Zeolites and Zeolite-like Materials

Zeolites and Zeolite-like Materials

Edited by

Bert F. Sels

Centre for Surface Chemistry and Catalysis, Leuven, Belgium

Leonid M. Kustov

N.D. Zelinsky Institute of Organic Chemistry, Russian Academy of Sciences; Moscow State University, Moscow, Russia

ELSEVIER

AMSTERDAM • BOSTON • HEIDELBERG • LONDON • NEW YORK • OXFORD
PARIS • SAN DIEGO • SAN FRANCISCO • SINGAPORE • SYDNEY • TOKYO

Elsevier
Radarweg 29, PO Box 211, 1000 AE Amsterdam, Netherlands
The Boulevard, Langford Lane, Kidlington, Oxford OX5 1GB, UK
50 Hampshire Street, 5th Floor, Cambridge, MA 02139, USA

Library of Congress Cataloging-in-Publication Data
A catalog record for this book is available from the Library of Congress

British Library Cataloguing-in-Publication Data
A catalogue record for this book is available from the British Library

ISBN: 978-0-444-63506-8

Publisher: John Fedor
Acquisition Editor: Kostas KI Marinakis
Editorial Project Manager: Sarah Jane Watson
Production Project Manager: Maria Bernard
Designer: Matthew Limbert

Contents

Contributors

J. Aguilhon Refining & Base Chemicals, Total Research & Technology Feluy, Seneffe, Belgium
Ph. Bodart Refining & Base Chemicals, Total Research & Technology Feluy, Seneffe, Belgium
J. Caro Leibniz University Hannover, Hannover, Germany
J.-P. Dath Refining & Base Chemicals, Total Research & Technology Feluy, Seneffe, Belgium
J. Dijkmans Centre for Surface Chemistry and Catalysis, Leuven, Belgium
M. Dusselier Centre for Surface Chemistry and Catalysis, Leuven, Belgium
T. Ennaert Centre for Surface Chemistry and Catalysis, Leuven, Belgium
E.J.M. Hensen Eindhoven University of Technology, Eindhoven, The Netherlands
V.I. Isaeva N.D. Zelinsky Institute of Organic Chemistry, Russian Academy of Sciences, Moscow, Russia
A. Kuperman Chevron Corporation, Richmond, United States
L.M. Kustov N.D. Zelinsky Institute of Organic Chemistry, Russian Academy of Sciences; Moscow State University, Moscow, Russia
A. Kustov N.D. Zelinsky Institute of Organic Chemistry, Russian Academy of Sciences; Moscow State University, Moscow, Russia
J. Liu Jilin University, Changchun, PR China
D. Minoux Refining & Base Chemicals, Total Research & Technology Feluy, Seneffe, Belgium
N. Nesterenko Refining & Base Chemicals, Total Research & Technology Feluy, Seneffe, Belgium
R. Oord Utrecht University, Utrecht, The Netherlands
E.A. Pidko Eindhoven University of Technology, Eindhoven, The Netherlands
W. Schutyser Centre for Surface Chemistry and Catalysis, Leuven, Belgium
B.F. Sels Centre for Surface Chemistry and Catalysis, Leuven, Belgium
E.F. Sousa-Aguiar Petrobras Research Centre (CENPES); Federal University of Rio de Janeiro (UFRJ), Rio de Janeiro, Brazil
A.G. Stepanov Boreskov Institute of Catalysis SB RAS, Novosibirsk, Russia
B.M. Weckhuysen Utrecht University, Utrecht, The Netherlands
J. Yu Jilin University, Changchun, PR China

Preface

Zeolites have been known for about 250 years, though their active life in science and industry started in late 1960s. The definition of zeolites experienced a few attacks from aluminophosphates (AlPOs, SAPOs, MeAPOs etc.), and metallosilicates (Fe-silicate, B-silicate, Ti-silicate, and others) in the 1980s, then from mesoporous oxide materials (silicate families MCM-41, SBA-15, mesoporous TiO_2, ZrO_2, and the like), and now from the organic-inorganic hybrid materials (metal-organic frameworks, MOFs, covalent organic frameworks, COFs, and porous coordination polymers, PCPs), though the hybrid materials may contain no traces of silicon or aluminum. All these new materials should be classified as zeolite-like materials. They are not true silica-alumina compositions and cannot be included in the family of zeolites. On the other hand, they do exhibit a clear molecular sieve effect and extended micro/meso porosity. In some aspects, these new materials are better or richer in properties and structures than the classical zeolites, and provide more opportunities for variation of the structure and composition. The new zeolite-like materials are not "boiling stones" anymore. Moreover, some of them are hydrophobic materials with preference for adsorption of non-polar molecules rather than water. But, the zeolite nature, or character, of these newcomers in the zeolite world can be easily recognized by the shape-selectivity effects, huge specific surface areas and pore volumes, and typical zeolite architecture.

The new wave of materials looks like a dangerous avalanche that can absorb the classical zeolites. But this will not happen, because the Si-Al zeolites have already inscribed their names (faujasites, mordenites, ZSM-5 zeolites, just to mention a few) in the history of modern industry—oil refining, petrochemicals synthesis, air separation, detergents, and other applications. Zeolites continue to be catalysts of choice in the conversion of light and heavy oil hydrocarbons, and the last decade offered a number of challenges for the use of zeolite-like materials. The conversion of biomass is another intriguing area where zeolites and zeolite-like materials will definitely show promising results. There is still a very large domain of fine chemicals synthesis, as well as other areas of potential applications, where zeolites do not provide an ideal solution. Yet, the time has not come to judge whether or not the little brothers of zeolites are better than their elder relatives.

This book is not a comprehensive handbook covering all the areas related to zeolites; it has been prepared as a kind of an introduction to zeolites for the Pre-Conference School preceding the 17th International Zeolite Conference (Moscow, July 2013). Also, the book is a contribution of the catalysis commission of the International Zeolite Association (IZA). The incentive was to show the progress in the key aspects of the zeolite arena. The lecturers of the Pre-Conference School were asked to prepare their chapters for this book, and they responded positively to this initiative. It was a difficult task to prepare chapters that should be, on the one hand, tutorial and understandable to the young generation of zeolite chemists (especially to those who just started the journey in this wonderful world) and, on the other hand, reflective of the modern state of the art, showcasing the hot spots in zeolite research. We intend to show that we are just in the middle of the road and the perception of the current state of the art will help to trace the path to new, yet undiscovered structures and functions of zeolite-like materials.

We hope that this book will be considered as a modest extension of the comprehensive encyclopedia of zeolites—*Introduction to Zeolite Science and Practice*, 2001. On the other hand, we were not going to be a shadow of this great book, and tried to identify those areas that are hot and need to be analyzed and updated after twelve years since 2001.

Every new material brings a new knowledge, not necessarily the promise of a new application, but this knowledge will serve as the foundation for the forthcoming applications of new generations of materials.

Bert F. Sels
Centre for Surface Chemistry and Catalysis, Leuven, Belgium

Leonid M. Kustov
N.D. Zelinsky Institute of Organic Chemistry, Russian Academy of Sciences; Moscow State University, Moscow, Russia

CHAPTER 1

Toward Greener and Designed Synthesis of Zeolite Materials

J. Liu, J. Yu
Jilin University, Changchun, PR China

1 Introduction

Zeolites are crystalline nanoporous inorganic materials formed by TO_4 tetrahedra (T = Si, Al, P, etc.), which show widespread applications in many industrial processes such as catalysis, adsorption, and separation [1–3]. In the 1940s, Barrer and Milton opened up the avenue to the synthesis of zeolites. Since then, there have been considerable efforts in the synthesis of new zeolite materials. Up to Jun. 2013, 206 types of zeolite materials have been identified by the Structure Commission of the International Zeolite Association (IZA), each of which has been named with a three-letter code [4]. Table 1 presents the new framework types approved since the 16th International Zeolite Conference (IZC-16) in 2010.

Zeolites are typically synthesized in a gel under hydrothermal/solvothermal conditions with alkali metal cations or organic amines/ammonium cations as the templates or structure-directing agents (SDAs). The reactive reagents usually include tetrahedral atom (Al, Si, P, etc.) sources, the SDA, the mineralizer (OH^- or F^-), the solvent, and so on. The mixture is then heated in a polytetrafluoroethylene-lined stainless-steel autoclave under autogenous pressure at temperature below 250°C for a period of time. Many synthetic factors influence the formation of zeolites, such as the source materials, solvent, SDA, gel composition, pH value, and crystallization temperature and time. So far, there is insufficient understanding of the formation mechanism of zeolite materials. Such hydrothermal/solvothermal synthesis process coupled with the unclear formation mechanism makes the synthesis of zeolites not in an environmentally friendly and rational way. Despite this, in recent decades, many research efforts have been made toward the greener and designed synthesis of zeolites. Traditionally, green chemistry is defined as *the utilization of a set of principles that reduces or eliminates the use or generation of hazardous substances in the design, manufacture, and application of chemical products* [17]. It emphasizes the economy of resource and energy, minimizing waste and cost, reducing chemical hazards, ensuring process safety, and the like [18]. In the aspect of zeolite synthesis, various approaches have been exploited toward greener synthesis by

Zeolites and Zeolite-like Materials. http://dx.doi.org/10.1016/B978-0-444-63506-8.00001-X

Table 1: New framework types approved by the IZA Structure Commission since IZC-16

Date Approved	Framework Type	Type Material	Tetrahedral Elements	Refs.
04/03/2011	-ITV	ITQ-37	Ge, Si	[5]
	IRR	ITQ-44	Ge, Si	[6]
	UWY	IM-20	Ge, Si	[7]
15/10/2011	JST	GaGeO-CJ63	Ga, Ge	[8]
	LTJ	Linde Type J	Si, Al	[9]
	NPT	Oxonitridophosphate-2	Ba, P	[10]
	*SFV	SSZ-57	Si	[11]
07/10/2012	BOZ	Be-10	Be, As(P)	[12]
	JOZ	LSJ-10	Be, Si	[13]
	JSN	MAPO-CJ69	Co(Zn), Al, P	[14]
	JSW	MAPO-CJ62	Co(Zn), Al, P	[15]
	PCR	IPC-4	Si	[16]

considering the green chemistry principles such as cost and environmental impact reduction, waste reusing, energy efficiency, process safety, and more. Consequently, some greener synthesis methods have been developed, such as organic-template-free synthesis, solvent-free synthesis, cheap or low-toxic organic-template-directed synthesis, waste reused synthesis, microwave-assisted synthesis, ionothermal synthesis, and more. Meanwhile, many attempts have been devoted to the rational synthesis of zeolites with specific framework structures and functions by novel SDA design, charge density mismatch (CDM), heteroatom substitution, and topotactic conversion [19,20]. In this chapter, we highlight some recent progress toward greener and rational synthesis of zeolite materials.

2 Green Synthesis

Currently, with increasing attention being paid to environmental and resource issues, green chemistry has been in the spotlight as a topic of research and has become one of the main objectives of by synthetic chemists. Anastas and Warner summarize 12 principles to guide the green chemical synthesis [17]. Many relevant factors should be taken into account in the process of green synthesis, including the concepts of the use and generation of safe and low or no-toxic substances, the design of energy-efficient processes, the best disposal of waste filtrate, and the like [18].

However, for zeolite synthesis, the traditional synthesis strategies more or less violate the principles of green chemistry. The hydrothermal or solvothermal synthesis method is usually conducted at high temperature and high pressure, undoubtedly increasing the hidden danger factors and causing energy consumption; the organic templates and organic solvents are usually toxic and expensive, thus increasing the synthesis costs; the discarded filtrates contain abundant inorganic/organic species that can be reused; and the resultant products need to be calcined at high temperature to remove the occluded organic templates, further causing environmental pollution. Chemists have therefore been making great efforts toward greener synthesis of zeolite materials. Herein, we will introduce the recent representative works toward greener synthesis of zeolites from the aspects of reduced cost and environmental impact, waste reused, energy efficiency, and process safety.

2.1 Synthesis for Reduced Cost and Environmental Impact

2.1.1 Organic-template-free synthesis

In the early time, zeolites are synthesized mainly in the basic $M_2O/MO–SiO_2 – Al_2O_3 – H_2O$ ($M = Li^+$, Na^+, K^+, Ca^{2+}, Ba^{2+}, etc.) systems with metal cations as templates. These cations play the role of directing the formation of structures and balancing the negative framework charges. Since Barrer and Denny first introduced organic templates into the synthesis system in 1961 [21], many novel zeolite structures have been prepared by using various types of organic amines/ammonium cations as templates. However, using organic templates also has disadvantages in zeolite synthesis. On the one hand, most organic amines/ammonium cations are toxic and expensive. The synthesis process will produce more or less poisonous vapor with a hazard to human health and the environment. And the posttreatment process requires a lot of water to wash reaction products and remove excessive raw materials, resulting in much toxic wastewater. On the other hand, the subsequent process of calcination to remove the organic amines/ammonium cations for opening up the channel system will be accompanied with the release of CO_2, NO_x, and other waste gases. Both aspects are against the environmentally friendly principles. To this end, the organic-template-free synthesis of zeolites becomes an effective strategy to avoid the environmental hazard, save sources, and reduce costs. To exclude organic templates from the synthesis, some crucial factors are needed to be considered [22]: (1) Metal cations. In replace of organic templates, inorganic metal cations (eg, Na^+) become an important factor in the formation of zeolites acting as the templates and framework charge balancers. (2) Molar ratio of the reactant gel. Because of the absence of organic templates, the phase regions of zeolite formation may be narrowed and different from the ordinary ones. So, careful exploration of the reactant gel molar ratio is needed. (3) Seeds. The addition of seeds may help to accelerate the growth of zeolite crystals. (4) Potential building units. The seed solution containing the potential building units of targeted zeolite products may direct the growth of zeolite crystals.

2.1.1.1 Metal-cation-directed synthesis

To overcome the disadvantages caused by organic templates, researchers attempted to directly omit the organic templates by carefully controlling the gel molar ratio in the synthesis. The aluminosilicate zeolite ZSM-5 (MFI) can be traditionally synthesized with the tetrapropylammonium cation as the organic template [23]. ZSM-5 with high crystallinity was first synthesized in the system of $Na_2O–SiO_2–Al_2O_3–H_2O$ in the 1970s [24,25]. Several years later, Clearfield and coworkers [26] further studied the organic-template-free synthesis of ZSM-5 by systematically adjusting the molar composition of the reaction gel. They found that the Si/Al ratio and Na/Al ratio were the most important factors affecting the synthesis. Only when the Si/Al ratio was 20 and Na/Al ratio was 2.25–3, pure ZSM-5 was crystallized; when the Si/Al ratio was lower (<15), the resulting product was mainly aluminosilicate zeolite mordenite (MOR) along with traces of ZSM-5; when the Si/Al ratio increased (>30), the amount of α-quartz increased coexisting with ZSM-5 and mordenite; when the gel contains no alumina, almost pure α-quartz could be obtained. Similarly, the increase of Na/Al ratio could also lead to the crystallization of mordenite and α-quartz along with ZSM-5. N_2 absorption data showed the as-synthesized materials possessed open channel system before calcination, which provided the successful synthesis of ZSM-5 in the organic-template-free system.

The aluminosilicate zeolite ECR-1 (EON) was first synthesized with the bis-(2-hydroxyethyl)-dimethyl-ammonium cations as the organic template [27]. As gallosilicate zeolite TNU-7 (EON) that has the same framework type as ECR-1 can be hydrothermally synthesized in the organic-template-free system [28], the organic-template-free synthesis of ECR-1 becomes possible. In 2006, Xiao and coworkers reported their work on the synthesis of ECR-1 without organic templates [29]. Through mixing NaOH, $NaAlO_2$, and water glass by stirring and aging at room temperature, a precursor solution (SiO_2: Al_2O_3: Na_2O: H_2O=15: 1.0: 16: 320) was prepared. Then, the precursor solution was mixed with H_2O, water glass, $Al_2(SO_4)_3$, $NaAlO_2$, and HCl to give the final mixture with a molar ratio of SiO_2: Al_2O_3: Na_2O: H_2O=10: 1.0: 2.5: 200. After crystallization at 100°C for 2 weeks, ECR-1 was obtained. By systemically studying the synthetic factors, it was found that alkalinity, temperature, and crystallization time played a pivotal role in the synthesis.

In 2012, Mintova and coworkers captured ultra-small EMT zeolite of 6–15 nm from the organic-template-free system at low temperature [30]. Traditionally, pure EMT could only be synthesized with an expensive organic template (18-crown-6 ether) [31]. The novel organic-template-free synthesis of EMT zeolite is from Na-rich precursor suspensions. Gel composition, nucleation temperature and time, and heating method are the key for the formation of ultra-small EMT zeolite, avoiding the conversion of zeolite EMT into zeolite FAU or SOD. The synthesis proceeded at near ambient temperature (30°C) with a very high yield. The successful synthesis of EMT zeolite provides a good example of zeolite synthesis by an energy-efficient and green route.

2.1.1.2 Seed-assisted synthesis

It is thought that the zeolite seed solution may contain the zeolite building units such as secondary building units (SBUs), which play a potential structure-directing role in the crystallization of zeolite crystals. The introduction of seed crystals into the starting aluminosilicate gels is considered to be an efficient way to accelerate the crystallization rate of zeolites. A typical example is shown by the aluminosilicate zeolite Beta (*BEA). In 2008,

Xiao and coworkers first reported the organic-template-free and fast synthesis route for Beta zeolite by seed-assisted synthesis [32]. The calcined Beta zeolite seeds with particle sizes of 60–100 nm were added to the aluminosilicate gel with a molar ratio of SiO_2: Al_2O_3: Na_2O: H_2O = 40: 1: 10: 570 and Beta zeolite with high crystallinity was successfully synthesized. Compared with the traditional synthesis of Beta zeolite with tetraethylammonium (TEA) cations as the template (three to four days, 140°C), the crystallization time of Beta crystals in seed-assisted synthesis (17–19 h, 140°C) was greatly decreased. Later in 2011 [33], they found the seed-directed synthesis of Beta exhibited larger textural parameters, more stable Al species, and an unprecedentedly higher density of active sites, giving rise to superior catalytic activity and selectivity for valuable products in catalysis. It is reported that borosilicate zeolite RUB-13 (RTH) [34], aluminosilicate zeolite ZSM-12 (MTW) [35], aluminosilicate zeolite RUB-50 (LEV) [33], aluminosilicate zeolite Heulandite (HEU) [33], and others can also be synthesized by the seed-assisted route.

In 2008, Xiao and coworkers also synthesized aluminosilicate zeolite ZSM-34 from a zeolite L seed solution without organic templates [36]. ZSM-34 is an intergrowth of offretite (OFF) and erionite (ERI) zeolites. Structurally, as with zeolite L, both OFF and ERI contain cancrinite cages. Therefore, they used zeolite L seed solution to directly induce and accelerate the formation of ZSM-34. ZSM-34 with heteroatom substitution by B, Ga, and Fe can also be synthesized by this organic-template-free route [37].

Although Al-rich aluminosilicate zeolite ferrierite (FER) could be synthesized by the cooperative effect of sodium and potassium cations in the absence of organic templates [38], it could not be applied to high-silica FER zeolite. In 2011, Xiao and coworkers developed a novel strategy to synthesize high-silica FER zeolite [39]. High-silica FER zeolite, named ZJM-2, was successfully prepared by introducing CDO SBUs resulting from the dissolution of aluminosilicate zeolite RUB-37 (CDO) under the organic-template-free conditions. The Si/Al ratio of ZJM-2 zeolite is as high as 14.5.

Recently, Okubo and coworkers built a working hypothesis in seed-assisted synthesis without organic templates, which may provide a new idea for synthesis and broaden framework types of zeolites [40]. By analyzing the examples of the seed-assisted and organic-template-free synthesis of zeolites in detail, they concluded that the composite building units that both the frameworks of zeolite seeds and the crystallized zeolites contained were the key to the organic-template-free synthesis.

2.1.2 Solvent-free synthesis

Traditional zeolite synthesis, for example, hydrothermal/solvothermal synthesis, necessarily is performed in the presence of solvents. The addition of solvents raises the cost of industrial production and also leads to waste solvent, which is hard to purify and recycle. Researchers should ask if they need a solvent at all in the green synthesis of zeolite materials. In recent work reported by Xiao and coworkers, the solvent-free route has been attempted in zeolite synthesis. The dry raw materials were grounded and heated to a properly high temperature [41]. Zeolites with framework types of MFI, SOD, MOR, *BEA, and FAU were successfully synthesized by the solventless synthesis. Heteroatoms (Al, Fe, B, Ga) could also be introduced into the frameworks, which enlightened their further applications in catalysis. Interestingly, hierarchical Si-ZSM-5 crystals could be obtained by adding nanosized $CaCO_3$ (average size of 100 nm) as a solid template. One factor to note here is that the raw materials must be hydrated. As the silicate condensation acts as the crucial role in zeolite synthesis, the existence of water will influence some hydration processes and kinetic pathways, which accelerates the crystallization [42]. Although any exploration of commercial applications is still to come, the solvent-free synthesis has already demonstrated some green features, such as the high yield of zeolite products associated with materials efficiency, reducing waste and cost, and reducing reaction pressure associated with safety concern.

2.1.3 Cheap or low-toxic organic-template-directed synthesis

The organic templates used in zeolite synthesis are usually expensive and toxic. Introducing cheap organic templates and lowering the toxicity are desired to reduce costs and environmental impact. With the cost-saving principle, Zones and coworkers produced aluminosilicate zeolite SSZ-25 (MWW) by using a cheaper isobutylamine together with a small amount of aminoadamantane to replace the more expensive hexamethyleneimine [43]. Xiao and coworkers also used low-toxic organic templates for zeolite synthesis [44]. They introduced tetramethylguanidine as the template, which was the product of animal metabolism, and successfully prepared aluminophosphate and heteroatom-containing aluminophosphate zeolites AlPO-5 (AFI), SAPO-5, MnAPO-5, and CoAPO-5.

2.2 Synthesis for Waste Reused

2.2.1 Organic-template-reused synthesis

Maximizing the materials efficiency and minimizing the generation of hazardous species are the aims of green chemistry. Davis and coworkers did excellent work by recycling organic amines/ammonium cations during zeolite synthesis. They reported a feasible method to extract TEA^+ cations from zincosilicate CIT-6 (*BEA) in the solution containing acetic acid [45]. The extraction process could simultaneously remove TEA^+ and Zn^{2+} cations, making CIT-6 a useful precursor to prepare novel *BEA-type zeolite materials. This work provided the possibility to reuse templates. Davis and coworkers further reported a facile method to recycle organic amines/ammonium cations during the synthesis of zeolite materials [46]. The organic templates could remain intact under zeolite synthesis conditions (high pH) and be cleavable at conditions that would not destroy the assembled zeolite. After the synthesis

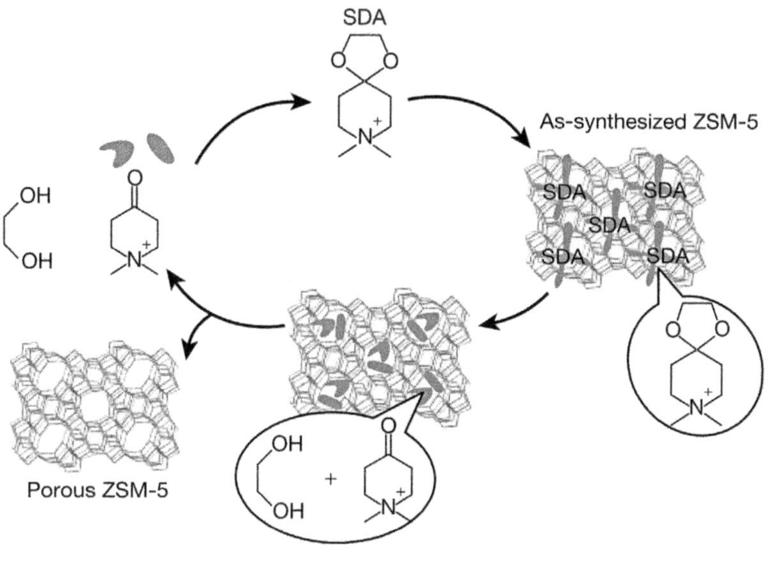

Fig. 1
Schematic of the organic-template-reused synthesis of ZSM-5 using a structure-directing agent (SDA) that contains a cyclic ketal: the as-synthesized ZSM-5 is first synthesized by assembling the SDA with silica precursor, H_2O, alkali metal ions, and so on; the organic templates cleave inside the zeolite pores; the fragments can be removed by ion-exchange treatment; the fragments recombine into the original template. *Reprinted by permission from Macmillan Publishers Ltd.: Lee H, Zones SI, Davis ME. A combustion-free methodology for synthesizing zeolites and zeolite-like materials. Nature 2003;425:385–388. Copyright (2003).*

of ZSM-5 with the organic template containing a cyclic ketal group, the organic template could be broken into smaller fragments with the aid of 1 M HCl solution at 80°C for 20 h. The fragments could be easily removed from the pores by the ion-exchange treatment with a mixture of 0.01 M NaOH and 1 M NaCl at 100°C for 72 h and the following step can recombine the fragments for repeated use (shown in Fig. 1). The organic-template-reused synthesis reduces the costs of templates and makes the synthesis recyclable.

2.2.2 Mother-liquor-reused synthesis

Usually, the synthesis of zeolite materials requires an excess amount of reactant reagents. After the resultant products are filtered from the reaction mixture, the filtrate (mother-liquor) containing abundant nutrients is discarded as a waste solution. With the concept of green chemistry, Yu and coworkers attempted a mother-liquor-reused route to synthesize cobalt aluminophosphate zeolites CoAPO-CJ51 (GIS) and CoAPO-CJ52 (LAU) by introducing Co^{2+} cations into the filtrated mother liquors of two layered aluminophosphates Mu-4 and AlPO-IM, respectively [47]. The mother-liquor-reused route provides a low-cost way for the synthesis of zeolite materials.

2.3 Synthesis for Energy Efficiency: Microwave-Assisted Synthesis

As a green chemistry process, microwave-assisted synthesis has become a promising alternative to conventional hydrothermal/solvothermal synthesis of zeolites. Because of the rapid nucleation of the initial crystallites and the uniform growth process during the microwave-assisted synthesis, it not only shortens crystallization time and decreases energy consumption but also favors the formation of products with uniform morphology and a wide variety of compositions [48]. The fast reaction rate and high-energy efficiency in a narrow temperature range change the association between reaction species, which finally influences the structure and morphology of zeolites. The microwave-assisted synthesis of zeolites originated from the synthesis of NaA and ZSM-5 in 1988 by Mobil [49]. Up to now, many zeolites such as LTA, SOD, FAU, EMT, and MFI have been synthesized under microwave irradiation [50,51]. Because of its superiority in saving time and energy, microwave-assisted synthesis has been developed along with other synthesis methods.

Recently, Yu and coworkers have successfully controlled the crystal size and morphology of the silicoaluminophosphate zeolite SAPO-34 (CHA) under microwave irradiation in the system of Al_2O_3–P_2O_5–SiO_2–TEAOH–H_2O by studying the synthetic factors such as the silica source, water content, crystallization time, and aging time [52]. Microwave-assisted synthesis proved to be an efficient approach to produce nanosized zeolite crystals [53]. As a comparison to traditional hydrothermal synthesis at 200°C for 72 h, the microwave-assisted synthesis of SAPO-34 needed only 1 h at 200°C, which greatly reduced the reaction time. The crystals obtained by conventional hydrothermal synthesis were much larger than the particles formed by microwave-assisted synthesis. Fig. 2 shows the nanosheet-like SAPO-34

Fig. 2
SEM images of SAPO-34 crystals: (A) nanosheet-like SAPO-34 (20 × 250 × 250 nm), (B) the nanospheres of SAPO-34 (dia. ~80 nm). *Reprinted with permission from Yang G, Wei Y, Xu S, Chen J, Li J, Liu Z, Yu J, Xu R. Nanosize-enhanced lifetime of SAPO-34 catalysts in methanol-to-olefin reactions. J Phys Chem C 2013;117:8214–8222. Copyright (2013) American Chemical Society.*

(20×250×250 nm) and the nanospheres of SAPO-34 (dia. ~80 nm). Studies on methanol-to-Olefin reactions demonstrated that the nanosized SAPO-34 catalysts exhibited enhanced catalyst lifetime and reduced coking rate.

2.4 Synthesis for Process Safety: Ionothermal Synthesis

As opposed to traditional hydrothermal/solvothermal synthesis, where the solvents are predominantly molecular, ionothermal synthesis is based on the use of ionic liquids as solvents, and in many cases also as templates. Ionic liquids are a class of molten salts, composed of organic cations and matching anions. At room temperature or near room temperature, ionic liquids are traditionally defined as liquids and can safely be used as solvents. In 2004, Morris and coworkers first developed the ionothermal synthesis approach to prepare zeolite materials [54]. Since then, ionothermal synthesis has received great attention due to its potential advantages over traditional hydrothermal synthesis.

Ionic liquids show a range of properties, making them suitable to be used as green and safe media for the preparation of zeolite materials [55]. For example, ionic liquids can be relatively polar solvents with excellent solvating properties. Their low melting points favor the formation of a homogeneous mixture of chemical reagents at room temperature. Their high thermal stability (stable below 300°C) provides a stable medium for high-temperature synthesis (usually below 250°C). Ionic liquids in particular exhibit low vapor pressure, so the synthesis can take place at ambient pressure, which not only avoids toxic volatile vapors but also eliminates safety concerns associated with the high pressure of hydrothermal/solvothermal reactions. Therefore, the ionothermal reactions can proceed in an open reaction vessel instead of a conventional Teflon-lined stainless-steel autoclave. Ionothermal synthesis undoubtedly provides a facile synthesis route for the preparation of zeolites at ambient pressure.

Ionic liquid 1-methyl-3-ethylimidazolium bromide ([EMim][Br]) is the most commonly used ionic liquid in the ionothermal synthesis of zeolite materials, leading to the formation of cobalt aluminophosphate zeolites SIZ-7 (SIV), SIZ-8 (AEI), and SIZ-9 (SOD) [56]. In ionothermal synthesis, the amount of water has a pronounced influence on the zeolite synthesis. Tian and coworkers investigated the effect of water on the ionothermal synthesis of aluminophosphate zeolites AlPO-5 and AlPO-11 (AEL), showing the important role of water [57]. The addition of organic amines/ammonium cations has also been found to affect the crystallization process of ionothermal reactions. Alternatively, it would be interesting to combine the ionic liquid and organic amine as the co-templates in the ionothermal synthesis of zeolite structures. Yu and coworkers reported the co-templating ionothermal synthesis of open-framework aluminophosphate JIS-1 by using the aromatic amine 1-methylimidazole (MIA) and [Emim][Br] [58]. $[Emim]^+$ cation and the protonated $[MIAH]^+$ cation were both found in the channels. In 2010, a 20-ring extra-large pore aluminophosphate zeolite DNL-1

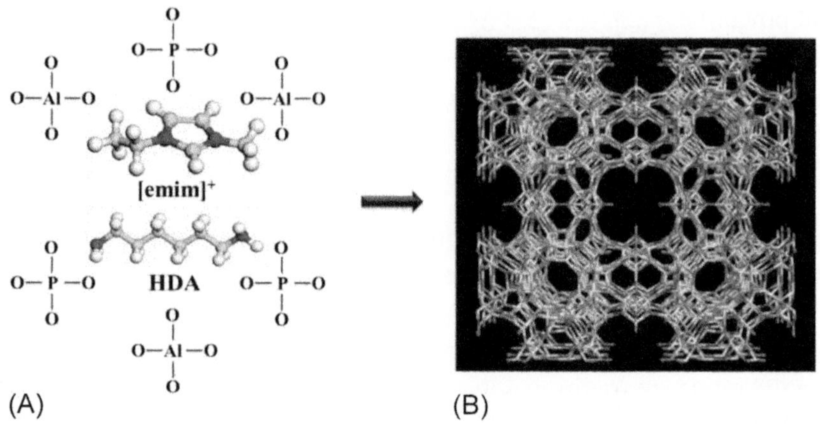

Fig. 3
(A) Co-templating of [Emim]+ and 1,6-hexanediamine (HDA) in ionothermal synthesis of DNL-1; (B) the framework structure of DNL-1 viewed along the [001] direction. *(B) is reprinted with permission from Wei Y, Tian Z, Gies H, Xu R, Ma H, Pei R, Zhang W, Xu Y, Wang L, Li K, Wang B, Wen G, Lin L. Ionothermal synthesis of an aluminophosphate molecular sieve with 20-ring pore openings. Angew Chem Int Ed 2010;49:5367–5370. Copyright (2010) John Wiley and Sons.*

(-CLO) was synthesized with the co-templates of [Emim]Br and 1,6-hexanediamine (shown in Fig. 3) [59]. Compared with the $GaPO_4$ analogue cloverite, DNL-1 has excellent stability.

Importantly, ionic liquids show excellent properties on the adsorption of microwave. The combination of microwave-assisted synthesis and ionothermal synthesis opens up an efficient and safe route to prepare zeolite materials [60]. Tian and coworkers reported the microwave-assisted ionothermal synthesis of AlPO-11 using [Emim]Br [61]. Silicon element was also successfully introduced to the framework forming SAPO-11 with potential catalytic applications. In addition, Yan and coworkers developed the ambient pressure synthesis method for silicate zeolites by combining the advantages of ionothermal synthesis, dry-gel conversion, and microwave irradiation [62]. With the assistance of microwave-assisted synthesis, the safe and fast process of ionothermal synthesis has shown to be a promising synthetic route for a variety of zeolite structures.

3 Designed Synthesis

The traditional way to seek new zeolites was primarily based on a trial-and-error synthesis process. With the growing needs for applications of zeolite materials, the rational synthesis of zeolites with desired structures and functions has become the objective of synthetic chemists. However, the rational synthesis of zeolites with desired structures is still a great challenge due to the unclear formation mechanism of zeolite crystals. With the advancement of understanding the synthesis chemistry and structure chemistry of zeolites in recent decades, several powerful strategies have been developed based on novel SDA design, CDM, heteroatom substitution, topotactic conversion, and the like.

3.1 Designing Novel SDAs

Various SDAs such as organic amines, quaternary ammonium cations, phosphonium cations, phosphazenes, and metal complexes have been employed in zeolite synthesis. SDAs are featured for their structure diversity and controllability to meet the objective of designed synthesis of zeolites with specific structures. Studies show that SDAs with proper shape, polarity, charges, and flexibility are the keys to achieve the design of novel zeolite structures [63]. Nowadays, many zeolite structures have been synthesized by using novel SDAs. Besides, hierarchical zeolites can be synthesized with bifunctional templates. Some representative works will be discussed later.

3.1.1 SDA design for novel zeolite structures

SDAs can act as space-filling agents, structure-directing agents and templates in the synthesis of zeolites. Even tiny differences in the structure of SDAs may lead to different products. For example, Fig. 4A shows the organic SDAs for the synthesis of different zeolites: silicate zeolite SSZ-74 (-SVR) [64], silicate zeolite IM-5 (IMF) [65], and aluminosilicate zeolite TNU-9 (TUN) [66]. These organic SDAs differ in the length of —CH_2— groups. SSZ-74 was prepared in fluoride media with 1,6-bis(*N*-methylpyrrolidinium)hexane as the SDA, showing a three-dimensional (3D) 10-ring channel system. IM-5 could be synthesized with 1,5-bis(*N*-methylpyrrolidinium)pentane as the SDA. The structure of IM-5 has a two-dimensional (2D) 10-ring channel system with a limited third dimension. The formation of TNU-9 was under the reaction system of SDA—Na_2O—Al_2O_3—SiO_2—H_2O with the SDA of 1,4-bis(*N*-methylpyrrolidinium)butane at 160°C [67]. The 3D pore system of TNU-9 consists of two parallel 10-ring channels along the [010] direction, and they are connected via 10-ring openings along the [10–1] direction. The analogue organic SDAs, trimethonium,

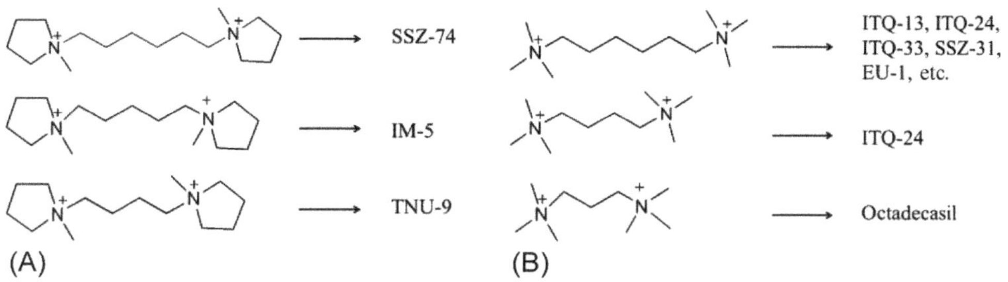

Fig. 4
(A) Organic SDAs for high-Si zeolites SSZ-74, IM-5, and TNU-9; (B) organic SDAs for zeolites ITQ-13, ITQ-24, octadecasil, and so on.

tetramethonium, and hexamethonium dications, with different lengths of —CH_2— groups can also direct a series of different zeolites such as silicate zeolite ITQ-13 (ITH), germanoaluminosilicate zeolite ITQ-24 (IWR), silicate zeolite octadecasil (AST), among others (shown in Fig. 4B).

Recently, Corma and coworkers explored the use of phosphonium cations as SDAs and successfully synthesized several germanosilicate zeolites containing double four rings (D4Rs), such as ITQ-26 (IWS) [68], ITQ-34 (ITR) [69], and ITQ-40 [70]. The aluminosilicate zeolite ITQ-27 (IWV) [71] can also be synthesized with dimethyldiphenylphosphonium. Notably, the extra-large pore germanosilicate zeolite ITQ-40 presents the first zeolite structure that contains D4Rs and double three rings (D3Rs) as SBUs [70]. It was synthesized with dimethyldiphenylphosphonium or diethyldiphenylphosphonium as SDAs in concentrated synthesis gels with the introduction of Ge atom and F^- ion. Corma and coworkers also reported the use of highly versatile organic SDAs based on phosphazenes for zeolite synthesis [72]. The aluminosilicate zeolite boggsite, which was a rare natural zeolite and difficult to synthesize experimentally, was prepared successfully by using phosphazenes as the organic SDAs. These phosphazene-derived molecules have adequate polarity and stability to mobilize silica species and offer more structural possibilities than quaternary ammonium or phosphonium cations.

Strikingly, a novel germanoaluminosilicate zeolite ITQ-43 with a hierarchical meso-microporous structure has been prepared with (2′R,6′S)-2′,6′-dimethylspiro[isoindoline-2, 1′-piperidin]-1′-ium hydroxide as the organic SDA [73]. Yu and coworkers further designed seven isoindoline-based organic SDAs with increasing sizes to systematically study the synthesis of extra-large pore zeolites by using the high-throughput synthesis technique [74]. Eight crystalline germanosilicates zeolite products have been synthesized by adjusting the synthesis parameters. It was found that when the organic SDAs are relatively small or they are large but very flexible, products are Beta (*BEA), ITQ-7 (ISV), or ITQ-17 (BEC). However, if the size of organic SDA increases above a certain dimension, four extra-large pore zeolites including ITQ-15 (UTL), ITQ-37 (-ITV), ITQ-44 (IRR), and ITQ-43 have been obtained (shown in Fig. 5).

More recently, as a newly developed organic SDA, bulky aromatic proton sponge, shows its advantage in structure-directing extra-large zeolites for its high basicity along with the large size and rigidity. A new silicoaluminophosphate zeolite ITQ-51 was synthesized with 1,8-bis(dimethylamino)naphthalene as the organic SDA (shown in Fig. 6) [75]. ITQ-51 has a unique zeolite framework with 1D 16-ring channels (9.9×7.7 Å) along the *c*-axis. Two different chains (the lau chain often found in zeolites and a unique helical four-ring chain) are connected with the common TO_4 tetrahedra, forming the framework of ITQ-51. Importantly, ITQ-51 is stable after calcination, being one of the very few examples of hydrothermally stable extra-large-pore zeolites.

Besides organic SDAs, metal complexes have also demonstrated their ability in directing the zeolite structures. The synthesis of silicate zeolite UTD-1 (DON) with

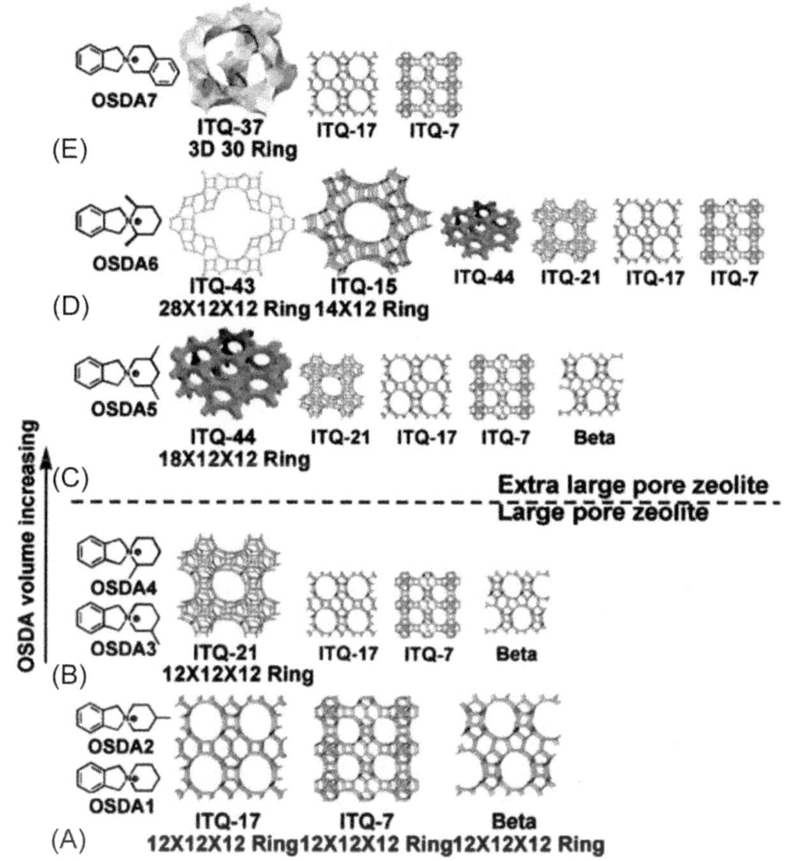

Fig. 5

Relationship between the size of organic SDAs and products in high-throughput synthesis of germanosilicate zeolites: (A) OSDA1,2 can only direct Beta, ITQ-7, and ITQ-17; (B) OSDA3,4 can produce ITQ-21 besides Beta, ITQ-7, and ITQ-17; (C) OSDA5 can direct extra-large pore zeolite ITQ-44 besides ITQ-21, Beta, ITQ-7, and ITQ-17; (D) OSDA6 can induce ITQ-15 and ITQ-43 besides ITQ-44, ITQ-21, ITQ-7, and ITQ-17; (E) OSDA7 can produce mesoporous chiral zeolite ITQ-37 besides ITQ-7 and ITQ-17. *Reprinted with permission from Jiang J, Xu Y, Cheng P, Sun Q, Yu J, Corma A, Xu R. Investigation of extra-large pore zeolite synthesis by a high-throughput approach. Chem Mater 2011;23:4709–4715. Copyright (2011) American Chemical Society.*

bis(pentamethylcyclopentadienyl) cobalt(III) ions presents an example of using metal complexes as the SDA [76]. UTD-1 has proved to be the first high-silica zeolite containing one-dimensional (1D) 14-ring channels. Recently, Yu and coworkers synthesized a new gallogermanate zeolite GaGeO-CJ63 ($[Ni(en)_3][Ga_2Ge_4O_{12}]$, en = ethylenediamine, JST) with chiral metal complex $[Ni(en)_3]^{2+}$ cations as the SDA [8]. Its framework is constructed exclusively by three rings (shown in Fig. 7A). GaGeO-CJ63 possesses 3D intersecting 10-ring channels with one of the lowest framework densities among the known zeolites.

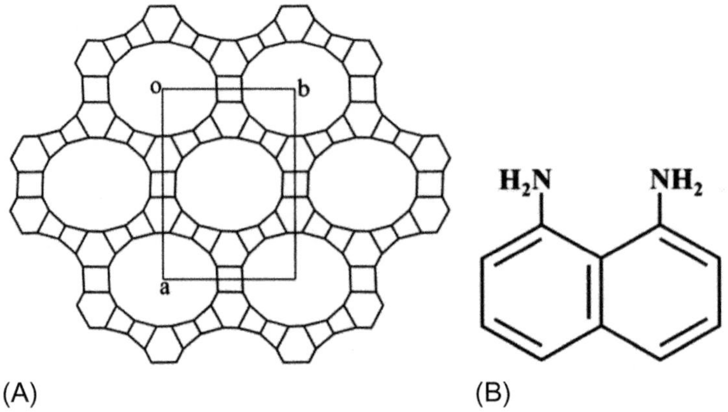

Fig. 6
(A) Structure of ITQ-51 zeolite viewed along the *c*-axis; (B) organic SDA, bulky aromatic proton sponge, used in the synthesis of ITQ-51.

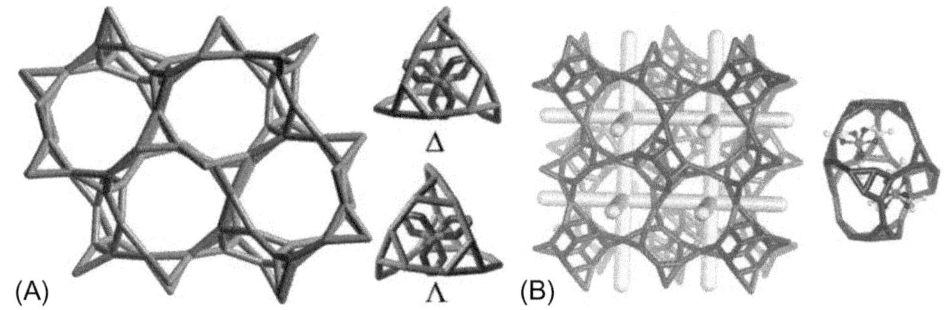

Fig. 7
(A) Framework of GaGeO-CJ63 and the chiral [$3^4.6.10^3$] cages with chiral $[Ni(en)_3]^{2+}$ cations; (B) the framework of GaGeO-JU64 showing the three-dimensional intersecting 11-ring channels and the chiral [$3^{12}.4^3.6^2.11^6$] cavity, induced by chiral $[Ni(1,2\text{-PDA})_3]^{2+}$ cations. *Panel (A): Reprinted with permission from Han Y, Li Y, Yu J, Xu R. A gallogermanate zeolite constructed exclusively by three-ring building units. Angew Chem Int Ed 2011;50:3003–3005. Copyright (2011) John Wiley and Sons. Panel (B): Reprinted with permission from Xu Y, Li Y, Han Y, Song X, Yu J. A gallogermanate zeolite with eleven-membered-ring channels. Angew Chem Int Ed 2013;52:5501–5503. Copyright (2013) John Wiley and Sons.*

The racemic $[Ni(en)_3]^{2+}$ cations are matched with chiral [$3^4.6.10^3$] cages, revealing that the chirality can be transferred from the chiral metal complex to the inorganic framework.

By using metal complexes as the SDA, Yu and coworkers also reported a novel gallogermanate zeolite GaGeO-JU64 ($|(Ni(C_3H_{10}N_2)_3)_{36}Ni_{4.7}|[Ga_{81.4}Ge_{206.6}O_{576}]$) with an unprecedented 3D intersecting 11-ring channel system (shown in Fig. 7B) [77]. GaGeO-JU64 has the lowest framework density of FD = 9.9 T/1000 Å^3 among all known oxide zeolites and features pairs of chiral [$3^{12}.4^3.6^2.11^6$] cavities, induced by chiral $[Ni(1,2\text{-PDA})_3]^{2+}$ cations formed in situ in the synthesis (1,2-PDA = 1,2-diaminopropane).

3.1.2 SDA design for hierarchical zeolites

Hierarchical zeolites with mesoporous and microporous channels are very promising for the catalysis of large molecular species. In 2006, Ryoo and coworkers developed an amphiphilic organosilane-directed method for the synthesis of tunable mesoporous zeolites [78,79]. The bifunctional template, [3-(trimethoxysilyl)propyl]hexadecyldimethylammonium chloride ([$(CH_3O)_3SiC_3H_6N(CH_3)_2C_{16}H_{33}$]Cl, TPHAC), contains three parts: the first is a hydrophobic alkyl chain to direct the formation of mesoporous channels; the second part is a quaternary ammonium group to direct the synthesis of zeolite structures; the third part is a hydrolysable methoxysilyl moiety, which is incorporated in the resultant zeolite frameworks. Fig. 8 shows the direct hydrothermal assembly process for the hierarchically porous aluminophosphates. Experimental results indicated that the methoxysilyl moiety could form covalent bonds with other SiO_2 and Al_2O_3 sources and interact with the growing zeolites. The mesopore diameters could be tuned by the length of the alkyl chain and hydrothermal synthetic conditions. By using TPHAC, they have successfully prepared hierarchical zeolites LTA, AlPO-*n*, and MFI.

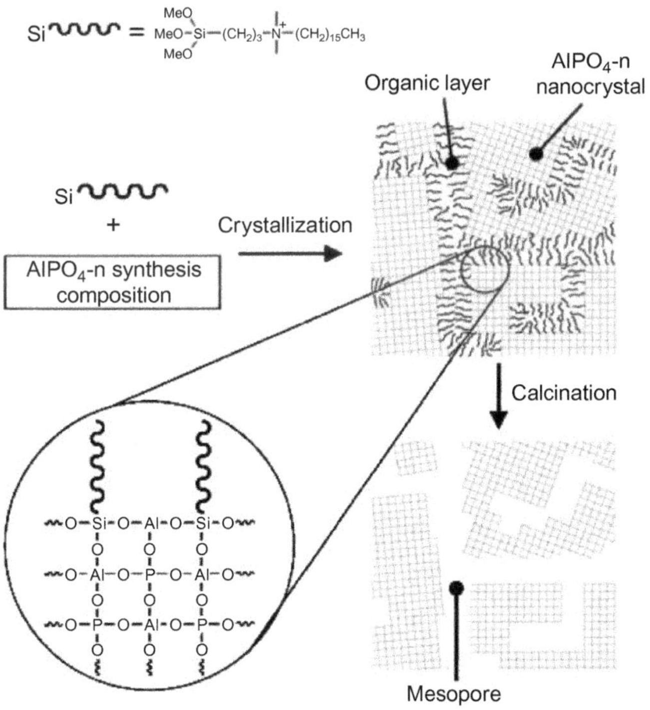

Fig. 8
Proposed synthesis mechanism for the hierarchically porous aluminophosphates (HP-AlPO-*n*) with TPHAC as the mesopore-directing agent. *Reproduced from Choi M, Srivastava R, Ryoo R. Organosilane surfactant-directed synthesis of mesoporous aluminophosphates constructed with crystalline microporous frameworks. Chem Commun 2006;4380–4382 with permission of The Royal Society of Chemistry.*

Ryoo and coworkers also used a bifunctional surfactant to direct the formation of hierarchical MFI zeolite nanosheets with the thickness of only 2 nm (Shown in Fig. 9) [80]. The surfactant was composed of two parts. Two quaternary ammonium groups separated by a C_6 alkyl chain act as an SDA and fit the *b*-axis dimension of a single MFI unit cell. The long-chain alkyl group (C_{22}) with hydrophobic interaction supports the formation of mesoporous structures. The MFI zeolite nanosheets presented advantages in catalytic applications for their high thermal stability, hydrothermal stability, and strong acidity. The authors also successfully synthesized zeolite MTW using the same method. The follow-up work presented a hierarchically mesoporous/microporous zeolite synthesized by pilling the multilamellar mesostructure [81]. Even after calcination to remove the surfactant, the mesopore between the lamellar space could remain with the support of silica pillar. By using well-designed bifunctional surfactant composed of long hydrophobic chains and hydrophilic zeolite structure-directing groups, much progress can be made in fabricating hierarchical zeolites.

Recently, Schwieger and coworkers reported the use of TPHAC to synthesize a house-of-cards-like arrangement of zeolite X (FAU) nanosheets with a unique, threefold (micro, meso,

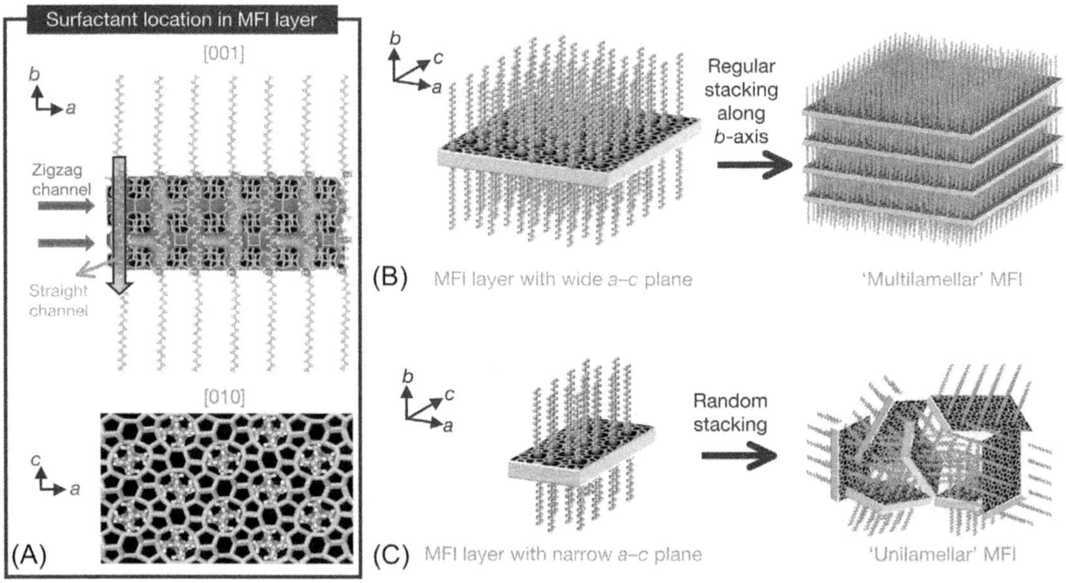

Fig. 9

Crystallization of MFI nanosheets (A) the proposed structure model for the surfactant located in the single MFI nanosheet. The red spheres indicate the quaternary ammonium groups in the channel intersections: one is inside the zeolite framework, and the other locates at the pore opening of the external surface. Both multilamellar stacking along the *b*-axis (B) and a random assembly of unilamellar structure (C) can be formed by the MFI nanosheets. *Reprinted by permission from Macmillan Publishers Ltd: Choi M, Na K, Kim J, Sakamoto Y, Terasaki O, Ryoo R. Stable single-unit-cell nanosheets of zeolite MFI as active and long-lived catalysts. Nature 2009;461:246–249. Copyright (2009).*

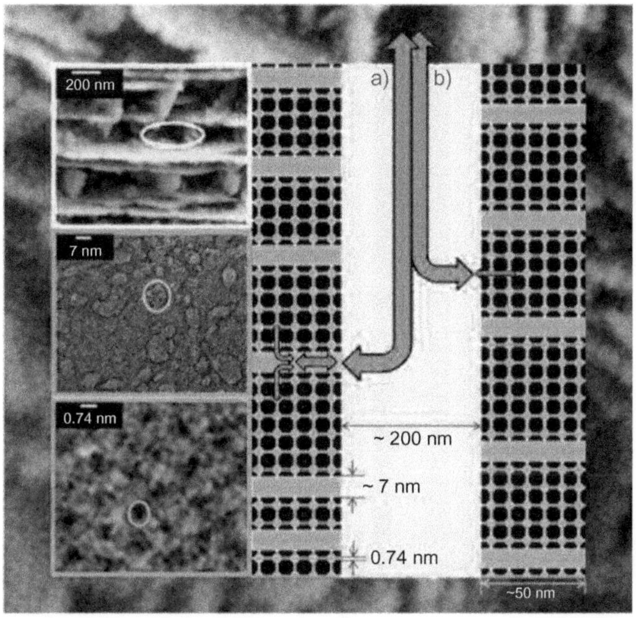

Fig. 10

Three hierarchy levels of the channel structure of the mesoporous zeolite X nanosheet sample, 200 nm = macropores; 7 nm = mesopores; 0.74 nm = micropores; possible flow paths in the hierarchical pore system are shown by the arrows on the right: route (a) macro-meso-micro, route (b) macro-micro. *Reprinted with permission from Inayat A, Knoke I, Spiecker E, Schwieger W. Assemblies of mesoporous FAU-type zeolite nanosheets. Angew Chem Int Ed 2012;51:1962–1965. Copyright (2012) John Wiley and Sons.*

and macropores) hierarchical pore system (shown in Fig. 10) [82]. The external surface of the hierarchical assembly is more than 10 times that of the conventional zeolite X. Such designed zeolite nanosystems may offer great potential for new applications [83].

More recently, an alternative and mesoporogen-free approach has been put forward by Okubo and coworkers for the construction of hierarchically organized MFI zeolites by sequential intergrowth (shown in Fig. 11) [84]. The simple $(C_3H_7)_3N^+$-$(CH_2)_n$-$N^+(C_3H_7)_3$ cation was used as the organic SDA to form plate-like MFI zeolites with enhanced 90 degree rotational intergrowths during crystal growth. The intergrowths formed a hierarchical structure with three classes of porosity: the intrinsic microporosity of the zeolite framework, mesoporosity existing within the zeolite plates, and macroporosity arising from the complex intergrown structure.

3.2 Charge Density Mismatch

Against the trend of using more complicated and expensive organic SDAs to produce new zeolite framework types, the CDM approach was initially proposed as a cheaper alternative for the designed synthesis of zeolites. The concept of CDM approach was first developed

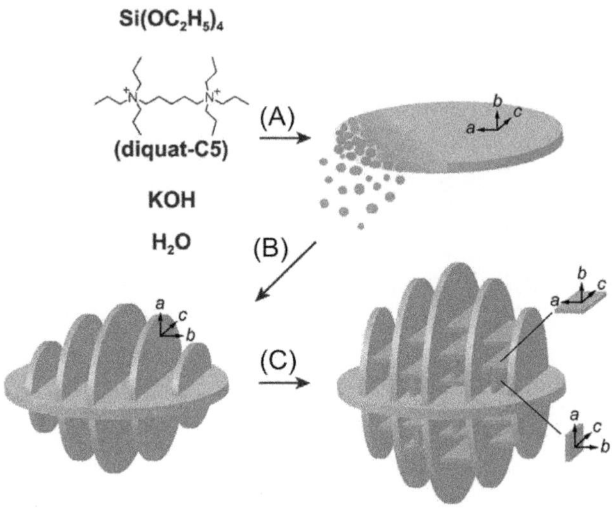

Fig. 11
Formation schematic of hierarchically organized MFI zeolites by sequential intergrowth, (A) generation of MFI zeolite plates by the aggregation of amorphous nanoparticles, which probably are hydrated diquat-C5-silica composites synthesized from a solution; (B) 90 degree rotational intergrowths developed by crystal growth also from nanoparticle aggregates; (C) formation of hierarchical structure with the nanoparticles and the soluble species in solution supplying the precursors for further crystal growth. *Reprinted with permission from Chaikittisilp W, Suzuki Y, Mukti RR, Suzuki T, Sugita K, Itabashi K, Shimojima A, Okubo T. Formation of hierarchically organized zeolites by sequential intergrowth. Angew Chem Int Ed 2013;52:3355–3359. Copyright (2013) John Wiley and Sons.*

by researchers at UOP in the 2000s, which was designed to develop cooperation between two or more different SDAs during the crystallization process of zeolites. Practice shows the use of multiple small SDAs tends to produce structures with lower Si/Al ratio [85]. Early in 2004, Wilson and coworkers demonstrated the template cooperation in the synthesis of aluminosilicate zeolite UZM-9 (LTA) [86,87]. They used the common organic SDAs, tetramethylammonium (TMA^+) and TEA^+ cations, together with Na^+ ion to direct the crystallization of UZM-9. In 2010, Hong and coworkers synthesized aluminosilicate zeolite UZM-12 (ERI) zeolite by the cooperative effects of TEA^+, alkali metal and diquaternary ammonium cations in crystallization to direct zeolite frameworks [88]. Later, they synthesized a series of aluminosilicates and gallosilicates using the CDM approach [89].

Recently, the mechanism study of the CDM approach by Hong and coworkers provided an explanation to the formation pathway for high-silica LTA zeolite crystals (shown in Fig. 12) [90]. In the $TEA^+/TMA^+/Na^+$ system, it was found that the nucleation began with the larger LTA cages instead of smaller SOD or D4R cages. In the initial condensation and nucleation process, both Na^+ and TMA^+ cations played an important role. Then the crystals grew with the incorporation of Na^+, TMA^+, and TEA^+ cations into the solid phase.

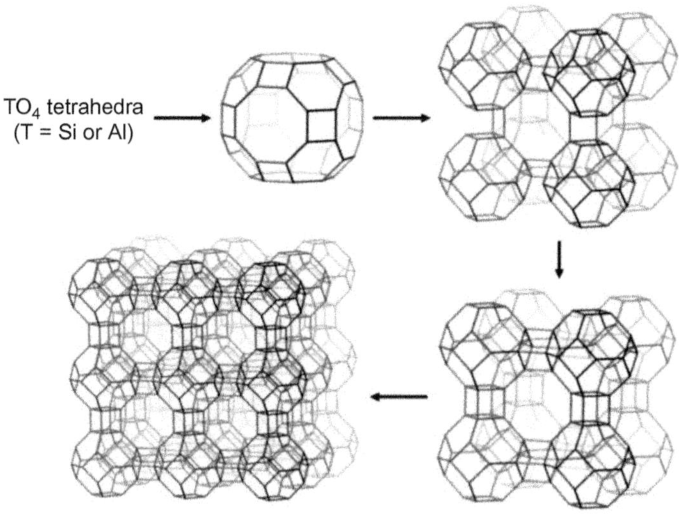

Fig. 12
Schematic illustration for the formation pathway of the high-silica LTA zeolite in the $TEA^+/TMA^+/Na^+$ system. *Reprinted with permission from Park MB, Lee Y, Zheng A, Xiao F-S, Nicholas CP, Lewis GJ, Hong SB. Formation pathway for LTA zeolite crystals synthesized via a charge density mismatch approach. J Am Chem Soc 2013;135:2248–2255. Copyright (2013) American Chemical Society.*

3.3 Heteroatom Substitution

Zeolites are typically constructed by TO_4 tetrahedra (T=Si, Al, P, etc.). The substitution of zeolite frameworks with heteroatoms may not only lead to special properties but also generate particular building units and influence the framework structures. In general, heteroatoms B, Ga, Be, and Zn may direct three rings, while Ge tends to direct D4Rs or sometimes D3Rs [20]. In this way, the substitution of heteroatom provides a promising way for designed synthesis of zeolites with specific structures.

3.3.1 Heteroatom substitution with trivalent and divalent metal ions

The metal heteroatoms play a significant role in the formation of various zeotype structures. On one hand, the negative charges introduced by the trivalent and divalent metal ions in the silicate-based zeolites must be balanced by the positive charges of protonated organic SDAs. Consequently, the existence of more trivalent and divalent metal ions needs more balancing SDAs, which may result in a larger pore volume of zeolites. On the other hand, the substituted elements causing different T—O bond lengths and T—O—T angles may be essential for some specific SBUs and structures [63]. In the case of heteroatom substitution by Ga atom, an 18R extra-large-pore gallosilicate ECR-34 (ETR) [91] is built of specific SBUs that have never been seen in the aluminosilicate frameworks.

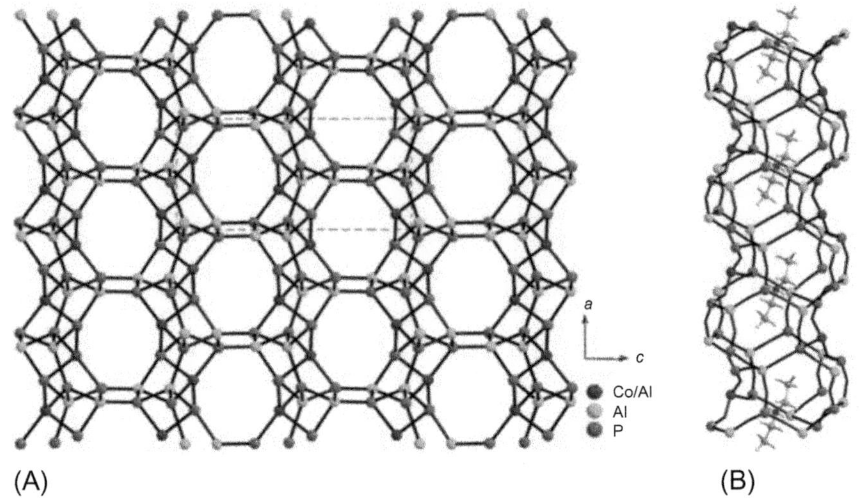

Fig. 13
(A) Framework structure of CoAPO-CJ40 viewed along the [010] direction; (B) helical 10-ring channel of CoAPO-CJ40 with helical arrangement of cobalt atoms shown in pink. *Reprinted with permission from Song X, Li Y, Gan L, Wang Z, Yu J, Xu R. Heteroatom-stabilized chiral framework of aluminophosphate molecular sieves. Angew Chem Int Ed 2009;48:314–317. Copyright (2009) John Wiley and Sons.*

Many examples can be found in the synthesis of heteroatom-containing aluminophosphate zeolites with new zeotype structures. Recently, Yu and coworkers synthesized the first heteroatom-containing chiral aluminophosphate zeolite MAPO-CJ40 (M = Co, Zn) with a new zeotype structure assigned as JRY by IZA, through the introduction of divalent metal ions to the aluminophosphate synthesis system [92]. It is made of strict alternation of MO_4/AlO_4 and PO_4 tetrahedra and the negative charges are balanced by protonated diethylamine cations. The framework contains 1D helical 10-ring channels (shown in Fig. 13). The heteroatom atoms in the framework of MAPO-CJ40 play a significant role in relaxing the high geometric distortion of chiral aluminophosphate framework based on the molecular simulations and geometric calculations. The pure chiral aluminophosphate analogue of MAPO-CJ40 could not be obtained in the absence of divalent metal ions. Strikingly, the heteroatoms in the framework display a helical array along the channel, which makes MAPO-CJ40 a potentially interesting single-site chiral solid catalyst.

Heteroatom-containing aluminophosphates MAPO-CJ62 (M = Co, Zn) have been hydrothermally synthesized by introducing heteroatoms Co and Zn into the framework, respectively [15]. This new zeolite framework, assigned as JSW, contains 1D eight-ring channels along the [010] direction. Interestingly, the metal atoms are orderly distributed and selectively occupy two of the three crystallographically distinct metal positions in the framework. It is quite different from the random distribution of heteroatoms observed in other aluminophosphate zeolites. A pure aluminophosphate analogue of MAPO-CJ62 without M-incorporation could not be obtained, suggesting the necessity of introducing M atoms in MAPO-CJ62. The reason has been explained by analyzing the distortions of Al-centered tetrahedra in the hypothetical framework of AlPO-CJ62.

By choosing Co and Zn as heteroatoms, two novel heteroatom-containing aluminophosphate zeolites MAPO-CJ69 (M = Co, Zn), with a new zeotype structure assigned as JSN, have been solvothermally synthesized [14]. The structure of MAPO-CJ69 is composed of the 4-4-SBUs and possesses two intersecting eight-ring channels along the [010] and [001] directions. The heteroatom and Al atom occupy the same position in the framework.

The preceding examples demonstrate that the incorporation of heteroatoms in the aluminophosphate frameworks can stabilize the zeolite framework and induce novel zeotype structures.

3.3.2 Heteroatom substitution with Ge atom to direct D4R

Because of their similarity germanium can be a substituent of silicon in zeolite frameworks. Because of the flexible Ge—O—Ge or Si—O—Ge angles, the introduction of Ge can stabilize the formation of D4R. Given that the framework density decreases with the size of the smallest rings [93], researchers have been exploring low-framework-density zeolites by Ge substitution approach, which may increase the possibility to form extra-large pore structures. A series of novel D4R-containing extra-large pore germanosilicates have been successfully prepared with designed organic SDAs and F^- ions in concentrated gels. Notable examples are the ITQ-*n* series of compounds synthesized by Corma and coworkers. Some zeolite structures resulting from heteroatom substitution with Ge atom are summarized in Table 2. Notably, with the preferential occupancy of germanium, ITQ-44 [6] was successfully synthesized which contains not only D4Rs but also the unusual D3Rs.

3.4 Topotactic Conversion of Framework

3.4.1 Topotactic condensation of layered structures (2D–3D)

In the search for new zeolite synthesis routes, the topotactic condensation of layered precursors into 3D structures has proven to be a promising method for generating novel zeolite materials with desired framework structures. Based on a solid-state reaction with thermal treatment, two silanol groups on the surface of two neighboring layers dehydrate spontaneously and high-silica zeolites can be formed while keeping the layered structures. In consideration of the structures, one must have proper numbers and intralayer distance of terminal silanol/siloxy groups on either side of the layer and suitable organic cations of low-charge density are also needed [107,108]. Up to now, several examples of high-silica silicates have been obtained by a topotactic conversion of layer silicates (shown in Table 3). Note that four new zeolite framework types, including CDO, NSI, RRO, and RWR, have been

Table 2: Some zeolite structures resulting from heteroatom substitution with Ge atom

Zeolite	Framework Type	Framework Density (T/1000 Å^3)	Channel System	Refs.
ITQ-15	UTL	15.4	2D, 14R×12R	[94]
IM-12	UTL	15.2	2D, 14R×12R	[95]
ITQ-17	BEC	14.6	3D, 12R×12R×12R	[96]
ITQ-21	-	13.6	3D, 12R×12R×12R	[97]
ITQ-22	IWW	16.1	3D, 12R×10R×8R	[98]
ITQ-24	IWR	15.5	3D, 12R×10R×10R	[99]
ITQ-26	IWS	14.3	3D, 12R×12R×12R	[68]
ITQ-29	LTA	13.8	3D, 8R×8R×8R	[100]
ITQ-33	ITT	12.3	3D, 18R×10R×10R	[101]
ITQ-34	ITR	17.4	3D, 10R×10R×9R	[69]
ITQ-37	-ITV	10.3	3D, 30R×30R×30R	[5]
ITQ-38	ITG	16.1	3D, 12R×10R×10R	[102]
ITQ-40	-IRY	10.1	3D, 16R×16R×15R	[70]
ITQ-43	-	11.4	3D, 28R×12R×12R	[73]
ITQ-44	IRR	10.9	3D, 18R×12R×12R	[6]
ITQ-49	IRN	15.5	1D, 8R	[103]
SU-15	SOF	15.5	3D, 8R×9R×12R	[104]
SU-32	STW	15.3	1D, 8R	[104]
SU-78A	-	15.3	3D, 12R×12R×12R	[105]
SU-78B	-	14.8	3D, 12R×12R×12R	[105]
PKU-12	-CLO	10.6	3D, 20R×20R×20R	[106]

Table 3: Examples of topotactic condensation of layered structures

Type Material	Framework Type	Precursors	Refs.
RUB-41	RRO	RUB-39	[109]
RUB-24	RWR	R-RUB-18	[110]
Zeolite RWR	RWR	AA-RUB-18	[111]
RUB-37	CDO	RUB-36, RUB-38, RUB-48	[112,113]
CDS-1	CDO	PLS-1	[114]
MCM-65[a]	CDO	MCM-65[b]	[115]
ZSM-52	CDO		[115]
ZSM-55	CDO		[115]
UZM-25	CDO	UZM-13, UZM-17, UZM-19	[116]
Unnamed	CDO	Several unnamed	[117]
Unnamed	CDO	PLS-4	[118]
RUB-15-SOD	SOD	HOAc-RUB-15	[119]
Silica-sodalite	SOD	HAc-RUB-15	[120]
CDS-3	FER	PLS-3	[118]
Ferrierite	FER	PREFER	[121]
ERS-12[a]	Incomplete condensation	ERS-12[b]	[122]
MCM-47[a]		MCM-47[b]	[123]
MCM-22	MWW	MCM-22-precursor	[124]
ERB-1[a]	MWW	ERB-1	[125]
ITQ-1[a]	MWW	ITQ-1[b]	[126]
Nu-6(2)	NSI	Nu-6 (1)	[127]
EU-20	CAS	EU-19	[128]
EU-20b	CAS	EU-19	[129]
IPC-4	PCR	IPC-1P	[16]

[a]The type material is calcined.
[b]The precursor is synthesized without calcination.

synthesized via topotactic condensation of layered structures, and these zeolites have not been achieved by direct hydrothermal synthesis so far.

Early in 1988, Lowe and coworkers prepared piperazine silicate EU-19 in the hydrothermal system and found it could convert into a thermally stable silica polymorph EU-20 (CAS/NSI) upon calcination [128]. However, the structure of EU-20 was not clear at that time and the condensation was only proposed conceptually. Boron aluminosilicate zeolite MCM-22 (MWW) provides the first unequivocal example, which was produced with a layered precursor, MCM-22P [124].

Ikeda and coworkers synthesized a novel silicate zeolite CDS-1 (CDO) with a cylindrically double saw-edged framework in 2004 [114]. CDS-1 was converted by dehydration-condensation between the layers of pentagonal cylinder-layered silicate (PLS-1). Then in 2008, Okubo and coworkers synthesized novel silicate sodalite (SOD) by topotactic conversion [119]. Another example is shown by Gies and coworkers with the preparation of silicate zeolite RUB-41 (RRO) [109]. By studying the synthesis of hydrous-layered silicates carefully, they used dimethyldipropylammonium hydroxide as the organic SDA to synthesize the layered silicate precursor (RUB-39), and RUB-41 was synthesized after the calcination of RUB-39.

Incorporating Si between layers may open the possibility to expand the pore size in 3D structures. For most zeolites synthesized by direct topotactic condensation without the addition of a silane agent between the layers, the layered silicate precursors usually contain pentagon cylinders and zeolites with 8R or 10R pores synthesized after calcination. Recently, Wu and coworkers found that the apertures formed between the layers could be expanded by inserting a monomeric Si source into the interlayer spaces through one-step dialkoxysilylation (shown in Fig. 14) [130]. By calcinating to remove the organic moieties, new structures with expanded pore windows, high crystallinity, and outstanding hydrothermal stability were formed. This interlayer-expanded zeolite exhibited higher catalytic activities in the redox and solid acid-catalyzed reactions of bulky molecules than the conventional MWW zeolite. The interlayer-expanded method and topotactic condensation of layered structures could be applied to other zeolite frameworks such as FER, CDO, and MCM-47 for new structures.

Notably, different condensations of the same layer may result in different zeolite structures. There are some structurally related pairs of framework types HEU-RRO, FER-CDO, and CAS-NSI, which may explain the stacking disorder in some cases. For example, the *fer* layers with stacking sequence ABAB along the *a*-axis can cause two different framework types of FER and CDO due to the different shift vector. As for the *cas* layers, the CAS framework type is connected by *cas* layers with stacking sequence ABAB, and the NSI framework type is connected by *cas* layers with stacking sequence AAA. Layers A and B are the mirror image of each other [107].

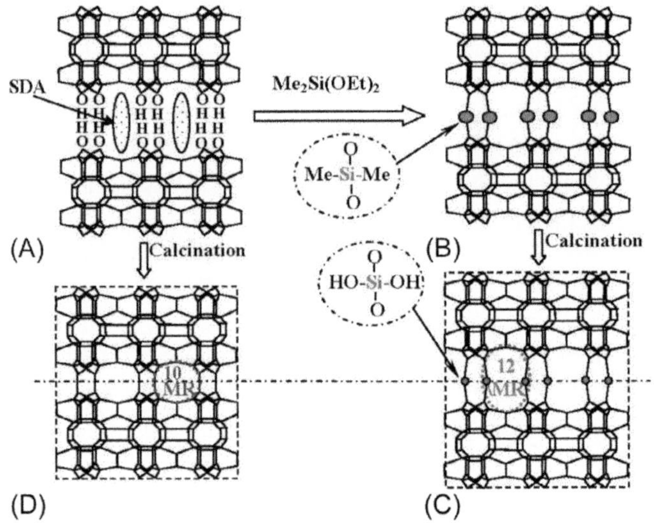

Fig. 14
Postsynthesis of interlayer-expanded zeolites through dialkoxysilylation of lamellar precursor (eg, MWW precursor): (A) MWW lamellar precursor; (B) interlayer expanded via the reaction of silane and silanols; (C) a novel 3D structure with expanded pore window after calcination; (D) common 3D MWW structure obtained through direct calcination. *Reprinted with permission from Wu P, Ruan J, Wang L, Wu L, Wang Y, Liu Y, Fan W, He M, Terasaki O, Tatsumi T. Methodology for synthesizing crystalline metallosilicates with expanded pore windows through molecular alkoxysilylation of zeolitic lamellar precursors. J Am Chem Soc 2008;130:8178–8187. Copyright (2008) American Chemical Society.*

3.4.2 Topotactic conversion of a 3D framework

Opposite to the 2D to 3D topotactic condensation, recent efforts have focused on the 3D to 2D framework conversion. In 2011, Čejka and coworkers first reported the 3D to 2D conversion of germanosilicate UTL zeolite with or without boron [131]. The conversion is based on that the structure of UTL can be viewed as dense layers pillared by D4R bridging units, and by mild treatment, the 3D structure can be degenerated to 2D layers, which can be transformed into novel 3D structures by further posttreatment. In the experiment, the calcined UTL crystalline samples were treated hydrothermally from room temperature up to 100°C, and the pH range was adjusted from neutral to acidic (0.1 M HCl). D4R units were removed from the structure; consequently, the 3D structure transformed into the 2D lamellar structure. Following the idea of topotactic condensation of layered structures, the lamellar germanosilicate zeolite precursor IPC-1P, which was obtained through UTL 3D to 2D conversion, was treated with a silylating agent in the nitric acid solution and the new germanosilicate zeolite called IPC-2 was synthesized after calcination.

Using the same UTL zeolite, Kirschhock and coworkers proposed the theoretical concept of inverse sigma transformation by removal of the germanium layer followed by calcination and generated a new zeolite structure experimentally (shown in Fig. 15) [132]. Since the D4R

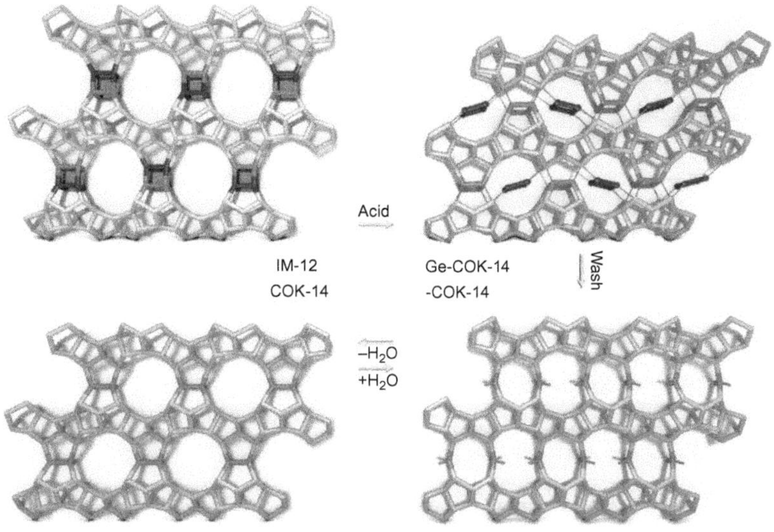

Fig. 15
Topotactic conversion of IM-12 zeolite to form a Ge-free framework COK-14. The Ge-4R units (*dark blue*) in IM-12 zeolite are shifted into the channels of Ge-COK-14 by acid leaching. By further leaching, a Ge-free interrupted framework -COK-14 is formed, which can be annealed on calcination (550°C). (Si-atoms in layers in *gray*, Si-atoms in layer-connecting Si-4Rs in *light blue*.) *Reprinted by permission from Macmillan Publishers Ltd: Verheyen E, Joos L, Van Havenbergh K, Breynaert E, Kasian N, Gobechiya E, Houthoofd K, Martineau C, Hinterstein M, Taulelle F, Van Speybroeck V, Waroquier M, Bals S, Van Tendeloo G, Kirschhock CEA, Martens JA. Design of zeolite by inverse sigma transformation. Nat Mater 2012;11:1059–1064. Copyright (2012).*

building units are the preferential location of germanium and the bond between the layers and D4R is very sensitive to be hydrolysed in the presence of water, germanium-containing D4R units in the germanosilicate IM-12 (UTL) can be removed in strong mineral acid. By leaching and calcining the products, the new all-silica zeolite COK-14 with an intersecting 12- and 10-ring channel system is synthesized.

4 Conclusion

Through decades of development and continuous efforts, there has been significant progress in zeolite synthesis. Some strategies toward greener and designed synthesis of zeolite materials have been developed. As for the greener synthesis, some principles, such as cost and environmental impact reduction, waste reused, energy efficiency, process safety, and the like, have been well applied in guiding the zeolite synthesis. Some effective synthetic routes have been explored such as organic-template-free synthesis, solvent-free synthesis, low-toxic and cheap organic-template-directed synthesis, organic SDAs reused and mother-liquor-reused synthesis, microwave-assisted synthesis, and ionothermal synthesis. However, some green chemistry principles still challenge the synthetic chemists to further develop new methods toward greener synthesis of zeolites. For example, catalytic reagents with high selectivity can reduce the side products and accelerate reactions, which meet the demand for atom-economy and energy efficiency. Given that zeolites can be applied in catalysis of many industrial reactions, however, the study of catalysts for zeolite synthesis is still unexplored. Accelerating zeolite synthesis selectively with catalysts may be a new route for synthetic chemists. In addition, analytical techniques for monitoring the zeolite synthesis process in situ are needed to be developed. Analytical methods may detect and quantify impurities in zeolite synthesis, thus optimizing production processes and minimizing waste as well as providing mechanistic information. Furthermore, the combination of several greener synthesis methods may extend the benefits of green routes and make zeolite synthesis faster, safer, more economic, and environmentally friendly.

Toward the designed synthesis of zeolites, great efforts have been paid to the rational synthesis of zeolite materials with desired structures and functions. Powerful strategies have been developed by novel SDA design, CDM, heteroatom substitution, topotactic conversion, and the like. Theoretically, desired zeolite structures can be designed with predefined building units and pore geometries by computer simulations [20]. Moreover, a set of reliable criteria has been proposed based on the observation that the local interatomic distances in all existing zeolites strictly obey several rules [133]. For example, there is a linear relationship between the average T—T and T—O distances (T = Si, Al, P, etc.) in existing zeolites. Using such criteria, the most realizable synthetic candidates can be selected from the huge pool of hypothetical zeolites, which provides a promising shortcut to future target synthesis. Yu and coworkers depicted a future blueprint for the tailor-made synthesis of zeolite materials [20]. The zeolite materials with desired structures and functions could eventually be accessed by the rational synthesis approaches with the assistance of computer simulations. It should be pointed out that establishing the correlation between synthesis parameters and the structures, as well as a deep understanding of the formation mechanism of zeolites at the molecular level, is the key for designed synthesis of zeolite materials.

In summary, the direction of synthesis chemistry has focused more and more on an atom-economic, a greener, and a rational synthesis of materials with desired structures and functions [134]. Driven by this direction, more research is expected in the future.

References

[1] Xu R, Pang W, Yu J, Huo Q, Chen J. Chemistry of zeolites and related porous materials synthesis and structure. Singapore: Wiley; 2007.
[2] Breck DW. Zeolite molecular sieves: structure, chemistry, and use. New York: Wiley; 1974.
[3] Čejka J, van Bekkum H, Corma A, Schüth F. Introduction to zeolite science and practice. 3rd ed. Studies in surface science and catalysis. vol. 168. Amsterdam: Elsevier; 2007.
[4] Baerlocher Ch, McCusker LB. http://www.iza-structure.org/databases/.
[5] Sun J, Bonneau C, Cantín Á, Corma A, Díaz-Cabañas MJ, Moliner M, et al. The ITQ-37 mesoporous chiral zeolite. Nature 2009;458:1154–7.

[6] Jiang J, Jordá JL, Díaz-Cabañas MJ, Yu J, Corma A. The synthesis of an extra-large-pore zeolite with double three-ring building units and a low framework density. Angew Chem Int Ed 2010;49:4986–8.
[7] Dodin M, Paillaud J-L, Lorgouilloux Y, Caullett P, Elkaïm E, Bats N. A zeolitic material with a three-dimensional pore system formed by straight 12-and 10-ring channels synthesized with an imidazolium derivative as structure-directing agent. J Am Chem Soc 2010;132:10221–3.
[8] Han Y, Li Y, Yu J, Xu R. A gallogermanate zeolite constructed exclusively by three-ring building units. Angew Chem Int Ed 2011;50:3003–5.
[9] Broach RW, Kirchner RM. Structures of the K^+ and NH^{4+} forms of Linde J. Micropor Mesopor Mater 2011;143:398–400.
[10] Sedlmaier SJ, Döblinger M, Oeckler O, Weber J, Schmedt auf der Günne J, Schnick W. Unprecedented zeolite-like framework topology constructed from cages with 3-rings in a barium oxonitridophosphate. J Am Chem Soc 2011;133:12069–78.
[11] Baerlocher Ch, Weber T, McCusker LB, Palatinus L, Zones SI. Unraveling the perplexing structure of the zeolite SSZ-57. Science 2011;333:1134–7.
[12] Littlefield BTR, Weller MT. Lightweight nanoporous metal hydroxide-rich zeotypes. Nat Commun 2012;3 [art. no. 1114.].
[13] Armstrong JA, Weller MT. Beryllosilicate frameworks and zeolites. J Am Chem Soc 2010;132:15679–86.
[14] Liu Z, Song X, Li J, Li Y, Yu J, Xu R. $|(C_4NH_{12})_4|[M_4Al_{12}P_{16}O_{64}]$ (M = Co, Zn): new heteroatom-containing aluminophosphate molecular sieves with two intersecting 8-ring channels. Inorg Chem 2012;51:1969–74.
[15] Shao L, Li Y, Yu J, Xu R. Divalent-metal-stabilized aluminophosphates exhibiting a new zeolite framework topology. Inorg Chem 2012;51:225–9.
[16] Roth WJ, Nachtigall P, Morris RE, Wheatley PS, Seymour VR, Ashbrook SE, et al. A family of zeolites with controlled pore size prepared through a top-down method. Nat Chem 2013;5:628–33.
[17] Anastas PT, Warner JC. Green chemistry: theory and practice. New York: Oxford University Press; 1998.
[18] Dahl JA, Maddux BLS, Hutchison JE. Toward greener nanosynthesis. Chem Rev 2007;107:2228–69.
[19] Yu J, Xu R. Rational approaches toward the design and synthesis of zeolitic inorganic open-framework materials. Acc Chem Res 2010;43:1195–204.
[20] Wang Z, Yu J, Xu R. Needs and trends in rational synthesis of zeolitic materials. Chem Soc Rev 2012;41:1729–41.
[21] Barrer RM, Denny PJ. Hydrothermal chemistry of the silicates. Part IX. Nitrogenous aluminosilicates. J Chem Soc 1961;971–82.
[22] Meng X, Xie B, Xiao F-S. Organotemplate-free routes for synthesizing zeolites. Chin J Catal 2009;30:965–71.
[23] Argauer RJ, Landolt GR. Crystalline zeolite ZSM-5 and method of preparing the same. US Patent 3 702 886; 1972.
[24] Li H, Xiang S, Wu D, Liu Y, Zhang X, Liu S. Study on the synthesis of zeolite ZSM-5. Chem J Chin Univ 1981;2:517–9.
[25] Grose RW, Flanigen EM. Novel zeolite compositions and processes for preparing and using same. US Patent 4 257 885; 1981.
[26] Shiralkar VP, Clearfield A. Synthesis of the molecular sieve ZSM-5 without the aid of templates. Zeolites 1989;9:363–70.
[27] Leonowicz ME, Vaughan DEW. Proposed synthetic zeolite ECR-1 structure gives a new zeolite framework topology. Nature 1987;329:819–21.
[28] Warrender SJ, Wright PA, Zhou W, Lightfoot P, Camblor MA, Shin C-H, et al. TNU-7: a large-pore gallosilicate zeolite constructed of strictly alternating MOR and MAZ layers. Chem Mater 2005;17:1272–4.
[29] Song J, Dai L, Ji Y, Xiao F-S. Organic template free synthesis of aluminosilicate zeolite ECR-1. Chem Mater 2006;18:2775–7.
[30] Ng E-P, Chateigner D, Bein T, Valtchev V, Mintova S. Capturing ultrasmall EMT zeolite from template-free systems. Science 2012;335:70–3.
[31] Dougnier F, Patarin J, Guth JL, Anglerot D. Synthesis, characterization, and catalytic properties of silica-rich faujasite-type zeolite (FAU) and its hexagonal analog (EMT) prepared by using crown-ethers as templates. Zeolites 1992;12:160–6.

[32] Xie B, Song J, Ren L, Ji Y, Li J, Xiao F-S. Organotemplate-free and fast route for synthesizing Beta zeolite. Chem Mater 2008;20:4533–5.
[33] Xie B, Zhang H, Yang C, Liu S, Ren L, Zhang L, et al. Seed-directed synthesis of zeolites with enhanced performance in the absence of organic templates. Chem Commun 2011;47:3945–7.
[34] Yokoi T, Yoshioka M, Imai H, Tatsumi T. Diversification of RTH-type zeolite and its catalytic application. Angew Chem Int Ed 2009;48:9884–7.
[35] Iyoki K, Kamimura Y, Itabashi K, Shimojima A, Okubo T. Synthesis of MTW-type zeolites in the absence of organic structure-directing agent. Chem Lett 2010;39:730–1.
[36] Wu Z, Song J, Ji Y, Ren L, Xiao F-S. Organic template-free synthesis of ZSM-34 zeolite from an assistance of zeolite L seeds solution. Chem Mater 2008;20:357–9.
[37] Zhang L, Yang C, Meng X, Xie B, Wang L, Ren L, et al. Organotemplate-free syntheses of ZSM-34 zeolite and its heteroatom-substituted analogues with good catalytic performance. Chem Mater 2010;22:3099–107.
[38] Suzuki Y, Wakihara T, Itabashi K, Ogura M, Okubo T. Cooperative effect of sodium and potassium cations on synthesis of ferrierite. Top Catal 2009;52:67–74.
[39] Zhang H, Guo Q, Ren L, Yang C, Zhu L, Meng X, et al. Organotemplate-free synthesis of high-silica ferrierite zeolite induced by CDO-structure zeolite building units. J Mater Chem 2011;21:9494–7.
[40] Itabashi K, Kamimura Y, Iyoki K, Shimojima A, Okubo T. A working hypothesis for broadening framework types of zeolites in seed-assisted synthesis without organic structure-directing agent. J Am Chem Soc 2012;134:11542–9.
[41] Ren L, Wu Q, Yang C, Zhu L, Li C, Zhang P, et al. Solvent-free synthesis of zeolites from solid raw materials. J Am Chem Soc 2012;134:15173–6.
[42] Morris RE, James SL. Solventless synthesis of zeolites. Angew Chem Int Ed 2013;52:2163–5.
[43] Zones SI, Hwang S-J. Synthesis of high silica zeolites using a mixed quaternary ammonium cation, amine approach: discovery of zeolite SSZ-47. Chem Mater 2002;14:313–20.
[44] Wang J, Song J, Yin C, Ji Y, Zou Y, Xiao F-S. Tetramethylguanidine-templated synthesis of aluminophosphate-based microporous crystals with AFI-type structure. Micropor Mesopor Mater 2009;117:561–9.
[45] Takewaki T, Beck LW, Davis ME. Zincosilicate CIT-6: a precursor to a family of *BEA-type molecular sieves. J Phys Chem B 1999;103:2674–9.
[46] Lee H, Zones SI, Davis ME. A combustion-free methodology for synthesizing zeolites and zeolite-like materials. Nature 2003;425:385–8.
[47] Duan F, Li J, Chen P, Yu J, Xu R. A low-cost route to the syntheses of microporous cobalt-substituted aluminophosphates by using the waste mother-liquor. Micropor Mesopor Mater 2009;126:26–31.
[48] Rao KJ, Vaidhyanathan B, Ganguli M, Ramakrishnan PA. Synthesis of inorganic solids using microwaves. Chem Mater 1999;11:882–95.
[49] Chu P, Dwyer FG, Vartuli JC. Crystallization method employing microwave radiation. US Patent 4 778 666; 1988.
[50] Kingstone HM, Haswell SJ. Microwave-enhanced chemistry. Washington, DC: American Chemical Society; 1997.
[51] Tompsett GA, Conner WC, Yngvesson KS. Microwave synthesis of nanoporous materials. ChemPhysChem 2006;7:296–319.
[52] Lin S, Li J, Sharma RP, Yu J, Xu R. Fabrication of SAPO-34 crystals with different morphologies by microwave heating. Top Catal 2010;53:1304–10.
[53] Yang G, Wei Y, Xu S, Chen J, Li J, Liu Z, et al. Nanosize-enhanced lifetime of SAPO-34 catalysts in methanol-to-olefin reactions. J Phys Chem C 2013;117:8214–22.
[54] Cooper ER, Andrews CD, Wheatley PS, Webb PB, Wormald P, Morris RE. Ionic liquids and eutectic mixtures as solvent and template in synthesis of zeolite analogues. Nature 2004;430:1012–6.
[55] Ma Z, Yu J, Dai S. Preparation of inorganic materials using ionic liquids. Adv Mater 2010;22:261–85.
[56] Parnham ER, Morris RE. The ionothermal synthesis of cobalt aluminophosphate zeolite frameworks. J Am Chem Soc 2006;128:2204–5.

[57] Ma H, Tian Z, Xu R, Wang B, Wei Y, Wang L, et al. Effect of water on the ionothermal synthesis of molecular sieves. J Am Chem Soc 2008;130:8120–1.
[58] Xing H, Li J, Yan W, Chen P, Jin Z, Yu J, et al. Cotemplating ionothermal synthesis of a new open-framework aluminophosphate with unique Al/P ratio of 6/7. Chem Mater 2008;20:4179–81.
[59] Wei Y, Tian Z, Gies H, Xu R, Ma H, Pei R, et al. Ionothermal synthesis of an aluminophosphate molecular sieve with 20-ring pore openings. Angew Chem Int Ed 2010;49:5367–70.
[60] Morris RE. Ionic liquids and microwaves—making zeolites for emerging applications. Angew Chem Int Ed 2008;47:442–4.
[61] Xu Y, Tian Z, Wang S, Hu Y, Wang L, Wang B, et al. Microwave-enhanced ionothermal synthesis of aluminophosphate molecular sieves. Angew Chem Int Ed 2006;45:3965–70.
[62] Cai R, Liu Y, Gu S, Yan Y. Ambient pressure dry-gel conversion method for zeolite MFI synthesis using ionic liquid and microwave heating. J Am Chem Soc 2010;132:12776–7.
[63] Jiang J, Yu J, Corma A. Extra-large-pore zeolites: bridging the gap between micro and mesoporous structures. Angew Chem Int Ed 2010;49:3120–45.
[64] Baerlocher Ch, Xie D, McCusker LB, Hwang S-J, Chan IY, Ong K, et al. Ordered silicon vacancies in the framework structure of the zeolite catalyst SSZ-74. Nat Mater 2008;7:631–5.
[65] Baerlocher Ch, Gramm F, Massüger L, McCusker LB, He Z, Hovmöller S, et al. Structure of the polycrystalline zeolite catalyst IM-5 solved by enhanced charge flipping. Science 2007;315:1113–6.
[66] Gramm F, Baerlocher Ch, McCusker LB, Warrender SJ, Wright PA, Han B, et al. Complex zeolite structure solved by combining powder diffraction and electron microscopy. Nature 2006;444:79–81.
[67] Hong SB, Lear EG, Wright PA, Zhou W, Cox PA, Shin C-H, et al. Synthesis, structure solution, characterization, and catalytic properties of TNU-10: a high-silica zeolite with the STI topology. J Am Chem Soc 2004;126:5817–26.
[68] Dorset DL, Strohmaier KG, Kliewer CE, Corma A, Díaz-Cabañas MJ, Rey F, et al. Crystal structure of ITQ-26, a 3D framework with extra-large pores. Chem Mater 2008;20:5325–31.
[69] Corma A, Díaz-Cabañas MJ, Jordá JL, Rey F, Sastre G, Strohmaier KG. A zeolitic structure (ITQ-34) with connected 9-and 10-ring channels obtained with phosphonium cations as structure directing agents. J Am Chem Soc 2008;130:16482–3.
[70] Corma A, Díaz-Cabañas MJ, Jiang J, Afeworki M, Dorset DL, Soled SL, et al. Extra-large pore zeolite (ITQ-40) with the lowest framework density containing double four- and double three-rings. Proc Natl Acad Sci U S A 2010;107:13997–4002.
[71] Dorset DL, Kennedy GJ, Strohmaier KG, Díaz-Cabañas MJ, Rey F, Corma A. P-derived organic cations as structure-directing agents: synthesis of a high-silica zeolite (ITQ-27) with a two-dimensional 12-ring channel system. J Am Chem Soc 2006;128:8862–7.
[72] Simancas R, Dari D, Velamazán N, Navarro MT, Cantín A, Jordá JL, et al. Modular organic structure-directing agents for the synthesis of zeolites. Science 2010;330:1219–22.
[73] Jiang J, Jordá JL, Yu J, Baumes LA, Mugnaioli E, Díaz-Cabañas MJ, et al. Synthesis and structure determination of the hierarchical meso-microporous zeolite ITQ-43. Science 2011;333:1131–4.
[74] Jiang J, Xu Y, Cheng P, Sun Q, Yu J, Corma A, et al. Investigation of extra-large pore zeolite synthesis by a high-throughput approach. Chem Mater 2011;23:4709–15.
[75] Martínez-Franco R, Moliner M, Yun Y, Sun J, Wan W, Zou X, et al. Synthesis of an extra-large molecular sieve using proton sponges as organic structure-directing agents. Proc Natl Acad Sci U S A 2013;110:3749–54.
[76] Balkus Jr KJ, Gabrielov AG. Synthesis of novel molecular sieves using a metal complex as template. US Patent 5 489 424; 1996.
[77] Xu Y, Li Y, Han Y, Song X, Yu J. A gallogermanate zeolite with eleven-membered-ring channels. Angew Chem Int Ed 2013;52:5501–3.
[78] Choi M, Srivastava R, Ryoo R. Organosilane surfactant-directed synthesis of mesoporous aluminophosphates constructed with crystalline microporous frameworks. Chem Commun 2006;4380–2.

[79] Choi M, Cho HS, Srivastava R, Venkatesan C, Choi D-H, Ryoo R. Amphiphilic organosilane-directed synthesis of crystalline zeolite with tunable mesoporosity. Nat Mater 2006;5:718–23.
[80] Choi M, Na K, Kim J, Sakamoto Y, Terasaki O, Ryoo R. Stable single-unit-cell nanosheets of zeolite MFI as active and long-lived catalysts. Nature 2009;461:246–9.
[81] Na K, Choi M, Park W, Sakamoto Y, Terasaki O, Ryoo R. Pillared MFI zeolite nanosheets of a single-unit-cell thickness. J Am Chem Soc 2010;132:4169–77.
[82] Inayat A, Knoke I, Spiecker E, Schwieger W. Assemblies of mesoporous FAU-type zeolite nanosheets. Angew Chem Int Ed 2012;51:1962–5.
[83] Pérez-Ramírez J. Zeolite nanosystems: imagination has no limits. Nat Chem 2012;4:250–1.
[84] Chaikittisilp W, Suzuki Y, Mukti RR, Suzuki T, Sugita K, Itabashi K, et al. Formation of hierarchically organized zeolites by sequential intergrowth. Angew Chem Int Ed 2013;52:3355–9.
[85] Wilson ST. Overview of zeolite synthesis strategies. In: Xu R, Gao Z, Chen J, Yan W, editors. From zeolites to porous MOF materials—the 40th anniversary of international zeolite conference Studies in surface science and catalysis, 170A. Amsterdam: Elsevier; 2007. p. 3–18.
[86] Lewis GJ, Miller MA, Moscoso JG, Wilson BA, Knight LM, Wilson ST. Recent advances in the science and technology of zeolites and related materials. In: van Steen E, Claeys M, Callanan LH, editors. Studies in surface science and catalysis, 154A. Amsterdam: Elsevier; 2004. p. 364–72.
[87] Moscoso JG, Lewis GJ, Gisselquist JL, Miller MA, Rohde LM. Crystalline aluminosilicate zeolitic composition: UZM-9. US Patent 6 713 041; 2004.
[88] Lee JH, Park MB, Lee JK, Min H-K, Song MK, Hong SB. Synthesis and characterization of ERI-type UZM-12 zeolites and their methanol-to-olefin performance. J Am Chem Soc 2010;132:12971–82.
[89] Park MB, Cho SJ, Hong SB. Synthesis of aluminosilicate and gallosilicate zeolites via a charge density mismatch approach and their characterization. J Am Chem Soc 2011;133:1917–34.
[90] Park MB, Lee Y, Zheng A, Xiao F-S, Nicholas CP, Lewis GJ, et al. Formation pathway for LTA zeolite crystals synthesized via a charge density mismatch approach. J Am Chem Soc 2013;135:2248–55.
[91] Strohmaier KG, Vaughan DEW. Structure of the first silicate molecular sieve with 18-ring pore openings, ECR-34. J Am Chem Soc 2003;125:16035–9.
[92] Song X, Li Y, Gan L, Wang Z, Yu J, Xu R. Heteroatom-stabilized chiral framework of aluminophosphate molecular sieves. Angew Chem Int Ed 2009;48:314–7.
[93] Brunner GO, Meier WM. Framework density distribution of zeolite-type tetrahedral nets. Nature 1989;337:146–7.
[94] Corma A, Díaz-Cabañas MJ, Rey F, Nicolopoulus S, Boulahya K. ITQ-15: the first ultralarge pore zeolite with a bi-directional pore system formed by intersecting 14- and 12-ring channels, and its catalytic implications. Chem Commun 2004;1356–7.
[95] Paillaud J-L, Harbuzaru B, Patarin J, Bats N. Extra-large-pore zeolites with two-dimensional channels formed by 14 and 12 rings. Science 2004;304:990–2.
[96] Corma A, Navarro MT, Rey F, Valencia S. Synthesis of pure polymorph C of Beta zeolite in a fluoride-free system. Chem Commun 2001;1486–7.
[97] Corma A, Díaz-Cabañas MJ, Martínez-Triguero J, Rey F, Rius J. A large-cavity zeolite with wide pore windows and potential as an oil refining catalyst. Nature 2002;418:514–7.
[98] Corma A, Rey F, Valencia S, Jordá JL, Rius J. A zeolite with interconnected 8-, 10- and 12-ring pores and its unique catalytic selectivity. Nat Mater 2003;2:493–7.
[99] Castañeda R, Corma A, Fornés V, Rey F, Rius J. Synthesis of a new zeolite structure ITQ-24, with intersecting 10-and 12-membered ring pores. J Am Chem Soc 2003;125:7820–1.
[100] Corma A, Rey F, Rius J, Sabater MJ, Valencia S. Supramolecular self-assembled molecules as organic directing agent for synthesis of zeolites. Nature 2004;431:287–90.
[101] Corma A, Díaz-Cabañas MJ, Jordá JL, Martínez C, Moliner M. High-throughput synthesis and catalytic properties of a molecular sieve with 18-and 10-member rings. Nature 2006;443:842–5.
[102] Moliner M, Willhammar T, Wan W, González J, Rey F, Jordá JL, et al. Synthesis design and structure of a multipore zeolite with interconnected 12-and 10-MR channels. J Am Chem Soc 2012;134:6473–8.

[103] Hernández-Rodríguez M, Jordá JL, Rey F, Corma A. Synthesis and structure determination of a new microporous zeolite with large cavities connected by small pores. J Am Chem Soc 2012;134:13232–5.
[104] Tang L, Shi L, Bonneau C, Sun J, Yue H, Ojuva A, et al. A zeolite family with chiral and achiral structures built from the same building layer. Nat Mater 2008;7:381–5.
[105] Yu Z-B, Han Y, Zhao L, Huang S, Zheng Q-Y, Lin S, et al. Intergrown new zeolite beta polymorphs with interconnected 12-ring channels solved by combining electron crystallography and single-crystal X-ray diffraction. Chem Mater 2012;24:3701–6.
[106] Su J, Wang Y, Lin J, Liang J, Sun J, Zou X. A silicogermanate with 20-ring channels directed by a simple quaternary ammonium cation. Dalton Trans 2013;42:1360–3.
[107] Marler B, Gies H. Hydrous layer silicates as precursors for zeolites obtained through topotactic condensation: a review. Eur J Mineral 2012;24:405–28.
[108] Li J, Corma A, Yu J. Synthesis of new zeolite structures. Chem Soc Rev 2015;44:7112–27.
[109] Wang YX, Gies H, Marler B, Müller U. Synthesis and crystal structure of zeolite RUB-41 obtained as calcination product of a layered precursor: a systematic approach to a new synthesis route. Chem Mater 2005;17:43–9.
[110] Marler B, Ströter N, Gies H. The structure of the new pure silica zeolite RUB-24, $Si_{32}O_{64}$, obtained by topotactic condensation of the intercalated layer silicate RUB-18. Micropor Mesopor Mater 2005;83:201–11.
[111] Ikeda T, Oumi Y, Takeoka T, Yokoyama T, Sano T, Hanaoka T. Preparation and crystal structure of RUB-18 modified for synthesis of zeolite RWR by topotactic conversion. Micropor Mesopor Mater 2008;110:488–500.
[112] Marler B, Wang Y, Song J, Gies H. RUB-36, a new layer silicate with ferrierite-type layers which can be condensated to form a highly ordered CDO-type zeolite. In: Book of Abstracts, 15th International Zeolite Conference, Beijing; 2007. p. 599–600.
[113] Marler B, Wang Y, Song J, Gies H. Condensation of layer silicates with ferrierite-type layers to form porous tectosilicates. Book of abstracts, III international workshop on layered materials. Germany: Bochum; 2010. p. 67–8.
[114] Ikeda T, Akiyama Y, Oumi Y, Kawai A, Mizukami F. The topotactic conversion of a novel layered silicate into a new framework zeolite. Angew Chem Int Ed 2004;43:4892–6.
[115] Dorset DL, Kennedy GJ. Crystal structure of MCM-65: an alternative linkage of ferrierite layers. J Phys Chem B 2004;108:15216–22.
[116] Knight LM, Miller MA, Koster SC, Gatter MG, Benin AI, Willis RR, et al. UZM-13, UZM-17, UZM-19 and UZM-25: synthesis and structure of new layered precursors and a zeolite discovered via combinatorial chemistry techniques. In: Xu R, Gao Z, Chen J, Yan W, editors. From zeolites to porous MOF materials—the 40th anniversary of international zeolite conference Studies in surface science and catalysis, 170A. Amsterdam: Elsevier; 2007. p. 338–46.
[117] Yamamoto K, Ikeda T, Onodera M, Muramatsu A. Layered silicate as precursor for zeolite. In: Book of abstracts, 15th international zeolite conference, Beijing, 2007. p. 452–3.
[118] Ikeda T, Kayamori S, Mizukami F. Synthesis and crystal structure of layered silicate PLS-3 and PLS-4 as a topotactic zeolite precursor. J Mater Chem 2009;19:5518–25.
[119] Moteki T, Chaikittisilp W, Shimojima A, Okubo T. Silica sodalite without occluded organic matters by topotactic conversion of lamellar precursor. J Am Chem Soc 2008;130:15780–1.
[120] Voss H, Therre J, Gies H, Marler B. Template-free clathrasils and clathrasil membranes. Intern. Patent WO/2010/037690 A1, Application No. PCT/EP2009/062479; 2010.
[121] Schreyeck L, Caullet P, Mougenel JC, Guth JL, Marler B. PREFER: a new layered (alumino) silicate precursor of FER-type zeolite. Micropor Mater 1996;6:259–71.
[122] Millini R, Carluccio LC, Carati A, Bellussi G, Perego C, Cruciani G, et al. ERS-12: a new layered tetramethylammonium silicate composed by ferrierite layers. Micropor Mesopor Mater 2004;74:59–71.
[123] Burton A, Accardi RJ, Lobo RF, Falcioni M, Deem MW. MCM-47: a highly crystalline silicate composed of hydrogen-bonded ferrierite layers. Chem Mater 2000;12:2936–42.

[124] Leonowicz ME, Lawton JA, Lawton SL, Rubin MK. MCM-22: a molecular sieve with two independent multidimensional channel systems. Science 1994;264:1910–3.
[125] Millini R, Perego G, Parker Jr WO, Bellussi G, Carluccio L. Layered structure of ERB-1 microporous borosilicate precursor and its intercalation properties towards polar molecules. Micropor Mater 1995;4:221–30.
[126] Camblor MA, Corell C, Corma A, Díaz-Cabañas MJ, Nicolopoulos S, González-Calbet JM, et al. A new microporous polymorph of silica isomorphous to zeolite MCM-22. Chem Mater 1996;8:2415–7.
[127] Zanardi S, Alberti A, Cruciani G, Corma A, Fornés V, Brunelli M. Crystal structure determination of zeolite Nu-6(2) and its layered precursor Nu-6(1). Angew Chem Int Ed 2004;43:4933–7.
[128] Blake AJ, Franklin KR, Lowe BM. Preparation and properties of piperazine silicate (EU-19) and a silica polymorph (EU-20). J Chem Soc Dalton Trans 1988;2513–2517.
[129] Marler B, Camblor MA, Gies H. The disordered structure of silica zeolite EU-20b, obtained by topotactic condensation of the piperazinium containing layer silicate EU-19. Micropor Mesopor Mater 2006;90:87–101.
[130] Wu P, Ruan J, Wang L, Wu L, Wang Y, Liu Y, et al. Methodology for synthesizing crystalline metallosilicates with expanded pore windows through molecular alkoxysilylation of zeolitic lamellar precursors. J Am Chem Soc 2008;130:8178–87.
[131] Roth WJ, Shvets OV, Shamzhy M, Chlubná P, Kubů M, Nachtigall P, et al. Postsynthesis transformation of three-dimensional framework into a lamellar zeolite with modifiable architecture. J Am Chem Soc 2011;133:6130–3.
[132] Verheyen E, Joos L, Van Havenbergh K, Breynaert E, Kasian N, Gobechiya E, et al. Design of zeolite by inverse sigma transformation. Nat Mater 2012;11:1059–64.
[133] Li Y, Yu J, Xu R. Criteria for zeolite frameworks realizable for target synthesis. Angew Chem Int Ed 2013;52:1673–7.
[134] Noyori R. Synthesizing our future. Nat Chem 2009;1:5–6.

CHAPTER 2

Metal-Organic Frameworks and Related Materials: Miles to Go

V.I. Isaeva, L.M. Kustov
N.D. Zelinsky Institute of Organic Chemistry, Russian Academy of Sciences, Moscow, Russia

Abbreviations

4-btapa	tris[*N*-(4-pyridyl)amide]
4-btapa	1,3,5-benzene tricarboxylic acid tris-*N*-(4-pyridyl)amide
ad	adeninate
adip	5,5′-(9,10-anthracenediyl)-di-isophthalate
bbc	4,4′4″-[benzene-1,3,5-triyl-tris(benzene-4,1-diyl)-tribenzoate]
bdc	benzene-1,4-dicarboxylate
bipy	bipyridine
bpdc	4,4′-biphenyldicarboxylate
bptc bpydc	biphenyl-3,3′,5,5′-tetracarboxylate
bpydc	2,2′-bipyridyl-5,5′-dicarboxylate
btb	benzene-1,3,5-tribenzoate
btc	benzene-1,3,5-benzenetricarboxylate
bte	4,4′4″-[benzene-1,3,5-triyl-tris(ethyne-2,1-diyl)tribenzoate]
btt	1,3,5-benzenetristetrazol-5-yl
CHA	cyclohexylamine
ctc	*cis,cis*-1,3,5-cyclohexanetricarboxylate
cyclam	1,4,8,11-tetraazacyclotetra decane
dabco	1,4-diazabicyclo[2.2.2]octane
ddbb	(R)-6,6′-dichloro-2,2′-dihydroxy-1,1′-binaphthyl-4,4-bipyridine
DEF	diethylformamide
DETA	diethylenetriamine
DMA	*N,N*-dimethylacetamide
DMF	*N,N*-dimethylformamide
DMSO	dimethyl sulfoxide
$e_3$hb	ethyl-3-hydroxybutyrate
EtOH	ethanol
H_2hfipbb	4,4′-(hexafluoroisopropylidene)-bis(benzoic acid)
H_4dhtp	2,5-dihydroxyterephthalate
hmmi	2-hydroxymethyl-1-methylimidazole
hmt	hexamethylenetetramine
hpdc	4,5,9,10-tetrahydropyrene-2,7-dicarboxylate
htdbd	4,4′-(6-hydroxy-1,3,5-triazine-2,4-diyl)-bis (azanediyl)dibenzoic acid
ica	imidazolate-2-carboxyaldehyde

Zeolites and Zeolite-like Materials. http://dx.doi.org/10.1016/B978-0-444-63506-8.00002-1

im	imidazolate
imdc	4,5-imidazoledicarboxylate
ipa	isopropanolamine
JUC	Jilin University China
l-lac	l-lactic acid
meim	methylimidazolate
MOF	metal-organic framework
ndc	naphthalenedicarboxylate
ntb	4,4′,4-nitrilotrisbenzoate
PCN	porous coordination network
PCP	porous coordination polymers
pda	*p*-phenylenediacrylate
pdc	pyrene-2,7-dicarboxylate
pip	piperazine
ptmtc	4,4′,4-methanetriyltris(2,3,5,6-tetrachlorobenzoate)
pydca	pyridine-2,5-dicarboxylate
pyz	pyrazine
pzdc	pyrazine-2,5-dicarboxylate
qs	8-hydroxyquinoline-5-sulfonate
R^6-bdc	1,2-dihydrocyclobutylphenylene-3,6-dicarboxylate
SBU	secondary building unit
$t_2$dc	thieno[3,2-b]thiophene-2,5-dicarboxylate
ta	tartaric acid
tatab	4,4′,4″-*s*-triazine-1,3,5-triyltri-*p*-aminobenzenecarboxylate
tba	tri-*n*-butylamine
tbip	5-tertbutylisophthalate
tea	triethylamine
ted	triethylenediamine
tpa	tri-*n*-propylamine
ttdc	thieno-[3,2]thiophene-2,5-dicarboxylate
ZIF	zeolitic imidazolate framework

1 Introduction

Metal-organic frameworks (MOFs) are crystalline coordination polymers built from organic linkers (bridging ligands) and inorganic nodes. MOFs are a novel class of microporous polymer materials, whose key structural elements are mono- or polynuclear coordination sites—metal ions interlinked by organic fragments. When researchers describe the synthesis of MOFs, they use such words as "construction," "logical synthesis," "molecular design," and "modular chemistry" to reflect the fact that such synthesis puts into practice the attractive but often hard to implement idea of assembling molecular units, "modules," into a new compound with controlled physical and chemical properties.

Although now we can witness the appearance of about 1200–1500 papers per annum related to MOFs and coordination polymers, the first work was published only in 1989 [1]. The authors predicted the fast growth of the number of these materials due to the variability of their compositions and structures. Later Yaghi coined the term MOF [2]. In 2002 covalent organic frameworks (COFs) were built on the basis of imidazolate structures that exhibit similarities with zeolites; they were therefore named ZIFs (zeolitic imidazole frameworks).

The modular nature (a combination of inorganic and organic components) of these new porous materials is perfectly suited for chemical manipulations aimed at fine-tuning of the structures and functions of MOFs in accordance with their specific application [3]. One of the most remarkable features of MOFs is their extremely high porosity. Depending on the sizes of ligands and inorganic building units, as well as the framework connectivity, open channels and pores with sizes ranging from a few angstroms to several nanometers are present in MOFs (Table 1).

In spite of a number of similarities to zeolites, MOFs exhibit also certain advantages and disadvantages compared to zeolites. The main advantage is the possibility of synthesis of a

Table 1: The MOF structures of different topology

MOF	Surface Area, BET (m^2/g)	Pore Volume (cm^3/g)	Pore Aperture (Å)	Refs.
IRMOF-3 $Zn_4O(NH_2\text{-bdc})_3$ 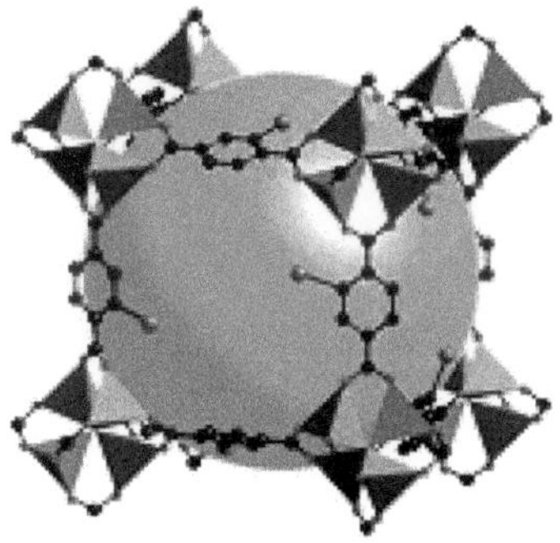	3062	1.07		[4]
IRMOF-8 $Zn_4O(ndc)_3$ 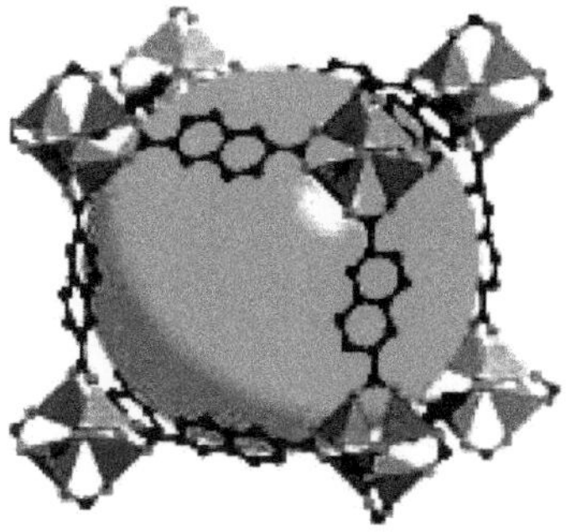	1466	0.52		[4]

(Continued)

Table 1: The MOF structures of different topology—cont'd

MOF	Surface Area, BET (m^2/g)	Pore Volume (cm^3/g)	Pore Aperture (Å)	Refs.
IRMOF-11 $Zn_4O(hpdc)_3$ 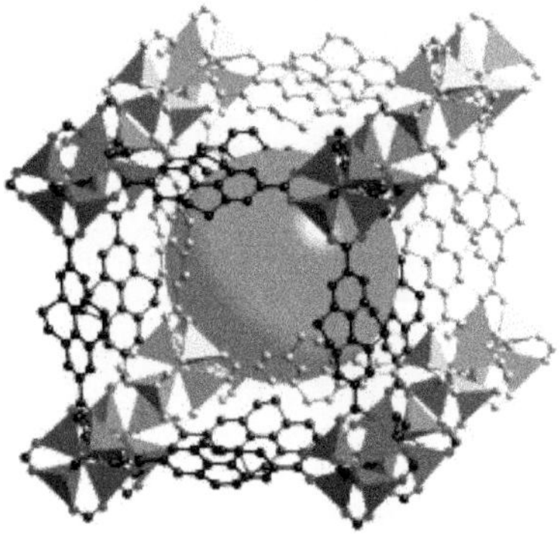	2100	0.73		[4]
MOF-74 $Zn_2(2,5\text{-}(OH)_21,4\text{-}(bdc)_2)$ 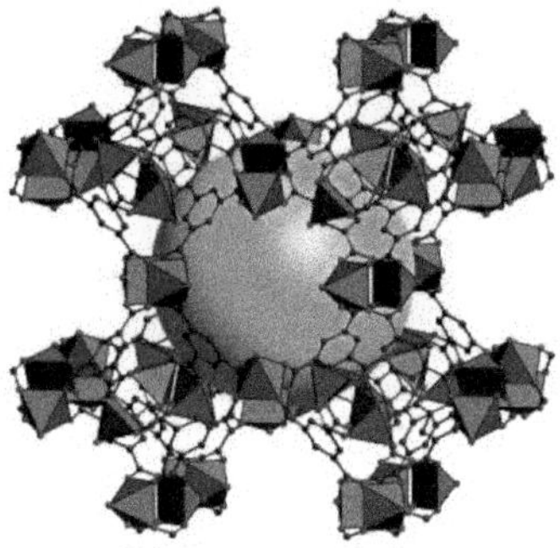	1132	0.39		[5]
MOF-177 $Zn_4O(btb)_3$ 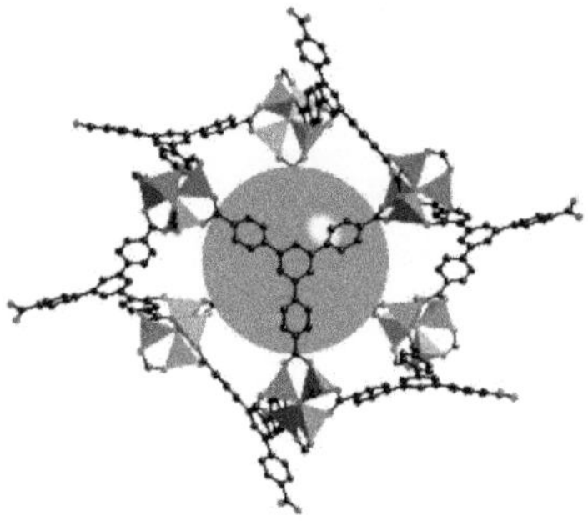	4526	1.61		[6]

Table 1: The MOF structures of different topology—cont'd

MOF	Surface Area, BET (m^2/g)	Pore Volume (cm^3/g)	Pore Aperture (Å)	Refs.
MOF-200 $Zn_4O(bbc)_2$ 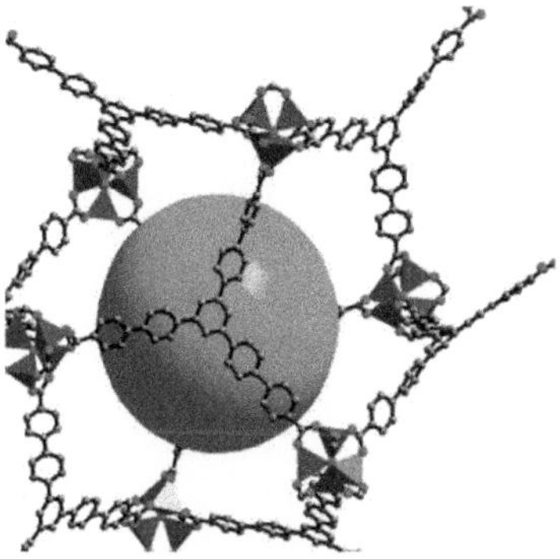	4530	3.59	18×28	[7]
MOF-210 $Zn_4O(bte)(bpdc)$ 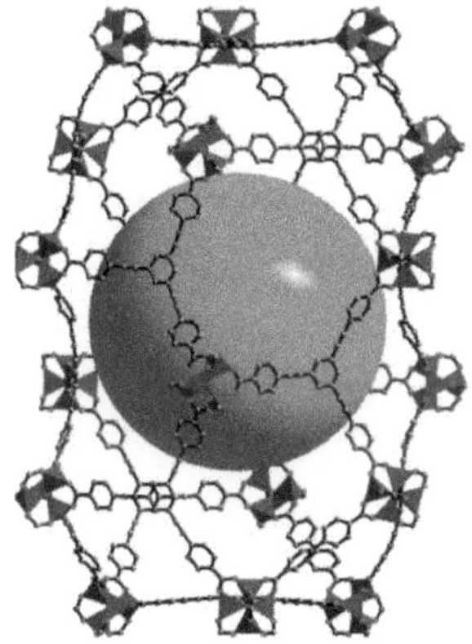	6240	3.6	26.9×48.3	[7]
DMOF-1 $Zn_2(1,4\text{-}(bdc)_2(dabco))$ 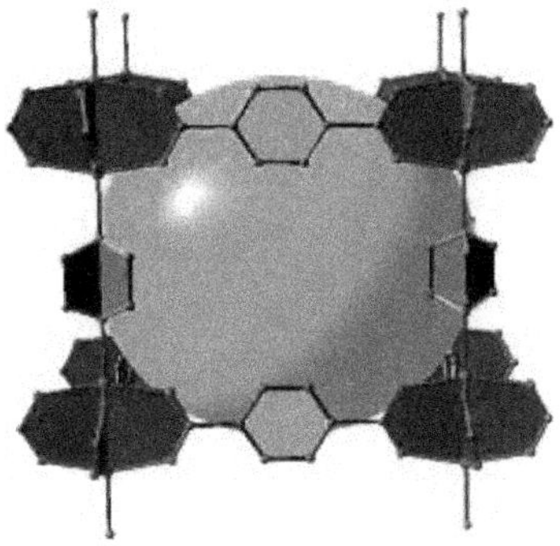	1450	-	7.5×7.5	[8]

(Continued)

Table 1: The MOF structures of different topology—cont'd

MOF	Surface Area, BET (m^2/g)	Pore Volume (cm^3/g)	Pore Aperture (Å)	Refs.
HKUST-1 (MOF-199) $Cu_2(btc)_{4/3}$ 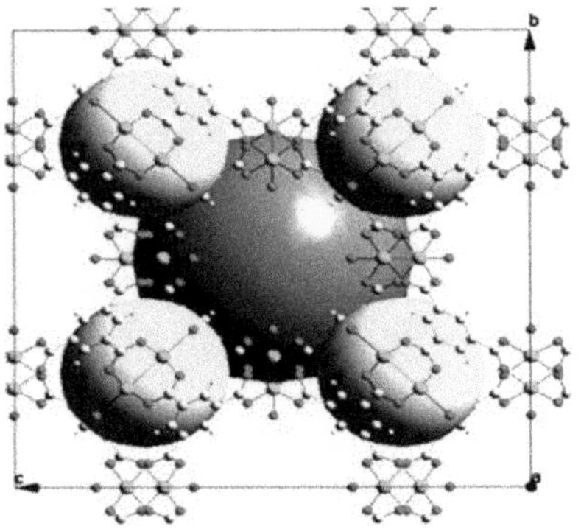	2175	0.75	7, 9	[9]
MIL-53(Cr) Al(OH)bdc 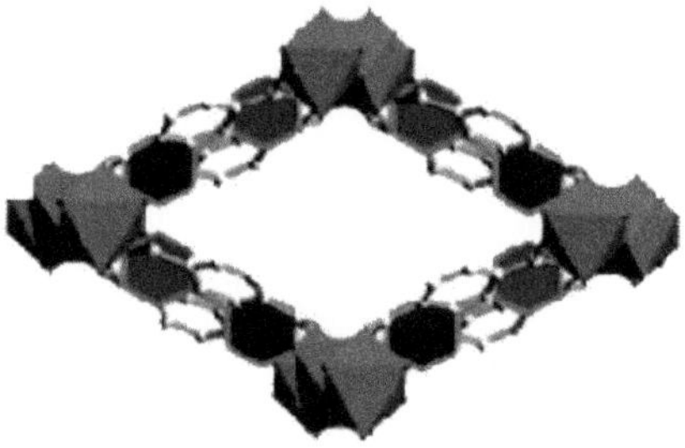	1100	-	8.5	[10]
MIL-100(Cr) $Cr_3O(H_2O)_3F(btc)_2$ 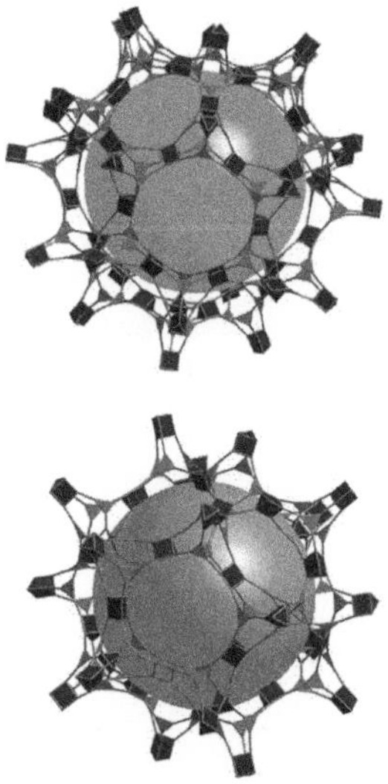	3100, Langmuir	1.16	25, 29	[11]

Table 1: The MOF structures of different topology—cont'd

MOF	Surface Area, BET (m^2/g)	Pore Volume (cm^3/g)	Pore Aperture (Å)	Refs.
MIL-101(Cr) $Cr_3O(H_2O)_2F(bdc)_3$ 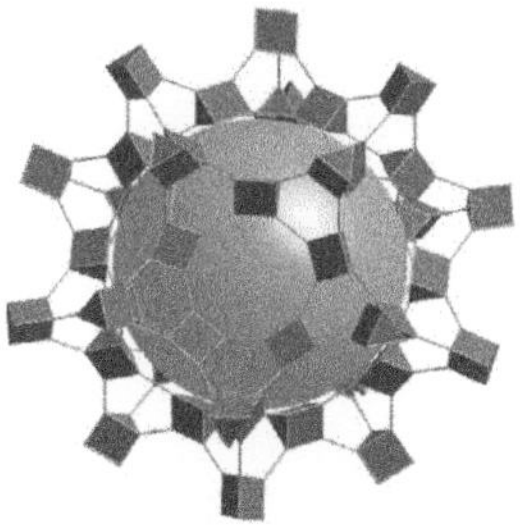 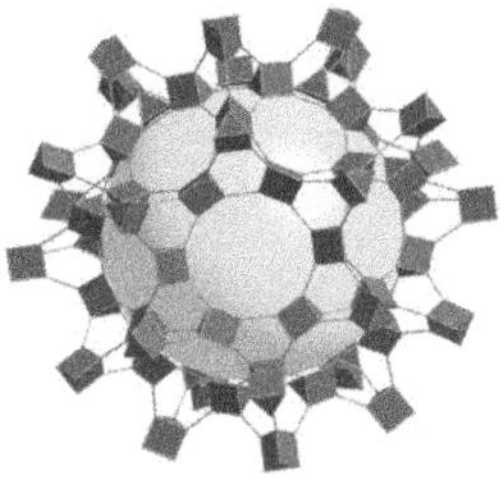	5900, Langmuir	1.9	29, 34	[12]
ZIF-8 $Zn(meim)_2$ 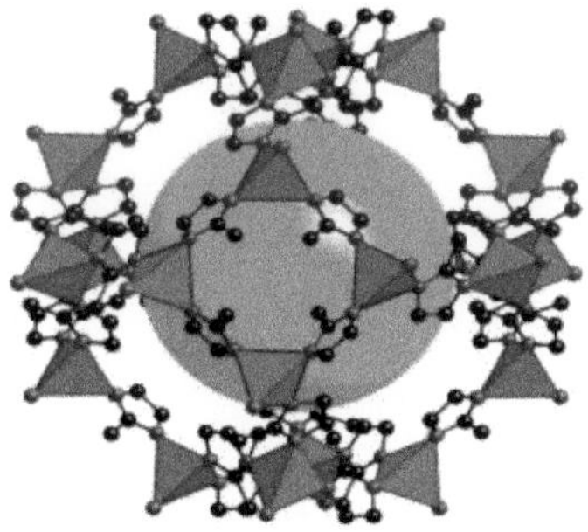	1630	0.64		[13]
ZIF-90 $Zn(ica)_2$ 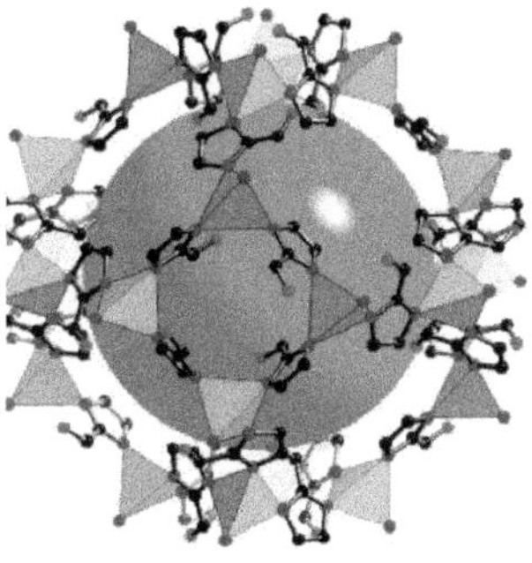	1270		3.5/11.2	[14]

(Continued)

Table 1: The MOF structures of different topology—cont'd

MOF	Surface Area, BET (m^2/g)	Pore Volume (cm^3/g)	Pore Aperture (Å)	Refs.
PCN-105 $Cd_4Na(H_2O)_2(htdbd)_3(tdbd)\cdot 10DMF\cdot 6EtOH\cdot 3H_2O$ 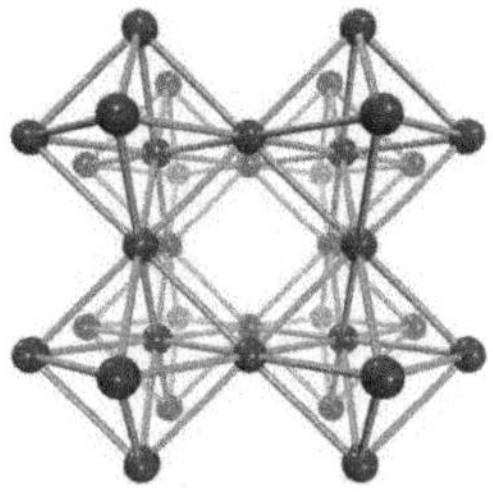	1067		9/21	[15]
MesoMOF-1 $Cu_3(tatab)_2(H_2O)_3\cdot 8DMF\cdot 9H_2O$ 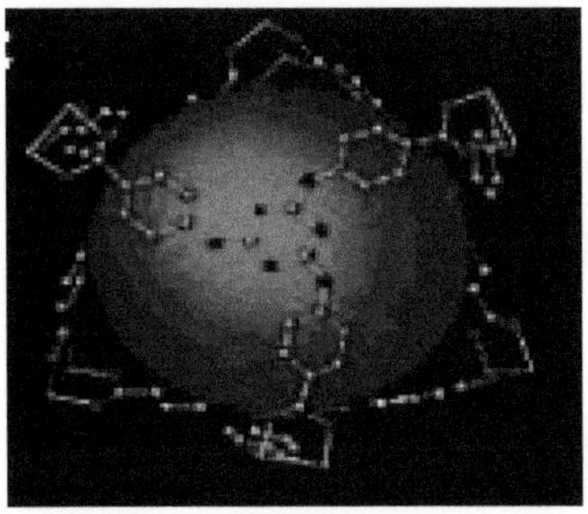	729		22.5×26.1	[16]
JUC-40 $Cd(btc)(H_2O)(H_2pip)_{0.5}\cdot H_2O$ 			5×10	[17]

Table 1: The MOF structures of different topology—cont'd

MOF	Surface Area, BET (m^2/g)	Pore Volume (cm^3/g)	Pore Aperture (Å)	Refs.
JUC-48 $Cd_3(bpdc)_3(DMF)\cdot 5DMF\cdot 18H_2O$ 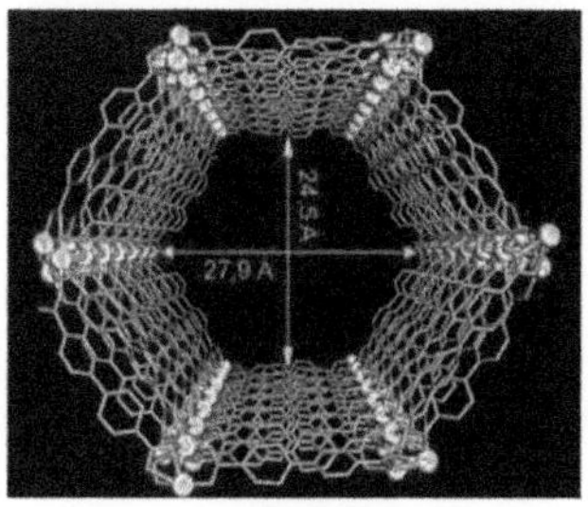	629	0.19	21.1×24.9	[18]
MOF-808 $Zr_6O_5(OH)_3(btc)_2(HCOO)_5\cdot(H_2O)_2$ 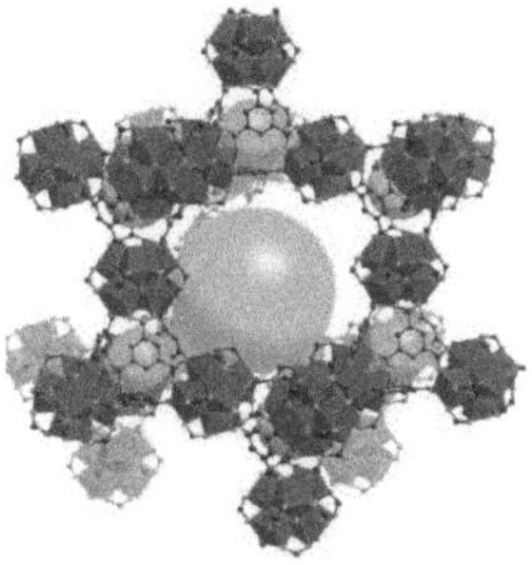	2060	0.84	18.4	[19]
bio-MOF-1 $Zn_8(ad)_4(bpdc)_6O$ 	1700			[20]

nearly infinite number of MOFs due to the combinations of diverse organic and inorganic blocks. A few thousand structures of MOFs and, in general, coordinated polymers are known, with more than a hundred being highly porous (large surface areas and pore volumes), that is, they are acceptable for catalysis, adsorption, and other applications related to the porosity. Second, the pore size of MOFs gently bridges the gap between classical zeolites (below 1 nm, even for the large-pore zeolites such as cloverite, VPI-5, IM-12, and ITQ-33) and mesoporous silicates (like MCM-41, usually around 3 nm). The pores of MIL-101 (3.4 nm) can accommodate bulky molecules such as polysaccharides and proteins [21]. The flexibility of MOFs may be beneficial in some applications (like shape-selective reactions of fine chemicals synthesis) or, on the contrary, may be a drawback, being the reason behind the poor hydrothermal and thermal stability of the framework. There are though a few examples of MOFs with a rather high hydrothermal and thermal stability extending up to 400–450°C. The specific surface areas of MOFs extend from a few m^2/g to over 5500 m^2/g (5640 m^2/g for MOF-177 [22] and 5900 m^2/g for MIL-101 [12]). Theoretically the surface areas up to 10,000 m^2/g are predicted [7].

The simplicity of synthesis with no necessity for the use of organic templates and other structure-directing agents, and for high-temperature calcinations to remove the template makes MOFs competitive in comparison with zeolites. The principles of the modular chemistry and methodology of synthesis of MOFs and, in general, porous coordination polymers (PCPs) are presented in a few very good recent reviews [23–25]. The synthesis and properties of the new JUC family of MOFs (JUC is the abbreviation of Jilin University China) is the focus of the review [26].

The synthesis in the absence of a solvent (solvent-free) has also became possible, either in a liquid ligand (eg, formic acid) or in water, which is a benign solvent. Also, the solvent-free synthesis can be carried out, at least in certain cases like Cu_3btc_2, by gently grinding the solid ligand and the salt as a precursor of the inorganic part of MOF [27,28]. The solvent-free approach was demonstrated in the case of Mg-MOF with formic acid as an organic linker [29]. The syntheses in water were also successful for aluminum fumarate MOF [30]. The latter, as well as aluminum terephthalate MOF, MIL-53(Al), MIL-101(Fe), MIL-101(Fe), and ZIF-8, ZIF-9, and ZIF-10 turned out to be highly stable under the hydrothermal conditions [31,32].

Ionic liquids (room-temperature molten salts) with chiral cations were efficiently used as reaction media to prepare homochiral MOFs [33]. But the use of homochiral ligands as precursors in the synthesis is even more efficient. For instance, Lin et al. [34] prepared MOFs with BINOL (1,1′-bi-2,2′-naphthol) and BINAP (2,2′-bis(diphenylphosphino)-1,1′-binaphthyl) as chiral linkers. Unfortunately, attempts to use such chiral catalysts in asymmetric catalysis showed some but not high enantiomeric excess in the studied reactions, perhaps the flexibility of the ligands in the MOF framework is not sufficient to induce the chirality effect in the reactions. The most illustrative example was presented by Lin [35]: the reaction of asymmetric hydrogenation of aromatic ketones demonstrated a 99.2% ee for Ru-BINAP/MOF systems.

The microwave-assisted (MW) synthesis also promised to provide a "green" solution to the problem of synthesis of MOFs [36]. Indeed, the use of MW radiation decreases dramatically the time required to complete the synthesis of a number of MOFs, produces particles with the distribution shifted toward the nanosize range and reduces the energy consumption, thereby improving the economics of the process. It is important that MW helps to accelerate the stage of nucleation of MOF crystals from solutions [37]. The small MOF particles are in strong demand for biomedical applications, magnetic resonance imaging, and in membrane or sensor systems.

Sonochemical treatment during the synthesis also speeds up the process due to the cavitation effect, also providing conditions for control of interpenetration of different frameworks (catenation) [38,39]. If separate frameworks are self-assembled within each other, this effect is known as framework catenation [40]. It is divided into interpenetration and interweaving. In interpenetrated MOFs, frameworks are maximally displaced from each other; in interwoven MOF structures, the distance between both frameworks is minimized (eg, IRMOF-13, Table 1). Catenation results in the enhancement of the rigidity of the material at the expense of its porosity, but increases the stability as well as may help in the case of adsorption and storage of small molecules due to the size selectivity.

Electrochemical approaches to the synthesis and postsynthetic modification of MOFs should also be mentioned here, for instance, the synthesis of Zn-EZIF [41] or HKUST-1 [41].

Novel MOFs/COFs structures appear at a high rate. We also discovered several new structures based on pyridine-2,5-dicarboxylate materials (Figs. 1 and 2). The novel metal-organic coordination networks comprised of Zn^{2+}-ions and multidentate pyridine-2,5-dicarboxylate linkers synthesized under mild conditions. Small variations in the parameters of the synthesis led to the formation of the metal organic systems of different topology, either three-dimensional (3D) MOFs or 0D molecular complex. The novel MOFs featured a permanent porosity and enhanced thermal stability (up to 350 °C). The specific surface BET areas for the synthesized 3D Zn-organic frameworks ranged within 330–380 m^2/g. The excess adsorption hydrogen capacity measured for pyridine-2,5-dicarboxylate MOF is about 1 wt.% at 77 K (1 atm). PXRD measurements revealed also the phase transformation as a function of the atmospheric water income for several samples without the crystallinity loss.

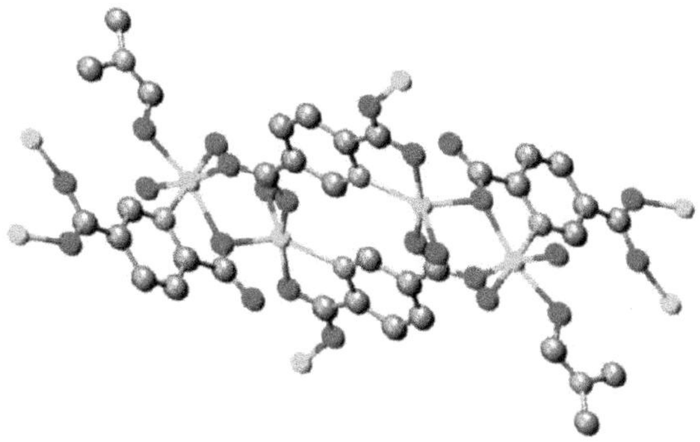

Fig. 1
3D structure of MOF (IOC-1).

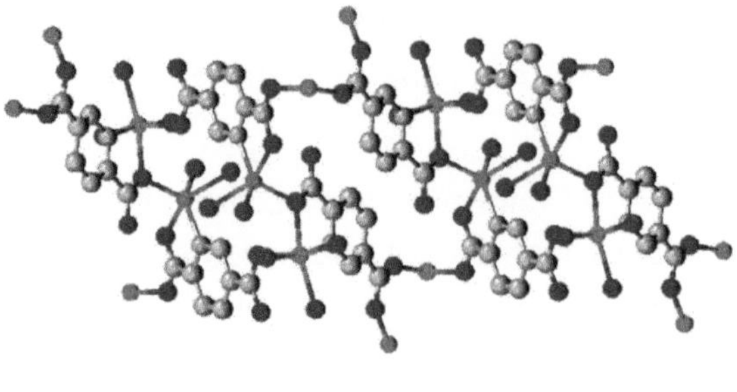

Fig. 2
3D structure of MOF (IOC-2).

The novel Zn(II)-based complex, [Zn(pz25dc)$(DMF)_2$] (**1**) (where pz25dc stands for pyrazine-2,5-dicarboxylate ligands) has been synthesized by the direct mixing method (Fig. 3). This structure confirmed by a combination of physicochemical methods features a 1D linear polymeric chain. The local structure of a Zn atom in the unit cell of [Zn(pz25dc)$(DMF)_2$] derived from EXAFS data is in a good agreement with that revealed from PXRD data [42].

Postsynthetic functionalization of MOFs, including the approaches of "click chemistry" is rationalized by Cohen et al. [43–45]. Unlike zeolites, for MOFs a variety of functional groups can be introduced by using the methodology of organic syntheses (amino, amido, merkapto groups, etc). Some of these functional groups improve the stabilization of encapsulated metal nanoparticles or guest molecules or increase the adsorption strength of adsorbates in gas storage and drug delivery systems, sensor materials, etc.

MOFs can be further coated with polymers or silica shells to bring them stability at certain pH values and prevent decomposition [46,47].

Of special interest are diverse composite materials based on MOFs, including host-guest assemblies, membranes (both ceramic and mixed matrix membranes), core-shell structures,

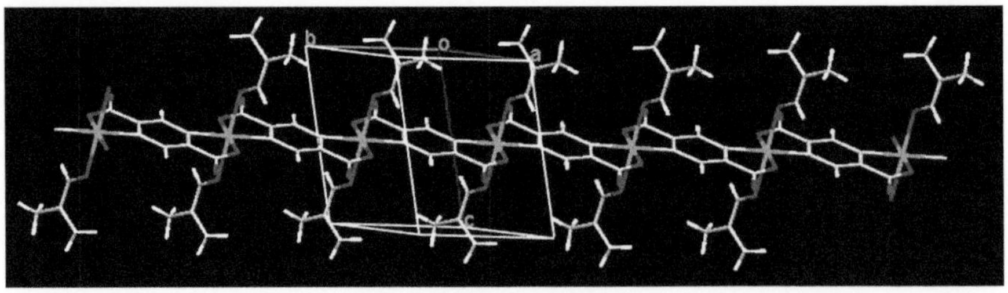

Fig. 3
[Zn(pz25dc)$(DMF)_2$] linear chain. *From Isaeva V, Chernyshev V, Afonina E, Tkachenko O, Klementiev K, Nissenbaum V, Grünert W, Kustov L. Inorg Chim Acta 2011;376:367–372 with permission.*

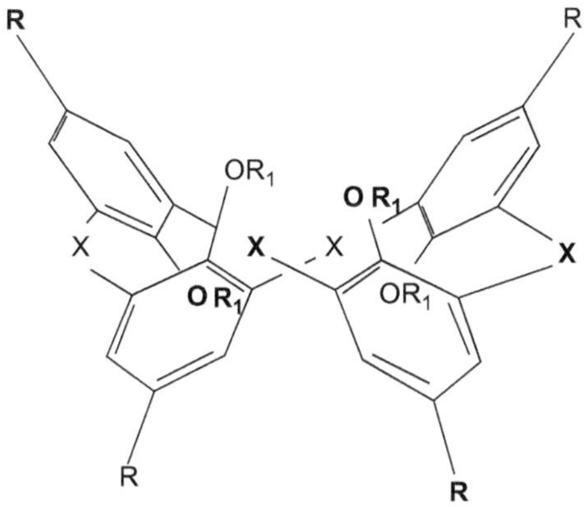

Fig. 4
Calix[4]arene.

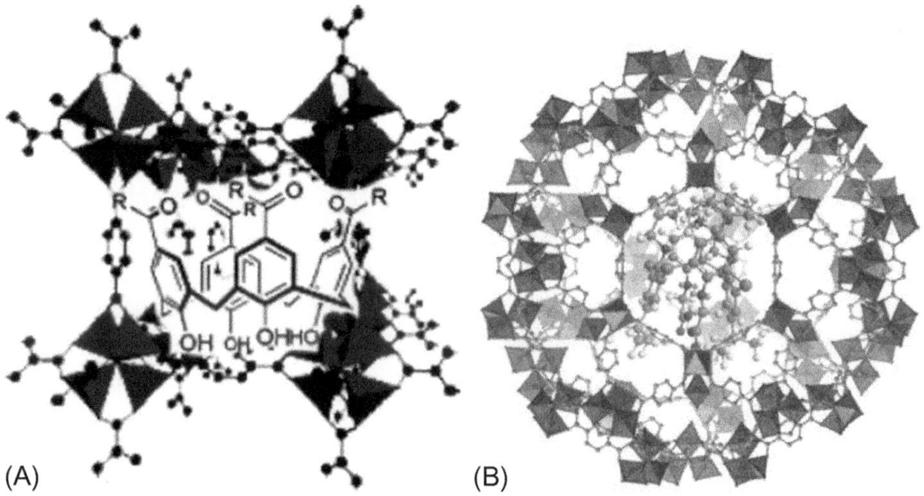

Fig. 5
(A) MOF-5/calix[4]arene composite. (B) NH_2-MIL-101(Al)/calix[4]arene composite.

and the like. Novel host-guest materials were synthesized by encapsulation of calix[4]arenes of different structures (Figs. 4 and 5) in situ during the synthesis of microporous MOF-5 and mesoporous MILs according to the procedure developed by us [48]. XRD analysis revealed the differences in the synthesized composite framework structures depending on the calix[4]arene content. N_2 adsorption measurements gave the BET surface areas ranging from 170 to 2560 m^2/g. The novel host-guest materials were used as the supports for Pd nanoparticles by incipient wetness deposition. The composite materials, Pd@MOF/calix[4]arene, and Pd@MILs/calix[4]arene have shown a high activity in the partial hydrogenation of the unsaturated compounds (1,3-cyclohexadiene, phenylacetylene, 1,4-butyne-2-diol). The selectivity in this reaction was directed by changing the substituent in the calix[4]arene molecule [49,50].

$$R_1 = CH_2COOH; \quad X = CH_2$$
$$C_s(\text{cone}) — R = H, \quad R_1 = CH_2COOH; \quad X = S$$
$$C_{s+}(\text{cone}) — R = t-Bu, \quad R_1 = CH_2COOH; \quad X = S$$
$$C_{CN}(\text{cone}) — R = t-Bu, \quad R_1 = CH_2CN; \quad X = S$$
$$C_{CONEt_2}(\text{cone}) — R = t-Bu, \quad R_1 = CH_2CONEt_2; \quad X = S$$

The possibilities of theoretical approaches to predict the structure and properties of MOFs, including the host-guest interactions were outlined in Ref. [51]. Modeling of MOFs definitely is characterized by some distinctive features compared to zeolites due to the less rigid structures of MOFs.

Their scalability is still in the early stage, but recent publications provide evidence in favor of improved space-time-yields of MOFs reaching 3000–5000 kg/m^3 per day, whereas zeolites are behind with some 50–150 kg/m^3 per day [31,52]. The ongoing research is directed toward solvent-free syntheses of MOFs or syntheses in water as a benign solvent, the use of microwave and sonochemical methods of activation to shorten the time of synthesis, as well as to produce nanoparticles of the materials.

The nature of a MOF as a solid determines not only catalytic and adsorptive but also other MOF physical properties, such as magnetic susceptibility, conductivity, and optical characteristics. Various particles can be encapsulated into the pores to bring about new properties to such composite materials. Also recently hierarchical and composite materials with host-guest structures are being developed quite extensively. The catalytic properties of a MOF relate not only to the presence of framework or extra-framework metal cations or reduced metals but also to the presence of functional groups on the inner surface of the MOF voids and channels [53]. To date, MOFs have also proved as promising systems in various applications, such as gas storage and separation, luminescence, and drug delivery [24,54]. Fig. 6A–C shows different ways of imaging the structure of MOF-5 (the numbering of MOFs is chronological in accordance with the time of their synthesis).

The first mentions of the use of MOFs in heterogeneous catalysis and gas adsorption and separation dated to the beginning of the past decade [1,55]. However, the systematic study of MOFs as catalytic and adsorptive systems began relatively recently. Lee et al. [56] noted that the backlog in the number of published works in the field of catalysis compared to works in the field of sorption is due to the fact that the development of synthetic methods for the preparation of MOFs took a long time, whereas obtaining the sorption characteristics of MOFs is an indispensable part of structural (textural) studies. Different approaches to reproducible, practicable synthesis of a great variety of MOFs have been developed to date. As a rule, they are synthesized by mixing organic ligands and metal salts under solvothermal reaction conditions at relatively low temperatures (typically, below 300 °C). The synthesis of one of the most well-known MOFs, MOF-5, is depicted later (Scheme 1):

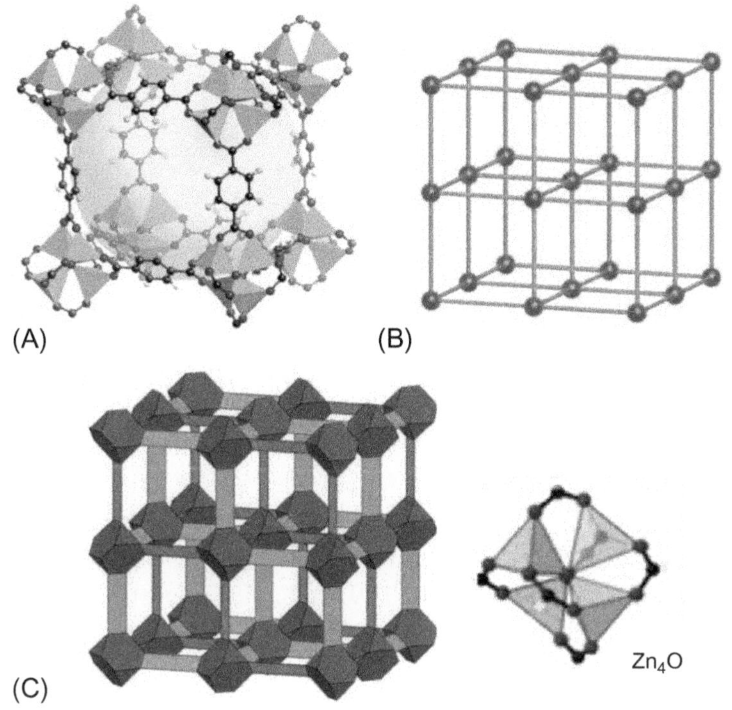

Fig. 6
Structure and topology of the MOF-5 MOF. (A) The structure of MOF-5 is represented as ZnO_4 tetrahedra connected by phenylenedicarboxylate linkers with the formation of a three-dimensional cubic framework, which contains interconnected small pores with a cross section of 0.8 nm, as well as a large spherical pore of 1.2 (1.5) nm in diameter. The large pore (or cavity) is represented as a sphere, whose diameter is determined by the distance between the van der Waals surfaces of the atoms of the framework, and which is inscribed into the cavity. (B) The topology of the MOF-5 structure (a primitive cubic lattice) represented as a ball and rod model. (C) The model shows a way of connection for $(OZn_4)O_{12}$ clusters and phenylenedicarboxylate ions. *From Tranchemontagne DJ, Hunt JR, Yaghi OM. Tetrahedron 2008;64:8553–8357 with permission.*

$$H_2BDC + Zn(NO_3)_2\,6H_2O \xrightarrow[\text{20°C, 150 h, } Et_3N]{DMF} Zn_4O(BDC)_3\,(DMF)_7 \longrightarrow Zn_4O(BDC)_3$$

H_2BDC is benzene-1,4-dicarboxylic acid.

Scheme 1
The synthesis of the framework MOF-5 [78]

The properties of organic ligands (bond angles, the ligand chain length, volume, chiral properties, etc.) play a key role in the formation of a particular type of MOF [54]. The topology of the structure of MOFs is determined by the coordination number of the metal ion (Fig. 6). The reagents are mixed in high-boiling polar solvents, such as water, *N*,*N*-dialkyl formamides, dimethyl sulfoxide, and acetonitrile. The most important parameters of the solvothermal synthesis are the temperature, the concentrations of the metal salt and organic ligand, the solubility of the reactants in the solvent, and pH of the solution. In addition, the direct mixing method has been developed, which is adapted for the synthesis of MOF-5 and a number of homologous MOFs (IRMOFs) as well as other MOFs and related materials [57]. A number of synthetic methods developed for the production of MOFs also include premixing of immiscible solvents [41,58]. One of the most promising techniques is the microwave-assisted solvothermal synthesis, which makes it possible to carry out the process over a wide temperature range, to reduce the time of crystallization, and to control the morphology and particle-size distribution [59–61].

Noteworthy, a broad range of conditions and procedures have been tried to obtain MOF-5 either in a large scale or with particular crystallite size or to reduce the synthesis time. Huang et al. reported a synthesis strategy for fast formation of nanocrystalline MOF-5 (30–150 nm in diameter) at room temperature [57]. "Soft" and "hard" template approaches were used to manipulate the morphology of MOFs. In further work, Yaghi with coworkers developed room-temperature or direct mixing methods for well-known MOFs: MOF-5, MOF-74, MOF-177, MOF-199, and a new isoreticular framework IRMOF-0 having the same cubic topology as MOF-5 (IRMOF-1) [5]. Ni et al. presented a microwave-assisted solvothermal synthesis, which allows one to synthesize MOF-5 crystals of the uniform size (about 4 μm) in less than a minute [37]. Synthesis procedures yielding nanocrystalline MOFs are interesting owing to their practical simplicity and because they present an opportunity to vary the geometrical properties of MOFs in adsorption or catalysis [57,62].

The size and chemical nature of the free volume of the pores formed due to this framework architecture are determined by the length and functionality of the organic linkers [63]. By selecting appropriate building blocks and bridges, one can control the architecture and properties of the material. Yaghi and coworkers introduced the notion of "reticular synthesis," with the integrity and rigidity of building blocks remaining unchanged during the entire process of "assembly" of the framework. "Reticular synthesis" profoundly differs from the synthesis of supramolecular ensembles because all the building blocks in the resulting MOF are linked with strong covalent bonds. Therefore, synthesis of MOFs requires not only selection of appropriate building blocks (modules) but also predicting how they will be linked in the final framework. The notion of a secondary building unit (SBU), originally used for describing the main zeolite fragments, helps to understand the structure of the framework being built [63]. In the context of MOFs, SBUs are simple geometric shapes, representing inorganic clusters and metal ion coordination spheres connected into the framework by organic linkers. Examples of secondary units in carboxylate MOFs are shown in Fig. 7. This approach provides a clear representation of the spatial structure of the resulting frameworks and allows a skillful researcher to use a large number of organic and inorganic SBUs with various geometries. The goal of the researcher is to determine the reaction conditions at

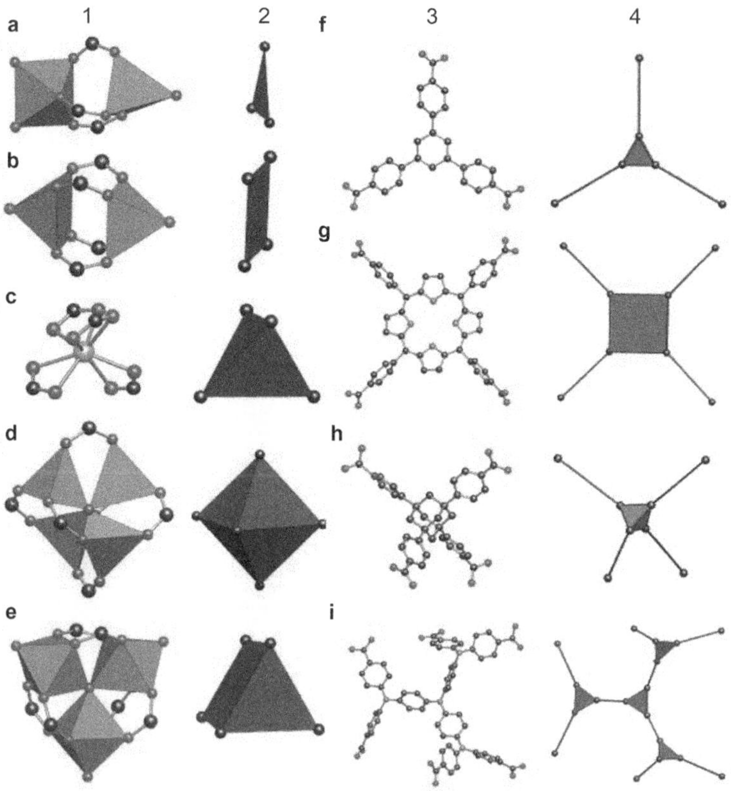

Fig. 7
The different SBUs in carboxylate MOF structures. *From Rowsell JLC, Yaghi OM. Micropor Mesopor Mater 2004;73:3–14 with permission.*

which a particular inorganic SBU will be formed in situ. Combining appropriate inorganic and fixed organic structural units results in the formation of MOFs with a predetermined structure [64].

As an example of this approach, we can consider the synthesis of 20 isoreticular 3D porous MOFs (IRMOFs) with the common formula $Zn_4O(O_2C)_6$ and identical lattice topology (Fig. 8) [4]. These frameworks are built from rigid organic linkers (derived from the completely deprotonized 1,4-phenylene dicarboxylic acid) and Zn^{2+} ions. Their structure comprises oxo-centered Zn_4O tetrahedra placed at the vertices of a cube and interlinked with six cabroxylate linkers, resulting in octahedral SBUs. The pore size was controlled by selecting cabroxylate bridges of a variable length, containing biphenyl, tetrahydropyrene, pyrene, and terphenyl fragments. Additionally, a number of IRMOFs were created containing various functional groups, such as $-Br$, $-NH_2$, $-C_2H_4-$, $-C_4H_6-$, $-OC_3H_7$, and $-OC_5H_{11}$ (Fig. 9).

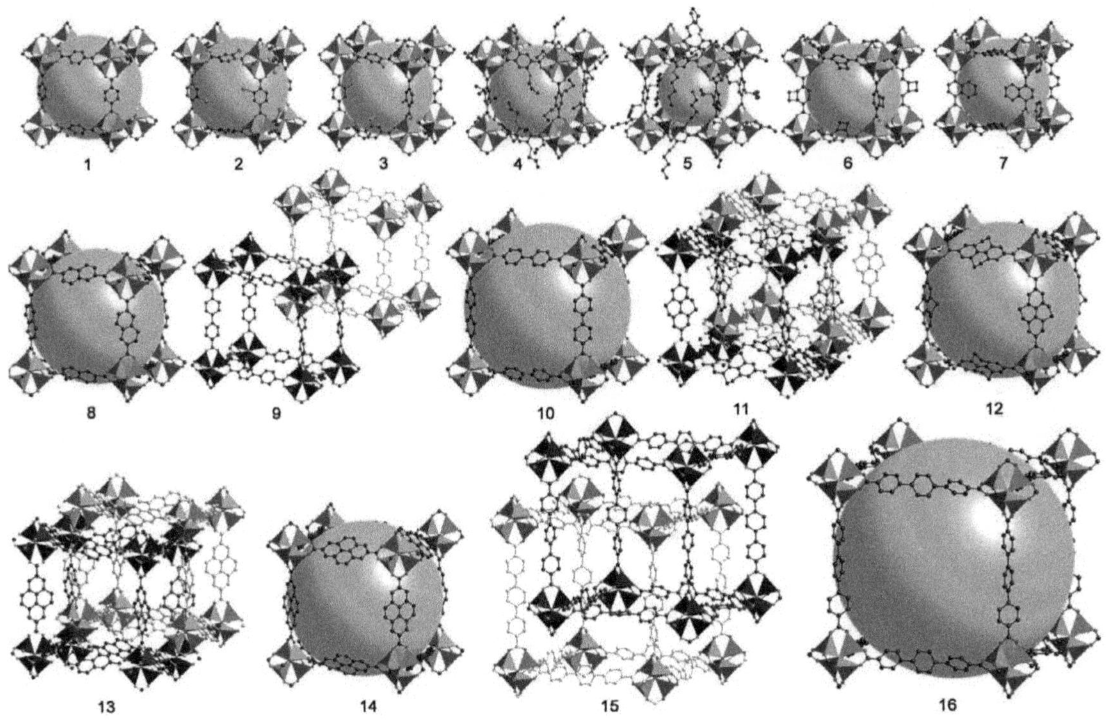

Fig. 8
IRMOF family. *From Eddaoudi M, Kim J, Rosi NL, Vodak D, Wachter J, O'Keeffe M. Science 2002;295:469–472 with permission.*

Approximately 11,000 MOFs have been synthesized recently [64–67]. Among these, c.3000 compounds have a 3D and c.6000, two-dimensional (2D) structure. Various organic components of MOFs can be classified as two main types: charged and uncharged. The first type includes, most importantly, carboxylate ions, and the second type includes heterocyclic linkers. The most commonly used metal ions are Cu(II), Mn(II), Zn(II), Ni(II), and lanthanides.

The disadvantage of using flexible heterocyclic fragments and aliphatic bridges, which allow free rotation around their axis, is the spatial uncertainty of the ligands surrounding the metal ion. This makes the control of the architecture of the synthesized MOF difficult. In contrast, rigid aromatic polydentate carboxylate fragments with chelating ability allow one to rigidly stabilize metal ions within metal-carboxylate lattice vertices or SBUs, providing for a rigid MOF with a predetermined topology [68].

An important distinction between porous MOFs and traditional porous crystalline materials, such as zeolite molecular sieves, stems from the fact that, in MOFs, the pores do not have walls. The free space inside MOFs is composed of open channels and cavities

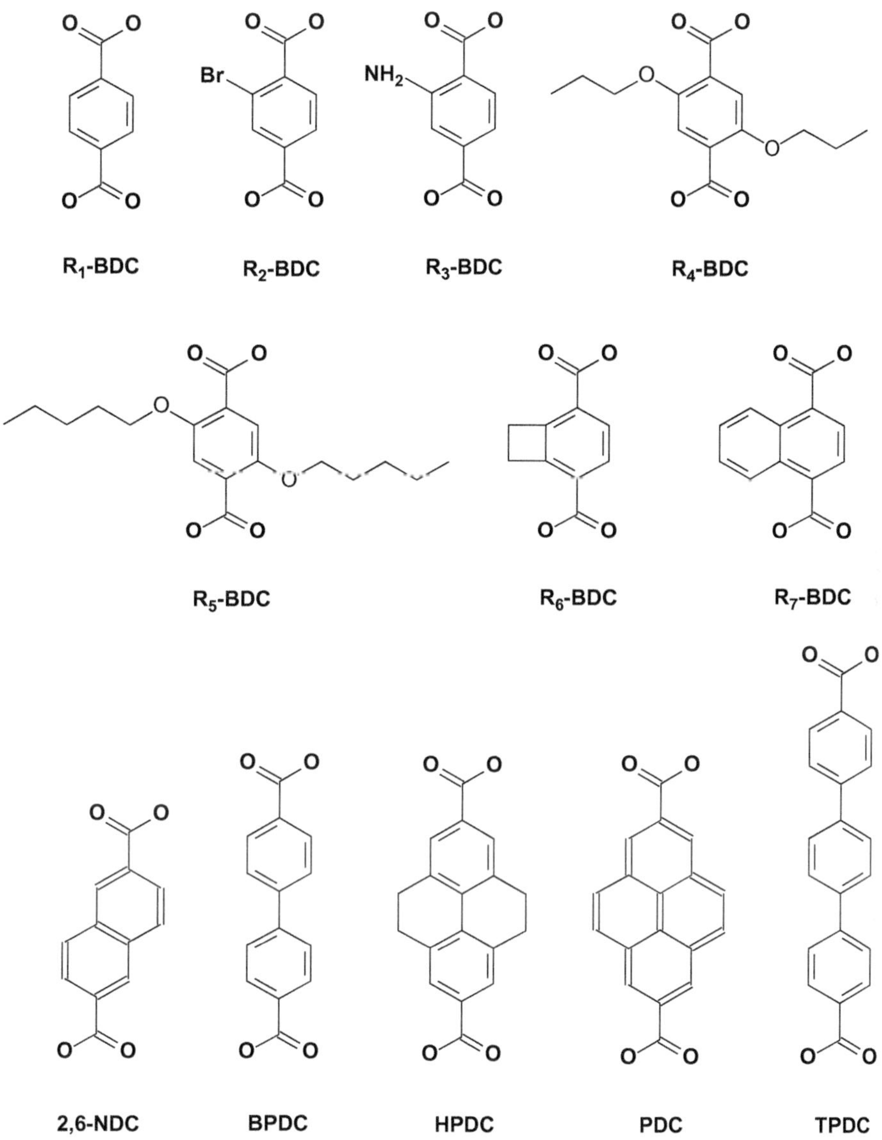

Fig. 9
Organic linkers used in the isoreticular MOF (IRMOF) series.

(<2 nm in diameter) where these channels intersect, which determines a large specific surface area and large volume of pores. Such cavities become accessible for the sorbate molecules after the molecules of solvent or remaining reagents, trapped during the synthesis of the MOF, are removed. In a number of MOFs, the total volume of pores (0.8–2.5 cm^3/g) is several times larger than in zeolites (the corresponding figures for the zeolite A, ZSM-5, and faujasite, 0.3–0.5 cm^3/g). The size of pores in MOFs can be controlled by selecting appropriate organic linkers [69]. For example, all the IRMOFs synthesized by Yaghi are highly porous materials with a very low crystalline density. The percent of free volume inside these MOFs with respect to the overall volume varies between 55.8% in IRMOF-5 (organic linker, R^5-bdc; for the names of organic linkers, see Abbreviations) to 91.1% in IRMOF-16 (organic linker, tpdc), while their density ranges from 1.0 to 0.21 g/cm^3. This is remarkable because the most open zeolites have no more than 30–40% of their overall volume accessible [68]. An important advantage of MOFs over traditional sorbents is the uniformity of the pore size. The shape of pores in MOFs can be highly diverse [69]. While zeolites typically have spherical, elliptical, and slit pores, MOFs can also have square, rectangular, and triangular pores [70–72]. This diversity of pore geometry is important for selective adsorption.

Descriptions of MOFs with rigid carboxylate linkers often use the term "permanent" porosity, because, in contrast to MOFs with heterocyclic linkers, such frameworks do not change after removal of "guest" molecules from their pores [73]. In other words, the "permanent" porosity of MOFs indicates their stability in vacuum. The notion of porosity is closely associated with the term "open metal organic framework," which is used to describe the MOFs containing freely movable solvent molecules within their framework [63]. Thermogravimetric analyses indicated that for most carboxylate MOFs the break-down temperatures lie in the range between 300°C and 500°C, which means that they are stable above the temperatures of desorption of the "guest molecules" [74]. This thermal stability is important for practical use of MOFs.

The majority of 3D MOFs have large specific surface area (up to 4500 m^2/g), which is an order larger than in zeolites, silica, and charcoals. For example, as it was already mentioned, the theoretical maximum for carbon-based adsorbents is 2630 m^2/g, which corresponds to covering both sides of a graphene sheet.

1.1 Synthesis of Carboxylate MOFs

In contrast to traditional porous catalysts and mesoporous materials like MCM-41, MOFs are synthesized in one step, under relatively mild conditions, and with a high yield (up to 95%). Usually it is done by precipitation from the solution of ingredients. Like with the synthesis of organic polymers, it is important that the insoluble product is formed at the rate high enough to avoid recrystallization.

Because currently the most popular among MOFs used as sorbents are the carboxylate MOFs, we will consider their synthesis in more detail. Several protocols for the synthesis of carboxylate MOFs exist, all of them requiring the presence of a base, usually amine, for deprotonizing the organic acid (the future linker fragment) and initiating the reaction. During this process it is important to preclude competitive coordination between the base and the organic acid.

Until recently, the most widespread was the so-called diffusion method. Its prerequisite is a high solubility of the reaction components. During the synthesis, volatile amine is slowly diffused into a dilute solution of a metal salt and organic acid at the ambient temperature and pressure [75–77]. The drawback of this method is as follows: it requires too much time (1–2 weeks). The "diffusion method" produces large crystals (several hundred micrometers in size), used mostly for structural analysis. These large crystals are typically useless for selective separation and catalysis due to unfavorable mass transfer.

The alternative, "solvothermal," method allows one to reduce the synthesis times to just 24 h. In this case, crystallization of MOFs from dilute solutions in polar solvents (dimethyl formamide, water, acetonitrile) usually takes place at moderately high temperatures (not exceeding 110°C) in an autoclave [70,78]. The solvothermal method allows one to overcome limitations due to low solubility of the components. Mixtures of solvents can be used to control the polarity of the solution and kinetics of the solvent-ligand exchange. As a rule, the MOF crystals produced by the solvothermal method are better suited for technological purposes. Recently, an express method was described which reduces the synthesis time to just 0.5–4 h [57]. It involves direct mixing of an amine into the reaction medium containing the solution of a metal salt and organic acid at room temperature. The process gives a high yield (c.90%) of high-purity nanocrystals. The one-step synthesis, using either of the three above methods, is highly reproducible and easily scalable [79].

A number of application areas of MOFs in catalysis proposed on the basis of the elaborated synthetic principles are being successfully developed at present; they include the heterogenization of the conventional homogeneous catalysts [80]; stabilization in MOF of catalytically active nanosized particles, which are unstable otherwise [80]; encapsulation of catalysts in the molecular framework [81]; the combination of catalysis with chemical separation [82], postsynthesis introduction of catalytic metal sites [9,83,84]; and catalysis with molecular sieve selectivity [55,80,83,85].

2 MOFs as Selective Heterogeneous Catalysts

Size and shape selectivity is extremely important for many industrial catalytic processes. Such selectivity can be achieved with the use of nanoporous catalysts. Selectivity can be based on the shape or size of reactants, products, or intermediates. To ensure the selective properties pertinent to the shape or size, the catalyst should have uniform pores with molecular dimensions and the architecture of the pore or cavity enabling a perfect accommodation of the reactant, product, or intermediate, in which case the mere effect of shape selective catalysis is realized. Selective heterogeneous catalysis of transformations of organic compounds has been considered from the early 1990s as one of the most important areas of the application of microporous MOFs [86]. As discussed earlier, the distinguishing feature of functional MOFs is their porosity; in this context, an analogy between MOFs and zeolites, another class of porous materials, was drawn [56]. Zeolites are the most important class of industrial heterogeneous catalysts [87]. Crystalline MOFs possess the features of zeolites important in heterogeneous catalysis, such as large specific surface areas and uniform size distributions of pores and voids. The uniformity of pores is an advantage of highly organized structures, such as MOFs, in which the pores or channels are the part of the building blocks (ie, repeating units). The system of channels with a strict geometry in MOFs allows for their use in size- and shape-selective catalysis, which previously was carried out exclusively over zeolites.

The capabilities of zeolites are largely limited by the almost exclusive use of SiO_4/AlO_4 tetrahedral building blocks (aluminophosphate and other structures have not yet found large-scale application in catalysis, except for SAPO-34 used for methanol to olefins conversion). The 3D framework of zeolites is microporous with channels or voids of up to 1.0 nm in size. It is for this reason that the field of catalytic applications of zeolites is limited to small organic molecules (usually not larger than *o*-xylene). In contrast, MOFs contain bulky organic components and can be formed from an infinite set of building blocks, which make it possible to finely tune their porous properties.

A unique feature of MOFs, which distinguishes these systems from other materials (such as metal oxides), is the absence of an inaccessible volume (so-called dead volume) [52]. Moreover, owing to their open architecture, the intrinsic diffusion coefficients of reactant molecules are only slightly less than in the bulk of the solvent, that is, mass transfer in the porous framework system of MOFs is not hindered. The pore sizes in MOFs vary in a tunable manner in a range from 0.4 to 3 nm (eg, for the IRMOF series), with the proper choice of the organic linker determining the pore diameter [4].

Another important characteristic of MOFs for the application in liquid-phase catalytic reactions is their hydrophilicity-hydrophobicity: whereas HKUST-1 is highly hydrophilic, MIL-101 and MIL-100(Fe) are moderately hydrophilic, and the ZIF-8 material exhibits a high hydrophobicity [88].

Note that many MOFs exhibit permanent porosity like zeolites, but it is not retained after the removal of the solvent for a number of systems (eg, clays, unless they are pillared by specially introduced agents). The preservation of the porosity after the removal of the solvent upon evacuation plays a decisive role for gas-phase catalytic reactions, as well as for the separation and storage of gases. However, for catalysis in condensed phases, this property is not significant.

Quite a few excellent reviews on catalysis by MOFs can be recommended to the reader [32,67]. One of the earliest reports on MOF-based catalysis by Fujita et al. [55] concerned the cyanosilylation of aldehydes in the presence of a 2D MOF based on 4,4′-bipyridine and cadmium nitrate $Cd(4,4'\text{-bpy})_2(NO_3)_2$, which does not possess permanent porosity. The substrate size and shape selectivity was observed in the presence of this catalyst. The active sites in the catalytic system are Cd(II) ions, which act as Lewis acids in the catalytic process. Catalysis occurs mainly through the substitution of the substrate for labile axially coordinated water molecules, rather than the equatorially bonded dipyridyl bridges.

Kitagawa et al. [89] synthesized a 3D PCP containing amide functions ([Cd-(4-btapa)$_2$(NO$_3$)$_2$]·6H$_2$O·2DMF)$_n$ by the reaction between Cd(NO$_3$)$_2$·4H$_2$O and a tridentate ligand (4-btpa). The amide groups are linked in a homogeneously ordered manner with the surface of channels. The system catalyzes the Knoevenagel reactions of benzaldehyde condensation with active methylene compounds (malonic acid dinitrile, ethyl, and *tert*-butyl esters of cyanoacetic acid) (Reaction 1):

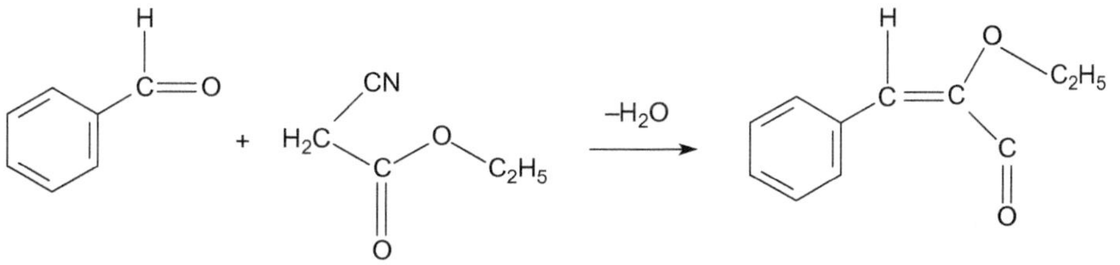

Reaction 1

The highest conversion (98%) was observed in the case of malonodinitrile. In the case of other substrates, the condensation almost did not proceed, thereby indicating the correlation between the size of the reactant molecules and the diameter of the "window" of the "host" pores.

Qiu [86] used the method of modular design to synthesize a series of supramolecular MOFs with 3D nanoscale channels containing catalytically active sites inside the channels. The MOFs corresponded to the formula [M$_1$(H$_2$O)$_6$] [M$_2$(phen)$_2$(H$_2$O)$_2$]$_2$ (2(btc) xH$_2$O), where M$_1$, M$_2$ = Co(II), Ni(II), Cu(II), Zn(II), and Mn(II); phen = 1,10-phenanthroline, btc = 1,3,5-benzene-tricarboxylate, X = 22–24. This approach based on the combination of a framework and a catalytic element (module) was used to obtain selective catalytic systems. The method of modular design implies a combination of a framework and a catalytic element in a precisely controlled process. To obtain MOFs as potential catalytic systems exhibiting size and shape selectivity, it is necessary by the proper choice of organic ligands and inorganic nodes to ensure the free internal space (ie, to prevent the intergrowth of the framework with the formation of a densely packed structure) and to preserve unsaturated metal sites (ie, to preclude the coordination of the catalytic building block with the organic ligand).

Hexa-aqua metal complexes and the btc ligand were chosen as the building blocks that form the MOF. The metal complex $\left[M_2\left(\text{phen}\right)_2\left(H_2O_2\right)\right]_2^+$ was used as a building unit. These supramolecular MOFs showed an appreciable substrate size and shape selectivity in the reaction of the catalytic oxidation of phenol to quinone (Reaction 2), indicating that the process proceeds in the framework channels:

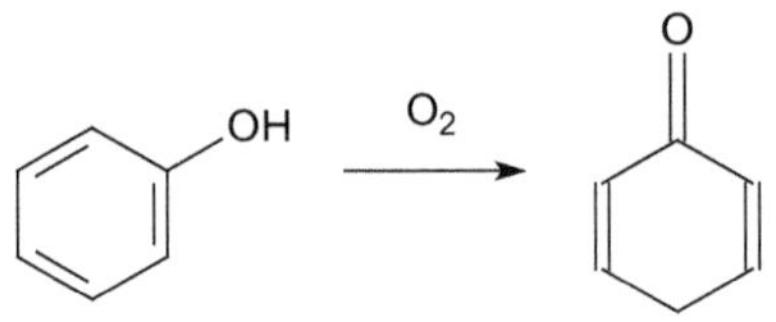

Reaction 2

From the economic and environmental standpoints, the replacement of homogeneous catalytic processes with effective processes to be run in the presence of heterogeneous and heterogenized catalysts is an extremely important task. The Friedel-Crafts alkylation of aromatic hydrocarbons is a very illustrative example of such an approach. In this process, a series of homogeneous catalysts (such as HF, H_2SO_4, and $AlCl_3$) have been replaced over time with zeolites in the H-form [90]. For example, the main product of the substrate shape-selective reaction of β-methylation of naphthalene with methanol on mordenite (0.6 nm) is 2,6-dimethylnaphthalene [91]. A high β-monomethylation and/or β,β′-dimethylation selectivity is due to the larger size of 1- and 3-oriented products, which limits their diffusion through the system of zeolite pores with an average pore size of 0.6–0.65 nm. Nevertheless, even large pores of zeolites cannot adsorb bulky organic molecules and/or alkylating agents, thereby impeding the catalytic processes inside the pores. Because most MOFs have a pore size larger than that of zeolites, such as zeolite Beta (BEA) and mordenite (MOR), the application of MOFs opens up new opportunities for the alkylation of large polyaromatic compounds. For example, Ferey et al. [92] reported on the catalytic activity of the MIL-101 (Material Institute Lavoisier) MOF structure, ($[M_3F(H_2O)_2O(bdc)_3]$, M = Fe, Cr), characterized by large cavity sizes, in the Friedel-Crafts benzylation (Reaction 3):

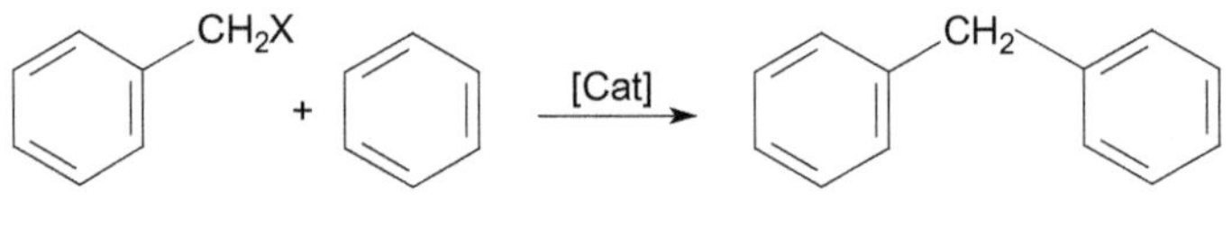

Reaction 3

Ravon et al. [93] found that zinc carboxylates exhibit an exceptionally high shape selectivity in a number of alkylation reactions. They studied catalysis on MOF-5, which is the archetype of a variety of MOFs with phenylenedicarboxylate-type ligands. This MOF structure consists of coordinatively saturated Zn_4O cluster units and completely complexed bdc ligands. However, MOF-5 exhibits catalytic activity in the Friedel-Crafts *tert*-butylation of toluene and biphenyl (*tert*-butyl chloride as a *tert*-butylating agent). Unlike the case of $AlCl_3$, mainly *para*-alkylation occurs on this catalyst, whereas *ortho*-alkylation proceeds at a low rate. This indicates the occurrence of the process within the porous structure.

The researchers describe a number of approaches to the design of new acidic MOFs with a selectivity pattern tuned for the shape of the substrate. The first approach is the so-called zeolite-mimetic by the cited authors. It includes the design of porous MOFs based on Zn or Al, which contain bridging OH groups similar to those present in the H-form of zeolites [Si–O(H)–Al]. Two compounds synthesized by the Ferey group were selected for tests, namely, MIL-53 (Al) [94] and MOF-69C [95] with structurally identical hydroxyl groups, as follows from the formulae reflecting their composition: Al(OH)(bdc) and $Zn_3(OH)_2(bdc)$, respectively. MIL-53 (Al) is built from infinite chains with common vertices (corner-sharing) based on $AlO_4(OH)_2$, whereas MOF-69C consists of $ZnO_2(OH)_2$ tetrahedra and $ZnO_4(OH)_2$ octahedra with $(\mu\text{-}OH)_2$ and μ_3-OH as the linkers.

The aim of the second approach is the development of MOFs with structural defects, which would play the role of catalytically active sites. In this case, it would be metal-containing nodes of the framework coordinated to terminal OH groups instead of bridging ligands, which takes place in the ideal structure. The authors tested two synthetic strategies: (1) rapid precipitation and (2) partial replacement of dicarboxylate with monocarboxylate bridging ligands. Since MOF-5, like all the family of IRMOFs (isoreticular MOFs, ie, with the same lattice topology), does not contain OH groups in its composition, this framework was chosen as the starting material for the development of OH groups as structural defects (the role of structural defects as active sites will be considered later).

The efficiency of the developed synthetic approaches was tested in the alkylation reactions of toluene and *tert*-butylbenzene, as well as large aromatic molecules, naphthalene and biphenyl, to better identify the shape selectivity of the pores. It was found that monoalkylation of biphenyl with *tert*-BuCl had a 100% *para*-selectivity, which exceeds the selectivity shown by H-MOR and H-BEA taken as model zeolite catalysts. It was concluded that, in the case of dicarboxylate MOF structures, the pore size can be "tuned" depending on the size of the substrate by an appropriate choice of the bridging ligand.

A highly organized structure of the MOF makes it possible to sharply separate active sites in space. Owing to a high specific surface area, MOF-based catalysts can have a high bulk concentration of active sites [96]. These fully open sites in the MOF structure provide an extremely high degree of dispersion of the metal. In addition to a high metal content in MOFs, one of the important advantages of these systems is the homogeneity of the active sites, which is provided by a high degree of crystallinity of these materials. Thus, the catalytic processes on MOFs can both be size- and shape-selective as a result of the porosity of these systems and be mediated by catalytically active sites based on transition metal ions.

The ability to incorporate functional groups into porous MOFs also makes them unique candidates for the fabrication of heterogeneous catalysts [97]. Mild conditions, generally used for the synthesis of MOFs, make it possible to directly introduce various functions into the framework structure. For example, enantiopure chiral ligands or their metal complexes can be directly introduced into the framework of MOFs with the formation of an effective asymmetric catalyst. This process cannot be used in the case of zeolites or other microporous oxide materials because severe conditions are usually practiced for their synthesis (eg, calcination at a high temperature for decomposition of an organic template).

The hydrolytic and thermal instability of MOFs limits their use in heterogeneous catalysis. Although many MOFs exhibit stability at high temperatures and some of them showed resistance up to 450 °C, these materials could not surpass zeolites in stability [98,99].

The reactions involving MOFs as catalysts and a number of basic approaches to create active sites of different nature in the catalytic systems based on MOFs are: the use of a MOF as a classical support in a heterogeneous catalytic system; the use of MOFs with an active site in the organic ligand; performing a catalytic process on the active sites placed on inorganic units of the framework; the introduction of an active site through the "host-guest" interaction; and postsynthetic modification of the framework, resulting in the formation of catalytically active sites. The nature of catalytically active sites in MOFs will be discussed later in more detail.

3 The Nature of Catalitically Active Sites in MOFs

As we have mentioned, MOFs are formed by metal ions or clusters as the lattice nodes and organic ligands as linkers. Due to the hybrid nature of MOFs, the active site can initially be present in the organic or inorganic component of the framework or can be introduced from the outside (host-guest interaction, a postsynthesized site, doping with catalytically active metals).

3.1 Catalysis on Open Catalytically Active Sites

The main issue in the design of MOF-based catalysts is how the coordinatively saturated metal-containing nodes of the framework can exhibit the catalytic effect. If the coordination sphere of the metal ions is completely blocked by organic bridges, the ions have no free positions available for interaction or activation of reactants. To avoid this limitation, different approaches were developed to produce MOFs with open metal sites. One of the approaches involves the use of labile ligands in the synthesis of MOFs; the ligands tend to be removed at the activation step preceding the catalytic reaction [9,85]. Most frequently, these labile ligands are solvent molecules, which open free coordination sites when removed. Catalytically active particles can be generated and stabilized by the framework; however, they are either unknown in homogeneous metal-complex systems or highly unstable. Free metal centers in the completely activated $[Cu_3(btc)_2]$ framework formed upon the detachment of water molecules from MOF-199 or HKUST-1 (Hong Kong University of Science and Technology) are an example of this approach. The latter MOF has the structure $[Cu_3(btc)_2(H_2O)_3]$, (Table 1). Coordination water is readily removed from the system, leaving open Cu(II) centers. When the apical aqua ligand is removed from the prototype of HKUST-1, the unsaturated Cu center of $Cu_2(CH_3COO)_4{\cdot}2H_2O$ coordinates oxygen as a donor from the second fragment of the complex with the formation of the μ_2-oxygen bridge. This way of the saturation is impossible in MOFs because of the existence of the framework; thus, open metal sites are stable and highly reactive.

This approach is exemplified by the protonation of a chiral MOF and its use in the enantioselective methanolysis of *cis*-2,3-epoxybutane [100] (Reaction 4):

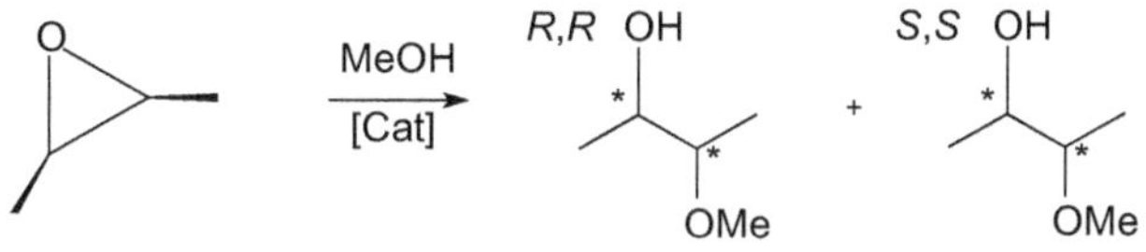

Reaction 4

In some cases, the stabilization of highly reactive particles, for example, open metal sites, is much more effective in the MOF structure than in the homogeneous phase [52].

Kitagawa et al. [101] proposed an alternative approach to the preparation of MOFs with open metal sites. Their strategy includes the use of the preliminarily "assembled" metal-containing ligand, which can react with a second metal, acting as a framework node.

3.2 Catalysis on Formally Saturated Catalytically Active Sites

There are published examples of catalysis on formally saturated sites (metal ions or clusters forming nodes of a MOF). Zou et al. [102] synthesized a 3D functional MOF based on the cubic metal organic building block $[Ni_8L_{12}]_{20}^{-}$ (H_3L=4,5-imidazoledicarboxylic acid). Alkali metal ions (Na^+) are placed in the nodes of the framework. This framework shows a stable catalytic activity in the oxidation of CO to CO_2.

Gandara et al. [103] synthesized a new In(III)-based framework formed by bulky layers containing channels of a square cross section [In(OH)L]×Py, L=4,4′-(hexafluoroisopropylidene)-bis-(benzoic acid). This microporous, thermally stable framework has proved to be an effective heterogeneous catalyst for the acetalization of aldehydes. The difference in the catalytic activity between the frameworks with filled and free channels is evidence that catalysis takes place inside the pores.

One more example is related to the linalool oxidation with hydrogen peroxide using a [Ln(OH)(H_2O)(naphthalenedisulfonate)], where Ln=Nd, Pr, La [104], though the nature of the active sites is not quite clear, since the MOF contains no transition metal ions, like Ti, Fe, Co, and so on. Similarly, oxidation of methylphenylsulfide to sulfoxide and sulfone with H_2O_2 is catalyzed by scandium ions in a [$Sc_2(bdc)_3$] MOF [105]. Among other oxidation reactions, the use of a Cu-MOF ([$Cu_3(btc)_2$]) in the conversion of phenols to higher phenols deserves attention and the nature of activity is not so elusive [106].

Last but not least, Ti-based MOFs were tried in the Ziegler-Natta polymerization of ethylene and propylene, though these systems were inferior to the well-known single-site metallocene polymerization catalysts with their tremendous turnover frequencies and space-time yields [107].

3.3 Catalysis on Framework Defects

The literature contains a number of examples of catalysis on active sites qualified as framework defects, for example, on open or uncoordinated carboxylic acids. The authors cited above [93], who studied the alkylation reaction in the presence of MOF-5, attribute the catalytic activity to encapsulated clusters of zinc hydroxide, which are formed as an impurity in the rapid synthetic routes, and may also be present in a partially hydrolyzed form of the framework structure, $[Zn_3(OH)_2(bdc)_2]$ [108,109].

4 Design of Catalysts Based on MOFs

There are several basic approaches to the design of heterogeneous catalysts based on MOFs: the use of MOFs as classical supports for an active metal, the immobilization of a catalytically active site in the framework per se (in the inorganic node or in the organic linker), "host-guest" interaction, postsynthetic modification of the framework, and so forth. These approaches will be discussed later in more detail.

4.1 MOF as a Support for Heterogeneous Catalysts

The use of MOFs as a support for active metals is an approach that is well known for typical heterogeneous catalysts, such as palladium/platinum on carbon or on zeolites [86]. Note that the introduction of ions of catalytically active transition metals into the framework of the synthesized MOFs can formally be considered as a postsynthetic modification as well, which we will discuss in more detail.

Note that different methods are used for the introduction of a metal into the framework: incipient wetness impregnation, MOCVD (introducing a metal into a PCP by metal-organic chemical vapor deposition), and coprecipitation. For example, Pd was introduced into the highly porous MOF of MOF-5 using $[Pd(acac)]_2$ (acac is acetylacetonate) as a precursor by incipient wetness impregnation [110]. After the introduction of Pd, the specific surface area decreased from 2885 to 985 m^2/g. Palladium-containing MOF-5 showed a high catalytic activity in the hydrogenation of styrene comparable to the activity of the classical Pd/C, but the hydrogenation of *cis*-cyclooctene proceeded at a much lower rate. Even at room temperature, the catalyst was not stable in air because of the low hydrothermal stability of the support, MOF. In general, the hydrothermal stability of this particular structure, MOF-5, is known to be lower than that of a number of other MOFs, especially MOF-177 [39], UiO-66 [98], Cr-MIL-100 [11], Cr-MIL-101 [12], MIL-125 [111], CAU-1 [112], ZIF-8 [113], and Ni-CPO-27 [114].

The coprecipitation method was successfully applied for the preparation of the 0.5% Pd/MOF-5 catalyst [115]. The specific activity of the system in the hydrogenation reaction was two times higher than that of the Pd/C industrial catalyst with the same metal content. It was shown that there was the "true" heterogeneous catalysis in this case, since no leaching of palladium into the reaction mixture took place. A decrease in the activity of MOF-5 supported palladium in repeated catalytic cycles was not observed, but a complete loss of the pore volume was marked. Consequently, it may be assumed that palladium is not localized inside the micropores and, probably, is connected with the outer surface of the crystals.

The Fisher group [116] has developed a metal-organic chemical vapor deposition procedure for metal introduction into the framework of MOFs. The general problem was to reveal whether the framework of the host, for example, MOF-5 or MOF-177 (a specific surface area of up to 4500 m^2/g and a pore volume of 0.69 cm^3/cm^3), is appropriate for the active and selective adsorption of vapors of metal organic chemicals used as precursors. A systematic study of doping MOFs with metal nanoparticles with the formation of composites of the general formula Metal@MOFs was conducted in refs. [116–119]. Microporous materials with a high internal surface area and a large pore volume, activated (ie, solvent-free) MOFs adsorb volatile metal organic complexes $[ML_n]$ of a proper size directly from the gas phase. As a result, relatively stable inclusion compounds $[ML_n]_a$@MOF can be obtained [119]. The parameter a refers to the number of adsorbed molecules per formula unit. As a rule, a is very close to an integer in the case of the limiting load, although these systems $[ML_n]_a$@MOF are not stoichiometric "compounds." The release of the metal atoms of the precursor introduced into the cavities results in the formation of "naked" nanoclusters. In the cited study, the highly PCP MOF-5 was modified with typical MOCVD precursor molecules containing Fe, Pt, Pd, Au, Cu, Zn, and Sn metals [110]. For example, MOF-5 was exposed to the vapor of the palladium-containing precursor $[(\eta^5\text{-}C_5H_5)Pd(\eta^3\text{-}C_3H_5)]$. Similarly, other typical organic precursors $[(\eta^5\text{-}C_5H_5)Cu(PMe_3)]$ and $[(CH_3)Au(PMe_3)]$ were used. The action of vapors of volatile organometallic compounds on $[Zn_4O(bdc)_3]$ results in the formation of the aforementioned compounds of the $[ML_n]_a$/MOF-5 type. The 3D crystal organization of the guest framework of MOF-5 remained unchanged after the introduction of the metal-containing precursor.

The compound Pd/MOF-5 exhibits a moderate catalytic activity in the hydrogenation of cyclooctene, and Cu/MOF-5 catalyzes the formation of methanol from the synthesis gas (220 °C, CO/H_2). In contrast, highly porous Au/MOF-5 is inactive in the catalytic reduction of CO (because of the lability of the system).

The introduction of Ru nanoparticles with retaining the MOF of MOF-5 was carried out in Ref. [120]. After the introduction of Ru in the form of $[Ru(cod)(cot)]_{3,5}$@MOF-5 into the framework, hydrogenolysis is carried out, which results in the formation of Ru nanoparticles inside the cavities with the formation of the Ru@MOF-5 material. This catalytic system shows a moderate activity in the oxidation of benzyl alcohol to benzaldehyde (Reaction 5):

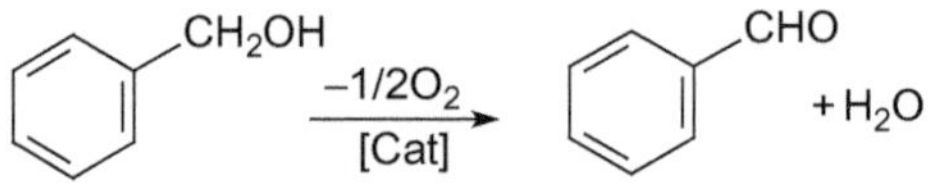

Reaction 5

The use of labile organometallic precursor complexes formed by the MOCVD technique offers new possibilities for mild and quantitative decomposition with the release of metal atoms, thereby making it possible to accomplish the formation of metal clusters or nanocrystallites inside the MOF channels or cavities. Physicochemical data showed that the host lattice and the guest molecule interact only through weak van der Waals forces without any changes in the MOF framework and the chemical nature of included molecules. For example, in the case of ferrocene $[Fe(\eta^5\text{-}C_5H_5)_2]$ as a precursor, a $[Fe(\eta^5\text{-}C_5H_5)_2]_7$@MOF-5 sample was obtained, which is characterized by an ordered superstructure of 56 molecules of ferrocene included in the unit cell of MOF-5. These molecules are distributed in two types of cavities in a ratio of 8:6 [121]. Similarly, $[Zn_4O(btb)_2]$ (MOF-177) was modified with ferrocene yielding an inclusion compound of the $[Fe(\eta^5\text{-}C_5H_5)_2]_{11}$@MOF-177 composition [122,123].

Fischer et al. [124] prepared MIL-53(Al) modified via the hydroxyl groups of MOF by 1,1′-ferrocenediyl-dimethylsilane. The grafted Fe complex turned out to be efficient in benzene hydroxylation with H_2O_2 to yield phenol (15%) at 330 K (Reaction 6):

Reaction 6

The related reaction is the synthesis of hydrogen peroxide from H_2 and O_2. Usually Pd or Au catalysts are the best choice for this catalytic process or industrial importance. The BASF team together with O. Yaghi performed this synthesis on a Pt@$[Zn_4O(bdc)_3]$ (MOF-5) [125].

On top of the oxidation reactions, the same team demonstrated that the silver-containing MOF-5 turned out to be active in propene epoxidation with molecular oxygen [126].

A systematic study of the limiting contents of Pd nanoparticles in the MOF structure of MOF-5 upon the introduction of various precursors from the vapor phase and solution, that is, by the MOCVD and incipient wetness impregnation techniques, respectively, was reported in Ref. [117]. The volatile organometallic precursor $[Pd(\eta^5\text{-}C_5H_5)(\eta^3\text{-}C_3H_5)]$ was used for the vapor deposition at 10^{-3} mbar to form the $[Pd(\eta^5\text{-}C_5H_5)(\eta^3\text{-}C_3H_5)]_x$@MOF-5 compound in the pores of the framework. In these conditions, the maximum content of the metal makes four molecules of $[Pd(\eta^5\text{-}C_5H_5)(\eta^3\text{-}C_3H_5)]$ ($x = 4$) per formula unit (35 wt.%). According to the second approach, a $[Pd(acac)_2]$ metal organic precursor was used. A series of samples $[Pd(acac)_2]_X$@MOF-5 with different metal contents was obtained by the incipient wetness impregnation method. The maximum palladium content in these samples was lower than in the first case, and the result was approximately two precursor molecules per formula unit of MOF-5. These two methods made it possible to obtain nanoparticles inside the porous guest matrix. Homogeneously distributed nanoparticles with a diameter of 2.4 (±0.2) nm were obtained by photolysis of the inclusion compounds $[Pd(\eta^5\text{-}C_5H_5)(\eta^3\text{-}C_3H_5)]_x$@MOF-5 ($x \leq 4$), whereas the hydrogenolysis of $[Pd(acac)_2]_x$@MOF-5 ($x \leq 2$) resulted in the formation of a mixture of small particles (<3 nm) inside the framework and large (~40 nm) Pd agglomerates on the outer surface of the MOF-5 particles. Thus, in the case of highly loaded Pd-containing samples of MOF-5, the vapor deposition method for the introduction of palladium results in monodisperse particles of a smaller size fixed within the framework. Apparently, the method of the metal deposition by impregnating the framework with the subsequent thermal decomposition in hydrogen results in a more inhomogeneous distribution of particles, including the formation of large agglomerates on the outer surface of the framework, even in the case of low metal contents (~1% of Pd). Systems based on the formed palladium nanoparticles, Pd_x@MOF-5, are stable in a hydrogen atmosphere (2 bar) at 150°C for a long time. Under these conditions, neither the hydrogenation of terephthalate linkers nor the growth of metal particles was observed. However, the collapse of the MOF-5 matrix was observed in the case of the repeated deposition, irrespective of the precursor and the metal introduction method.

One of the model reactions used to test encapsulated metal nanoparticles in MOFs is the reduction of nitrophenol into aminophenol [127] with the nanoparticles of about 2 nm being active in this reaction (Reaction 7):

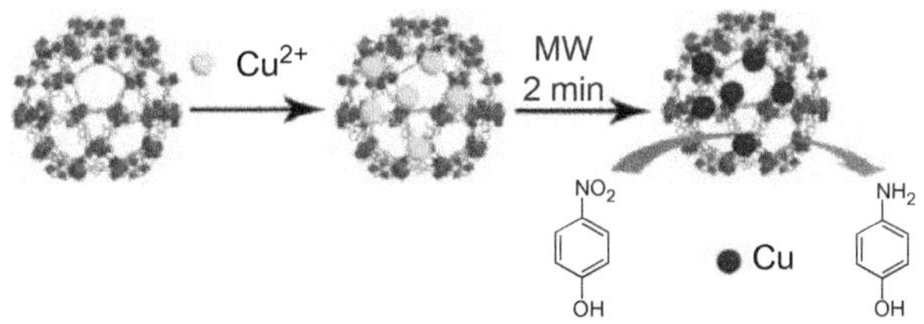

Reaction 7

Thus, the careful conduction of the hydrolysis/calcination operations opens up new perspectives for the development of new nanocomposite materials based on MOFs as guest matrices and on MOCVD precursors.

The hydroformylation of 1-alkenes on Rh@MOF-5 [16] results in the formation of a relatively large amount of *i*-aldehydes by the double bond isomerization and further hydroformylation. *n*-Octene-1 was found to be less active than the larger *n*-decene-1 and *n*-dodecene-1 molecules in the hydroformylation of *n*-alkenes-1 over Rh@MOF-5 and Rh@IRMOF-3 [128]. The reason behind this effect may be the constraints for the relatively long and straight *n*-octene-1 molecule (about 10 Å in length) to be accommodated at the active sites in the limited space of the pores of these MOFs.

4.2 Bridging Ligand as an Active Site

4.2.1 Metal-containing ligands

Kitagawa et al. [129] proposed a strategy based on the use of known homogeneous catalysts as ligands in MOFs, thus realizing their heterogenization. They noted that the introduction of coordinatively unsaturated metal sites into the organic ligand of MOFs opens up broad prospects for various applications, such as chemical sorption, separation, catalysis, and the development of chemical sensors. The work of Cho et al. [80,130] on the synthesis of a double-intergrown MOF structure based on a Mn(salen) complex and its application in catalysis illustrates this approach. The introduction of 4,4′-biphenyldicarboxylate (bpdc) ligands into the MOF made it possible to develop an analog of the known catalyst of the Katsuki-Jacobsen epoxidation reaction, (1,2-cyclohexanediamino-*N*,*N*′-bis(3-*tert*-butyl-5(4-pyridyl)-salicylidene)Mn(III)Cl) [131,132]. The study of the epoxidation reaction of olefins in the presence of the MOF showed that the substrate size selectivity substantially increases with the progress of the reaction, from the ratio of 2:1 at the beginning of the reaction to 18:1 in the case of a small or large substrate at a 45% conversion. This indicates a mixed "inside/outside" type of catalysis at the beginning of the reaction and then, with the progress of the reaction, the catalysis occurs predominantly "inside" the framework structure. The partial decomposition of the external ligands of the framework occurs during the reaction, which "forces" the substrate to enter into the MOF to reach the active sites and results in the enhancement of the selectivity for small substrates. Note that the reaction rate was only slightly lower than in the case of the homogeneous catalyst, and the deactivation of the catalyst was not observed. The active sites in the catalytic system based on the MOF are spatially well separated from one another in fixed positions, an arrangement that precludes the occurrence of side reactions, including those resulting in deactivation. These phenomena were first called "the MOF effect in catalysis." Kitaura et al. [133] showed that Mn(salen) complexes encapsulated into the MOF framework demonstrated an extremely high shape selectivity in the interaction with phenols and dihydroxybenzenes using hydrogen peroxide as an oxidant.

Metal porphyrins are widely used as molecular catalysts both in synthetic systems and as active sites in many metal enzymes. The first report on the use of porphyrins as ligands for catalytically active MOFs refers to the PIZA-3 (Porphyrinic Illinois Zeolite Analogue) system [134]. In this MOF structure, trinuclear manganese clusters are the nodes, and [tetrakis(phenylcarboxylate)porphyrin]Mn(III) complexes are the ligands. PIZA-3 has a permanent porosity and is active in the reactions of the hydroxylation of linear and cyclic alkanes, as well as epoxidation of olefins (iodosobenzene as an oxidant). Studies of the catalytic action of the system in the gas phase have been conducted, and it was concluded on the basis of the results on the selectivity for the substrate size and shape that catalysis occurs mainly on the outer surface of the framework. In another study, a highly porous MOF, ZnPO-MOF based on $Zn(NO_3)_2 \times 6H_2O$, dpydphf$_5$por (5,15-dipyridyl-10,20-bis-(pentafluorophenyl)-porphyrin) and a tetradentate ligand (H_4tcpb, 1,2,4,5-tetrakis(4-carboxyphenyl)-benzene) was synthesized [130]. This framework structure shows activity in the intramolecular acyl-transfer reaction [135]. This MOF is of interest from the standpoint that the porphyrin ligand, which enters into the synthesis, is spontaneously metallated with Zn(II) upon the formation of the MOF. The tetradentate ligand (tcpb) provides for the structural stability and, hence, permanent porosity—the lack of the latter is a drawback of porphyrin-based MOFs. It is the tetradentate ligand that prevents the lattice from the intergrowth (both in this framework and in other MOFs) [84].

4.2.2 Catalysis by metal-free organic ligands

Most catalytic systems based on MOFs exhibit catalytic activity due to metal ions (or atoms) as the active sites. However, there are reported examples of catalysis by organic ligands that contain no metal.

Kitagawa et al. [89] reported the synthesis of a 3D MOF with the formula [Cd(4-btpda)$_2$(NO_3)$_2$]. Although the MOF structure consists of identical pairs of intergrown frameworks, it contains pores of a molecular size. The nodes in the framework are composed of cadmium ions octahedrally coordinated to pyridyl nitrogen atoms. It was shown that the amide functions are active in the Knoevenagel condensation of benzaldehyde with malonodinitrile. A sharp decrease in the conversion during the condensation with the use of bulkier nitriles indicates that catalysis takes place only inside the channels of the framework, rather than on its outer surface. It was shown that in a homogeneous solution with free molecules of the catalyst, the potential ligand, catalysis does not occur, apparently because of the formation of intermolecular hydrogen bonds between btpda molecules [136].

MOF structures with uncoordinated amino groups, for example, IRMOF-3 and MIL-53, are stable base catalysts for the Knoevenagel condensation of benzaldehyde with ethyl cyanoacetate or ethyl acetoacetate [137]. IRMOF-3 shows the same activity as the known active catalysts. The selectivity for the condensation products is 100%. The catalytic activity of IRMOF-3 correlates with the degree of accessibility to basic sites. Note that there are no diffusion constraints in this case. In contrast, when using the MIL-53 framework, a low activity is associated with strong adsorption and constrained diffusion in unidimensional pores. The catalytic properties exhibited by IRMOF-3 indicate that the basicity of aniline-type amino groups increases after their introduction into the framework (an increase in pK_b is observed). Consequently, the aromatic amino group of the MOF is more active than that of the homogeneous catalyst (aniline). As a catalyst, IRMOF-3 is stable under the reaction conditions and, therefore, can be reused. The activity of the MOF structure is comparable to the activity of the best known solid base catalysts. Note that the catalytic properties of IRMOF-3 in various solvents indicate that the framework (with an open, accessible, and geometrically well-defined structure) is similar to homogeneous base catalysts, rather than to solid heterogeneous systems.

4.3 Inorganic Node as an Active Site

The Corma group [138] has demonstrated an approach comprising the synthesis of a MOF based on a catalytically active metal, that is, its use as an inorganic component in the SBU. It was shown that Pd-MOF (Pd-containing MOF with a molecular formula of $[Pd(2\text{-pymo})_2]_n$ (2-pymo = 2-hydroxypyrimidinolate)) displays a marked catalytic activity due to the active site, tetracoordinated palladium ion. This material is structurally related to the type of 3D lattices of the sodalite type with two different types of hexagonal windows with a free aperture of 0.48 and 0.88 nm and the fractional free volume of the crystal available for guest molecules of 42%. This framework structure showed properties typical of Pd-based heterogeneous catalysts. In particular, it accelerated the C–C cross-coupling reaction, aerobic oxidation, and hydrogenation reactions. The oxidation of aromatic enol to enal is an example of aerobic oxidation (Reaction 8):

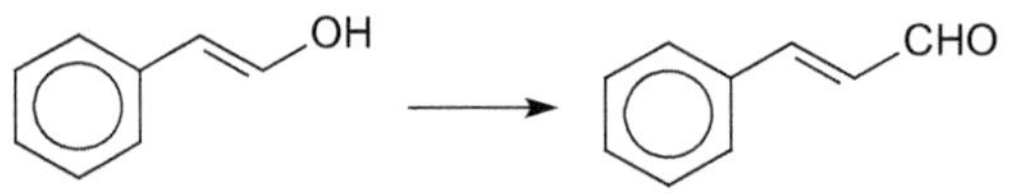

Reaction 8

The conversion reached 100% at a sufficiently high selectivity for the aldehyde (75%) under relatively mild conditions. The presence of palladium atoms in the microporous channels opens the possibility of substrate shape-selective catalysis. It was found that the Pd-MOF catalyst is active and recyclable in the Suzuki cross-coupling, even at room temperature. The authors conclude that the Pd-MOF is an active heterogeneous catalyst free of phosphine ligands. The conversion was 85% with an extremely high selectivity (~99%), and the turnover frequency was 1250 h^{-1} (Reaction 9):

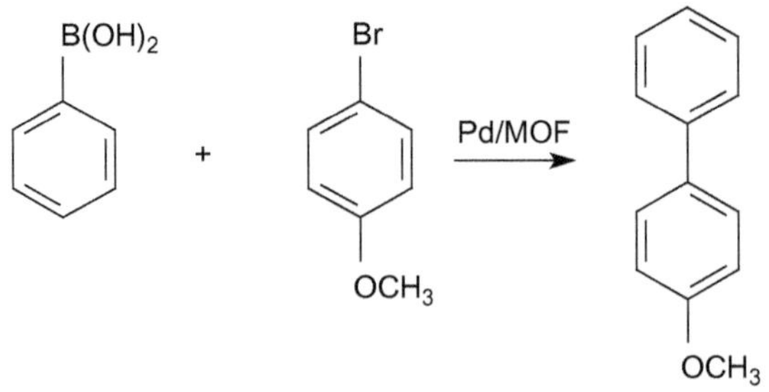

Reaction 9

Dispersed Pd nanoparticles supported on amino functionalized MOFs MIL-53(Al)-NH_2 (Al(OH)[H_2N-bdc], H_2N-bdc = 2-aminoterephthalic acid, MIL = Materials of Institut Lavoisier) were prepared by direct anionic exchange and further reduction with $NaBH_4$ [139]. The Pd/MIL-53(Al)-NH_2 catalyst exhibited a high activity and good stability for the Suzuki-Miyaura cross-coupling reaction.

Corma et al. note that the cross-coupling reactions can occur because of the Pd(II)–Pd(0) transitions in the metal nodes, processes that are accompanied by changes in coordination numbers. However, such sharp changes in the electronic state of the metal can result in the destabilization and potential degradation of the original MOF. The observed substrate shape and size selectivity indicates that the heterogeneous catalysis takes place inside the framework. Note that, in the case of the hydrogenation reaction, the possibility of the occurrence of the catalytic reaction on the sites formed by the partial degradation of the framework (palladium nanoparticles) and on the lattice defects is not ruled out. It seems more likely that catalysis proceeds on coordinatively saturated palladium atoms in the MOF nodes.

Similar systems ([Cu(2-pymo)$_2$] and [Co(Phim)$_2$], where PhIM is phenylimidazolate) containing, respectively, Cu^{2+} and Co^{2+} ions and anionic diazaheterocyclic ligands as organic linkers, have been used successfully for the aerobic oxidation of tetralin yielding α-tetralone as the main product [140]. Both of the frameworks are stable and recyclable under the reaction conditions. Nevertheless, kinetic studies showed substantial differences between the two MOFs because of the different nature of the central ions. [Cu(2-pymo)$_2$] is highly effective in the activation of tetralin with oxygen giving tetralin hydroperoxide (T–OOH) and is less active in reactions involving peroxide. At the same time, the use of the cobalt-based catalyst is accompanied by a long induction period. Once T–OOH is formed, Co^{2+} rapidly and efficiently transforms it to T=O with a high tetralon to tetralol ratio. The combination of both systems is a promising way of making a highly effective, selective, and recyclable catalyst for the liquid-phase aerobic oxidation of tetralin. The Cu-MOF system with bipyridine ligands [Cu(bpy)(H_2O)$_2$(BF_4)$_2$(bpy)] also shows a unique catalytic activity [141]. In the case of the methanolysis reaction of epoxides, the catalytic activity of the Cu-MOF was comparable to the activity of the homogeneous catalyst [Cu(BF_4)$_2$(H_2O)$_2$]. However, the relative activity in this case substantially decreased with the increasing steric volume of alcohol, indicating the localization of catalytically active sites in the micropores. The catalyst can be completely recycled without any loss of the activity.

The De Vos's group [142] studied [Cu_3(btc)$_2$], anhydrous MOF-199 or HKUST-1 as a selective catalyst exhibiting Lewis acidity in the isomerization reactions of terpene derivatives, pinene oxide rearrangement to campholene aldehyde, and cyclization of citronellal to isopulegol (Reaction 10):

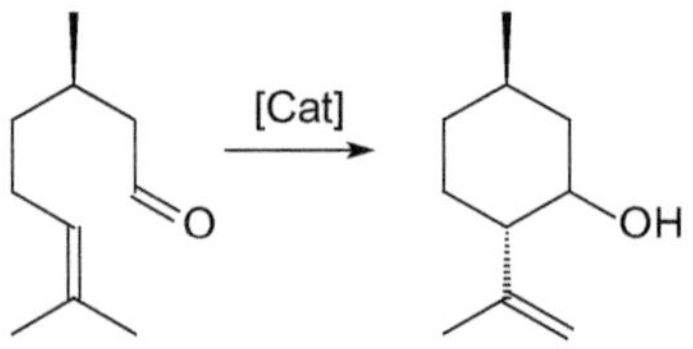

Reaction 10

The choice of the reactions is caused by the fact that their products differ depending on whether catalysis occurs on Broensted or Lewis acid sites. On the basis of the experimental data, the conclusion was made on the Lewis acid catalysis.

The same MOF shows a moderate catalytic activity in "wet" catalytic oxidation with hydrogen peroxide using *p*-coumaric acid as a model phenolic compound [143]. Structural and morphological changes of the Cu-MOF were observed during the catalytic reaction. Note that the activity of the reused catalyst is higher than that of the fresh sample.

A similar MOF $[Cu_3(btc)_2(H_2O)_3 \times H_2O]$ catalyzes the reaction of cyanosilylation of aldehydes with a yield of 50–60% and a selectivity of 60–90% (Reaction 11); the study of this reaction indicates the presence of relatively strong Lewis acid sites, due to the presence of an inorganic component of the MOF structure [144]:

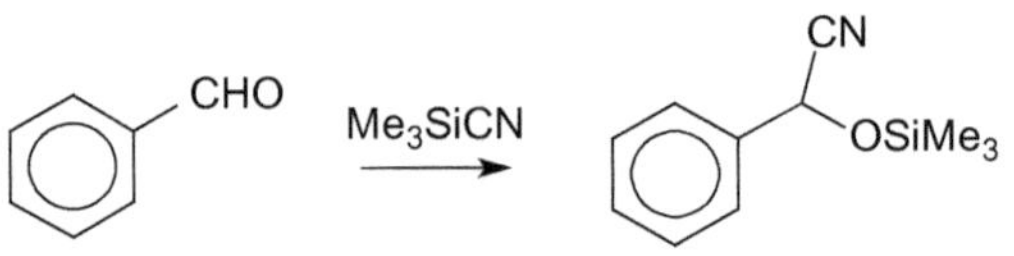

Reaction 11

Kaskel et al. [145] also studied the catalytic action of MIL-101 in the cyanosilylation reaction. The coordinated water molecules are easily detached, thereby a coordination unsaturation at the Cr(III) sites is created. Owing to the fact that the Lewis acidity of the Cr(III) sites is much higher than that of Cu(II), MIL-101 turned out to be much more active than HKUST-1 in this reaction. In addition, the catalytic sites of MIL-101 are resistant to the side reaction of benzaldehyde reduction, unlike HKUST-1 sites.

The same Lewis acid-catalyzed reaction of the cyanosilylation of aromatic aldehydes was investigated by Long et al. [85]. The catalytic activity of the sodalite-type compound $Mn_3[Mn_4Cl_3(btt)_8(CH_3OH)_{10}]$ (BTT is 1,3,5-benzene-tris(tetrazol-5-yl)) was studied. This material is a 3D porous structure with a pore diameter of 1 nm. It was assumed that each of the two types of Mn(II) sites exhibits catalytic activity. The distinctive features of this catalyst are the high conversions in the case of small substrates and a good substrate size selectivity, properties that are consistent with the occurrence of the catalytic process in the channels. The surface of the framework bears open Mn^{2+} ions, which act as Lewis acid sites.

Hereupon, the cyanosilylation of aromatic aldehydes and ketones, as well as the Mukayama aldol condensation, which is more sensitive to acidity (this reaction requires stronger Lewis acid sites than cyanosilylation), become feasible. In the presence of the Mn-MOF, a pronounced molecular-sieve effect is observed, which correlates well with the size of pores. This manifestation of the substrate size selectivity indicates the occurrence of the true heterogeneous catalysis.

There are also published data on benzaldehyde acetalization with trimethyl orthoformate on the In(III) Lewis acid sites of a 2D MOF accessible to the substrate [146]. As noted in Ref. [54], the nodes of most MOFs are Lewis acid sites. These open acid sites are located on the surface of MOFs and can be good catalysts, while the terminal ligands can behave as Lewis bases.

Lu et al. [147] carried out the oxidation of alkenes over a 3D porous MOF with the general formula [Co(bpb)]·3DMF, (bpb is 1,4-bis(4′-pyrazolyl)-benzene). Cobalt(II) ions in this structure are organized as one-dimensional (1D) chains through tetrahedral coordination with pyrazolyl nitrogen. It is assumed that the treatment with *tert*-butyl hydroperoxide transforms Co(II) into Co(III). The addition of cyclohexene results in the formation of *tert*-butyl-2-cyclohexenyl peroxide with a yield of 83%, but some degradation of the structure of MOFs is observed under these conditions after several catalytic cycles.

Fan et al. [148] have recently shown that MOFs with copper active sites can catalyze the oxidative addition of 2,6-dimethylphenol yielding poly(1,4-phenylene ether) as a commercially important polymer. These systems can perform C–O/C–C oxidative coupling with a selectivity comparable to the selectivity obtained for other homo- and heterogeneous catalysts (up to ~90% in optimized conditions) at a low catalyst/substrate ratio and a short reaction time.

4.4 Introduction of Guest Molecules Containing Catalytically Active Sites

It has been noted in the literature that the supramolecular chemistry of Zn carboxylate-based "host-guest" MOFs pertinent to catalysis is still in its infancy. The use of zeolite-like rho-ZMOF based on In-HImDC as the "host" for large catalytically active molecules, such as metal porphyrins, resulting in the enhancement of the catalytic activity has been demonstrated [81]. To synthesize the system, 4,5-imidazoledicarboxylic acid (H_3imdc) and $In(NO_3)_3$ were used to obtain a stable anionic framework rho-ZMOF with the general formula $[In_{48}(Himdc)_{96}]$. In this case, pores framing large cavities are small enough to freely pass porphyrins. Owing to this fact, cationic porphyrin can be electrostatically incorporated into the framework during the synthesis of the MOF. Under optimal conditions, the porphyrin content reached 60%. Porphyrin as the free base was postsynthetically modified with different ions of transition metals (manganese, cobalt, nickel, and zinc) to give a wide range of encapsulated metal porphyrins. Hydrocarbons were subjected to oxidation in the presence of Mn-tmpyp (5,10,15,20-tetrakis(1-methyl-4-pyridinium)porphyrin-tetra-(*p*-toluenesulfonate)) included into the rho-ZMOF. Cyclohexane was oxidized to cyclohexanol and cyclohexanone with *tert*-butyl hydroperoxide (tbhp) as an oxidant. This option of encapsulation into MOF made it possible to compare the results with those of oxidative catalysis on the Fe(porphyrin) and Mn(porphyrin) systems [12,149].

A controlled growth of catalytically active clusters in sufficiently large MOF cavities is also an interesting way of introducing the "guest" nanoparticles into MOF [54]. There is an example of the encapsulation of polyoxometalates into the cavities of MIL-101 [12]. Kholdeeva et al. [150] showed that cluster@MOF systems based on substituted titanium or cobalt ions can exhibit a catalytic activity. For instance, the $[PW_{11}TiO_{39}]^{5-}$@MIL-101 framework accelerates the oxidation of α-pinene to the corresponding alcohol and ketone with hydrogen peroxide as an oxidant. $[PW_{11}CoO_{39}]^{5-}$@MIL-101 is also active in this reaction with the participation of O_2. In the latter case, it was found that, unlike oxygen, hydrogen peroxide destroys the MOF.

4.5 Active Sites Introduced by the Postsynthetic Modification of the Framework

A new synthetic strategy, which includes the postsynthetic modification of an as-formed MOF by appropriate functional groups, has been recently proposed as an alternative to the previously proposed "presynthetic" approaches. Cohen et al. [151,152] described the modification of amino groups in IRMOF-3 with alkylanhydrides. This idea was used by Costa et al. [153] for the preparation of an MOF based on Gd(III) and aminoterephthalate ligands. The amino group of the organic ligand was postsynthetically modified with the use of isocyanate or carboxylic acid to obtain various functions.

Ferey [154] has shown that the presence of coordinatively unsaturated Cr(III) sites in the inorganic nodes of MIL-101 makes it possible to introduce thermally stable amino groups with the chelating ability to the MOF. This ability can open new possibilities for the selective introduction of framework-linked organic functions. In this example, Cr(III) in inorganic nodes of the framework was coordinated to one of the two nitrogen atoms of a diamine molecule (ethylenediamine and diethylenetriamine) to obtain modified MIL-101 with grafted amino groups (ED-MIL-101 and APS-M1L-101). Free terminal amines show appreciable activity as Broensted base catalysts in the Knoevenagel condensation with a selectivity of 99.3% for the condensation product. Because the reaction takes place inside the pores, the formation of bulky condensation products is hindered. Therefore, high selectivity is directly related to the structural features of the MOF itself. Palladium-containing APS-MIL-101 and ED-MIL-101 also showed an increased activity in the catalysis of the Heck cross-coupling reaction (393 °C).

Ingleson et al. [100] carried out a rational postsynthetic modification of a series of MOFs, $M(asp)L_{0.5}$ ($M=Ni^{2+}$, Cu^{2+}; asp = L- or D-aspartate; L = bipy or bpe). After protonation of the synthesized frameworks by their treatment with HCl in diethyl ether, carboxylic acids in the frameworks act as Broensted acid catalysts, accelerating the methanolysis of small epoxides that can penetrate into MOF cavities. The yield of the reaction was 65%.

The postsynthesis approach was used for the introduction of an Au(III) complex with a Schiff base into the walls of MOF pores [155]. The Au(III)-containing MOF showed high activity and selectivity in domino coupling and cyclization reactions. Gold(III) particles remain unreduced after the catalytic reaction; thus, the catalyst is completely recyclable. This resulted in a higher activity than that cited earlier for gold-based homogeneous and supported catalysts. In addition, properly localized Au(III) active sites are active in the dissociation of hydrogen (H_2), and they accelerate the selective gas-phase hydrogenation of 1,3-butadiene to butane as well.

5 Catalysis in the Presence of Chiral MOFs

Enantioselective interaction in the presence of metal organic systems occupies a special place. In order for homochiral MOFs to act as efficient heterogeneous asymmetric catalysts, it is necessary to retain the structure of the framework during the catalytic reaction and open channels for transport of reactants and products must also be present [97]. Chiral MOFs with large open channels (typically more than 1.5 nm) have the advantage for this type of catalysis. This is caused by the large size of prochiral substrates and formed chiral products. Unfortunately, MOFs with large open channels tend to undergo substantial distortions of the lattice after the removal of solvent molecules, thus, generating a material with characteristics that differ from the original ones. In particular, a discrepancy in X-ray diffraction patterns and a decrease in the specific surface area are observed. Consequently, the stability of a homochiral framework during the catalytic reaction and the accessibility of its channels to large organic molecules are of substantial significance.

There are several ways for the preparation of enantioselective materials based on MOF. For the synthesis of homochiral open MOFs, the following approaches are used: (1) the use of rigid homochiral organic ligands as linkers for binding a suitable active site based on a metal or SBUs [67]; (2) the use of homochiral organic ligands containing an auxiliary functional group, which directly does not participate in the formation of the framework but causes the lattice to take the specific chiral topology [156]; (3) another rational approach to the synthesis of homochiral MOFs includes the use of a metal ion and a homochiral organic ligand to form SBUs, which are connected by rigid links with the formation of a lattice structure in a one-pot reaction [157]. Kim et al. [158] reported the synthesis of a homochiral porous material $[Zn_3(3\text{-}O)(L_6)]\cdot 2H_3O\cdot 12H_2O$ (D-Post-1, L is D-tartaric acid), in which enantiopure metal organic clusters are used as SBUs. The presence of pyridyl groups in the channels also offers great opportunities for catalysis in the presence of POST-1. The catalytic trans-esterification of ethanol in the presence of this material occurs with the conversion of 77%. However, in the case of bulkier alcohols (isobutanol, neopentanol, and 3-phenyl-1-propanol), the reaction is much slower or almost does not proceed.

Lin et al. [159] synthesized a homochiral porous MOF, $[Cd_3L_4(NO_3)_6 \cdot 7MeOH] \cdot 5H_2O$ (L = (*R*)-6,6′-dichloro-2,2′-dihydroxy-1,1′-binaphthyl-4,4′-bipyridine) in DMF/$CHCl_3$/MeOH. The treatment of this material with excess $Ti(OiPr)_4$ in toluene makes it possible to obtain an active catalyst for the addition of diethylzinc to aromatic aldehydes to produce chiral secondary alcohols after hydrolysis. The study also exemplifies the postsynthetic modification of the framework. Chiral secondary alcohols were obtained, in this case, with a high enantioselectivity and a good yield. For example, the addition of diethylzinc to (*R*)-1-naphthaldehyde allows for the synthesis of (*R*)-1-(1-naphthyl)-propanol with 90.0% ee and complete conversion.

6 Other Types of Catalytic Processes on MOFs

Because many MOFs are semiconductors with a band gap of 1–4 eV, one may expect photocatalytic activity. The band gap may be changed by varying the extent of conjugation in the system [160].

A few examples of polymerization reactions catalyzed by MOFs are reported [161]. Zn carboxylate [162] was shown to catalyze the polymerization of propylene with CO_2 to polycarbonates (M_w = 75,000 g/mol). Alkoxylation of propylene glycol or acrylic acid with ethylene/propylene oxide resulted in polyols. Radical polymerization of divinylbenzene was performed using $[M_2(bdc)_2(ted)_2]$ (M = Cu, Zn) [163–165].

To conclude, the MOF materials by themselves offer new opportunities in catalysis, especially in fine chemicals synthesis. However, the expectations have not yet been proved perfectly with fascinating experimental evidence for the significant advantages of MOFs compared to zeolites and other materials. On the average, with a few exceptions, they are similar to zeolitic catalysts. The realization of host-guest interactions and stabilization of metal or oxide nanoparticles of precise size and morphology would help to foster the development and application of these revolutionary materials.

7 Adsorption on MOFs

7.1 Hydrogen Storage

The unique porous structure, high surface area, and high pore volume of MOFs allow one to use them as selective adsorbents and gas storage systems. One of the most challenging areas so far is hydrogen storage [166]. Hydrogen can be used as a fuel in internal combustion engines or in fuel cells. The materials for hydrogen storage are the key issue in developing hydrogen energetics. The U.S. Department of Energy developed the targets for hydrogen storage materials [167] aiming at the weight capacity of 9 wt.% by 2015. There are several methods of hydrogen storage: high-pressure gas cylinders (up to 80 MPa); cryogenic tanks at about 20 K; adsorbed hydrogen in materials with a high surface area at a temperature close to or below room temperature, or in the chemically bound state at atmospheric pressure using intermetallides, covalent or ionic compounds, including organic molecules that can be reversibly hydrogenated and dehydrogenated to accumulate and release hydrogen [168].

7.1.1 Physical sorption of hydrogen

The mass density of liquid hydrogen is only 70.8 kg/m^3 at 20 K and atmospheric pressure [169], while the systems for physical adsorption are characterized by low storage capacities, from 1 to ~4.5 wt.% and low desorption temperature (below 273 K, typically 77 K). An ideal adsorbent should demonstrate a high gravimetric and volumetric hydrogen storage capacity and should be capable of fast saturation and release of hydrogen. The energy of interaction between H_2 and adsorbent should be about 5 kJ mol^{-1} [170]. The common problem of the physical adsorption of hydrogen by materials with a high specific surface area (eg, MOFs, zeolites, carbon materials) is the very low energy of hydrogen adsorption to provide for the sufficient sorption at $T > 77$ K. Therefore, the efficient adsorbents should exhibit a high energy of the interaction between H_2 and the surface [171].

This part of the review presents the analysis of the literature data on physical adsorption of hydrogen by MOFs. We considered separately the spillover effect as an efficient approach to enhance the adsorption capacity for hydrogen. It should be noted that all the microporous materials, including MOFs, are characterized by I type isotherms with an abrupt saturation at low p/p_s and a flat region at higher p/p_s.

7.1.2 H_2 storage in zeolites

Physical sorption of H_2 on diverse zeolites with different composition, pore sizes, and architecture (LTA, Rho, X, Y) has been studied in a wide temperature and pressure ranges (77–573 K, 0.1–10 MPa) [172–175]. Two factors govern the hydrogen adsorption capacity: the structure of the zeolite providing the free space for adsorption and the nature of exchanged cations responsible for the interaction and binding with hydrogen. Usually the amount of adsorbed hydrogen increases almost linearly with the surface area. Quite high adsorption capacity (Table 2) was found at 77 K for CaX zeolite (2.19 wt.%), whereas NaY shows a somewhat lower capacity (1.8 wt.%) (77 K, 1.5 MPa) [174]. The H_2 adsorption in zeolites is limited, because the content of cations cannot be increased significantly over the stoichiometric ratio, whereas extra-framework cations introduced over the stoichiometry block the pores, especially for LTA and Rho zeolites unlike X or Y zeolites [177].

Table 2: Hydrogen uptake by zeolites

Material	Surface Area, BET (m^2/g)	H_2 Uptake (wt.%)	Conditions	Refs.
ZSM-5	n/a	0.7	77 K, 0.1 MPa	[176]
NaY	725	1.8	77 K, 1.5 MPa	[174]
CaX	n/a	2.19	77 K	[175]

The further gain in the hydrogen adsorption may be achieved by introducing metals with enhanced affinity to hydrogen like Pd(0). In this case atomically adsorbed hydrogen forms a metal hydride [177]. In some cases, the use of hydrogen spillover increases the adsorption capacity of such systems.

7.1.3 H_2 storage in MOFs

MOFs provide new opportunities for gas storage due to their increased surface area and pore volume [62]. As we have mentioned, the chemical properties and spatial structure of MOFs can be tailored for a specific application. Such purpose-oriented molecular-level engineering enabled successful use of MOFs in various fields. In particular, MOFs are widely used in adsorption and selective separation, including separation of gases (N_2, Ar, CO_2 [176,178] and methane [4,70]).

In order to bring the enhanced adsorption function to MOFs, a metal (Li or other metals) can be topologically accommodated in this highly porous and easily accessible structure [5].

MOFs have already demonstrated excellent hydrogen storage capacity under the cryogenic conditions (excess uptake of over 8 wt.% at 5 MPa [7]), whereas at room temperature the hydrogen uptake is lower; however, systematic studies at high pressures are missing and an option of a high-pressure tank filled with MOF as an adsorbent is promising. The theoretical estimate of the highest possible H_2 storage capacity for COFs is at the level of 20 wt.% [179].

So, there are no principle constraints to reach the values of 6–7 wt.% or higher capacities at pressures in the range of 35–70 MPa. The tuning of the structure of MOFs/COFs by choosing appropriate organic linkers (di- or tricarboxylic acids) and the inorganic cations, as well as the synthetic approaches governing the formation of interwoven structures will be efficient in reaching the capacity lead. The efficiency of the MOF materials may be enhanced by using loaded metals such as Li, Mg, Al, and others.

Although a large number of MOFs have been synthesized, only a few have been evaluated from the perspective of hydrogen adsorption [170,180,181]. We have already emphasized that the most attractive feature of MOFs is the possibility to control the relation between their structure and properties. The structure of MOFs has a distinct geometry; moreover, it is hybrid, in that it combines an organic component and an inorganic component (metal ion or metal oxide cluster). This determines two possible directions for enhancement of hydrogen adsorption by MOFs: increasing the inner surface area and/or volume of pores, and increasing the energy of hydrogen binding to the MOF [182]. Below we will consider the corresponding main factors that affect the adsorption of hydrogen by MOFs.

Let us discuss the data on H_2 adsorption using most typical MOF structures, like MOF-5. It is noteworthy that a broad range of conditions and procedures have been tried to obtain MOF-5 with particular crystallite size or to reduce the synthesis time. Synthesis procedures yielding nanocrystalline MOF-5 are interesting owing to their practical simplicity and because they present an opportunity to vary the geometrical properties of MOFs in adsorption [108].

Among the hierarchically ordered systems for H_2 storage, MOFs show significant promise. MOFs represent a new class of porous crystalline materials for which it is possible to design organic linkers and inorganic joints [32,144,183].

To enhance the H_2 capture function of the system, a specific porous structure is needed with specific porous features and specific adsorption sites such as metal ions or metal oxides or metal nanoparticles that can be structurally or topologically accommodated in this highly porous and easily accessible structure. It is quite clear that the opportunities afforded by this new class of inorganic-organic materials are unique, as the choice of the organic linker provides the possibility to tune both textural parameters and design of metal complexes. The former includes a specific surface area extending from hundreds to about 5000 m^2/g (for MOF-177), which is far higher compared to zeolites (about 500 m^2/g) and a pore opening diameter ranging from the dimension typical for zeolites (1 nm) to larger diameters allowing for accommodation of supramolecular structures, with a fraction of free volume reaching 90% (30–50% for zeolites) and pore volume close to 1.5–1.6 cm^3/g. The nature of the inorganic counterpart of the MOF architecture may be responsible for the H_2 storage properties of these materials, while the presence of functional groups at the linkers, the inorganic moieties, and introduced metal ions provides new opportunities in tailoring specific adsorption sites.

Different approaches to introduce the metal in the framework ("incipient wetness" impregnation, MOCVD loading of highly PCPs host lattices by metal organic vapor deposition, etc.) provide wide opportunities for the modification of the adsorption properties of MOFs/COFs [32,110,115,184–186].

Among the factors influencing hydrogen storage efficiency, the following parameters should be mentioned:

- Specific surface area,
- Shape and size of the pores,
- Availability of niches or side pockets,
- Ratio of micro-to-meso porosity,
- Presence of specific adsorption sites, like metal ions and atoms or NPs,
- Presence of intergrown guest structures like calixarenes and dendrimers,
- Possibility of formation of interwoven (two or more different MOFs) structures (catenation).

7.1.4 Specific surface area

It was mentioned previously that carboxylate MOFs, most intensively studied as potential sorbents of hydrogen, are characterized by extremely large values of the formal specific surface area. How does this parameter affect the physical adsorption of hydrogen? It turns out that the MOFs with a large specific surface area (>1000 m^2/g) show no direct correlation between the size of their specific surface area and hydrogen adsorption [173]. This is an important distinction between MOFs and the majority of porous carbon-based compounds (except nanofibers) and zeolites, which demonstrate higher hydrogen adsorption as their specific surface area increases, irrespectively of their morphology. The lack of such linear correlation is typical of MOFs and some other microporous materials. What affects the adsorption in these cases is not the formal specific surface area but the total volume of micropores [187].

This is well illustrated by studies on the synthesis and adsorption properties of MOF-5. This metal organic system has been synthesized using the diffusion, solvothermal, and express methods [4,181,188–190]. It turned out that the method of synthesis strongly affects the specific surface area of the MOF-5 preparations, although XRF and X-ray crystallographic analyses found no structural differences between them. The specific surface of the sample prepared by Yaghi using the solvothermal method (3362 m^2/g) [181] was significantly larger than that of the sample synthesized by Panella and Hirsher using the express method (572 m^2/g estimated with the BET method or 1014 m^2/g estimated with the Langmuir method) [188]. Nevertheless, the hydrogen adsorption capacities of these preparations were similar. The hydrogen adsorption capacity (1.6 wt.%, 77 K, 1 MPa) of the sample obtained using the express method was almost identical to that of the sample synthesized by the solvothermal method (1.32 wt.%, 77 K, 0.1 MPa, Table 3). This appears to be a direct proof that hydrogen adsorption is determined primarily by other factors, which will be discussed later.

7.1.5 Pore shape and size

The shape and size of the pore is not less important than the surface area and pore volume. Although MOFs are extremely highly porous systems, the volume of pores is not the only factor that determines the amount of absorbed hydrogen. Which pores are ideal for an efficient hydrogen adsorption by MOFs? The optimal material must have pores of the diameter as close as possible to that of the sorbate gas molecules and, simultaneously, a large overall pore volume. The smaller pore size facilitates interactions between the hydrogen molecule and multiple aromatic nuclei of the organic component of the MOF, which increases the energy of hydrogen binding [192] Therefore, the efficient adsorption requires not the presence of large pores, but the large number of small pores, close in size to the size of the hydrogen molecule [193]. This can be achieved by adding to the MOF cavities of organic or inorganic guest compounds, which will form a secondary microporous structure and create additional sites of hydrogen adsorption.

Table 3: Hydrogen uptake by MOFs

MOF	Surface Area, BET (m^2/g)	Pore Volume (cm^3/g)	H_2 Uptake (wt.%)	Conditions	Refs.
$Zn_4O(bdc)_3$, MOF-5	3362	1.19	1.32	77 K, 0.1 MPa	[181]
	2630	0.93	1.0	25 °C, 2 MPa	[79]
	1466	0.52	1.65	25 °C, 4.8 MPa	[79]
	1014	-	1.6	25 °C, 1.0 MPa	[188]
$Zn_4O(Br\text{-}bdc)_3$IRMOF-2	2544	0.88	1.2	77 K, 0.1 MPa	[4]
$Zn_4O(NH_2\text{-}bdc)_3$ IRMOF-3	3062	1.07	1.4		
$Zn_4O(R^6\text{-}bdc)_3$ IRMOF-6	3263	0.93	1.0		
$Zn_4O(ndc)_3$ IRMOF-8	1466	0.52	1.5	77 K, 0.1 MPa	[182]
$Zn_4O(bpdc)_3$ IRMOF-9	2613	0.9	1.0	77 K, 0.1 MPa	[4]
$Zn_4O(hpdc)_3$ IRMOF-11	1911	0.68	1.62	77 K, 0.1 MPa	[181]
$Zn_4O(pdc)_3$ IRMOF-13	2100	0.73	1.73	77 K, 0.1 MPa	[4]
$Zn_4O(tmbdc)_3$ IRMOF-18	1501	0.53	0.89	77 K, 0.1 MPa	[181]
$Zn_4O(ttdc)_3$ IRMOF-20	4346	1.53	1.31	77 K, 0.1 MPa	[79]
$Zn_4O(btb)_3$ IRMOF-177	4526	1.61	1.25	77 K, 0.1 MPa	[181]
Al(OH)bdc MIL-53(Al)	1590	-	3.8	77 K, 1.6 MPa	[180]
Cr(OH)bdc MIL-53(Cr)	1590	-	3.1		
Cu(hfipbb) $(H_2hfipbb)_{0.5}$	-	-	1.0	25 °C, 4.8 MPa	[189]
$Cu_3(tatb)_2(H_2O)_3$	3800	-	1.9	77 K	[182]
$Mn(HCO_2)_2$	240[a]	-	0.9	77 K, 0.1 MPa	[176]
Zn_2(1,4-$(bdc)_2$ (dabco))	1450	-	2.0	77 K, 0.1 MPa	[8]
$Cu_2(bptc)$, MOF-505	1646	0.63	2.48	77 K, 0.1 MPa	[191]
$Cu_2(btc)_{4/3}$, HKUST-1	2175	0.75	2.5	77 K, 0.1 MPa	[79]
Zn_2(2,5-$(OH)_2$1,4-$(bdc)_2$) MOF-74	1132	0.39	1.7		
$[Zn_4(\mu_4\text{-}O)(L_1)_3(DMF)_2]$	502[a]	0.20	1.12	298 K, 4.8 MPa	[183]
$[Zn_4(\mu_4\text{-}O)(L_2)_3]$	396[a]	0.13	0.98		

[a]L_1 = 6,6′-dichloro-2,2′-diethoxy-1,1′-binaphthyl-4,4′-dibenzoate and L_2 = 6,6′-dichloro-2,2′-dibenzyloxy-1,1′-binaphthyl-4,4′-dibenzoate.

In most MOFs, the free space between organic linkers is too large for an efficient hydrogen adsorption (Table 4). For example, the distance between phenylene planes in MOF-5 (Fig. 6) can be described as an approximately spherical pore, 12–15 Å in diameter, which is much larger than the kinetic diameter of the hydrogen molecule (2.8 Å). Therefore, even monolayer adsorption will be leaving some unused volume in the center of each pore. The pore size can be reduced to the optimum value by changing the shape of the pores, for example, from spherical to ellipsoid [170].

The role of pore geometry can be illustrated by the high hydrogen adsorption capacity of MIL-53, based on phenylenedicarboxylates (M(OH)(bdc)) (M = Al^{3+}, Cr^{3+}) [186]. The preparation with chromium showed the hydrogen adsorption capacity of c.3.1 wt.%, and that with aluminum, c.3.8 wt.% (77 K, 1.6 MPa). However, almost no hydrogen uptake was observed at room temperature, which is similar to the case of the MOF-5 preparation synthesized by Panella and Hirsher [188] (Table 3).

Table 4: Geometrical characteristics of MOFs

MOF	Free/Fixed Pore Diameter (Å)[a]	Accessible Pore Volume Fraction[b]	Free Pore Volume (%)	Refs.
$Zn_4O(bdc)_3$ IRMOF-1, MOF-5	7.8/15.2	0.59	79.2	[181]
$Zn_4O(Br\text{-}bdc)_3$ IRMOF-2	7.5/16.4	n/a	78.4	[4]
$Zn_4O(NH_2\text{-}bdc)_3$ IRMOF-3	9.6/18.6	n/a	78.7	[79]
$Zn_4O(C_3H_7O\text{-}bdc)_3$ IRMOF-4	5.4/14.1	n/a	64.6	[4]
$Zn_4O(C_3H_7O\text{-}bdc)_3$ IRMOF-5	3.6/12.3	n/a	55.8	[4]
$Zn_4O(R^6\text{-}bdc)_3$ IRMOF-6	5.9/15.2	0.50	77.5	[79]
$Zn_4O(R^7\text{-}bdc)_3$ IRMOF-7	5.5/13.6	n/a	76.6	[4]
$Zn_4O(ndc)_3$ IRMOF-8	8.4/18.0	0.66	n/a	[182]
$Zn_4O(bpdc)_3$ IRMOF-9	10.6/14.5	n/a	74.7	[79]
$Zn_4O(bpdc)_3$ IRMOF-10	15.4/24.5	n/a	87.0	[4]
$Zn_4O(hpdc)_3$ IRMOF-11	6.0/12.4	0.4	69.1	[181]
$Zn_4O(hpdc)_3$ IRMOF-12	13.0/24.5	n/a	84.3	[4]
$Zn_4O(pdc)_3$ IRMOF-13	10.2/15.8	n/a	70.2	[79]
$Zn_4O(pdc)_3$ IRMOF-14	13.8/24.5	n/a	85.3	[4]
$Zn_4O(tpdc)_3$ IRMOF-15	8.1/12.8	n/a	79.8	[4]
$Zn_4O(tpdc)_3$ IRMOF-16	19.1/28.8	n/a	91.1	[4]
$Zn_4O(tmbdc)_3$ IRMOF-18	5.4/13.8	0.42	n/a	[181]
$Zn_4O(btb)_3$ IRMOF-177	9.6/11.8	0.63	n/a	[181]
Al(OH)bdc MIL-53(Al)	6.4/6.4	0.29	n/a	[180]
Cr(OH)bdc MIL-53(Cr)	6.6/6.6	0.29		[180]
$Cu(hfipbb)(H_2hfipbb)_{0.5}$	3.0/4.7	0.03	11.6	[189]
$Mn(HCO_2)_2$	3.0/4.7	0.10	33.0	[176]
$Zn_2(1,4\text{-}bdc)_2$ (dabco)	7.8/9.5	0.45		[8]
$Cu_2(bptc)$, MOF-505	6.7/10.1	0.37	37.1	[191]
$[Zn_4(\mu_4\text{-}O)(L_1)_3(DMF)_2]$	3.8/7.8	0.21		[192]
$[Zn_4(\mu_4\text{-}O)(L_2)_3]$	3.8/5.4	0.17		[192]

[a]The free pore diameter is the diameter of the largest sphere that can pass through the small pore of the MOF; the fixed diameter is the diameter of the largest sphere that can fit into its largest pore [190].
[b]The fraction of the accessible pore volume was calculated based on the model radius, 1.45 Å, corresponding to the kinetic diameter of the hydrogen molecule, 2.8 Å [4, 63].

MIL-53 materials provide a convenient model for comparison with traditional sorbents with similar geometric characteristics, such as activated charcoals and zeolites. The specific surface area estimated for MIL-53, 1100 m^2/g (BET method), is similar to the average value for nanocarbon materials and exceeds that for zeolites. The framework of MIL-53 comprises unidimensional channels of 8.5 Å in diameter, which is similar to the pore diameter in zeolites (6–12 Å) and smaller than that in the IRMOF series, including MOF-5 (12–15 Å). At the same time, the hydrogen adsorption capacity of MIL-53 is somewhat higher than that of the zeolite CaX (2.19 wt.%, Table 2) and activated charcoals (2.15 wt.%) [179]. It is possible that this parameter is affected by the channel geometry, because zeolites more often have a 3D channel system, as opposed to the unbranched unidimensional channels in MIL-53. However, additional experiments on the adsorption mechanism are required to draw more definitive conclusions.

An interesting model to assess the roles of the specific surface area and pore size is provided by manganese formate, $Mn(HCO_2)_2$ [176]. Its specific surface area (c.240 m^2/g, BET method) is relatively small and similar to the average value for zeolites. X-ray crystallographic analysis suggested that its structure comprises unidimensional channels with the inner diameter of 5.5 Å, as in the case of MIL-53 (Fig. 10). Each octahedron shown in Fig. 10 represents a Mn(II) cation connected to six formate linkers. The volume of the channels is 33% of the overall volume (Table 4). The hydrogen adsorption capacity of $Mn(HCO_2)_2$, 0.9 wt.%, turned out to be somewhat larger than that of the zeolite ZSM-5 (0.7 wt.%, Tables 2 and 3). Interestingly, this value in manganese formate is only c.35% lower than in MOF-5, in spite of the fact that its specific surface area is almost 15 times smaller. This can be explained by the pore size in $Mn(HCO_2)_2$ better matching the kinetic diameter of the hydrogen molecule.

This idea about the proper fit of the channel geometry to the H_2 molecule can also be illustrated with the example of MOF-177 ($Zn_4O(btb)_2$), which has the largest specific surface

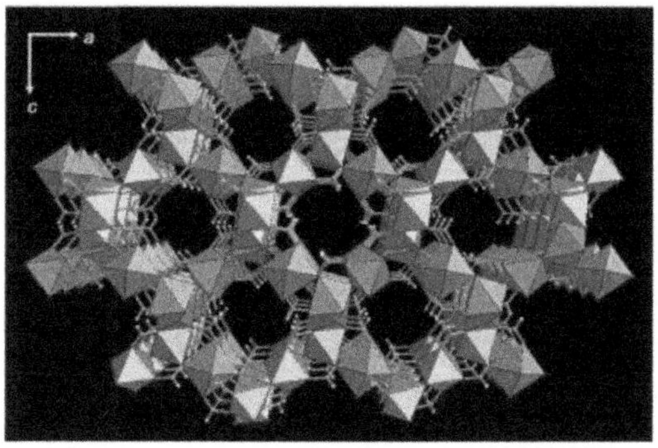

Fig. 10
Crystal structure of $Mn(HCO_2)_2$. *From Dubtsev D, Chun H, Yoon SH, Kim D, Kim K. J Am Chem Soc 2004;126:32–33 with permission.*

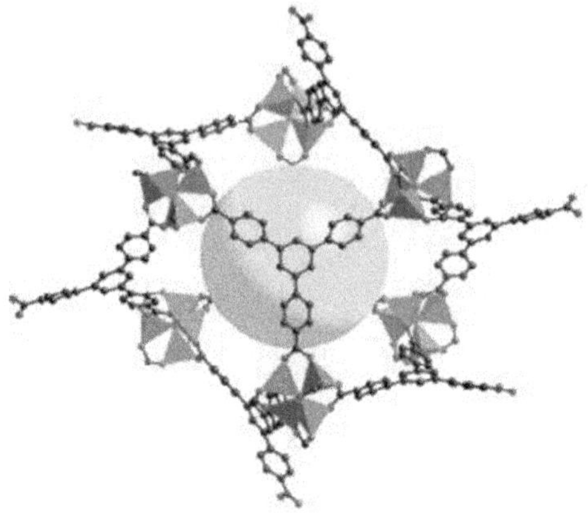

Fig. 11
MOF-177. *From Chae HK, Siberio-Perez DY, Kim J, Go YB, Eddaoudi M, Matzger AJ, O'Keeffe M, Yaghi OM. Nature 2004;427:523–527 with permission.*

area known so far, up to 5600 m^2/g. X-ray crystallographic analyses indicated that the structure of MOF-177 comprises a (6,3) coordinated net formed by Zn^{2+} oxide clusters and triangular carboxylate secondary structure (Fig. 11). Here the large pore is shown as a sphere inscribed into the framework; its diameter is determined by the distance between van der Waals surfaces of the framework atoms.

Although the pore volume is 0.69 Cm^3/g, the material has the channels and cavities with a diameter of 10.9–11.8 Å [6]. However, the H_2 capacity for MOF-177 is lower than that for the materials with a lower surface area, like IRMOF-8 (Table 3). Only some portion of the pore volume is filled with hydrogen. Hydrogen uptake by MOF-177 does not reach saturation at 77 K and atmospheric pressure because, at these conditions, hydrogen is absorbed to a lesser extent than nitrogen (used for estimation of the specific surface area). A systematic study of hydrogen uptake by representatives of the aforementioned IRMOF series (Table 3) also shows no adsorption saturation at these conditions (77 K, 0.1 MPa) [184]. Calculations demonstrated that only a small portion of the overall pore volume becomes filled with hydrogen molecules. Therefore, in order to improve the hydrogen adsorption capacity by decreasing the unfilled fraction of the inner volume, the geometry of pores needs to be optimized.

7.1.6 MOFs with catenane structures

One way to optimize the pore size in MOFs to match the size of the hydrogen molecule is creation of catenanes. A catenane structure is formed by "fusion" of two or more identical frameworks, provided that the necessary free pore volume is available. The so-called catenation can be in the form of interpenetration, when the two or more identical frameworks are set as far apart as possible, or in the form of "interweaving" when the distance between the frameworks is minimized and the latter are in close contact [63,68,194]. For instance,

IRMOF-11 (hpdc—4,5,9,10-tetrahydropyrene-2,7-dicarboxylate) is formed by two interwoven MOF structures. Although the pore volume usually decreases in this case, the narrow pores are formed that are capable of accommodating small H_2 molecules, especially if specific adsorption sites (metal ions and atoms) are present; this also leads to enhanced thermal and hydrothermal stability [79].

Traditionally, interpenetration of two or more systems is considered to be unfavorable for preparation of high-porosity frameworks because of the pore volume reduction. However, "interweaving" of two identical frameworks with minimal displacement reduces the pore size, improves the ratio between the surface area and pore volume, and increases the thermal stability [79]. These factors facilitate hydrogen uptake. Indeed, IRMOF-9, -11, and -13 have the "interwoven" structure (Fig. 8) and better hydrogen adsorption characteristics than MOF-5 (Table 3). For example, the hydrogen adsorption capacity of IRMOF-13, with pyrene linkers (Fig. 8), is almost twice as high as that of MOF-5, although the latter has a larger pore volume. This is a direct consequence of the pore size reduction due to the formation of a catenane structure.

An advantage of the "interwoven" structures is an increased proportion of the overall pore volume accessible for hydrogen (Table 3). This can be shown by comparing the hydrogen adsorption capacities of IRMOF-13 ("interwoven" structure), 0.89%, and IRMOF-20 (standard structure), 1.25%, the latter framework having the formal specific surface area three times larger than the former.

A significant improvement of the hydrogen adsorption capacity due to optimization of the pore size by formation of interpenetrating structures is evidenced by comparison of MOF-5 and [Cu(hfipbb)(H_2hfipbb)$_{0.5}$] (where H_2hfipbb = 4,4-(hexafluoroisopropylidene)bis(benzoic acid)). The latter MOF consists of two interpenetrating 3D MOFs [189]. Due to the catenane structure, the accessible pore volume in [Cu(hfipbb)(H_2hfipbb)$_{0.5}$] is significantly reduced and is estimated at 11.6% of the overall pore volume (Table 4). This value is 6.6 times lower than in MOF-5 (76.8%). However, at room temperature, [Cu(hfipbb)(H_2hfipbb)$_{0.5}$] absorbs only 1.8 times less hydrogen than MOF-5 (1.0%, 4.8 MPa, Table 3). This can be explained, in particular, by the size of pores: [Cu(hfipbb)(H_2hfipbb)$_{0.5}$] contains smaller (c.3.5 × 3.5 Å) and larger (c.5.1 × 5.1 Å) pores, but all of them are significantly smaller than the pores in MOF-5 (7.7 × 7.7 Å). The specific H_2 uptake for [Cu(hfipbb)(H_2hfipbb)$_{0.5}$] per 1 m^2 is higher compared to MOF-5.

The hydrogen uptake of a pair of isostructural compounds, catenated PCN-6 and noncatenated PCN-6′, was found to be more effective in catenated PCN-6 than that in noncatenated PCN-6′ [195].

The hydrogen storage was also studied on a microporous MOF, $[Mn(DMF)_6]_3[(Mn_4Cl)_3(btt)_8(H_2O)_{12}]_2 \cdot 42(DMF) \cdot 11(H_2O) \cdot 20(CH_3OH)$ with exposed Mn^{2+}

Table 5: Spillover effect during hydrogen uptake

Material	Surface Area, BET Method (m^2/g)	H_2 Uptake (wt.%)	Conditions	Refs.
CNF + 2% Pd		1.38	293 K, 7.7 MPa	[197]
SWNTs/5%Pd/C/carbon bridges 8:1:1	820	0.038	298 K, 100 kPa	[198]
AX-21/5%Pd/C/carbon bridges 8:1:1	2880	0.04	298 K, 100 kPa	[198]
AX-21/5%/Pd/C/carbon bridges 8:1:1	2880	1.8	298 K, 10 MPa	[198]
MWNTs		0.6	298 K, 10.73 MPa	[199]
MWNTs/20%Pd/plasma treatment		4.5	298 K, 10.72 MPa	[199]
$Zn_4O(bdc)_3$ + 5%Pt/C/carbon bridges, IRMOF-1, MOF-5	3362	3.0	298 K, 10 MPa	[200]
$Zn_4O(ndc)_3$ + 5%Pt/C/carbon bridges, IRMOF-8	1466	4.0	298 K, 10 MPa	[200]

SWNTs, single-walled nanotubes; MWNTs, multiwalled nanotubes; AX-21, superactivated charcoal; CNF, carbon nanofibers.

coordination sites [196]. The form of this compound after removal of the solvent $Mn_3[(Mn_4Cl)_3(btt)_8(CH_3OH)_{10}]_2$ shows a total H_2 adsorption capacity of 6.9 wt.% at 77 K and 90 bar and exhibits a maximum isosteric heat of adsorption of 10.1 kJ/mol among MOF materials (Table 5).

$Cu_3(tatb)_2(H_2O)_3$ (where tatb = 4,4′,4″-*s*-triazine-2,4,6-triyl-tribenzoate) can serve as an example of an MOF with the pore size close to the size of the hydrogen molecule [182]. In this system, the pore size is efficiently reduced due to the "interweaving" of two frameworks (specific surface area estimated using the Langmuir method is c.3800 m^2/g, the overall pore volume is 1.45 cm^3/g). Table 3 shows that the hydrogen adsorption capacity of $Cu_3(tatb)_2(H_2O)_3$ at 77 K and atmospheric pressure is c.1.9 wt.% of hydrogen (10.6 mg/cm^3), which exceeds the corresponding capacities of MOFs with a "normal" structure and larger specific surface area, such as IRMOF-20 (4346 m^2/g, 1.35 wt.%) and MOF-177 (4526 m^2/g, 1.25 wt.%).

7.1.7 Chemistry of organic linkers

Organic linkers with highly polarized aromatic fragments, such as phenylene, naphthylene, bipyridine, and biphenylene, are widely used for synthesis of MOFs because they make the framework rigid. The chemical nature of these molecular structural units is similar to that of carbon nanomaterials and, in particular, nanotubes, formed by sp^2-hybridized carbon atoms. An important difference between MOFs and carbon nanomaterials lies in the fact that chemical modification of the latter does not significantly affect their physical adsorption energy [170]. In contrast, modification of the organic components of MOFs provides for a stronger interaction between the framework's surface and hydrogen.

The effect of chemical modification of the organic component while preserving the topology of an MOF can be illustrated by the synthesis of the IRMOF series. It turned out that hydrogen adsorption per formula unit, both at 77 K and room temperature, increased with the increase of the number of aromatic rings in the organic fragment. Therefore, the length and width of the organic linker affect significantly the adsorption properties of the system (Table 3) [4,181]. The adsorption maximum increased from 4.2 molecules of H_2 per formula unit in IRMOF-8 (organic linker, 2,6-naphthalenedicarboxylate) to 8.9 in IRMOF-11 (organic linker, 4,5,9,10-tetrahydropyrene-2,7-dicarboxylate). At the same time, the gravimetric hydrogen capacity of IRMOF-11 is almost twice as high as that of IRMOF-18, formed by 2,3,5,6-tetramethylbenzene-1,4-dicarboxylate linkers (Table 3). A further increase of the physical adsorption of hydrogen was achieved by using biphenyldicarboxylate linkers (IRMOF-13, Table 3). These results demonstrate that the increased aromaticity of the organic linker significantly increases the adsorption capabilities of MOFs. One has to remember that bulkier and better branched organic linkers also strongly reduce the diameter of pores. Because of this, the question whether the increased adsorption results from the optimized pore sizes or optimized surface energetics remains open.

The effect of the organic linker on hydrogen adsorption was confirmed by studies using inelastic neutron scattering (INS) of several IRMOFs: IRMOF-1 (MOF-5), IRMOF-8, IRMOF-11, and IRMOF-177 [201]. It was found that the pores of MOFs contain numerous distinctly localized hydrogen adsorption sites. INS studies demonstrated that the mechanism of hydrogen binding depends on the type of MOF; given the same inorganic component (Zn_4O), it is directly affected by the chemistry of the organic linkers. Therefore, including a higher proportion of polarized polyaromatic linkers, which increase the energy of interaction between the hydrogen molecule and the surface, is an efficient way to synthesize MOFs with a high hydrogen adsorption capacity. This approach resulted in the creation of MOFs formed by polyaromatic binaphthyl compounds $[Zn_4(\mu_4\text{-}O)(L_1)_3(DMF)_2]$ and $[Zn_4(\mu_4\text{-}O)(L_2)_3]$ (where L_1 = 6,6′-dichloro-2,2′-diethoxy-1,1′-binaphthyl-4,4′-dibenzoate, and L_2 = 6,6′-dichloro-2,2′-dibenzyloxy-1,1′-binaphthyl-4,4′-dibenzoate) [191].

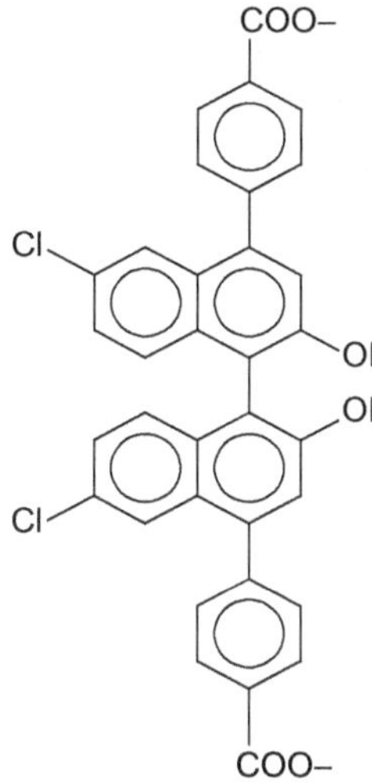

Although the geometrical characteristics, the specific surface area, and overall pore volume of $[Zn_4(\mu_4\text{-}O)(L_1)_3(DMF)_2]$ (502 m^2/g, 0.20 cm^3/g) and $[Zn_4(\mu_4\text{-}O)(L_2)_3]$ (396 m^2/g, 0.13 cm^3/g) are significantly lower than those of MOF-5, the hydrogen adsorption capacities of these three materials are very similar (Table 3). Therefore, "interpenetrating" MOFs with polyaromatic organic linkers show better hydrogen uptake capacity at room temperature. On the other hand, the uptake values for $[Zn_4(\mu_4\text{-}O)(L_1)_3(DMF)_2]$ and $[Zn_4(\mu_4\text{-}O)(L_2)_3]$ are very close to the hydrogen adsorption capacity of single-walled nanotubes. Apparently, the interpenetration of frameworks in $[Zn_4(\mu_4\text{-}O)(L_1)_3(DMF)_2]$ and $[Zn_4(\mu_4\text{-}O)(L_2)_3]$ facilitates the interaction between the hydrogen molecules and the framework surface. This is due to the multiple simultaneous contacts with multiple aromatic nuclei of both "interpenetrating" frameworks, which enhances physical adsorption of hydrogen. Similar to nodes in the "bundles" of nanotubes, these adsorption sites inside "interpenetrating" MOFs have a higher energy of hydrogen binding.

It turned out that incorporating functional groups into organic linkers of MOFs do not significantly affect the hydrogen adsorption capacity or, in some cases, even lowers it (see Table 3) [79]. This is explained by a combination of several antagonistic effects. In the MOFs containing various functional groups, the pore size is reduced from 12.1 Å (IRMOF-1) to c.9.6–9.8 Å (IRMOF-2, linker, 2-bromophenylenedicarboxylate; IRMOF-3, linker, 2-aminophenylenedicarboxylate; IRMOF-6, linker, 1,2-dihydrocyclobutylphenylene-3,6-dicarboxylate, Table 4). Calculations have demonstrated that the addition of electron-donor groups, such as amine or methyl, can increase the energy of interaction between hydrogen and the phenylene ring by c.15%. However, the hydrogen adsorption capacity of IRMOF-2 is only slightly higher than that of MOF-5 (Table 3). The weak electron-acceptor effect of bromine in the phenylene ring (IRMOF-2) is compensated by its strong interaction with the hydrogen molecule; nevertheless, the adsorption capacity of IRMOF-3 is lower than that of MOF-5. In the case of IRMOF-20, which has two intersecting spherical pores in its crystalline lattice (Fig. 12), the polarizability of the thieno-[3,2b]thiophene groups significantly improves the hydrogen uptake due to a stronger interaction between the hydrogen molecule and organic linker, but the presence of the heavy sulfur atom lowers the gravimetric capacity of this system. As a result, the adsorption capacities of MOF-5 and IRMOF-20 are nearly identical. These data suggest that the afore-mentioned functional groups facilitate hydrogen adsorption, but heavy substituents decrease the gravimetric hydrogen uptake (Table 3).

7.1.8 Flexibility of organic linkers

Until recently, the flexibility of organic linkers in MOFs was regarded as a disadvantage because it makes the configuration of the resulting structure less certain and can lower the thermal stability of the system. Recently, flexible heterocyclic linkers have been used in MOFs besides the rigid polyaromatic carboxylate linkers [202]. In this type of frameworks, pores can be narrowed after taking up guest molecules and can be widened after their desorption; the "response" of the porous framework is determined by the chemistry of the "guest" molecule.

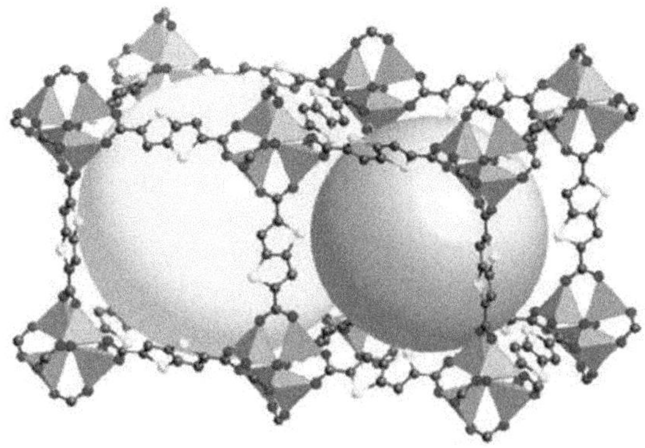

Fig. 12
Structure of IRMOF-20 formed by thieno[3,2-*b*]-thiophene-2,5-dicarboxylate fragments. *From Rowsell JLC, Yaghi OM. J Am Chem Soc 2006;128:1304–1315 with permission.*

An example of such systems is $[Zn_2(1,4\text{-bdc})_2(\text{dabco})]_4$ [8]. This MOF has a strong but flexible MOF, which dynamically responds to the presence of "guest" molecules. Its hydrogen adsorption capacity is 2.0 wt.% (77 K, 0.1 MPa), which is close to the highest known values of this parameter, measured in MOF-505 and HKUST-1 (Table 3). Yet, additional studies are needed to better understand the advantages of such systems for the purpose of hydrogen storage.

7.1.9 Presence of coordinatively unsaturated (open) metal centers

As hydrogen sorbents, MOFs have an important advantage over carbon-based materials because the metal ions in their composition can bind hydrogen more strongly that graphene-like carbon. Indeed, the inorganic part of the framework, that is, the chemical nature of the metal cation or metal oxide, plays an important role in hydrogen uptake. INS studies indicated the presence of two different types of hydrogen-binding sites in MOFs: metal ions and organic linkers [201]. It is well known that metal ions play the important role of active sites in a number of processes, including adsorption. Principles of modular chemistry, used for the synthesis of MOFs, allow one to stabilize available coordinatively unsaturated sites inside the porous crystalline matrix [191]. The main approach for engineering an open metal center is as follows. The SBU is formed from binuclear carboxylate clusters with labile ligands in the axial positions. Vacuum treatment eliminates these ligands, thereby creating an open coordinatively unsaturated site, while the porosity of the framework is preserved due to its rigidity. Because such coordinatively unsaturated sites are highly reactive, it is important to avoid the formation of dimeric clusters or polymeric structures after elimination of the axial ligands. Rigid carboxylate linkers lock the spatial configuration and stabilize the active centers.

Such coordinatively unsaturated metal centers significantly increase the hydrogen adsorption capacity [79]. This is convincingly illustrated by MOF-505 being used as a sorbent; its hydrogen adsorption capacity is the highest known so far (2.47 wt.% at 77 K and 0.1 MPa, Tables 3 and 5) [191]. The formula of this compound is $[Cu_2(bptc)(H_2O)(DMF)_3(H_2O)]$. The nodes of the lattice are formed by square clusters $Cu_2(O_2C\text{-})_4$. This MOF contains two types of pores, 8.30 and 10.10 Å in diameter (Table 4). Its overall accessible pore volume, 37.1%, is lower than in many Zn^{2+}-based IRMOFs and other MOFs (Table 4). Apparently, its extremely high hydrogen adsorption capacity is achieved due to its inorganic cluster, based on the coordinatively unsaturated Cu^{2+} ion site, which binds hydrogen with a higher energy than Zn^{2+}. The small pore size in this MOF also facilitates hydrogen uptake. The active sites of this MOF become open after vacuum treatment and elimination of axial ligands, molecules of water.

The idea of hydrogen physical adsorption being affected by the presence of coordinatively unsaturated centers is confirmed by measurements of the isosteric adsorption heat, which allows one to quantitatively estimate the strength of the interaction between the hydrogen molecule and the active center at the metal ion [79]. Such studies of MOF-74 (based on 2,5-dihydroxyphenylenedicarboxylates of Zn(II)) and HKUST-1 (based on phenylene-1,3, 5-tricarboxylates of Cu(II)) helped to understand the effect of the nature of the coordinatively unsaturated center. The gravimetric hydrogen adsorption capacities of both these frameworks are higher than in other members of the IRMOF series (Table 3). A stronger interaction of the H_2 molecule with the $Cu_2(O_2C\text{-})$ than with the $Zn_4O(O_2C\text{-})_6$– cluster is evidenced by the isosteric adsorption heat values measured for HKUST-1 and MOF-5. At a low surface coverage, this value for HKUST-1 is 2.0 kJ/mol higher than for MOF-5. The isosteric heats of adsorption for MOF-74 (2,5-dihydroxyphenylenedicarboxylate of Zn (II)) and HKUST-1 (phenylene-1,3,5-tricarboxylate of Cu(II)) that demonstrate a higher H_2 capacity than other IRMOFs are somewhat increased (by 2–3 kJ/mol) to reach 10–12 kJ/mol. This agrees with the hydrogen adsorption capacity of HKUST-1, which is almost twice as high as that of MOF-5 (Tables 3 and 6) and slightly higher than that of MOF-505, which also contains coordinatively unsaturated Cu^{2+} centers (Tables 3 and 6) [191].

Therefore, more efficient hydrogen uptake at low pressures can be achieved by stabilization of coordinatively unsaturated centers in MOFs and by choosing metal ions to be more efficient than Zn^{2+}.

This comparison of the effects of "organic" and "inorganic" components of MOFs on their hydrogen adsorption capacity shows that the role of the organic linker in hydrogen uptake is much less significant. The most recent theoretical and experimental studies demonstrated that inorganic clusters based on metal ions serve as more efficient binding sites for hydrogen and significantly improve its physical uptake.

7.1.10 Hydrogen spillover

One of the currently most promising approaches to enhancing the hydrogen adsorption capacity is the use of secondary hydrogen spillover [198,200,205]. According to one

Table 6: The best known results on H_2 storage on MOF materials

MOF	Hydrogen Uptake (wt.%)	H_2 Uptake (g/L)	Conditions	Refs.
MOF-5	1.65		295 K; 4.8 MPa	[79]
MOF-505; $Cu_2(bptc)$	2.48		77 K, 0.1 MPa	[191]
HKUST-1; MOF-199; $Cu_2(btc)_{4/3}$	3.6	22	77 K, 10 bar	[79]
$[Mn(DMF)_6]_3[(Mn_4Cl)_3(btt)_8(H_2O)_{12}]_2$	6.9	12.1 (298 K, 9 MPa)	77 K; 90 bar	[196]
MOF-177; $Zn_4O(btb)_3$	7.5	48.3[a]	77 K; 60 bar	[203]
	11.3[a]		77 K; 70 bar	[203]
MOF-646; $[Zn_4O(L)_3]$ (L = 1,3-azulenedicarboxylate)		20.9	77 K, 1 bar	[19]
MOF-210 $Zn_4O(bte)(bpdc)$	8.6	44	77 K, 80 bar	[7]
	2.9		298 K, 100 bar	[204]
MOF-200 $Zn_4O(bbc)_2$	7.4	36	77 K, 80 bar	[7]
	3.24		298 K, 100 bar	[204]

[a]Absolute predicted value of the adsorbed amount.

definition, the spillover effect is the migration of an active particle adsorbed onto one site to a different site, which under normal conditions does not itself adsorb such particles [198,206]. As a result, new sites in the immediate proximity to the surface of the catalyst become available for adsorption. The catalytic site (particle of metal) facilitates dissociation of hydrogen molecules into atoms, and monoatomic hydrogen "spills over" across the surface toward the receptor. To produce this effect, close mechanical mixtures are typically used that contain a catalyst (the source of monoatomic hydrogen) and a receptor, which is inert and has a low adsorption capacity [207,208]. In some cases, it is necessary to create physical "bridges" to facilitate spillover of hydrogen from dissociation sites to receptors [209]. Incorporating such "bridges" is very important to overcome the energy barrier that hinders diffusion of hydrogen atoms across the surface from the catalytic site to the receptor. One currently used solution is the creation of carbon "bridges," originating as a result of carbonization of the surface [198]. If the source of monoatomic hydrogen is a metal particle on a carrier with a low capacity, then the overall adsorption capacity can be increased by adding a receptor with a high capacity. In this case, migration of monoatomic hydrogen from the metal particle to the carrier is the primary spillover, and its transfer to the receptor is the secondary spillover (Fig. 13). Therefore, secondary spillover is achieved by dissociative chemisorption of hydrogen onto the metallic catalyst, subsequent migration of monoatomic hydrogen onto the surface of the carrier (the primary receptor of a hydrogen atom) located in a contact with the metal, and then onto the second carrier (the secondary receptor). Carbon bridges (*black areas* in Fig. 13) facilitate the spillover.

The hydrogen adsorption capacity of single-walled nanotubes (specific surface area, 820 m^2/g, pore diameter, 12 Å) and the superactivated charcoal AX-21 (specific surface area, 2820 m^2/g) was significantly enhanced by the addition of a Pd/C catalyst (5 wt.%) and carbon "bridges."

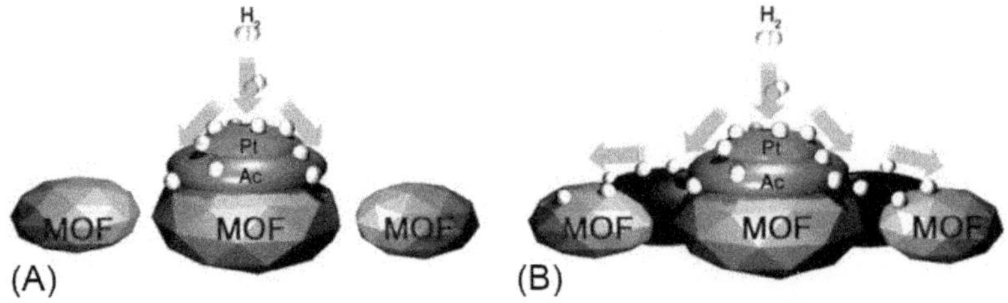

Fig. 13
(A) Primary spillover of atomic hydrogen from Pt metal to the activated carbon support and secondary spillover to the MOF receptor that has limited contacts with the support. (B) Facilitated primary and secondary spillover by using carbon bridges (*dark black zone*). *From Srinvas ST, Rao PK. J Catal 1994;148:470–477 with permission.*

As was indicated earlier, the catalyst catalyzes dissociation of hydrogen molecules into atoms, which facilitates penetration of hydrogen between the layers of graphene or into the inner space of nanotubes [198].

The carbon "bridges" are produced by adding a carbon precursor (D-glucose) to the mechanical mixture of components (the receptor:precursor:source weight ratio is 8:1:1), with subsequent thermal treatment at 673 K. The presence of such "bridges" increases the hydrogen adsorption capacity, at room temperature and at both low (100 kPa) and high (10 MPa) hydrogen pressures, 2–3 times for both superactivated charcoal and nanotubes. Under these conditions, the superactivated charcoal AX-21 demonstrated a higher hydrogen adsorption capacity than single-walled nanotubes (Table 5).

The uptake properties of multiwalled nanotubes (MWNTs) were enhanced through the combination of the hydrogen spillover effect and treating the nanotubes with hydrogen plasma in a microwave generator to create defects in their walls, which facilitates penetration of hydrogen into the tube's inner space between graphene layers (Table 5) [210]. Thus, the original hydrogen adsorption capacity of intact MWNTs was 0.6 wt.% (at 298 K, 10.7 MPa), plasma treatment increased this value more than two times, to 1.4 wt.%. Subsequent addition of Pd resulted in a further increase of the hydrogen adsorption capacity, to 4.5 wt.%, which is one of the best results achieved so far.

Hydrogen spillover turned out to be a highly efficient way of increasing the hydrogen adsorption capacity also for carbon nanofibers (CNF). For example, the addition of 2 wt.% of Pd allowed one to enhance the hydrogen uptake by CNF at room temperature (Table 5) [210].

Recently, the secondary hydrogen spillover effect was demonstrated also in MOFs [200]. The hydrogen uptake significantly increased when MOFs are mechanically mixed with a supported catalyst (Fig. 13). For example, the hydrogen adsorption capacity of IRMOF-8 at

298 K and 10 MPa was increased to 4% by addition of 5% of Pt/C (see Table 5). This is eight times the adsorption capacity of pristine IRMOF-8 at the same conditions (0.5%, Table 3). The same treatment increased the adsorption capacity of MOF-5 to 3% (298 K, 10 MPa). At these conditions, physical uptake of hydrogen is completely reversible. In the example discussed earlier, the activated charcoal is the primary and IRMOF-1 (MOF-5) and IRMOF-8 are the secondary receptors of hydrogen. For the optimal effect, the system was supplemented with carbon "bridges." To make such "bridges," a three-component mixture was prepared, which contained an MOF, a catalyst, and cellulose. The mixture was then carbonized as described earlier for carbon-based carriers. According to calculations, the amount of hydrogen taken up should correspond to c.34 hydrogen atoms per a $Zn_4O(C_{12}H_6O_4)$ formula unit. Therefore, the theoretical maximum of hydrogen adsorption is c.6.5% (298 K, 10 MPa), which is already not too far from actual demands of the automotive industry.

So, in 2003, Yaghi et al. [190] reported the first measurements of hydrogen adsorption on $Zn_4O(bdc)_3$ (MOF-5) and reached 4.5 wt.% at 77 K and 1 atm and 1.0 wt.% at room temperature and 20 bar. In 2006, Yaghi discovered another highly porous MOF, $Zn_4O(btb)_2$ (MOF-177) and reached an even higher total gravimetric uptake, 7.5 wt.% at 77 K and 78 bar [22]. Since 2003, at least 150 unique MOFs have been evaluated for their ability to store hydrogen.

MOF-210 and MOF-200 are believed to be candidates for the high performance in hydrogen storage [211]. The expected H_2 storage capacities for these materials are 2.9 and 3.24 wt.% at 298 K and 100 atm for MOF-210 and MOF-200, respectively. The metalation with alkali metals is supposed to further enhance the H_2 uptake. The calculation results provide an estimate of about 6.34 wt.% at 298 K for Li/MOF-200 and 5.94 wt.% for Na/MOF-200 at 100 atm. One of the JUC materials (JUC-48, $[Cd_3(bpdc)_3(DMF)]\cdot 5DMF\cdot 18H_2O$, where bpdc stands for 4,4′-biphenyldicarboxylate) also demonstrate quite high H_2 adsorption at 77 K and 40 atm (2.8 wt.%) [18].

The theoretical evaluation of MOFs as hydrogen storage materials predict the optimal size of the pore to be about 0.7 nm, when exactly two layers of hydrogen molecules adsorb on opposite surfaces [204]. As for today, the highest reachable hydrogen capacities for MOFs are listed in Table 6.

MOF-177 was evaluated by BASF (Basolite Z377 material) by loading the material in the high-pressure tank at 77 K and 50 bar [31]. The Basolite M050 material, a small-pore Mg formate, though demonstrating only modest H_2 adsorption capacity at 77 K (about 1 wt.% at 1 atm), shows a very good selectivity in separation of H_2 from N_2, with the adsorption of the latter being very small even at 77 K [52].

Thus, although there are some remarkable features related to the enhancement of the H_2 adsorption capacity of MOFs compared to conventional adsorbents, this field is definitely underexplored. Future research will probably be guided by the following hints to design the most efficient MOF-based hydrogen storage materials:

1. The high surface area itself is a prerequisite, but, in general, there should be no linear correlation between the H_2 uptake and the surface area at ambient temperatures (cf. the example of carbon materials with surface areas up to 3000 m^2/g but with a very low H_2 capacity, below 0.5%);
2. The high pore volume is more important than the surface area, but again the overall pore volume does not reflect the pore distribution among micro/meso/macropores; for high H_2 capacities, we need more micropores, less mesopores, and no macropores at all;
3. The synthesis should be directed toward the formation of intergrown (or interwoven or catenated) MOF structures, that is, two or more MOF architectures built-in each other;
4. The use of other guest structures with nanopores or nanocavities, like calixarenes or first or second generation dendrimers built in the MOF structure, especially into the interwoven structures, will enhance the H_2 uptake for sure;
5. The incapsulation of light metals inside the nanopores or nanocavities of MOFs or inside the calixarene baskets, or inside the branches of the dendrimer will result in the hydrogen spill-over and further enhancement of the H_2 capacity;
6. The use of functional groups at the organic linkers forming the MOF/COF structure (for instance, NH_2) will increase somewhat the hydrogen adsorption capacity;
7. Finally, all the approaches sketched earlier do not result in any high heats of hydrogen adsorption, and the heat of adsorption is typically not higher than 10 kJ/mol, which excludes the necessity of using heat exchangers in gas tanks filled with MOF.

MOFs have some properties in common with zeolites and modern carbon materials: large surface area, porosity (open channels), and presence of hydrogen-binding sites (metal ions). The physical characteristics and structure of MOFs and single-walled nanotubes are also similar. Both of these are systems with low density due to open channels or inner tube spaces; moreover, most MOFs contain aromatic carbon. As hydrogen sorbents MOFs have a number of advantages over carbon materials. Their channels are perfectly ordered, which facilitates more efficient penetration of hydrogen into their inner space. The inner surface of channels can be easily modified to change their curvature, which allows one to optimize hydrogen-sorbent interactions. MOFs are obtained by uncomplicated and repeatable synthesis, while cost-effective production of large quantities of nanotubes is not currently practical [199]. MOFs do not require severe activation conditions like nanotubes (nitric acid, CO_2, baking at 700 °C) or template removal like zeolites. As a rule, vacuuming at temperatures between 25°C and 190°C is sufficient to free the channels and pores of MOFs from solvent molecules or surplus reagents.

Literature data indicate that the values of hydrogen adsorption capacity for MOFs are, on average, intermediate between those for nanocarbon materials and zeolites. Indeed, MOFs, which are sometimes referred to as organic zeolites, have hydrogen adsorption capacity similar to or exceeding that of zeolites (Tables 2 and 3). Although the hydrogen adsorption capacity of current MOFs is lower than that of modern nanocarbon materials (see Tables 1 and 3), the diversity of the chemical nature of MOFs gives reason for optimism. Zeolites and carbon materials offer a limited number of structural parameters that can be modified to increase their hydrogen adsorption capacity. For zeolites, this can be achieved by supplementing their framework with extraframework cations, which serve as additional hydrogen binding sites, and increasing the surface area. These zeolite parameters have an upper limit. Hydrogen uptake by carbon materials can be enhanced by increasing their surface area, from which their hydrogen adsorption capacity depends linearly; such an increase is impossible above the theoretical limit. Numerical studies which simulated interactions between hydrogen and a wide range of nanotubes, of various geometries and prepared using various techniques, indicated that this class of nanomaterials has the limit of hydrogen adsorption equal to just several wt.%. In contrast, the number of parameters to modify the properties of MOFs is practically unlimited: volume, size, and shape of the pores, polarizability of the organic linker, nature of inorganic SBUs, number of coordinatively unsaturated sites within the framework, framework topology, etc. This offers hope for further enhancement of physical hydrogen adsorption properties of crystalline microporous MOFs.

7.1.11 Adsorption and storage of other molecules

Methane storage is another target for MOFs. Cu(II) benzene-1,3,5-tricarboxylate (MOF-177) was tested at BASF (Basolite C300 material produced by BASF on a large scale) on board of a car as a fuel, methane was stored at 200 bar [http://www.ecofuel-asia-tour.com]. Even in the case of a lower pressure of 38 bar, the capacity was enhanced (80 g/L) compared to the compressed cylinders of methane. Some of COFs (basically imidazolate-based) turned out to be even more efficient in methane storage compared to MOFs: COF-102 and COF-103 and later postsynthetically modified COF-103-ETH-trans and COF-102-ANT showed the CH_4 uptake of 190 v/v (STP) which is significantly higher compared to 112 v/v for MOF-177, with 137 v/v for COF-102 [211].

BASF reported on the pilot-scale purification of the natural gas and separation of a Kr-Xe mixtures by using their Basolite products [52,212,213].

Separation of acetylene from mixtures with other gases is a difficult problem. CuII coordination polymers of the type $[Cu_2(pzdc)_2(L)]n$ (L = pillar ligand) called "CPL" (coordination pillared layers) were demonstrated to adsorb selectively acetylene from the mixtures with CO_2 (CPL-1 with 4,4′-bpy as a ligand) [214]. Acetylene adsorption increased drastically at low pressures, whereas CO_2 adsorption followed the slowly rising curve, so that at $P = 1.1$ kPa the ratio of the adsorbed amount of C_2H_2 to that of CO_2 was about 26 at 270 K. The density of adsorbed acetylene was 0.434 g/cm^3, which is 200 times larger than the compression limit for the safe use of acetylene at 300 K and 0.2 MPa.

The separation of alkenes from the alkanes with the same number of carbon atoms is still more challenging compared with acetylene. The kinetic diameters of ethylene and ethane or propylene and propane are very close to each other, and the attempt to play with the size effect of the pore will fail. However, the olefins are quite strong bases, and this difference can be used for the separation of this pair of adsorbates. Also, there are examples of the formation of quite strong complexes of alkenes with coordinatively unsaturated metal ions (they also act as Lewis acid sites). The Fe_2(dobdc) MOF (dobdc is 2,5-dioxido-1.4-benzenedicarboxylate) managed to separate selectively ethylene and propylene from their saturated counterparts at 318 K [215]. Furthermore, the authors consider the possibility of fractionation of a mixture containing methane, ethane, ethylene, and acetylene, with the removal of acetylene from the mixture at the first stage and MOF/membrane separation at the second stage. Iron(II) coordinatively unsaturated ions were shown to act as adsorption sites for unsaturated molecules.

NO adsorption is of particular interest because of the relation to the problem of car exhaust control and drug delivery. Indeed, the too low concentration of NO in the human body results in hypertension, atherosclerosis, Parkinson disease, fibrosis, and sexual impotence, whereas the very high content of NO in the body may lead to septic shock, hypotension, excessive bleeding, meningitis, rheumatoid arthritis, ischemia, cancer, diabetes, and asthma. Therefore, the materials capable of strongly adsorbing NO at room temperature in large quantities are in strong demand. Morris and coworkers developed two coordination polymers, CPO-27-Co and CPO-27-Ni, that showed excellent performance of solids (ie, about 21 wt.%) for NO adsorption [216]. The activated CPO-27 material can adsorb ~7.0 mmol NO/g. This amount is three orders of magnitude greater than that found for HKUST-1 and seven times greater than that for the best zeolites.

CO_2 capture also calls for the efficient materials and MOFs show a significant promise in this application. The CO_2 adsorption capacity as high as 25–30 wt.% was found for several MOFs, including MOF-177 and MIL-53(Cr) [217,218]. However, their benefits seem to be better realized at high pressures. The recent review [219] summarizes the achievement made while using MOFs for CO_2 capture in three largest areas: postcombustion capture, precombustion capture, and oxy-fuel treatment. The highest adsorption capacities for CO_2 at elevated pressures (15–50 atm) were found to decrease in the raw:

MOF-210 > MOF-200 > CPO-27 > MOF-5 > MOF-205 > DUT-9 > MOF-177 > PCN-68 > MIL-101(Cr) > UMCM-1.

Other MOFs are inferior to this top 10, although the data set is not complete for many MOFs, especially at high pressures over 20–30 atm. For the atmospheric pressure the order of capacities is a little different:

CPO-27 > HKUST-1 > PCN-6 > MOF-505 > MIL-53 > MIL-101

Noteworthy that under the flow-through conditions the CO_2 capture capacity rarely exceeds 15–18 wt.%, which is about two times higher compared with liquid amines.

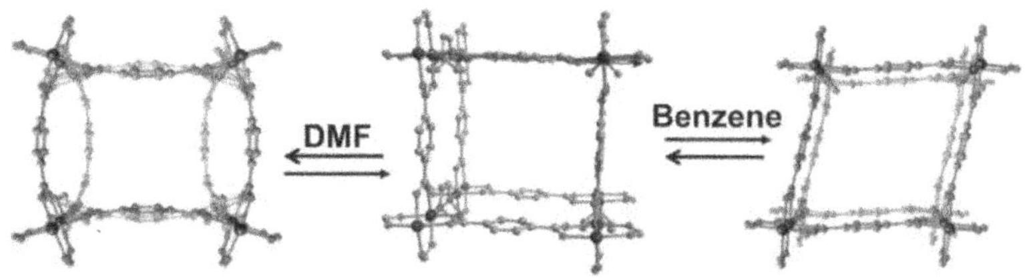

Fig. 14
Breathing effect. *From Schneemann A, Bon V, Schwedler I, Senkovska I, Kaskel S, Fischer RA. Chem Soc Rev 2014;43:6062–6096 with permission.*

De Vos and coauthors [220] studied the differences in adsorption of linear and branched hydrocarbons in the Cu-based MOF structure. The data obtained provide evidence for definite preferences of the linear alkanes in the adsorption. Quite a few examples of the so-called molecular recognition by MOFs were reported in the literature, the predominant adsorption of certain isomers over the other, such as *p*-dichlorobenzene or *p*-xylene compared to *m*-isomers or linear alkanes compared to the branched molecules is meant by this term. Another interesting feature of the MOF materials is the "breathing" effect, which is revealed as the changes in the pore diameter depending on the nature of the adsorbate (Fig. 14) [8,221].

8 MOFs as Sensor Materials

Diverse MOF materials are being explored from the point of view of their application in adsorption, catalytic, and separation processes [113,152] with limited but steadily growing citations of the use of MOF materials in sensor applications. It is quite clear that the opportunities afforded by this new class of inorganic-organic materials are unique, since the choice of the organic linker provides the possibility to tune both textural parameters and design of metal complexes. The nature of the inorganic part of the MOF structure may be responsible for the sensor properties of these materials, whereas the interplay of the functional groups located at the linkers, the inorganic moieties and introduced metal ions create new opportunities in designing unique single-site sensoring centers. MOF can be potentially used as both optical (luminescence-based) or electrochemical (based on the variation of the electroresistivity) due to adsorption and/or catalytic reaction occurring on the surface of the electroconductive plate (typically SnO_2) coated with an adsorbent or a catalyst.

The research focused on MOFs as sensors is still in the infant state and the number of publications on the use of MOFs in sensors is very limited and nonsystematic. We may consider MOFs as a meeting point of organic luminescent materials and inorganic luminophores. Due to the tailor-wise design and variability in combining diverse organic and inorganic structures in one material, MOFs seem to be the most promising alternative to other types of luminescent materials in many applications. On top of that, their intrinsic porosity (actually beating the champions in porosity like zeolites and carbon materials) adds one more dimension in their variability: one could also place a third luminescent material inside the cavity or channel of the size tunable from 0.3 to 3 nm with an increment of 0.1 nm and this cavity or channel can be also made of luminescent materials. This resembles a 3D sandwich or roll, which looks very tasty for the researchers.

The research focused on the use of MOFs in sensors is limited by the very recent (2006–2013) publications related to the luminescence, such as quenching of the nanosheets of Zn(BDC) MOF upon submission of ethylamine [222,223]. The emission at 446 nm was ascribed to the fluorescence from intraligand emission excited states. The morphology of such an MOF sensor (nanobelts, nanosheets, single crystals) was varied by applying different ultrasonic conditions in the synthesis. In general, d^{10} elements (like Zn^{+2} or Cd^{+2}) deserve special attention in this concern, as they are known to exhibit strong fluorescence in the solid state.

Strong blue fluorescent emission at 499 nm (extremely high intensity) and 550 nm (a satellite) of a new fourfold interpenetrated diamondoid MOF of Zn(II) urocanate (4/5-imidazoleacrylate) with 1D open channels was reported in Ref. [224]. A very similar MOF composition, Zn(II) imidazole-4,5-dicarboxylate, but having a helical structure, shows a strong luminescence at 420 nm upon excitation at 310 nm [225].

2,2′-Diimidazole-derived dicarboxylic acid was used for the synthesis of a Zn-MOF material [226]. The MOF structure exhibits a fluorescence at 414 nm. Other optical properties of Zn-MOF materials are reported in Ref. [227].

The luminescent properties of Cd(II) are very much similar to Zn(II) in the MOF materials. A series of coordination polymers based on rigid Cd(II) and Zn(II) MOFs with a tricarboxylic acid as a linker and flexible imiidazole ligands was synthesized [228] and the fluorescence at 425 nm (the same band for the three materials) was assigned to the π^*-π transitions in the coordinating ligands. The luminescent properties of binuclear and pentanuclear (based on adipic acid and triazole derivatives) structures of Cd(II)-MOFs are documented [229,230]. An analogous ligand-to-metal-charge-transfer emission at 364 nm (blue-shifted as compared to the free acid) and 460–495 nm was observed for the Cd-MOF prepared with cyclohexanetricarboxylate and pyrazinedicarboxylate as linkers [231,232]. The band position exhibits the dependence on the presence of a bipyridyl moiety in the structure.

Lanthanides present another group of luminescent materials suitable for sensors. Dysprosium and samarium pyridinedicarboxylates prepared by hydrothermal synthesis [233] exhibited a set of 3–4 emission bands assigned to different transitions of Dy and Sm with maxima ranging from 480 to 623 nm for Dy and 560 to 710 nm for Sm. The conclusion has been drawn that the energy transfer from the organic ligand to the lanthanide ion is effective.

The study of Eu and Tb MOF coordination polymers showed that they are characterized by strong photoluminescence peaks at 580–695 nm for Eu and 490–620 nm for Tb due to f*-f transitions [234]. The luminescent properties of Tb-, Yb, and Eu-MOFs were also reported in Ref. [235], and they were shown to be sensitive to the molecules present in the MOF cavities.

The luminescent properties of Pb-pyridineacrylate MOF materials are studied [236], and two emission bands at 418 and 438 nm were found due to CT interactions in the MOF structure. A series of MOF materials containing Pb(II), Co(II), Ni(II) pyridinedicarboxylates was synthesized and the Pb-sample was investigated for their phosphorescence properties (the red-shifted band at 475 nm) [237].

Finally, although the solid-state luminescence of silver complexes is typically observed only at low temperatures due to the intense spin-orbital coupling of Ag(I), several Ag-MOF materials with different organic pyrimidine linkers were studied [238] for their luminescent properties and the emission bands at 327–360 nm were revealed at room temperature, but these bands were assigned to the luminescence of the organic moieties.

A novel Zn(II) MOF $[Zn_3(tib)_2(L)_3]$ was synthesized with mixed ligands of 1,3,5-tris(1-imidazolyl)benzene (tib) and 4-carboxycinnamate (L^{2-}) exhibiting interesting twofold interpenetration, where triple-stranded homochiral left-(L-) and right-(R-) handed helical chains were constructed by linear rigid ligand L^{2-} and metal ions [239]. The complex shows high thermal stability and strong photoluminescent emission in the solid state at room temperature (Fig. 15).

The JUC-48 material also demonstrated quite good luminescent properties [203]. Also, JUC-48 assembled with a nanosized die (Rh_6G, a xanthene derivative) showed strong luminescence

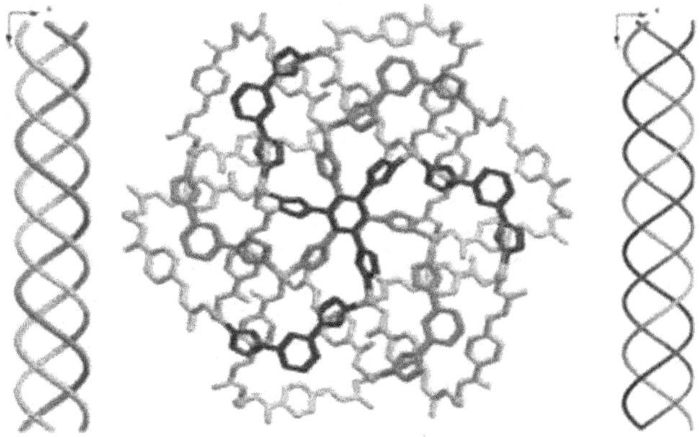

Fig. 15
Helical structure of $Zn_3(tib)_2(L)_3$. *From Su Z, Fan J, Sun W-Y. Inorg Chem Commun 2013;27:18–21 with permission.*

at 563 nm, with the peak position remaining unchanged with decreasing temperature, whereas the intensity increases in parallel with the temperature rise [240]. Such materials can be used in temperature-sensing devices.

The band gap of MOFs can be tuned to allow the materials to absorb visible light via substitution of oxygen in IRMOF-1 for sulfur and selenium [241].

Quite a comprehensive review of the luminescent properties of MOFs is presented by Cui et al. [242]. The outlook is broader than the sensor materials and covers the areas of lighting (MOF-based OLEDs), display materials, and optical devices.

The luminescence-based sensors prepared from MOFs are just one group of sensors, the second group is based on semiconducting MOFs, which exploit the semiconducting properties of Zn, In, W, Sn, and Ti oxides, with electroresistivity or electroconductivity being the detection parameter. Unfortunately, in spite of a number of publications related to such sensors based on bulk oxides (ZnO, SnO_2, In_2O_3) [243,244], there is only one paper by Corma et al. [245] describing the semiconductor properties of the conventional Zn-MOF-5 (terephthalate).

Another method that can be used in sensors is based on monitoring the magnetic properties of MOFs. Some of these materials exhibit unique magnetic features that are sensitive to the host-guest interactions, temperature, and presence of adsorbed molecules. For instance, an open-framework $Cu_3(ptmtc)_2(py)_6(CH_3CH_2OH)_2(H_2O)$ (MOROF-1), combining very large pores (2.8–3.1 nm) with bulk magnetic ordering shows a reversible and highly selective solvent-induced "shrinking-breathing" process that strongly influences the magnetic properties of the material, this magnetic spongelike behavior could be a route toward magnetic solvent sensors [246].

9 Other Potential Applications

The unique properties of MOFs (porosity, flexibility, possibility of combination of nearly any metal ion and any ligand in the composition) stimulated the quest for new applications of these materials in the fields that have been never considered for classical zeolites. We are not going to dwell upon the detailed discussion of these areas, since the research has just started and the number of publications grows exponentially. But we can present a list of priorities for the forthcoming applications of MOFs:

- Deodorant materials and humidity control systems; these applications are obviously based on adsorption properties of MOFs. One example of removal of tetrahydrothiophene (10–15 ppm) from natural gas by using electrochemically prepared Cu-btc-EMOF is described in Ref. [41]. The materials are capable of adsorbing up to 70 g of the sulfur compound per liter of MOF, which is an order of magnitude larger than the values obtained for carbon adsorbents.

- Second-order nonlinear optics, with 3D diamondoid MOFs with elongated unsymmetrical linear ligands in combination with optically transparent Zn^{2+} and Cd^{2+} ions being most reliable and predictable for engineering of noncentrosymmetric solids [247].
- Nanomaterials for drug delivery [248,249]. Quite a number of medicines have been tested in drug delivery experiments with PCPs and MOFs (ibuprofen, caffeine, doxorubicin, procainamide, vitamin B3, cisplatin complexes, NO, and others). Of course, for the efficient delivery the porous carrier should possess optimal porosity. The material should be characterized by a significant pore volume, whereas the pore size should match the size of the drug molecule to provide the proper rates of charging and discharging, (ie, delivery), keeping in mind that some drugs should act quickly and some of them should work for days or weeks, and the delivery process should fit into the prescribed rate. Not less important is the composition of MOFs usable in drug delivery, the presence of toxic components must be excluded. Even the conventional, considered nontoxic metal ions and organic moieties should be present in amounts well below the LD_{50} value, for instance 1 g/kg for Ca, 0.45 g/kg for Fe, 8.1 g/kg for Mg, 1.5 g/kg for Mn, and 0.35 g/kg for Zn. The best solution would be the MOFs consisting of amino acids and Fe or Ca as endogeneous components (ie, naturally present in the human body). Fortunately, the LD_{50} values for the typical ligands used in building MOFs, like terephthalic acid, imidazole, and the like are not very high (~1–10 g/kg). The "breathing" effect of MOFs is very important in designing efficient drug delivery materials based on MOFs. Long-term tests and nanotoxicity studies are required to conclude on the efficiency and safety of using MOFs for biomedicine.
- Ferroelectric materials with piezoelectric and pyroelectric properties that may find applications in diverse gauges, field-effect transistors, random access memory devices [250].
- Electrochemical devices and batteries, for instance, the use of MIL-53(Fe) in Li-ion batteries [251].
- Magnetic resonance imaging [252].
- Scintillation counters based on stilbene dicarboxylic acid containing MOFs [253] for measuring the radiation quanta.

10 Conclusions

The current status of research shows that the areas for the application of MOFs and their classical counterparts—zeolites—are different. When a catalytic reaction requires severe conditions, MOFs still are unable to compete with zeolites. Their scope is primarily fine organic synthesis, enantioselective synthesis, and preparation of valuable substances, because such catalytic processes can be carried out under mild conditions.

The analysis of published data shows that the use of the variety of properties of MOFs for the development of catalytic systems requires the complete realization of all possible ways of synthesis. The future prospect is to develop catalysts capable of a high molecular-sieve selectivity. Such systems must possess polyfunctional architecture, not achievable outside the area of the MOF structures. In our view, fine arrangement of the environment of active sites, which determines the reactivity, also seems an interesting problem. Thus, the future will demonstrate the unique catalytic properties inherent exclusively to MOFs.

Their application in gas storage and separation, especially coupled with the use of membranes, is another promising alley for the future research. In the case of H_2 storage or light hydrocarbons, the proper design of the nanopore will provide the solution to enhance the adsorption capacity.

The sensor area will start from scratch and will definitely develop into a group of interesting applications owing to the unique properties of MOFs, like extremely high specific surface area and pore volume, molecular sizes of the pores, and flexibility of the framework that is responsive (adaptive) toward the adsorbate.

Acknowledgment

This work was supported by the Ministry of Science and Education of Russian Federation (Contract no. RFMEFI61614X0014).

References

[1] Hoskins BF, Robson RJ. Infinite polymeric frameworks consisting of three dimensionally linked rod-like segments. J Am Chem Soc 1989;111:5962–4.
[2] Yaghi OM, Li G, Li H. Selective binding and removal of guests in a microporous metal–organic framework. Nature 1995;378:703–6.
[3] Wang Z, Cohen SM. Postsynthetic modification of metal–organic frameworks. Chem Soc Rev 2009;38:1315–29.
[4] Eddaoudi M, Kim J, Rosi NL, Vodak D, Wachter J, O'Keeffe M. Exceptional H_2 saturation uptake in microporous metal organic frameworks. Science 2002;295:469–72.
[5] Tranchemontagne DJ, Hunt JR, Yaghi OM. Room temperature synthesis of metal-organic frameworks: MOF-5, MOF-74, MOF-177. Tetrahedron 2008;64:8353–7.
[6] Chae HK, Siberio-Perez DY, Kim J, Go YB, Eddaoudi M, Matzger AJ, et al. A route to high surface area, porosity and inclusion of large molecules in crystals. Nature 2004;427:523–7.
[7] Furukawa H, Ko N, Go YB, Aratani N, Choi SB, Choi E, et al. Ultrahigh porosity in metal-organic frameworks. Science 2010;329:424–8.
[8] Dybtsev DN, Chun H, Kim K. Rigid and flexible: a highly porous metal-organic framework with unusual guest-dependent dynamic behavior. Angew Chem Int Ed 2004;43:5033–6.
[9] Chiu SS, Lo SM, Charmant JP, Orpen AG, Williams ID. Metal-organic frameworks with high capacity and selectivity for harmful gases. Science 1999;283:1148–50.
[10] Millange F, Serre C, Ferey G. Synthesis, structure determination and properties of MIL-53as and MIL-53ht: the first Cr^{III} hybrid inorganic–organic microporous solids: $Cr^{III}(OH)\cdot\{O_2C–C_6H_4–CO_2\}\cdot\{HO_2C–C_6H_4–CO_2H\}_x$. Chem Commun 2002;822–3.
[11] Ferey G, Serre C, Mellot-Draznieks C, Millange F, Dutour J, Margiolaki I. A hybrid solid with giant pores prepared by a combination of targeted chemistry, simulation, and powder diffraction. Angew Chem Int Ed 2004;43:6296–301.

[12] Ferey G, Mellot-Draznieks C, Serre C, Millange F, Dutour J, Surble S, et al. A chromium terephthalate-based solid with unusually large pore volumes and surface area. Science 2005;309:2040–2.
[13] Park KS, Ni Z, Côté AP, Choi JY, Huang R, Uribe-Romo FJ, et al. Exceptional chemical and thermal stability of zeolitic imidazolate frameworks. Proc Natl Acad Sci U S A 2006;103:10186–91.
[14] Morris W, Doonan CJ, Furukawa H, Banerjee R, Yaghi OM. Crystals as molecules: postsynthesis covalent functionalization of zeolitic imidazolate frameworks. J Am Chem Soc 2008;130:12626–7.
[15] Fang Q-R, Yuan D-Q, Sculley J, Lu W-G, Zhou H-C. A novel MOF with mesoporous cages for kinetic trapping of hydrogen. Chem Commun 2012;254–6.
[16] Wang X-S, Ma S, Sun D, Parkin S, Zhou H-C. A mesoporous metal–organic framework with permanent porosity. J Am Chem Soc 2006;128:16474–5.
[17] Fang Q-R, Zhu GS, Xue M, Wang Z, Sun J, Qiu S. Amine-templated assembly of metal–organic frameworks with attractive topologies. Cryst Growth Des 2008;8:319–29.
[18] Fang Q-R, Zhu GS, Jin Z, Ji Y-Y, Ye J-W, Xue M, et al. Mesoporous metal–organic framework with rare etb topology for hydrogen storage and dye assembly. Angew Chem Int Ed 2007;46:6638–42.
[19] Furukawa H, Gándara F, Zhang Y-B, Jiang J, Queen WL, Hudson MR, et al. Water adsorption in porous metal–organic frameworks and related materials. J Am Chem Soc 2014;136:4369–81.
[20] An J, Geib SJ, Rosi NL. Cation-triggered drug release from a porous zinc–adeninate metal–organic framework. J Am Chem Soc 2009;131:8376.
[21] Sonnauer A, Hoffmann F, Fröba M, Kienle L, Duppel V, Thommes M, et al. Giant pores in a chromium naphthalene-2,6-dicarboxylate open-framework structure with MIL-101 topology. Angew Chem Int Ed 2009;48:3791–4.
[22] Wong-Foy AG, Matzger AJ, Yaghi OM. Exceptional H_2 saturation uptake in microporous metal–organic frameworks. J Am Chem Soc 2006;128:3494–5.
[23] Noro S-i, Kitagawa S, Akutagawa T, Nakamura T. Coordination polymers constructed from transition metal ions and organic N-containing heterocyclic ligands: Crystal structures and microporous properties. Prog Polym Sci 2009;3:240–79.
[24] Kitagawa S, Kitaura R, Noro S-i. Functional porous coordination polymers. Angew Chem Int Ed 2004;43:2334–75.
[25] Stock N, Biswas S. Synthesis of metal-organic frameworks (MOFs): routes to various MOF topologies, morphologies, and composites. Chem Rev 2012;112:933–69.
[26] Qiu S, Zhu G. Molecular engineering for synthesizing novel structures of metal-organic frameworks with multifunctional properties. Coord Chem Rev 2009;253:2891–911.
[27] www.moftechnologies.com.
[28] Pichon A, James SL. Syntheses, crystal structures, and magnetic properties of one-dimensional oxalato-bridged Co(II), Ni(II), and Cu(II) complexes with n-aminopyridine ($n = 2–4$) as terminal ligand. CrystEngComm 2008;10:1839–47.
[29] Leung E, Müller U, Cox G. Solvent-free production of magnesium formate based porous metal-organic frame material. Patent WO 2010/106121 A1; 2010.
[30] Leung E, Müller U, Cox G, Mattenheimer H, Blei S. Process for preparing porous metal-organic frameworks based on aluminum fumarate. EP Patentanmeldung 10183283.0; 2010.
[31] Yilmaz B, Trukhan N, Mueller U. Industrial outlook on zeolites and metal organic frameworks. Chin J Catal 2012;33:3–10.
[32] Farrusseng D, Aguado S, Pinel C. Metal-organic frameworks: opportunities for catalysis. Angew Chem Int Ed 2009;48:7502–13.
[33] Lin Z, Slawin AMZ, Morris RE. Chiral induction in the ionothermal synthesis of a 3-D coordination polymer. J Am Chem Soc 2007;129:4880–1.
[34] Hu A, Ngo H, Lin WB. Remarkable 4,4′-Substituent effects on binap: highly enantioselective Ru catalysts for asymmetric hydrogenation of β-aryl ketoesters and their immobilization in room-temperature ionic liquids. Angew Chem Int Ed 2004;43:2501–4.
[35] Hu A, Ngo H, Lin WB. Chiral porous hybrid solids for practical heterogeneous asymmetric hydrogenation of aromatic ketones. J Am Chem Soc 2003;125:11490–1.

[36] Choi JS, Son WJ, Kim J, Ahn WS. Facile syntheses of metal-organic framework $Cu_3(BTC)_2(H_2O)_3$ under ultrasound. Micropor Mesopor Mater 2008;116:727–31.
[37] Ni Z, Masel RI. Metal–organic framework MOF-5 prepared by microwave heating: factors to be considered. J Am Chem Soc 2006;128:12394–5.
[38] Li ZQ, Qiu LG, Xu T, Wu Y, Wang W, Wu ZY, et al. Rapid production of metal–organic frameworks via microwave-assisted solvothermal synthesis. Mater Lett 2009;63:78–80.
[39] Jung DW, Yang DA, Kim J, Kim J, Ahn WS. Ultrasonic synthesis of the microporous metal–organic framework $Cu_3(BTC)_2$ at ambient temperature and pressure: an efficient and environmentally friendly method. Dalton Trans 2010;39:2883–7.
[40] Han SS, Mendoza-Cortes JL, Goddard 3rd WA. Recent advances on simulation and theory of hydrogen storage in metal–organic frameworks and covalent organic frameworks. Chem Soc Rev 2009;38:1460–7.
[41] Müller U, Schubert M, Teich F, Puetter H, Schierle-Arndt K, Pastré J. Metal–organic frameworks—prospective industrial applications. J Mater Chem 2006;16:626–36.
[42] Isaeva V, Chernyshev V, Afonina E, Tkachenko O, Klementiev K, Nissenbaum V, et al. Novel metal-organic 1-D oordination polymer based on pyrazine-2,5-dicarboxylate ligands: Synthesis and structure investigation. Inorg Chim Acta 2011;376:367–72.
[43] Cohen SM. Postsynthetic methods for the functionalization of metal–organic frameworks. Chem Rev 2012;112:970–1000.
[44] Cohen SM. Modifying MOFs: new chemistry, new materials. Chem Sci 2010;1:32–6.
[45] Tanabe K, Cohen SM. Postsynthetic modification of metal–organic frameworks – a progress report. Chem Soc Rev 2011;40:498–519.
[46] Rowe MD, Thamm DH, Kraft SL, Boyes SG. Polymer-modified gadolinium metal-organic framework nanoparticles used as multifunctional nanomedicines for the targeted imaging and treatment of cancer. Biomacromolecules 2009;10:983–93.
[47] Taylor-Pashow KML, Rocca JD, Xie Z, Tran S, Lin W. Postsynthetic modifications of iron-carboxylate nanoscale metal–organic frameworks for imaging and drug delivery. J Am Chem Soc 2009;131:14261–3.
[48] Isaeva VI, Tkachenko OP, Brueva TR, Kapustin GI, Afonina EV, Mishin IV, et al. Composite materials on the basis of phenylenecarboxylate framework MOF-5 and calix[4]arenes with various structures. Russ J Phys Chem A 2011;85:293–7.
[49] Isaeva VI, Tarasov AL, Tkachenko OP, Kapustin GI, Mishin IV, Solov'eva SE, et al. 1,3-cyclohexadiene hydrogenation in the presence of a palladium-containing catalytic system based on an MOF-5/calixarene composite. Kinet Catal (Russ) 2011;52:95–8.
[50] Isaeva V, Tkachenko O, Afonina E, Kozlova LM, Gruenert W, Solov'eva SE, et al. 2-butyne-1,4-diol hydrogenation over palladium supported on Zn^{2+}-based – MOF and host–guest MOF/calix[4]arene materials. Micropor Mesopor Mater 2013;166:167–75.
[51] Tafipolsky M, Amirjalayer S, Schmid R. Atomistic theoretical models for nanoporous hybrid materials. Micropor Mesopor Mater 2010;129:304–18.
[52] Czaja AU, Trukhan N, Mueller U. Industrial applications of metal–organic frameworks. Chem Soc Rev 2009;38:1284–93.
[53] Ferrey G, Mellot-Draznieks C, Millange F. Crystallized frameworks with giant pores: are there limits to the possible?. Acc Chem Res 2005;38:217–25.
[54] Kuppler RJ, Jimmons TD, Fang Q-R, Li J-R, Makala TA, Young MD, et al. Potential applications of metal-organic frameworks. Coord Chem Rev 2009;253:3042–66.
[55] Fujita M, Kwon YJ, Washizu S, Ogura K. Preparation, clathration ability, and catalysis of a two-dimensional square network material composed of cadmium(II) and 4,4'-bipyridine. J Am Chem Soc 1994;116:1151–2.
[56] Lee JY, Farha OK, Roberts J, Scheidt KA, Nguyen ST, Hupp JT. Metal–organic framework materials as catalysts. Chem Soc Rev 2009;38:1450–9.
[57] Huang L, Wang H, Chen J, Wang Z, Sun J, Zhao D, et al. Synthesis, morphology control, and properties of porous metal-organic coordination polymers. Micropor Mesopor Mater 2003;58:105–14.

[58] Forster PM, Thomas PM, Cheetham AK. Biphasic solvothermal synthesis: a new approach for hybrid inorganic–organic materials. Chem Mater 2002;14:17–20.
[59] Ferëy G. Hybrid porous solids. Stud Surf Sci Catal 2007;168:327–74.
[60] Tompsett GA, Conner WC, Yngvesson KS. Microwave synthesis of nanoporous materials. ChemPhysChem 2006;7:296–319.
[61] Jhung SH, Yoon JW, Hwang J-S, Cheetham AK, Chang J-S. Facile synthesis of nanoporous nickel phosphates without organic templates under microwave irradiation. Chem Mater 2005;17:4455–60.
[62] Klinowski J, Paz FA, Silva AP, Rocha J. Microwave-assisted synthesis of metal–organic frameworks. Dalton Trans 2011;40:321–30.
[63] Rowsell JLC, Yaghi OM. Metal–organic frameworks: a new class of porous materials. Micropor Mesopor Mater 2004;73:3–14.
[64] Yaghi OM, O'Keefi M, Ockwig NW, Chae H-K, Eddaoudi M, Kim J. Reticular synthesis and the design of new materials. Nature 2003;42:705–14.
[65] James SL. Metal-organic frameworks. Chem Soc Rev 2003;32:276–88.
[66] Papaefstathiou GS, MacGillivray LR. Inverted metal–organic frameworks: solid-state hosts with modular functionality. Coord Chem Rev 2003;246:169–84.
[67] Kesanli B, Lin W. Chiral porous coordination networks: rational design and applications in enantioselective processes. Coord Chem Rev 2003;246:305–26.
[68] Rosi NL, Eddaoudi M, Kim J, O'Keefi M, Yaghi OM. Advances in the chemistry of metal–organic frameworks. CrystEngComm 2002;4:401–4.
[69] Hagrman PJ, Hagrman D, Zubieta J. Organic-inorganic hybrid materials: from "Simple" coordination polymers to organodiamine-templated molybdenum oxides. Angew Chem Int Ed 1999;38:2638–84.
[70] Fletcher AJ, Thomas KM, Rosseinsky MJ. Flexibility in metal-organic framework materials: impact on sorption properties. J Solid State Chem 2005;178:2491–510.
[71] Noro S-i, Kitagawa S, Kondo M, Seki K. A new, methane adsorbent, porous coordination polymer. Angew Chem Int Ed 2000;39:2081–4.
[72] Paz FA, Klinowski J. Two-and three-dimensional cadmium-organic frameworks with trimesic acid and 4, 4'-trimethylenedipyridine. Inorg Chem 2004;43:3882–93.
[73] Eddaoudi M, Li H, Yaghi OM. Highly porous and stable metal–organic frameworks: structure design and sorption properties. J Am Chem Soc 2000;122:1391–7.
[74] Serre C, Millange F, Thouvenot C, Noguès M, Marsolier G, Louër D, et al. Very large breathing effect in the first nanoporous chromium(III)-based solids: MIL-53 or $Cr^{III}(OH)\cdot\{O_2C–C_6H_4–CO_2\}\cdot\{HO_2C–C_6H_4–CO_2H\}_x\cdot H_2O_y$. J Am Chem Soc 2002;124:13519–26.
[75] Yaghi OM, Davis CE, Li G, Li H. Selective guest binding by tailored channels in a 3-D porous zinc(II)–benzenetricarboxylate network. J Am Chem Soc 1997;119:2861–8.
[76] Li H, Davis CE, Groy TL, Kelley DG, Yaghi OM. Coordinatively unsaturated metal centers in the extended porous framework of $Zn_3(BDC)_3$,6CH_3OH(BDC)1,4-benzenedicarboxylate). J Am Chem Soc 1998;120:2186–7.
[77] Li H, Eddaoudi M, Groy TL, Yaghi OM. Establishing microporosity in open metal-organicframeworks: gas sorption isotherms for Zn(BDC)(BDC)1,4-benzenedicarboxylate). J Am Chem Soc 1998;120:8571–2.
[78] Li H, Eddaoudi M, O'Keefi M, Yaghi OM. Design and synthesis of an exceptionally stable and highly porous metal-organic framework. Nature 1999;402:276–9.
[79] Rowsell JLC, Yaghi OM. Effects of functionalization, catenation, and variation of the metal oxide and organic linking units on the low-pressure hydrogen adsorption properties of metal–organic frameworks. J Am Chem Soc 2006;128:1304–15.
[80] Cho SH, Ma B, Nguen ST, Hupp JT. A metal-organic framework material that functions as an enantioselective catalyst for olefin epoxidation. Chem Commun 2006;25630–5.
[81] Alkordi MHY, Liu RW, Larsen JF, Eubank JF, Eddaoudi M. Zeolite-*like* metal–organic frameworks as platforms for applications: on metalloporphyrin-based catalysts. J Am Chem Soc 2008;130:12639–41.
[82] Nuzhdin AL, Dybtsev DN, Bryliakov KP, Talsi EP, Fedin VP. Enantioselective chromatographic resolution and one-pot synthesis of enantiomerically pure sulfoxides over a homochiral Zn–organic framework. J Am Chem Soc 2007;129:12958–9.

[83] Wu CD, Hu A, Zhang L, Lin W. A homochiral porous metal–organic framework for highly enantioselective heterogeneous asymmetric catalysis. J Am Chem Soc 2005;127:8940–1.
[84] Farha OK, Mulfort KL, Hupp JT. An example of node-based postassembly elaboration of a hydrogen-sorbing, metal–organic framework material. Inorg Chem 2008;47:10223–5.
[85] Horike S, Dinca M, Tamaki K, Long JR. An example of node-based postassembly elaboration of a hydrogen-sorbing, metal–organic framework material. J Am Chem Soc 2008;130:5854–5.
[86] Qui L-G, Gu L-N, Hu G, Zhang L-D. Synthesis, structural characterization and selectively catalytic properties of metal–organic frameworks with nano-sized channels: a modular design strategy. J Solid State Chem 2009;182:502–8.
[87] Davis ME. New vistas in zeolite and molecular sieve catalysis. Acc Chem Res 1993;26:111–5.
[88] Kusgens P, Rose M, Senkovska I, Frode H, Henschel A, Siegle S, et al. Characterization of metal-organic frameworks by water adsorption. Micropor Mesopor Mater 2009;120:325–30.
[89] Hasegawa S, Horike S, Matsuda R, Furukawa S, Mochizuki K, Kinoshita Y, et al. Three-dimensional porous coordination polymer functionalized with amide groups based on tridentate ligand: selective sorption and catalysis. J Am Chem Soc 2007;129:2607–14.
[90] Ravon U, Savonnet M, Aguado S, Domine ME, Janneau E, Farrusseng D. Engineering of coordination polymers for shape selective alkylation of large aromatics and the role of defects. Micropor Mesopor Mater 2010;129:319–29.
[91] Corma A. State of the art and future challenges of zeolites as catalysts. J Catal 2003;216:298–312.
[92] Horcajada P, Surble S, Serre C, Hong D-Y, Seo Y-K, Chang J-S, et al. Synthesis and catalytic properties of MIL-100(Fe), an iron(III) carboxylate with large pores. Chem Commun 2007;2820–2.
[93] Ravon U, Domine ME, Gaudillere C, Desmartin-Chomel A, Farrusseng D. MOFs as acid catalysts with shape selectivity properties. New J Chem 2008;32:937–40.
[94] Loiseau T, Cerre S, Huguenard C, Fink G, Taulelle F, Henry M, et al. A rationale for the large breathing of the porous aluminum terephthalate (MIL-53) upon hydration. Chem Eur J 2004;10:1373–82.
[95] Loiseau T, Muguerra H, Ferey G, Haouas M, Taulelle F. Synthesis and structural characterization of a new open-framework zinc terephthalate $Zn_3(OH)_2(bdc)_2{\cdot}2DEF$, with infinite Zn–($\mu_3$-OH)–Zn chains. J Solid State Chem 2005;178:621–8.
[96] Müller U, Ceder C. A density functional theory study of hydrogen adsorption in MOF-5. J Phys Chem B 2005;109:17974–83.
[97] Ma L, Abney C, Lin W. Enantioselective catalysis with homochiral metal–organic frameworks. Chem Soc Rev 2009;38:1248–56.
[98] Cavka JH, Jakobsen SR, Olsbye U, Guillou N, Lamberti C, Bordiga S, et al. A new zirconium inorganic building brick forming metal organic frameworks with exceptional stability. J Am Chem Soc 2008;130:13850–1.
[99] Farha OK, Spokoyny AM, Mulfort KL. Synthesis and hydrogen sorption properties of carborane based metal–organic framework materials. J Am Chem Soc 2007;129:12680–1.
[100] Ingleson JM, Barrio JP, Bacsa J. Generation of a solid Brønsted acid site in a chiral framework. Chem Commun 2008;1287–9.
[101] Noro S-I, Kitagawa S, Yamashita M, Wada T. New microporous coordination polymer affording guest-coordination sites at channel walls. Chem Commun 2002;222–3.
[102] Zou RQ, Sakurai H, Xu Q. Preparation, adsorption properties, and catalytic activity of 3D porous metal–organic frameworks composed of cubic building blocks and alkali-metal ions. Angew Chem Int Ed 2006;45:2542–6.
[103] Gandara F, Gómez Lor B, Gutierrez Puebla E, Iglesias M, Monge MA, Proserpio DM, et al. An indium layered MOF as recyclable lewis acid catalyst. Chem Mater 2008;20:72–6.
[104] Gutiérrez Puebla E, Cáscales Sedano C, Gómez Lor B, Iglesias Hernández MM, Monge Bravo MA, Ruiz Valero C, et al. Organometallic framework materials of transition group III. Patent ES 2,200,681 A1; 2004.
[105] Perles J, Iglesias M, Martin-Luengo MA, Monge MA, Ruiz Valero C, Snejko N. Metal–organic scandium framework: useful material for hydrogen storage and catalysis. Chem Mater 2005;17:5837–42.

[106] De Rosa S, Giordano G, Granato T, Katovic A, Siciliano A, Tripicchio F, et al. Chemical pretreatment of olive oil mill wastewater using a metal-organic framework catalyst. J Agric Food Chem 2005;53:8306–9.
[107] Tanski JM, Wolczanski PT. Covalent titanium aryldioxy one-, two-, and three-dimensional networks and their examination as ziegler–natta catalysts. Inorg Chem 2001;40:2026–33.
[108] Havicovic J, Bjorgen M, Olsbye U, Dietzel PD, Bordiga S, Prestipino C, et al. The inconsistency in adsorption properties and powder XRD data of MOF-5 is rationalized by framework interpenetration and the presence of organic and inorganic species in the nanocavities. J Am Chem Soc 2007;129:3612–20.
[109] Kaye SS, Dailly A, Yaghi OM, Long JR. Impact of preparation and handling on the hydrogen storage properties of $Zn_4O(1,4\text{-benzenedicarboxylate})_3$ (MOF-5). J Am Chem Soc 2007;129:14176–7.
[110] Sabo M, Henschel A, Frode H, Klemm E, Kaskel S. Solution infiltration of palladium into MOF-5: synthesis, physisorption and catalytic properties. J Mater Chem 2007;17:3827–32.
[111] Dan-Hardi M, Serre C, Frot T, Rozes L, Morin G, Sanchez K, et al. Facile synthesis of MOF-177 by a sonochemical method using 1-methyl-2-pyrrolidinone as a solvent. J Am Chem Soc 2009;131:10857–9.
[112] Ahnfeldt T, Guillou N, Gunzelmann D, Margiolaki I, Loiseau T, Férey G, et al. A new photoactive crystalline highly porous titanium(IV) dicarboxylate. Angew Chem Int Ed 2009;48:5163–6.
[113] Banerjee R, Phan A, Wang B, Knobler C, Furukawa H, O'Keefi M, et al. $[Al_4(OH)_2(OCH_3)_4(H_2N\text{-bdc})_3]$ x xH(2)O: a 12-connected porous metal-organic framework with an unprecedented aluminum-containing brick. Science 2008;319:939–43.
[114] Dietzel PDC, Panella B, Hirscher M, Blom R, Fjellvåg H. High-throughput synthesis of zeolitic imidazolate frameworks and application to CO_2 capture. Chem Commun 2006;959–61.
[115] Opelt S, Turk S, Dietzsch E, Henschel A, Kaskel S, Klemm E. Hydrogen adsorption in a nickel based coordination polymer with open metal sites in the cylindrical cavities of the desolvated framework. Catal Commun 2008;1286–90.
[116] Hermes S, Schröter M-K, Schmid R, Khodeir L, Muhler M, Tissler A, et al. Preparation of palladium supported on MOF-5 and its use as hydrogenation catalyst. Angew Chem Int Ed 2005;44:6237–41.
[117] Esken D, Zhang X, Lebedev OL, Schröder F, Fischer RA. Metal@MOF: loading of highly porous coordination polymers host lattices by metal organic chemical vapor deposition. J Mater Chem 2009;19:1314–9.
[118] Turner S, Lebedev OL, Schröder F, Esken D, Fischer RA, Tendeloo GV. Pd@MOF-5: limitations of gas-phase infiltration and solution impregnation of $[Zn_4O(bdc)_3]$ (MOF-5) with metal–organic palladium precursors for loading with Pd nanoparticles. Chem Mater 2008;20:5622–7.
[119] Hermes S, Schröder F, Amirjalayer S, Schmid R, Fischer RA. Direct imaging of loaded metal–organic framework materials (Metal@MOF-5). J Mater Chem 2006;16:2464–72.
[120] Schröder F, Esken D, Sokoja M, van den Berg MWE, Lebedev OI, Van Tendeloo G, et al. Loading of porous metal–organic open frameworks with organometallic CVD precursors: inclusion compounds of the type $[L_nM]_a$@MOF-5. J Am Chem Soc 2008;130:6119–30.
[121] Kim H, Chun H, Kim G-K, Lee H-S, Kim K. Ruthenium nanoparticles inside porous $[Zn_4O(bdc)_3]$ by hydrogenolysis of adsorbed [Ru(cod)(cot)]: a solid-state reference system for surfactant-stabilized ruthenium colloids. Chem Commun 2006;2759–61.
[122] Park YK, Choi S, Kim H, Kim K, Won B-H, Choi K, et al. Vapor phase inclusion of ferrocene and its derivative in a microporous metal–organic porous material and its structural characterization by single crystal X-ray diffraction. Angew Chem Int Ed 2007;46:8230–3.
[123] Müller M, Lebedev OI, Fischer RA. Crystal structure and guest uptake of a mesoporous metal-organic framework containing cages of 3.9 and 4.7 nm in diameter. J Mater Chem 2008;18:5274–81.
[124] Meilikhov M, Yusenko K, Fischer RA. Gas-phase loading of $[Zn_4O(btb)_2]$ (MOF-177) with organometallic CVD-precursors: inclusion compounds of the type $[L_nM]_a$@MOF-177 and the formation of Cu and Pd nanoparticles inside MOF-177. J Am Chem Soc 2009;131:9644–5.
[125] Müller U, Metelkina O, Junicke H, Butz T, Yaghi OM. Turning MIL-53(Al) redox-active by functionalization of the bridging OH-group with 1,1′-ferrocenediyl-dimethylsilane. Patent US 2004/081611; 2004.

[126] Müller U, Lobree L, Hesse M, Yaghi OM, Eddaoudi M. Process for preparing hydrogen peroxide from the elements. Patent WO 03/101975; 2003.
[127] Wu F, Qiu L-G, Ke F, Jiang X. Method of using a metal organic frameworks based on aluminum fumarate. Inorg Chem Commun 2013;32:5–8.
[128] Vu TV, Kosslick H, Schulz A, Harloff J, Paetzold E, Radnik J, et al. Copper nanoparticles embedded in metal–organic framework MIL-101(Cr) as a high performance catalyst for reduction of aromatic nitro compounds. Micropor Mesopor Mater 2013;177:135–42.
[129] Kitagawa S, Noro S-I, Nakamura T. Hydroformylation of olefins over rhodium supported metal-organicframework catalysts of different structure. Chem Commun 2006;701–7.
[130] Ma B-Q, Mulfort KL, Hupp JT. Pore surface engineering of microporous coordination polymers. Inorg Chem 2005;44:4912–4.
[131] Katsuki T. Microporous pillared paddle-wheel frameworks based on mixed-ligand coordination of zinc ions. Coord Chem Rev 1995;140:189–214.
[132] Palucki M, Finney NS, Pospisil PJ, Güler ML, Ishida T, Jacobsen EN. Catalytic asymmetric oxidations using optically-active (salen)manganese(III) complexes as catalysts. J Am Chem Soc 1998;120:948–54.
[133] Kitaura R, Onoyama G, Sakamoto H, Matsuda R, Noro S-i, Kitagawa S. The mechanistic basis for electronic effects on enantioselectivity in the (salen)Mn(III)-catalyzed epoxidation reaction. Angew Chem Int Ed 2004;43:2684–7.
[134] Suslick KS, Bhyrappa P, Chou JH, Kosal M, Nakagaki S, Smithenry DW, et al. Immobilization of a metallo schiff base into a microporous coordination polymer. Acc Chem Res 2005;38:283–91.
[135] Schultz AM, Farha OK, Hupp JT, Nguen ST. Microporous porphyrin solids. J Am Chem Soc 2009;131:4204–5.
[136] Kumar DK, Jose DA, Dastidar P, Das A. A catalytically active, permanently microporous MOF with metalloporphyrin struts. Chem Mater 2004;16:2332–5.
[137] Gascon J, Hernandez-Alonso MD, Almeida AR, van Klink GPM, Kapteijn F, Mul G. Nonpolymeric hydrogelators derived from trimesic amides. ChemSusChem 2008;1:981–3.
[138] Llabres i Xamena FX, Abad A, Corma A, Garcia H. Isoreticular MOFs as efficient photocatalysts with tunable band gap: an operando FTIR study of the photoinduced oxidation of propylene. J Catal 2007;250:294–8.
[139] Huang Y, Zheng Z, Liu T, Lü J, Lin Z, Li H, et al. MOFs as catalysts: activity, reusability and shape-selectivity of a Pd-containing MOF. Catal Commun 2011;14:27–31.
[140] Llabres i Xamena FX, Casanova O, Galiasso Tailleur R, Garcia H, Corma A. Palladium nanoparticles supported on amino functionalized metal-organic frameworks as highly active catalysts for the suzuki–miyaura cross-coupling reaction. J Catal 2008;255:220–7.
[141] Jiang D, Mallat T, Krumeich F, Baiker A. Metal organic frameworks (MOFs) as catalysts: a combination of Cu^{2+} and Co^{2+} MOFs as an efficient catalyst for tetralin oxidation. J Catal 2008;257:390–5.
[142] Alaerts L, Thibault-Starzyk F, Séguin E, Denayer JFM, Jacobs PA, De Vos DE. Copper-based metal-organic framework for the facile ring-opening of epoxides. Stud Surf Sci Catal 2007;170:1996–2003.
[143] Caudo S, Centi G, Genovese C, Giordano G, Granato T, Katovic A, et al. MOF materials as catalysts for organic transformations and as selective hosts in recognition *of* organics. Stud Surf Sci Catal 2007;170:2054–9.
[144] Schlichte K, Kratzke T, Kaskel S. Cu-MOF: a new highly active catalyst for WHPCO of waste water from agro-food production. Micropor Mesopor Mater 2004;73:81–8.
[145] Henschel A, Gedrich K, Kraehnert R, Kaskel S. Improved synthesis, thermal stability and catalytic properties of the metal-organic framework compound $Cu_3(BTC)_2$. Chem Commun 2008;4192–4.
[146] Gandara F, Garsia-Cortes A, Cascales C, Gómez Lor B, Gutiérrez Puebla E, Iglesias M, et al. Catalytic properties of MIL-101. Inorg Chem 2007;46:3475–84.
[147] Lu Y, Tonigold M, Bredenkötter B, Volkmer D, Hitzbleck J, Langstein G. Crystal structure and low-temperature magnetic ordering in rare earth iron germanates $RFeGe_2O_7$, R = Y, Pr, Dy, Tm, and Yb. Z Anorg Allg Chem 2008;634:2411–7.
[148] Xiao B, Hou H, Fan Y. A cobalt(II)-containing metal-organic framework showing catalytic activity in oxidation reactions. J Organomet Chem 2007;692:2014–20.

[149] Kalilur Rahiman A, Rajesh K, Shanmuga Barathi K, Sreedaran S, Narayanan V. Catalytic applications of Cu^{II}-containing MOFs based on N-heterocyclic ligand in the oxidative coupling of 2,6-dimethylphenol. Inorg Chim Acta 2009;362:1491–500.
[150] Maksimchuk NV, Timofeeva MN, Melgunov MS, Shmakov AN, Chesalov YuA, Dybtsev DN, et al. Catalytic oxidation of alkenes by manganese(III) porphyrin-encapsulated Al, V,Si-mesoporous molecular sieves. J Catal 2008;257:315–23.
[151] Wang ZQ, Cohen SM. Heterogeneous selective oxidation catalysts based on coordination polymer MIL-101 and transition metal-substituted polyoxometalates. J Am Chem Soc 2007;129:12368–9.
[152] Tanabe KK, Wang Z, Cohen SM. Postsynthetic covalent modification of a neutral metal–organic framework. J Am Chem Soc 2008;130:8508–17.
[153] Costa JS, Gomez P, Black CA, Roubeau O, Teat SJ, Reedijk J. Systematic functionalization of a metal–organic framework via a postsynthetic modification approach. Eur J Inorg Chem 2008;1551–4.
[154] Hwang YK, Hong D-Y, Chang J-S, Jhung SH, Seo Y-K, Kim J, et al. Chemical modification of a bridging ligand inside a metal–organic framework while maintaining the 3D structure. Angew Chem Int Ed 2008;47:4144–8.
[155] Zhang X, Llabres i Xamena FX, Corma A. Amine grafting on coordinatively unsaturated metal centres of MOFs: consequences for catalysis and metal encapsulation. J Catal 2009;265:155–60.
[156] Bradshaw D, Prior TJ, Gussen EJ, Claridge JB, Rosseinsky MJ. Gold(III) – metal organic framework bridges the gap between homogeneous and heterogeneous gold catalysts. J Am Chem Soc 2004;126:6106–14.
[157] Dybtsev DN, Nuzhdin AL, Chun H, Bryliakov KP, Talsi EP, Fedin VP, et al. Permanent microporosity and enantioselective sorption in a chiral open framework. Angew Chem Int Ed 2006;45:916–20.
[158] Seo JS, Whang D, Lee H, Jun SI, Oh J, Jeon YJ, et al. A homochiral metal-organic material with permanent porosity, enantioselective sorption properties, and catalytic activity. Angew Chem Int Ed 2000;45:916–20.
[159] Wu CD, Lin W. Topotactic linear radical polymerization of divinylbenzenes in porous coordination polymers. Angew Chem Int Ed 2007;46:1075–8.
[160] Civalleri B, Napoli FY, Noel C, Roetti C, Dovesi R. Ab-initio prediction of materials properties with CRYSTAL: MOF-5 as a case study. CrystEngComm 2006;8:364–71.
[161] Müller U, Luinstra G, Yaghi OM. Process for producing polyalkylene carbonates. US Patent 6,617,467; 2003.
[162] Müller U, Stober M, Ruppel R, Baum E, Bohres E, Sigi M, et al. (BASF AG, University of Michigan), 2010-ACIE-metal-organic frameworks@opportunities for catalysis. PL370429; 2003.
[163] Uemura T, Hiramatsu D, Kubota Y, Takata M, Kitagawa S. Angew Chem 2007;119:5075–8.
[164] Uemura T, Hiramatsu D, Kubota Y, Takata M, Kitagawa S. Topotactic linear radical polymerization of divinylbenzenes in porous coordination polymers. Angew Chem Int Ed 2007;46:4987–90.
[165] Uemura T, Kitagawa K, Horike S, Kawamura T, Kitagawa S, Mizuno M, et al. Radical polymerisation of styrene in porous coordination polymers. Chem Commun 2005;5968–70.
[166] Dillon AC, Heben MJ. Hydrogen storage using carbon adsorbents: past, present and future. Appl Phys A 2001;172:133–42.
[167] Ning GQ, Wei F, Luo GH, Wang QX, Wu YL, Yu H. Hydrogen storage in multi-wall carbon nanotubes using samples up to 85 g. Appl Phys A 2004;78:955–9.
[168] Züttel A. Hydrogen storage methods. Naturwissenschaften 2004;91:157–72.
[169] Poirier E, Chaine R, Bernard PS, Cossement D, Lafi L, Melancon E, et al. Storage of hydrogen on single-walled carbon nanotubes and other carbon structures. Appl Phys A 2004;78:961–7.
[170] Rowsell JLC, Yaghi OM. Strategies for hydrogen storage in metal–organic frameworks. Angew Chem Int Ed 2005;44:4670–9.
[171] Ross DK. Hydrogen storage: the major technological barrier to the development of hydrogen fuel cell cars. Vacuum 2006;80:1084–9.
[172] Weitkamp J, Fritz M, Ernst S. Zeolites as media for hydrogen storage. Int J Hydrogen 1995;20:967–70.

[173] Schlapbach L, Züttel A. Hydrogen-storage materials for mobile applications. Nature 2001;414:353–8.
[174] Langmi HW, Walton A, Al-Mamouri MM, Johnson SR, Book D, Speight JD, et al. Hydrogen adsorption in zeolites A, X, Y and RHO. J Alloys Compd 2003;356–357:710–5.
[175] Langmi HW, Walton A, Book D, Johnson SR, Al-Mamouri MM, Speight JD, et al. Hydrogen storage in ion-exchanged zeolites. J Alloys Compd 2005;404-406:637–42.
[176] Dubtsev D, Chun H, Yoon SH, Kim D, Kim K. Microporous manganese formate: a simple metal–organic porous material with high framework stability and highly selective gas sorption properties. J Am Chem Soc 2004;126:32–3.
[177] Scarano D, Bordiga S, Lamberti C, Ricchiardi G, Bertarione S, Spoto G. Hydrogen adsorption and spill-over effects on H–Y and Pd-containing Y zeolites: an experimental and theoretical investigation. Appl Catal A 2006;307:3–12.
[178] Skoulidas AI. Molecular dynamics simulations of gas diffusion in metal–organic frameworks: argon in CuBTC. J Am Chem Soc 2004;126:1356–7.
[179] Tylianakis E, Klontzas E, Froudakis GE. Multi-scale theoretical investigation of hydrogen storage in covalent organic frameworks. Nanoscale 2011;3:856–69.
[180] Férey G, Latroche M, Serre C, Millange F, Loiseau T, Percheron-Guégan A. Hydrogen adsorption in the nanoporous metal-benzenedicarboxylate $M(OH)(O_2C–C_6H_4–CO_2)(M = Al^{3+}, Cr^{3+})$, MIL-53. Chem Commun 2003;2976–7.
[181] Rowsell JLC, Millward AR, Park KS, Yaghi OM. Hydrogen sorption in functionalized metal–organic frameworks. J Am Chem Soc 2004;126:5666–7.
[182] Sun D, Ma S, Ke Y, Kollins DJ, Zhou H-C. An interweaving MOF with high hydrogen uptake. J Am Chem Soc 2006;128:3896–7.
[183] Férey G. Metal-organic frameworks: the young child of the porous solids family. Stud Surf Sci Catal 2007;170:66–84.
[184] Bux H, Liang F, Li Y, Cravillon J, Wiebcke M, Caro J. Zeolitic imidazolate framework membrane with molecular sieving properties by microwave-assisted solvothermal synthesis. J Am Chem Soc 2009;131:16000–1.
[185] Gascon J, Aguado S, Kapteijn F. Manufacture of dense coatings of $Cu_3(BTC)_2$ (HKUST-1) on α-alumina. Micropor Mesopor Mater 2008;113:132–8.
[186] Aguado S, Nicolas C-H, Moizan-Basler V, Nieto C, Amrouche H, Bats N, et al. Facile synthesis of an ultramicroporous MOF tubular membrane with selectivity towards CO_2. New J Chem 2011;35:41–4.
[187] Nijkamp MG, Raayamakers JEM, van Dillen AJ, De Jong KP. Hydrogen storage using physisorption – materials demands. Appl Phys A 2001;72:619–23.
[188] Panella B, Hirsher M. Nanostructures with high surface area for hydrogen storage. J Alloys Compd 2005;404–406:399–401.
[189] Pan L, Sander MB, Huang X, Li J, Smith M, Bittner E, et al. Microporous metal organic materials: promising candidates as sorbents for hydrogen storage. J Am Chem Soc 2004;126:1308–9.
[190] Rosi NL, Eckert J, Eddaoudhi M, Vodak DT, Kim J, O'Keeffe M, et al. Hydrogen storage in microporous metal-organic frameworks. Science 2003;300:1127–9.
[191] Chen B, Ockwig NW, Millward AR, Contreras DS, Yaghi OM. High H_2 adsorption in a microporous metal-organic framework with open-metal sites. Angew Chem Int Ed 2005;44:4745–9.
[192] Kesanli B, Cui Y, Smith MR, Bittner EW, Bockrath BC, Lin W. Highly interpenetrated metal–organic frameworks for hydrogen storage. Angew Chem 2005;117:74–7.
[193] Chun H, Dubtsev DN, Kim H, Kim K. Synthesis, X-ray crystal structures, and gas sorption properties of pillared square grid nets based on paddle-wheel motifs: implications for hydrogen storage in porous materials. Chem Eur J 2005;11:3521–9.
[194] Batten SR, Robsom R. Interpenetrating nets: ordered, periodic entanglement. Angew Chem Int Ed 1998;37:1460–94.
[195] Ma SQ, Sun DF, Ambrogio M, Fillinger JA, Parkin S, Zhou H-C. Framework-catenation isomerism in metal–organic frameworks and its impact on hydrogen uptake. J Am Chem Soc 2007;129:1858–9.

[196] Dinca M, Long JR. High-enthalpy hydrogen adsorption in cation-exchanged variants of the microporous metal–organic framework $Mn_3[(Mn_4Cl)_3(BTT)_8(CH_3OH)_{10}]_2$. J Am Chem Soc 2007;129:11172–6.
[197] Marella M, Tomaselli M. Synthesis of carbon nanofibers and measurements of hydrogen storage. Carbon 2006;44:1404–13.
[198] Lachawiec Jr AJ, Qi GS, Yang RT. Assembly of metal ions and ligands with adaptable coordinative tendencies as a route to functional metal-organic solids. Langmuir 2005;21:11418–24.
[199] Dickie D, Shatte GM, Jennings CF, Jenkins HA, Khoo SYL, Clyburne JAC. A zinc carboxylate network containing metal sites with low coordination numbers. Inorg Chem 2006;45:1646–55.
[200] Li Y, Yang RT. Hydrogen storage in nanostructured carbons by spillover: bridge-building enhancement. J Am Chem Soc 2006;128:8136–7.
[201] Rowsell JLC, Eckert J, Yaghi OM. Characterization of H_2 binding sites in prototypical metal–organic frameworks by inelastic neutron scattering. J Am Chem Soc 2005;127:14904–10.
[202] Shimizu KH. Assembly of metal ions and ligands with adaptable coordinative tendencies as a route to functional metal-organic solids. J Solid State Chem 2005;178:2519–26.
[203] Furukawa H, Miller MA, Yaghi OM. Spillover in heterogeneous catalysis adsorption in metal–organic frameworks. J Mater Chem 2007;17:3197–204.
[204] Murray LJ, Dinca M, Long JR. Hydrogen storage in metal-organic frameworks. Chem Soc Rev 2009;38:1294–314.
[205] Yang FH, Yang RT. Ab initio molecular orbital study of adsorption of atomic hydrogen on graphite: insight into hydrogen storage in carbon. Carbon 2002;40:437–44.
[206] Conner Jr WC, Falconer JL. Spillover in heterogeneous catalysis. Chem Rev 1995;95:759–88.
[207] Srinvas ST, Rao PK. Direct observation of hydrogen spillover on carbon-supported platinum and its influence on the hydrogenation of benzene. J Catal 1994;148:470–7.
[208] Lueking AD, Yang RT. Hydrogen spillover to enhance hydrogen storage—study of the effect of carbon physicochemical properties. Appl Catal A 2004;265:259–68.
[209] Levy RB, Boudart MJ. The kinetics and mechanism of spillover. J Catal 1974;32:304–14.
[210] Mu S, Tang H, Qian S, Pan M, Yuan R-z. Hydrogen storage in *carbon* nanotubes modified by microwave plasma etching and Pd decoration. Carbon 2006;44:762–7.
[211] Mendoza-Cortes JL, Han SS, Goddard III WA. High H_2 uptake in Li-, Na-, K-metalated covalent organic frameworks and metal organic frameworks at 298 K. J Phys Chem A 2012;116:1621–33.
[212] Müller U, Hesse M, Puetter H, Wessel H. Adsorptive anreicherung von methan in methan-haltigen gasgemischen. Patent EP 1,674,555; 2005.
[213] Müller U, Hesse M, Puetter H, Kamieth M, Schierle-Arndt K. Adsorptive gewinnung von xenon aus krypton-xenon gasgemischen. Patent WO 2006072573 A2; 2006.
[214] Matsuda R, Kitaura R, Kitagawa S, Kubota Y, Belosludov RV, Kobayashi TC, et al. Highly controlled acetylene accommodation in a metal–organic microporous material. Nature 2005;436:238–41.
[215] Block ED, Queen WL, Krishna R, Zadrozny JM, Brow CM, Long JR. Hydrocarbon separations in a metal-organic framework with open iron(II) coordination sites. Science 2012;335:1606–10.
[216] McKinlay AC, Xiao B, Wragg DS, Wheatley PS, Megson IL, Morris RE. Exceptional behavior over the whole adsorption–storage–delivery cycle for NO in porous metal organic frameworks. J Am Chem Soc 2008;130:10440–4.
[217] Arstad B, Fjellva H, Kongshaug KO, Swang O, Blom R. Amine functionalised metal organic frameworks (MOFs) as adsorbents for carbon dioxide. Adsorption 2008;14:755–62.
[218] Llewellyn PL, Bourrelly S, Serre C, Filinchuk Y, Férey G. How hydration drastically improves adsorption selectivity for CO_2 over CH_4 in the flexible chromium terephthalate MIL-53. Angew Chem 2006;118:7915–8.
[219] Sumida K, Rogow DL, Mason JA, McDonald TM, Bloch ED, Herm ZR, et al. Carbon dioxide capture in metal–organic frameworks. Chem Rev 2012;112:724–81.
[220] Finsy V, De Bruyne S, Alaerts L, De Vos D, Jacobs PA, Baron GV, et al. Shape selective adsorption of linear and branched alkanes in the $Cu_3(BTC)_2$ metal-organic framework. Stud Surf Sci Catal 2007;170:2048–53.

[221] Schneemann A, Bon V, Schwedler I, Senkovska I, Kaskel S, Fischer RA. Flexible metal–organic frameworks. Chem Soc Rev 2014;43:6062–96.
[222] Li ZQ, Qiu LG, Wang W, Xu T, Wu Y, Jian X. Fabrication of nanosheets of a fluorescent metal–organic framework $[Zn(BDC)(H_2O)]_n$ (BDC = 1,4-benzenedicarboxylate): Ultrasonic synthesis and sensing of ethylamine. Inorg Chem Commun 2008;11:1375–7.
[223] Wong K-L, Law G-L, Yang Y-Y, Wong W-T. A highly porous luminescent terbium–organic framework for reversible anion sensing. Adv Mater 2006;18:1051–4.
[224] Zou RQ, Yamada Y, Xu Q. Diamondoid metal-organic framework of Zinc(II) urocanate with one-dimensional open channels. Micropor Mesopor Mater 2006;91:233–7.
[225] Zhong RQ, Zou RQ, Xu Q. Microporous metal-organic framework Zinc(II) imidazole-4,5-dicarboxylate: four-fold helical structure and strong fluorescent emission. Micropor Mesopor Mater 2007;102:122–7.
[226] Xu YH, La YQ, Zhao YH, Du D-Y, Xu G-J, Shao K-Z, et al. An unprecedented $[Cu_7I_4]^{3+}$ cationic cluster-based metal–organic framework (MOF) including 1D nanochannels. Inorg Chem Commun 2009;12:169–72.
[227] Chandra D, Kasture MW, Bhaumik A. A new microporous MOF material based on Zn(II)-polycarboxylate coordination polymer synthesized with the aid of 1,6-diaminohexane as template. Micropor Mesopor Mater 2008;116:204–9.
[228] Zhu S, Zhang H, Zhao Y, Shao M, Wang Z, Li M. Synthesis, structures and luminescence of three coordination polymers constructed from rigid 1,3,5-benzenetricarboxylic acid and flexible bis(imidazol-1-ylmethyl)-benzene. J Mol Struct 2008;892:420–6.
[229] Basak S, Sen S, Marschner C, Baumgartner J, Batten SR, Turner DR, et al. Synthesis, crystal structures and fluorescence properties of two new di- and polynuclear Cd(II) complexes with N_2O donor set of a tridentate Schiff base ligand. Polyhedron 2008;27:1193–200.
[230] Yang E-C, Liang Q-Q, Wang P, Zhao X-J. A 3D photoluminescent cadmium(II)-organic framework with unusual pentanuclear cluster as secondary building unit. Inorg Chem Commun 2009;12:211–3.
[231] Jin Z, Zhu G-S, Zou Y-C, Fang Q-R, Xue Mi, Li Z-Y, et al. Synthesis, structure and luminescent property of a new 3D porous metal organic framework with rutile topology. J Mol Struct 2007;871:80–4.
[232] Chen LF, Li ZJ, Qin YY, Cheng J-K, Yao Y-G. Syntheses, crystal structures and photoluminescence of two new pyrazinecarboxylate-based cadmium(II) coordination polymers. J Mol Struct 2008;892:278–82.
[233] Huang YG, Yuan DQ, Gong YQ, Jiang F-L, Hong M-C. Synthesis, structure and luminescent properties of lanthanide–organic frameworks based on pyridine-2,6-dicarboxylic acid. J Mol Struct 2008;872:99–104.
[234] Li X, Zhang Y-B, Shi M, Li P-Z. (4, 4)- and (6, 3)-2-D luminescent lanthanide(III) metal-organic frameworks constructed from tetrafluorosuccinate and 1,10-phenanthroline. Inorg Chem Commun 2008;11:869–72.
[235] Zhang ZH, Song Y, Okamura T, Hasegawa Y, Sun WY, Ueyama N. Syntheses, structures, near-infrared and visible luminescence, and magnetic properties of lanthanide-organic frameworks with an imidazole-containing flexible ligand. Inorg Chem 2006;45:2896–902.
[236] Zhu Y-J, Ren Z-G, Zhang W-H, Chen Y, Li H-X, Zhang Y, et al. Efficient mechanochemical synthesis of tris(pyrazolylborate) complexes of manganese(II), cobalt(II) and nickel(II). Inorg Chem Commun 2007;10:485–8.
[237] Zhao Y-H, Su Z-M, Fu Y-M, Shao K-Z, Li P, Wang Y, et al. Syntheses and characterizations of four metal coordination polymers constructed by the pyridine-3,5-dicarboxylate ligand. Polyhedron 2008;27:583–92.
[238] Luo G-G, Huang R-B, Zhang N, Lin L-R, Zheng L-S. Structural diversity in the (derivatives) system: new zero-, one-, and two-dimensional inorganic–organic hybrids. Polyhedron 2008;27:3231–33238.
[239] Su Z, Fan J, Sun W-Y. Novel two-fold interpenetrated Zn-based metal-organic framework with triple-stranded right- and left-handed helical chains. Inorg Chem Commun 2013;27:18–21.
[240] Marlow F, McGehee MD, Zhao D, Chmelka BF, Stucky GD. Doped mesoporous silica fibers: a new laser material. Adv Mater 1999;11:632–6.
[241] Choi JH, Jeon HJ, Choi KM, Kang JK. A series of s-block (Ca, Sr and Ba) metal–organic frameworks: synthesis and structure–property correlation. J Mater Chem 2012;22:10144–7.

[242] Cui Y, Yue Y, Qian G, Chen B. Luminescent functional metal-organic frameworks. Chem Rev 2012;112:1126–62.
[243] Obvintseva LA. Metal oxide semiconductor sensors for determination of reactive gas impurities in air. Ross Khim Zh (Russ) 2008;52:113–21.
[244] Rumyantseva MN, Makeeva EA, Gas'kov AM. Influence of the microstructure of semiconductor sensor materials on oxygen chemisorption on their surface. Ros Chem J (Russ) 2008;52:122–9.
[245] Llabres i Xamena FX, Corma A, Garcia H. Applications for metal–organic frameworks (mofs) as quantum dot semiconductors. J Phys Chem 2007;111:80–5.
[246] Maspoch D, Ruiz-Molina D, Wurst K, Domingo N, Cavallini M, Biscarini F, et al. A nanoporous molecular magnet with reversible solvent-induced mechanical and magnetic properties. Nat Mater 2003;2:190–5.
[247] Wang C, Zhang T, Lin W. Rational synthesis of noncentrosymmetric metal–organic frameworks for second-order nonlinear optics. Chem Rev 2012;112:1084–104.
[248] Ma Z, Moulton B. Recent advances of discrete coordination complexes and coordination polymers in drug delivery. Coord Chem Rev 2011;255:1623–41.
[249] Horcajada P, Gref R, Baati T, Allan PK, Maurin G, Couvreur P, et al. Metal–organic frameworks in biomedicine. Chem Rev 2012;112:1232–68.
[250] Zhang W, Xiong R-G. Ferroelectric metal–organic frameworks. Chem Rev 2012;112:1163–95.
[251] Ferrey G, Millange F, Morcrette M, Serre C, Doublet M-L, Grenèche J-M, et al. Mixed valence Li/Fe-based MOFs with both reversible redox and sorption properties. Angew Chem 2007;119:3323–7.
[252] Della Rocca J, Lin W. Nanoscale metal–organic frameworks: magnetic resonance imaging contrast agents and beyond. Eur J Inorg Chem 2010;3725–34.
[253] Doty FP, Bauer CA, Skulan AJ, Grant PG, Allendorf MD. Scintillating metal-organic frameworks: a new class of radiation detection materials. Adv Mater 2009;21:95–101.

CHAPTER 3

Computational Chemistry of Zeolite Catalysis

E.A. Pidko, E.J.M. Hensen

Eindhoven University of Technology, Eindhoven, The Netherlands

1 Introduction

Understanding the origin of reactivity of zeolite materials necessitates the development of the molecular-level understanding of the structure and chemical nature of catalytic centers. The direct information on the state of the intrazeolite active sites under the reaction conditions is seldom available from experiments. Furthermore, practical zeolite catalysts are usually characterized by a high heterogeneity of potentially reactive species in their micropores. This additionally complicates the identification of the active site, the reaction mechanism as well as the role of other zeolite components in catalytic reactions. In previous years, computational methods proved to be extremely useful in developing a deep fundamental understanding of catalytic reactivity, assisting the interpretation of experimental data, and increasingly in predicting of the properties and behavior of zeolite catalysts at the molecular level. The current advances in computational soft- and hardware make it possible to perform analysis of complex reaction networks using realistic zeolite models by means of quantum chemical calculations, thereby providing a framework for theoretical studies of zeolite reactivity. The state-of-the-art computational methodologies allow a realistic description of the reactive environment and its effect on the chemical transformations occurring within the zeolite micropores.

By using computational techniques the structural complexity of zeolites can be addressed by predicting the structural properties and stabilities of different potential species in their confined space under catalytic conditions. To understand further the relationship between structural and physicochemical properties of zeolite catalysts and their reactivity, macroscopic reaction rates for specific sites have to be predicted. This requires unraveling the complex network of elementary chemical transformations occurring in the zeolite pores during the catalytic reaction. Because the mechanism of most reactions, with the exception probably of the most trivial ones, is generally not known, the main aim of computational catalysis studies is to identify the preferred reaction mechanism by calculating and comparing energetics of different elementary reaction steps. The information on the energetics of

Zeolites and Zeolite-like Materials. http://dx.doi.org/10.1016/B978-0-444-63506-8.00003-3

reaction steps can then be used to compute rate constants of the relevant elementary steps providing a framework for the experimental validation of the computational results. The latter is a challenge in itself. To unambiguously verify the proposed catalytic sites and reaction mechanism, advanced in situ characterization studies complemented by careful kinetic experiments on well-defined catalytic materials have to be performed.

To create a comprehensive theoretical description of a catalytic reaction, one has to solve a multiscale problem that stems from the very different characteristic time and length scales of the key processes underlying a catalytic reaction (Fig. 1A). Whereas a characteristic timescale of a reaction in heterogeneous catalysis is 1 s, the elementary reactions inside the zeolite pores take place at a timescale of ca. 10^{-6} s, while the timescale of electronic motion that governs the intrinsic chemistry of such processes is 10^{-16} s. Therefore, to create a detailed picture of a catalytic reaction, a wide range of computational techniques covering different time and length scales has to be used in a complementary manner. Chemically realistic structural models of intrazeolite active sites and elementary reaction steps catalyzed by them can now be studied with high accuracy by performing electronic structure calculations. Such methodologies allow predicting structures and energies of different reaction intermediates and transition states. Reaction entropies can also be computed from ab initio calculations, thus allowing the prediction of rates at finite temperature. This information can then be included into kinetic schemes using, for example, a microkinetic modeling (MKM) approach [3] or kinetic Monte Carlo methods [4] to ultimately arrive at macroscopic rate constant of a catalytic reaction.

The computational chemistry toolbox [5] includes a wide variety of methods useful for the description of different physical and chemical processes in zeolites. They are usually classified on the basis of the size of the chemical system that they can simulate and the

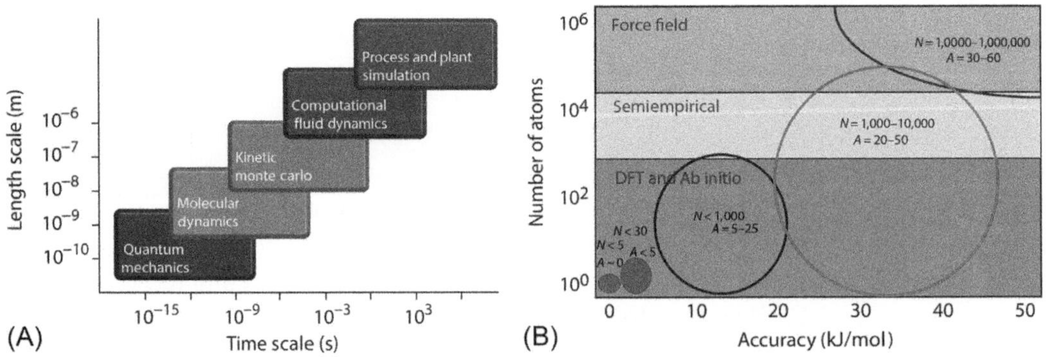

Fig. 1

(A) Time and length scales [1] and (B) hierarchy of different computational methods (*N* stands for the number of atoms, *A* is achievable accuracy in kJ/mol). *Figure from Ref. [2] with permission.*

accuracy that they can achieve (Fig. 1B). The processes controlled by very large ensembles of atoms can be routinely modeled in the framework of classical mechanics by using interatomic potentials (force fields) that are often based on empirical parameters. The results of such simulations can be used to estimate stabilities and predict structural properties of zeolite frameworks, to study adsorption properties by computing ensemble averages using Monte Carlo simulations, or to model diffusion inside zeolite pores via molecular dynamics simulations using classical Newton's law of motion. A large number of reliable force fields specifically designed for modeling different phenomena in zeolites are available. When no bond breaking or making processes need to be modeled, these methodologies are very useful for simulating various properties of zeolites [6,7].

The second large class of computational methods that is most useful for predicting reactivity of zeolites is based on the quantum mechanical description of a chemical system. Quantum mechanics represents the highest level in the hierarchy of computational methods. By solving the electronic structure problem they provide us the energy and wave function of a system, from which all properties of all atoms in it can be derived. In practice, however, the exact solution of the electronic structure problem cannot be obtained for any realistic system. Thus, a number of simplifying approximations must be made that lead to different levels of quantum chemical methods. The most drastic approximations to the Schrödinger equation and extensive parameterization with experimental values are made in the so-called semiempirical methods. Although these methods are relatively fast, they are still ca. 100–1000 times slower than the force-field methods and give low accuracy. Moreover, the application of these methods to systems that were not included in the parametrization is quite risky and may lead to unreliable results.

The highest accuracy can be achieved by using computational methods that solve the Schrödinger equation in an approximate manner without direct fitting to experiment. This can be done by solving the quantum mechanical equations directly to derive a many-electron wave function using various ab initio methods ("from the beginning," in other words, "from first principles"). An alternative approach is to use methodologies based on the density functional theory (DFT), in which the electron density is the key property of a system instead of the many-electron wave function. Both approaches have their advantages and limitations and can be efficiently used for the description of different chemical processes in zeolites. Nevertheless, because of their robustness, high efficiency, and high accuracy, the DFT methods have predominated in recent applications.

In this chapter we present an introductory overview of important theoretical concepts and practical tools essential for computational modeling of catalytic reactions in zeolites using quantum chemical calculations. The power and capabilities of state-of-the-art computational catalysis methodologies will be illustrated by discussing selected relevant examples from the recent literature.

2 Methodology

2.1 The Concept of the Potential Energy Surface

Computational studies on catalysis by zeolites begin with the need to identify the structures of the most probable active sites as well as different species formed within the zeolite pores in the course of the reaction. Such species correspond to specific stationary points on a so-called potential energy surface (PES) that is one of the core concepts in computational chemistry. The first derivative of the potential energy at these points is zero with respect to every degree of freedom in the system.

A catalytic reaction usually proceeds via the formation of a number of stable reaction intermediates representing energy minima on PES. At such points the second derivative of the potential energy is positive for all degrees of freedom. The lowest-energy path connecting the minima forms a reaction coordinate or intrinsic reaction coordinate, along which the chemical system evolves in the course of a reaction. For most of catalytic reactions, the degree of freedom corresponding to the reaction coordinates is usually composed of many individual geometrical parameters such as bond distances, angles, and dihedrals. The transition state is defined as an energy maximum along the reaction coordinate linking two minima. It represents a saddle point on the PES for which the first derivative of the potential energy is zero and the second derivative is positive for all degrees of freedom but the one corresponding to the reaction coordinate, along which it is negative.

Determining structures of stable reaction intermediates, finding the minimum energy paths, and locating the transition states on the PES are not trivial processes. There are different methods available for accomplishing these tasks that differ in their accuracy and efficiency. Generally, to find the stationary points and the reaction paths, the knowledge of the potential energy of a chemical system and its first derivative with respect to the atomic coordinates is required.

2.2 The Many-Body Problem and Born-Oppenheimer Approximation

A wide variety of chemical and physical properties, in the first place the energy, of a chemical system can be obtained by solving the many-body Schrödinger equation:

$$\hat{H}\psi = E\psi \tag{1}$$

where E is the total energy of the N-particle system, $\hat{H}$ is the Hamiltonian operator, and ψ is the many-body wave function. For any chemical system with the number of electrons exceeding 1, the exact solution of the Schrödinger equation is not possible. Therefore, a number of simplifying approximations resulting in different quantum chemical methods are required to solve it for realistic systems. In the following we provide an overview of the most important approximations and limitations of the resulting quantum chemical methods as applied to the zeolite chemistry. More detail and in-depth discussion on the electronic structure calculations can be found in a number of very good references [5,8].

The first approximation that is made in all practical quantum chemical methods is the Born-Oppenheimer approximation that was introduced in 1927 by Max Born and J. Robert Oppenheimer [9]. It assumes that because the nuclei are much heavier than the electrons, they move relatively slowly and therefore are assumed to be stationary with the electrons moving in their field. In other words, this approximation decouples the electronic and nuclear motion. This allows one to reduce the many-body problem to only the electronic problem. As a result, the total energy of a chemical system and its wave function can be determined by solving only the Schrödinger equation for a set of nuclear positions representing a particular geometry of a chemical system. Nevertheless, the exact solution of this equation still exists only for one-electron systems because it is not possible to resolve the electron-electron repulsion term analytically. To describe many-electron systems further approximations are necessary.

2.3 Ab Initio Methods

Although the positions of nuclei become now only a parameter, the electronic wave function depends on the coordinates of all electrons in the system and their spin. Traditionally, the derivation of an electronic wave function and energy calculation follows a consecutive procedure that involves the introduction of crucial physical phenomena step-by-step. All ab initio wave function-based methods are based on the Hartree-Fock (HF) approximation that yields an approximate zero-order electronic wave function. This method treats the electron-electron interaction within an N-electron system by a mean-field approximation in which each electron is considered as moving in the field generated by the nuclei and the other $N-1$ electrons. As a result, the motion of electrons is not correlated. This means that the probability of an electron having a particular set of spatial coordinates at a particular time does not depend on the coordinates of other electrons at that moment. In reality, the electronic motion is correlated. This property results in an effective "improvement" of their ability to avoid each other. The mean-field approximation in the HF treatment leads to a very strong overestimation of the electron-electron repulsion and accordingly gives too high energy of the chemical system.

The energy that cannot be accounted for within the HF procedure is then defined as the correlation energy. It consists of two components, namely static and dynamic correlation. Dynamic correlation energy is associated with the fact that within HF, electrons are not kept sufficiently apart; in other words it relates to the interaction between electrons in the system. Static correlation is associated with the use of a single determinant. The underestimation of the static correlations arises from the fact that there may be several almost degenerate frontier orbitals for which the assignment of occupancy cannot be unambiguously done within HF.

The HF approach cannot adequately describe either component of the electron correlation energy. In practical terms the neglect of electron correlation in HF can cause very serious errors in geometries, vibrational, and other molecular properties as well as in predicted energies of chemical transformations.

To improve the description of the correlated motion of electrons different post-HF methods were developed that introduce additional corrections to the results of HF calculations. An in-depth description of the theoretical basis of these methodologies as well as other computational techniques can be found in Refs. [5,8]. The dynamic correlation energy can be efficiently recovered by using the CI (configurational interaction), Møller-Plesset perturbation theory (MP*n*), and coupled cluster (CC) methods, whereas multireference CI methods such as complete active space self-consistent field (CASSCF) allow an adequate description of the static correlations.

One of the most widely used approaches to recover correlation energy in wave function-based methods is based on perturbation theory. This approach is based on the idea that the correct description of a complex system can be achieved by describing a perturbed version of its simplified model. In the Møller-Plesset methods the correlation energy and the CIs are treated as small perturbations to the HF Hamiltonian. Within this methodology, the basic HF approach is denoted as MP1, a method that does not yet account for electron correlation. The first MP level that goes beyond HF is the second-order Møller-Plesset perturbation theory (MP2). The MP2 energy (E_{MP2}) can be expressed as a sum of HF energy (E_{HF}) and the perturbation correction E^2. The latter term is purely electrostatics and represents a sum of terms describing double excitations from occupied to formally unoccupied molecular orbitals. The second order Møller-Plesset perturbation theory (MP2) typically recovers 80–90% of the correlation energy, while MP4 provides a reliably accurate solution to most system.

Another approach to improve the description of correlated motion of electrons originates from the proposal that the HF wave function can be improved by including additional terms due to electron excitations from occupied to virtual molecular orbitals. This approach forms a basis for the CI methods, in which a wave function is represented by a linear combination of the ground-state and excited-state wave functions. Similar to the MP case, *the partial population of the excited states makes it easier for electrons to avoid each other and thus decreases the electron-electron repulsion*. Although the basic concept for the recovery of electron correlation in MP and CI methods is similar, the mathematical approach is very different. In CI, the first term of the trial wave function is the single-determinant HF wave function describing the ground state of the system. Further terms represent additional determinants representing excited state configurations of the initial HF configuration. When all possible electronic configurations are included in the expansion, the full CI wave function is constructed that is an exact solution for the electronic problem. Unfortunately, the full CI calculation can be applied only to very small molecules. For realistic chemical problems, the CI expansion has to be truncated so that only a limited number of determinants are considered. The trial wave function can include the exchange of 1, 2, or 3 electrons from the valence band into unoccupied orbitals. The corresponding methods are known as CI singles, CI doubles, and CI triples and allow for single, single/double, and single/double/triple excitations. A number of mathematical treatments have been developed to facilitate the recovery of correlation energy within CI calculations despite the necessity of truncation of the CI expansion. The most widely used modifications of CI are multiconfigurational SCF, CASSCF, and CC methods. The widely used CASSCF method is based on a careful selection of the orbitals to create the active space that is used for constructing the CI determinants. In this approach, the active space includes only those orbitals that directly "influence" the chemical process under consideration.

The CC method combines the basic ideas of both the perturbation methods and the CI treatment. Here the trial wave function includes a series of excitation operators to act on the ground-state HF wave function promoting thus electrons into unoccupied orbitals. This method again requires a truncation of the number of included excitations resulting in the following practical methods: CC doubles, CC singles and doubles or CC singles, doubles and triples (CCSDT). To decrease the computational demands of the latter, triple excitations are usually considered in an approximate way. The resulting CCSD(T) method represents the current golden standard of computational chemistry. With this method, moderate size practical chemical systems can be treated with an absolute accuracy of 1 kcal/mol.

Although the post-HF methods allow achieving a very high accuracy, almost all of them, with the exception of MP2, are prohibitively expensive for calculation of realistic models of heterogeneous catalysts including of course zeolites. It is also important to realize that for studying reactivity, the absolute accuracy of a computational methodology is usually of little importance. The correct description of qualitative trends and a good estimation of relative energies between different stationary points on the PES can provide a basis for a prediction of the chemical reactivity of complex chemical systems.

2.4 Density Functional Theory

A much more attractive and popular methodology to solve the electronic structure problem of complex chemical systems such as zeolites and other microporous materials is based on the DFT. DFT is not based on wave functions, but on electron density function (electron probability density function, electron density, charge density), $\rho(\mathbf{r})$. This is the main conceptual difference between the approaches discussed earlier and the DFT methodologies. The attractiveness of these methods is partially owing to the fact that a wave function cannot be experimentally measured. It is still debated whether a wave function is just a convenient mathematical model or a true physical entity. On the contrary, the electron density is a directly measurable property that can be obtained from, for example, X-ray diffraction experiments. Furthermore, the electron density is much easier to treat in a calculation. Electron density is a function of position making it dependent only on only three spatial coordinates. Wave function, on the other hand, is a function of three spatial coordinate plus spin coordinate for *each electron*. Thus, to describe an N-electron system, $4N$ variables have to be considered in the wave function-based method, whereas in DFT the electron density of any systems remains a function of only three spatial coordinates. Thus, the electron density has three important advantages compared to the wave function: it is easily comprehensible, it is directly measurable, and it is more mathematically tractable [5].

The foundation of DFT has been introduced through the work of Hohenberg and Kohn [10], in which it was formally demonstrated that any ground-state property of a system is a function of its electron density. Kohn and Sham [11] extended the theory to practice by showing how the energy can be portioned into kinetic energy of the motion of the electrons, potential energy of the nuclear-electron attraction, electron-electron repulsion, which involves coulomb as well as self-interactions, and exchange correlation, which covers all other electron-electron interactions. The energy of an N-particle system can then be written as

$$E[\rho]=T[\rho]+U[\rho]+E_{XC}[\rho] \tag{2}$$

Kohn and Sham demonstrated that the N-particle system can be written as a set of n-electron problems (similar to the molecular orbitals in wave function methods) that could be solved self-consistently [11].

DFT is then in principle an "ab initio" method because it is derived without using any adjustable parameters. However, the exact mathematical form of the DFT functional is not known, unlike the mathematically correct fundamental Schrödinger equation in the wave function-based methods. This implies that whereas in the latter methods, the quality of wave function and therefore the quality of the calculation can be improved by employing higher-level energy corrections, the systematic improvement of the DFT is not possible.

The last term in Eq. (2) represents the main problem of the DFT. In DFT methodologies the optimal electron density (ρ_0) is computed following the variational principle. Kohn and Sham proposed that a real electron density can be represented by a "fictitious" noninteracting reference system. Electrons in the latter system do not interact, but its ground-state electron density distribution ρ_r is exactly the same as ρ_0 corresponding to the real system under consideration. The deviation in the behavior of noninteracting electrons from that of the real ones is then taken into account by the unknown XC functional that is included in DFT methods in an approximate form. The development of such approximate functional is a very active research topic in theoretical chemistry.

A good functional should not only account for errors in electron exchange and correlation, but also for self-repulsion errors in the average electron density cloud and kinetic energy error. The simplest approximation to E_{XC} is the local density approximation (LDA) that assumes that at every point in a chemical system the energy density corresponds to that of a homogeneous electron gas with the density at that point. For most chemical systems this approximation leads to large errors in bond energies and cannot adequately describe weak bonds such as hydrogen bonds. These phenomena are the key for understanding the reactivity of zeolites thereby making the LDA approach not an optimal choice for computational catalysis studies.

The description of LDA can be substantially improved by taking into account not only the density at a given point in space but also its gradient, the first derivative of ρ with respect to its position, $\nabla\rho(r)$. This, however, cannot be performed by a simple gradient expansion, and a number of constraints have to be imposed to derive the exchange-correlation functional in the generalized gradient approximation (GGA). Different GGA functionals have been introduced so far. Among the most popular GGA functionals are plane wave (PW), PW91 [12], PBE [13], and B88 [14]. Their use allows one to drastically improve the accuracy of DFT calculations compared to the LDA approach. The accuracy of calculations of the atomization energy improves to about 20 kJ/mol compared to the accuracy of about 180 kJ/mol obtained with LDA [15].

In view of the dramatic improvement of accuracy achieved by the inclusion of the information about the spatial variation of density, it is reasonable to propose that the DFT results can be further improved by introducing additional terms that depend on the second derivative of ρ, $\nabla^2\rho(r)$. Such functionals are usually named meta-GGA (MGGA). Indeed, the inclusion of the additional parameter allows some improvements of the results of calculations, although this improvement over the GGA approximation is relatively small.

The exchange-correlation energy can be expressed as a sum of the exchange and correlation components, where the former term dominates. Therefore, the introduction of the exact HF exchange to the exchange-correlation functional is an attractive approach to improve the GGA approximation. The applicability of this approach follows from the adiabatic connection method that implies that the exchange-correlation energy can be expressed as a linear combination of the DFT exchange-correlation energy and HF exchange energy. The resulting hybrid DFT functional is contributed by a HF-type electron exchange energy computed from the Kohn and Sham wave functions for noninteracting electrons. This way an effective correction to the classical coulomb repulsion for Pauli repulsion is introduced. The most important parameter for hybrid functional is the amount of exact HF exchange used. This parameter is usually expressed in a mixing coefficient that is usually fitted to various parameters of large sets of molecules upon constructing the hybrid functional. One of the most popular hybrid functional is B3LYP [16], for which the exchange-correlation energy is given as:

$$E_{XC}^{B3LYP} = (1-a)E_X^{LDA} + aE_X^{HF} + bE_X^{GGA} + cE_C^{LYP} + (1-c)E_C^{LDA} \quad (3)$$

where $a=0.2$, $b=0.72$, and $c=0.81$.

The attractiveness of DFT methods is mainly due to their high accuracy and at the same time very high efficiency. DFT calculations provide results of accuracy close to that of the MP2 method, while requiring much less computational time. DFT methods formally scale with the increase in the number of basis functions (electrons) as N^3 and therefore can be easily applied to studying much more realistic structural models compared to the higher-level post-HF methods, which usually scale as N^5 for MP2 and up to N^7 for CCSD(T). The geometry and energetics calculated using hybrid functionals (eg, B3LYP and B3PW91 [16], PBE0 functional [17] etc.) are usually in a good agreement with the experimental results and with those obtained using the post-HF methods. Because of the approximate nature of the exchange-correlation functional, DFT methods still fail in describing effects originating from correlated motion of electrons such as, for example, dispersion interactions [18,19]. The importance of these effects for chemical reactions in confined space and the applicability of available DFT techniques for their description are discussed in Section 1.

2.5 Basis Sets

Another general approximation made in all electronic structure methods including DFT and ab initio methods is the representation of molecular orbitals by a finite number of simple mathematical functions. Indeed, in spite of the fact that formally the energy in DFT is a function of electron density, in practice the latter is represented by the Kohn-Sham orbitals, which are represented in calculations by a specific orthonormal basis set. In a limiting case, when an infinite number of basis function is used the orbitals and accordingly the true electronic structure can be described perfectly. For practical applications these functions have to be truncated. The resulting description of the orbitals by a finite number of basis functions leads to a loss in accuracy. A wide range of different basis sets has been constructed so far. The actual choice of the basis set for particular calculations is determined by the computational method used, the nature of the chemical system studied and the accuracy needed. Most commonly, basis sets are represented as a linear combination of local atomic orbitals described by Gaussian-type functions (GTOs), or a linear combination of plane-waves (PWs). Localized GTO basis sets are best suited for modeling of molecular systems, which also include cluster zeolite models (see Section 2.6). Such basis sets are implemented in quantum chemical programs including Gaussian [20], Tubomole [21], NWChem [22], among others. PWs are, on the other hand, more naturally applied to periodic systems such as zeolite crystals. The respective calculations are usually faster both for computing energies and gradients as compared to the approaches employing GTO basis sets. The PW approach is implemented in such computer programs as CASTEP [23], CPMD [24], VASP [25], and more recently in NWChem [22]. A special case is the CRYSTAL [26] program that utilizes GTO basis sets for describing both periodic and molecular systems within the same formalism. Because true electronic structure can be modeled by an infinite number of basis functions, very similar results can be obtained when a sufficiently large GTO or PW basis sets are used [27].

One more important difference between the GTO and PW approaches is that whereas in the former case the representation of both core and valence orbitals is the same, in the PW formalism the number of PW components needed to correctly describe the behavior of the wave function near the nucleus is prohibitively large. This problem is solved by modeling the core electrons using the pseudopotential approximation, in which it is assumed that the core electrons do not significantly influence the electronic structure and chemistry of atoms. The valence electrons of a particular atom are then considered to move in an effective ionic potential due to the core electrons and the nucleus. Same approach can also be applied to reduce the size of the GTO basis set in calculations without dramatic loss of accuracy. Furthermore, the use of pseudopotentials allows the inclusion of nonrelativistic effects in the calculations, which are particularly important for the chemistry of heavy elements.

2.6 Models

The use of different quantum chemical methods for studying chemical processes is inevitably associated with the introduction of different errors due to the various approximations made to solve the electronic problem in a particular computational method. When reactivity of such complex molecular systems as zeolites is considered, these are not the only sources of errors leading to limited accuracy. The complexity of realistic industrial zeolite catalysts containing different defects, interphases, and crystal intergrowths is prohibitive for modeling using any electronic structure method. Moreover, even ideal zeolite crystals often have very large unit cells (Fig. 2A and B). A comprehensive study of complex reaction networks inside such systems is very computationally demanding. Therefore, one may choose to select only

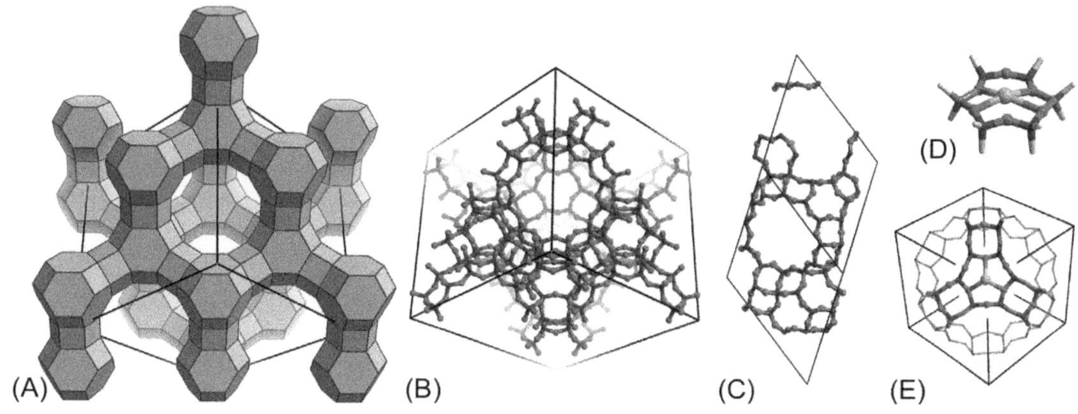

Fig. 2
Structural models of zeolite with faujasite topology: (A) crystallographic unit cell with the $Fd\bar{3}m$ symmetry containing 576 atoms (192 T atoms); (B) a smaller rhombohedral lower-symmetry unit cell with 144 atoms (48 T); (C) a cluster model containing six T-atoms; (D) representing a local structure of a single S_{II} cation site and a respective hybrid 12 T cluster embedded in the rhombohedral faujasite unit cell; (E) *(adapted from Ref. [28])*.

a limited subset of the atoms of the zeolite to construct an atomistic model of the reactive site and its environment. The size of the model can be critical for obtaining reliable results. Although when using a very small cluster model, high-level post-HF methods can be applied, important effects due to confinement and coordination environment of the reactive species can be neglected leading to fundamentally wrong predictions. This represents a so-called competition of "model" versus "method" accuracy.

The state-of-the-art methodologies in computational zeolite catalysis are based on DFT calculations with periodic boundary conditions. These allow theoretical analysis of structure and chemical properties of zeolites with moderate-sized unit cells using a real crystal structure as a model (Fig. 2C). Such periodic DFT calculations of zeolites are mostly limited to the GGA density functionals.

An alternative method for modeling zeolites is a so-called cluster approach, in which only a part of the zeolite framework, containing finite number of atoms, is considered explicitly, whereas the influence of other atoms of the zeolite lattice is neglected (Fig. 2D). Typically, a representative model of a zeolite active site should include the active site itself and a part of the structure near it to mimic the local properties of zeolite lattice and preferably its confinement space. Obviously, this approach leads to a substantial loss of "model" accuracy. Nevertheless, it can be very useful for the analysis of different local properties of porous materials such as elementary reaction steps, specific adsorption of small molecules, and the like. The effect of the extended zeolite structure can be further verified by comparing the results obtained for cluster models of increasing size with those from periodic calculations. Furthermore, the cluster modeling approach allows the application of higher-level wave function-based methods as well as hybrid density functionals to study the chemistry of zeolitic active sites. For small cluster models even the CCSD(T) method can be used, thereby providing benchmark data for validating the accuracy of more efficient methodologies used for realistic zeolite systems.

The main drawbacks of the cluster modeling approach can be partially overcome by using hybrid models based on quantum chemical embedding schemes. The associated methodologies combine two or more computational techniques in one calculation and make it possible to investigate the chemistry of such systems as zeolites with high precision. In such schemes, a local region inside the zeolite where chemical reaction takes place is treated with an appropriately accurate method, while the remainder of the system is described at a lower level of theory (Fig. 2E). The energy in this case is then calculated as

$$E_{\text{hybrid}}(\text{System}) = E_{\text{high}}(\text{Model}) + E_{\text{low}}(\text{System}) - E_{\text{low}}(\text{Model}) \tag{4}$$

where E_{high} (Model) is the energy calculated at higher level for the inner core region only. E_{low} (System) $- E_{\text{low}}$ (Model) corresponds to the difference in energy between the full system and the core region both calculated at the low level of theory. Usually quantum chemical methods

(QMs) are used for the description of the core region and the rest atoms of the system are treated by force-field methods. Nevertheless, the latter can be any lower-level method, which is faster than that used for the core.

3 Reactions in Confined Space: The Role of Dispersion Interactions

Because of the approximate nature of exchange-correlation functionals, standard DFT methodologies are generally not able to adequately describe chemical phenomena dominated by effects due to the electron correlation. In the zeolite chemistry, the most important physicochemical effects of this class are the so-called London dispersion or van der Waals interactions. The correct description of such interactions is crucial for the correct representation of reactions in zeolite pores. Accordingly, one of the main challenges in modeling chemical processes in zeolites is to predict accurately energies of adsorption and reaction profiles for transformations of different molecules in confined space. Although DFT methodologies are the state of the art in the computational zeolite catalysis, most of common exchange-correlation functionals fail to describe correctly the long-range dispersion interactions between confined molecules and walls of the zeolite channels [18,19]. The dominant interactions between the hydrocarbon species and the zeolite walls correspond to weak van der Waals interaction of dispersive nature. Such interactions originate from the attraction between induced dipoles formed due to charge fluctuations in the interacting species. This effect is the example of the instantaneous electron correlation. All semilocal and conventional hybrid functionals cannot correctly predict the asymptotic $-C_6/R^6$ behavior of the dispersive interactions with increasing interatomic distance. The incorrect description of the corresponding effects by conventional DFT methodologies may not only lead to decreased accuracy of the computed energetics of chemical reactions but even to wrong predictions about the stability or reactivity trends for systems with substantially different properties of confinement spaces.

The origin of dispersion interactions lies in the correlated motion of electrons in the interacting fragments. As has been outlined earlier, the simplest method that can explicitly account for such effects is Møller-Plesset perturbation theory (MP2). However, MP2 calculations for large chemical systems are very demanding. To describe periodic systems they are presently feasible only for the smallest zeolite unit cells. Recently, an embedding scheme to introduce local corrections at post-HF level to DFT calculations on a periodic structural model has been proposed [29]. Within this approach, the microporous reaction environment is accurately modeled by periodic DFT calculations. The DFT energies thus obtained are further corrected for the self-interaction effects and van der Waals interactions by applying a modified MP2 method to a cluster model representing the essential part of the framework inside the periodic model of a zeolite. Such MP2:DFT approach is suitable for studying reactions between small or medium-sized molecules and very large chemical

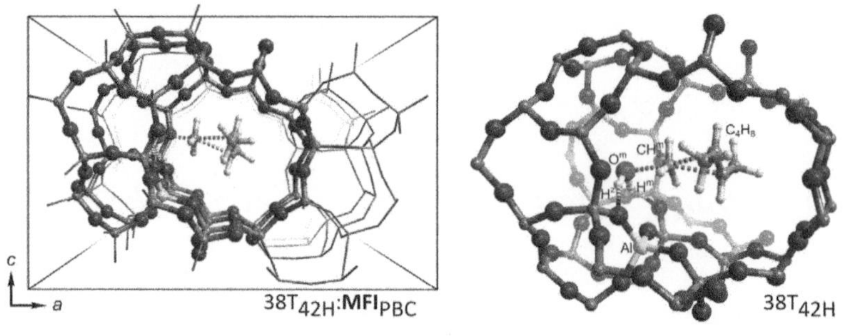

Fig. 3
Transition state structure for *t*-2-butene methylation with methanol over HZSM-5 zeolite. Left panel depicts the corresponding periodic MFI model shown along the straight channel. Highlighted atoms correspond to the largest 38 T embedded cluster (enlarged in right panel, boundary H atoms are omitted for clarity) treated at the MP2 level of theory in Ref. [30]. *Figure from Ref. [28] with permission.*

systems such as zeolites. A representative example of the successful application of this methodology is a computational study by Sauer and coworkers [30] on the methylation of small alkenes with methanol over zeolite HZSM-5. In this work, the zeolite was represented by a periodically repeated MFI unit cell. The interactions due to the confinement of the reactants in the zeolite channel and the energy profile for their transformations over a zeolite Brønsted acid site were refined by applying the MP2 correction to an extended cluster model embedded in the periodic structure (Fig. 3).

The methylation reaction is one of the key chemical transformations in the industrially important methanol to olefins (MTO) process. The growing interest this reaction promoted over acidic zeolites is determined by the fact that MTO provides a route to convert almost any carbon-containing feedstock (eg, natural gas, coal, biomass) into key intermediates for chemical industry such as ethylene and propylene. From previous experimental studies [31] on the methylation of small alkenes over HZSM-5 reaction rates and activation barriers are directly available. As a result, this reaction represents an attractive model for evaluating the predictive power and accuracy of the popular and more advanced computational methodologies. Fig. 4A depicts a simplified reaction energy diagram for this reaction. To compare directly the experimental and computational results several important assumptions with respect to the reaction kinetics had to be made. Experiments demonstrated that that the methylation reaction shows the first order behavior with respect to alkene concentration and zero order with respect to CH_3OH. This implies that the experimental apparent activation barriers [31] correspond to the energy difference between the transition state of the reaction and the state, in which methanol is adsorbed at the Brønsted acid site of a zeolite and alkene is in the gas-phase (Fig. 4A). The second assumption relates to the reaction mechanism. The methylation reaction is postulated to occur via an associative one-step mechanism without the intermediate formation of a methoxy group grafted to the zeolite lattice.

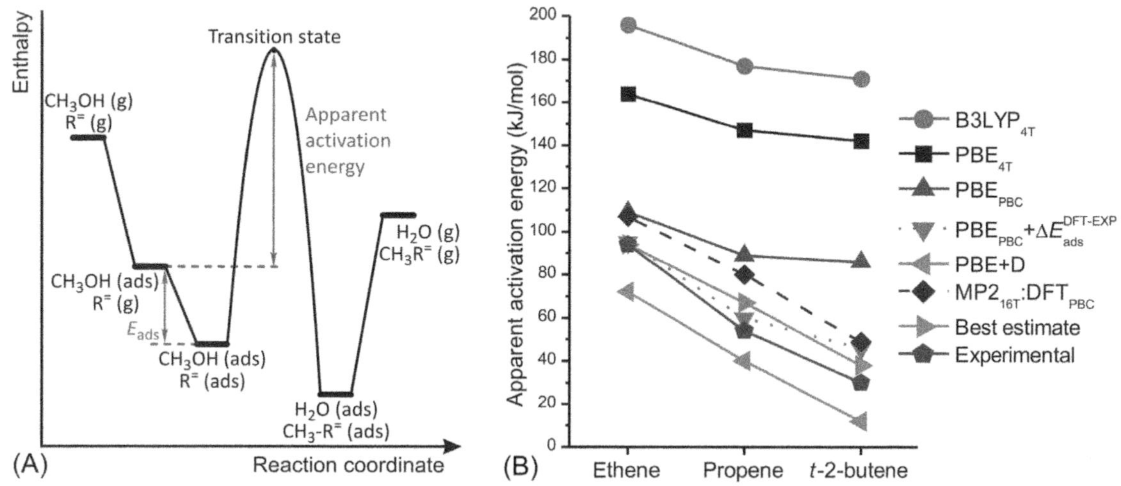

Fig. 4
Simplified reaction energy diagram for an alkene methylation with methanol over acidic zeolite (A) and the respective apparent activation barriers computed using various methods (B) [30]. *Figure from Ref. [28] with permission.*

When the reaction is modeled using a small 4 T cluster model [32] that completely neglects the electrostatic as well as van der Waals stabilization of the reactive species in the zeolite pores, the computed activation energies were strongly overestimated. More importantly, such a severe oversimplification of the zeolite active site led to a wrong prediction of the dependency of the apparent activation energy on the length of alkene chain (B3LYP$_{4T}$ and PBE$_{4T}$ in Fig. 4B). The electrostatic stabilization of polar transition state structures inside zeolite pores is correctly reproduced when the reaction is simulated within a more realistic zeolite model (periodically repeated MFI unit cell) resulting in much lower apparent activation barriers (PBE$_{PBC}$ in Fig. 4B). The barrier computed for methylation of ethane is only 15 kJ/mol higher than the experimental value. This error as well as the apparent absence of the hydrocarbon chain-length dependency of the apparent activation barriers calculated in the framework of periodic DFT calculations is mainly due to the inaccurate description of the dispersion interactions. The experimental trends are qualitatively reproduced when a local post-HF correction is applied to the DFT results using the MP2:DFT methodology. Furthermore, quantitative agreement between theory and experiment can be obtained by extrapolating the high-level correction results to the periodic and complete basis set (CBS) limits ("best estimate" in Fig. 4B). The MP2:DFT method by Sauer and coworkers [29,30] allows one to reproduce the experimental apparent enthalpy barriers with nearly chemical accuracy.

Despite that this hybrid MP2:DFT scheme gives access to very accurate adsorption and reaction energies in zeolites, it is very demanding and cannot be readily applied to investigate mechanisms of chemical reaction. It is mostly applicable as a benchmark to validate the results of more efficient computational methods. Moreover, in view of the long-range nature of van der Waals interactions, to model them correctly using this method one has to treat at the MP2 level an excessively large part of the zeolite if DFT is used for the periodic model [33].

To study catalytic reactivity of zeolites, more efficient and robust computational methodologies suitable for treating dispersion-dominated systems are needed. In recent years there was a great progress in improvement of DFT toward a better description of nonbonding interactions. The currently available dispersion-corrected DFT methods, their capabilities, and limitations have been recently reviewed by Grimme [19].

The understanding that standard DFT methods generally fail in reproducing the asymptotic r^{-6} behavior of PES for weakly interacting species have led to a proposal that their quality can be improved by including a semiclassical dispersion correction into the quantum chemical approach. This resulted in a scheme that is usually denoted as DFT-D. In this scheme, the DFT energy E(DFT) is "corrected" for dispersion interactions by adding a semiempirical term E(D) expressed as a sum over pairwise interactions represented as a truncated force-fieldlike potential:

$$E(\mathrm{D}) = -s_6 \sum \frac{C_{ij}}{r_{ij}^6} f_{\mathrm{D}}\left(r_{ij}\right) \tag{5}$$

where c_{ij} are the dispersion coefficients, the damping function $f_{\mathrm{D}}(r_{ij})$ removes contributions for short-range interactions, while the scaling parameter s_6 depends on the particular choice of the exchange-correlation functional. The DFT-D approach has been parameterized for a wide range of chemical system and can be used in a combination with a wide variety of exchange-correlation functionals. It has been implemented in many commercial quantum chemical programs. When applied to chemical processes in microporous materials, this approach has been shown to provide realistic adsorption energies for hydrocarbons in zeolites [33,34]. Although the DFT-D substantially improves the results of pure DFT (Fig. 4B), its accuracy is still lower than that of the benchmark MP2:DFT. The DFT-D apparent activation energies are systematically underestimated by ~20–30 kJ/mol (Fig. 4B), while the qualitative trend in the hydrocarbon chain dependency is predicted perfectly. Further improvement of this methodology can be achieved by further optimization of the empirical parameters in Eq. (5). The construction of hybrid schemes by applying post-HF local correction to periodic DFT-D models will not lead to the increased accuracy but rather will result in overestimation of dispersion interactions due to double-counting effect. Nevertheless, DFT-D is currently considered a "standard" methodology for studying chemical transformations in a confined space. It is a powerful, efficient, and robust computational tool for investigating different chemical processes and account for London dispersion interactions in molecular systems and solids.

Besides introducing external corrections, the performance of DFT itself can be improved via parameterization of the exchange-correlation functionals. Zhao and Truhlar have developed a family of MGGA functionals (M05 [35], M06 [36], and related functionals) showing much better accuracy in description of nonbonding interactions and prediction of reaction energetics compared to the conventional GGA and hybrid functionals.

In conclusion of this section, we would like to emphasize that one has to be very careful when applying these methodologies to specific chemical problems without additional validation. Crucial approximations such as the assumption of pairwise additivity of van der Waals interactions, the presence of empirically fitted parameters both in the force fields and in the parameterized density functionals can produce unreliable results for systems different from the training set. The development of novel ab initio and semiempirical schemes for modeling dispersion-dominated systems is currently a hot topic in theoretical chemistry. Many new methods are currently being introduced and tested for their applicability to all fields of chemistry including catalysis by microporous materials.

4 Structural Complexity of Active Sites in Zeolites

Catalytic properties of zeolite catalysts can be widely adjusted by introducing different metal cations into their micropores. However, unlike active sites in their parent hydrogen forms, in which (in a general case) the catalytic Brønsted acid sites are represented as proton compensating for the negative lattice charges due to isomorphous substitution of part of the silicon atoms in the framework by aluminum, the structural properties of active sites in transition-metal-modified zeolites are usually much less defined. Most of the synthetic catalyst preparation procedures lead to the highly heterogeneous speciation of metal inside the micropores. As a result, unambiguous identification of the true catalytic site for a particular chemical reaction and, accordingly, the reaction mechanism under the catalytically relevant conditions based solely on the experimental results is very difficult. Computational chemistry provides useful tools to unravel the structural complexity of such catalysts. We will illustrate this below on the example of high-silica zeolites modified with iron.

Iron-modified zeolites (Fe/ZSM-5) are highly efficient catalysts for a wide range of important processes than include the selective oxidation of benzene with N_2O (the Panov reaction) [37], catalytic decomposition of N_2O [38], selective catalytic reduction (SCR) of NOx [39], and many others. These unique properties stem from the presence of specific extra-framework iron-containing cationic species in the micropores of ZSM-5 zeolite [40]. Because of a very heterogeneous iron speciation in the zeolite catalyst, the direct determination of the catalytically active iron species and the mechanism of the catalytic reaction by experimental methods was not possible. The exact speciation will obviously depend on such parameters as the Fe loading, the method of iron introduction, and the history of the sample (calcination, reduction, etc.). Many studies have indicated that the reactivity of Fe/ZSM-5 for the selective benzene oxidation is mainly associated with the presence of highly dispersed Fe^{2+} extra-framework cations [41]. On the contrary, the high catalytic activity in the N_2O decomposition to N_2 and O_2 is related to the presence of more aggregated iron species in the micropores [42]. The questions on the nuclearity of iron species in high-silica zeolites, the nature of catalytic sites for a particular reaction and the role of alternative iron-containing extra-framework complexes, as well as the mechanisms of these reactions were under continuous debates.

Only recently, it was possible to shed light onto these questions by performing a comprehensive DFT study on the stability and reactivity of different iron-containing species in a realistic model of ZSM-5 zeolite [43–46]. The stability of a wide range of mono- and binuclear Fe complexes at different positions inside zeolite channels was investigated by periodic DFT calculations [43,45]. It was demonstrated that isolated exchangeable Fe^{2+} cations can be accommodated only within a very limited number of specific cation sites in ZSM-5 with a highly symmetric distribution of lattice Al (Fig. 5). It was also found that although the local zeolite environment strongly influences the stability of exchangeable Fe^{2+} sites, its effect on their intrinsic reactivity is very small. A direct correlation between the stability of Fe^{2+} sites and their oxygenated FeO^{2+} form (Fig. 5) evidences the independence of the associated reaction energy from the coordination of exchangeable Fe^{2+} cations.

The preferred location of isolated Fe^{2+} are six-membered zeolite rings with a symmetric distribution of framework Al atoms (most stable configurations in Fig. 5). Only a very limited number of cation sites in high-silica zeolite are expected to fulfill these requirements. The

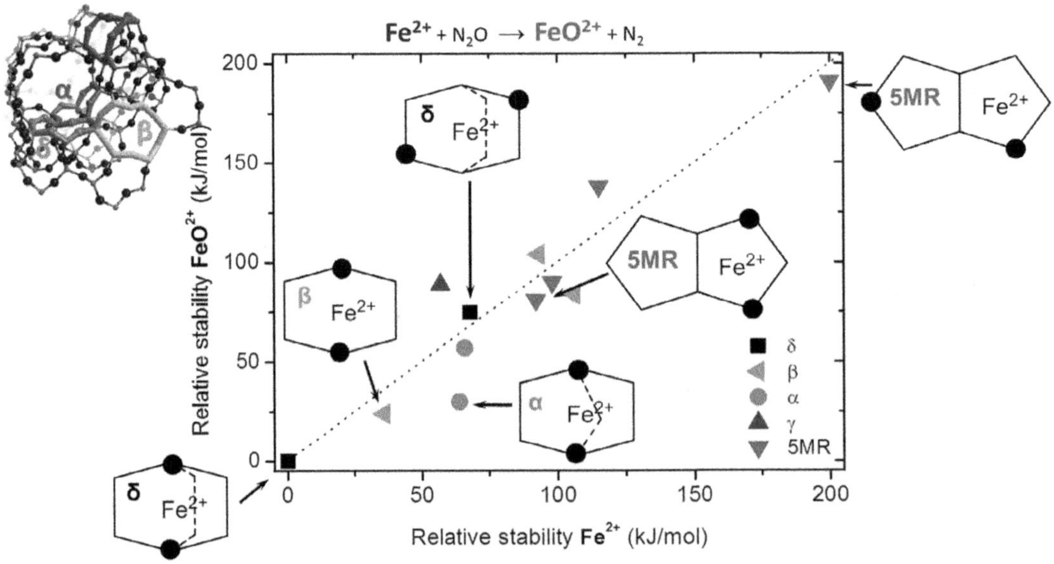

Fig. 5

The relationship between the relative stabilities of isolated exchangeable Fe^{2+} cations at different sites in ZSM-5 and that of the corresponding oxygenated FeO^{2+} species. *Figure from Ref. [43] with permission.*

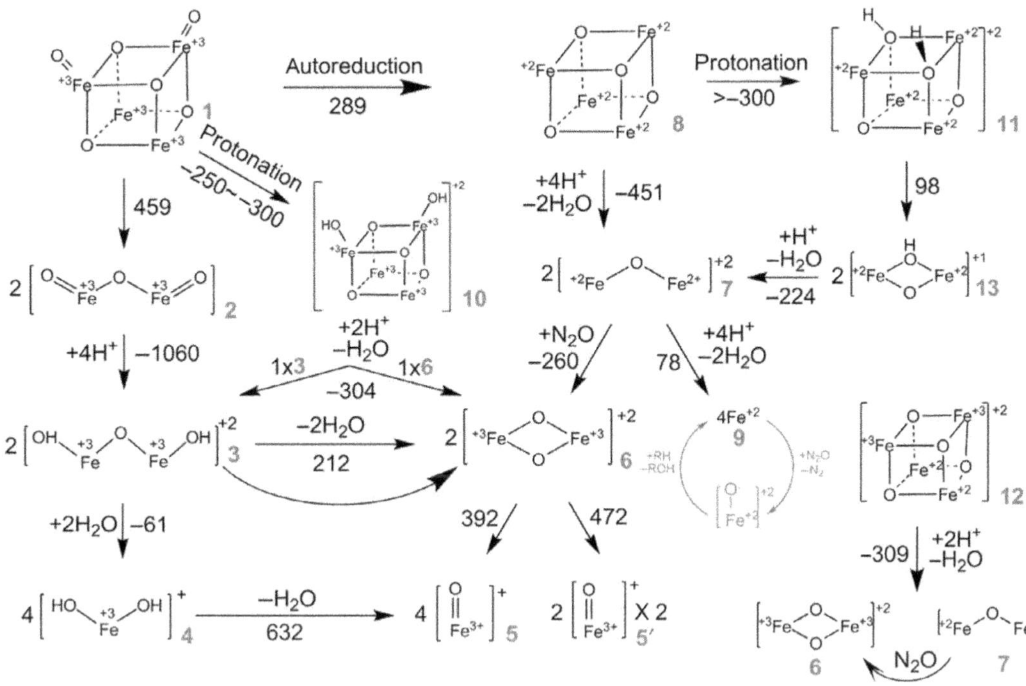

Fig. 6
Reaction paths for interconversion of iron-containing complexes confined in ZSM-5 zeolite. Reaction energies are given in kJ/mol. *Figure from Ref. [45] with permission.*

largest fraction of iron in Fe/ZSM-5 catalyst will be present as multinuclear oxygenated or hydroxylated species, the stability of which is less sensitive to the local environment. These may take such forms as isolated $[FeO]^+$ or various oxygen/hydroxy-bridged binuclear Fe^{2+} and Fe^{3+} complexes.

To establish which extra-framework complexes can be potentially formed in Fe/ZSM-5, a reaction scheme for the transformation of a hypothetical iron (III) oxide precursor in the presence of H_2O, O_2 and zeolitic protons was constructed and energetics of each reaction steps was predicted by periodic DFT calculations (Fig. 6, [45]). These results support making a number of very important conclusions. First of all, mononuclear oxygenated Fe(III) species are highly unstable and tend to self-organize into binuclear species. This phenomenon appears to be general for cation-exchanged high-silica zeolites [47]. Secondly, there are two potential paths for the conversion of the initial iron oxide precursor proceeding either via protonation over zeolitic Brønsted acid sites and ultimately leading to the formation of binuclear Fe(III) species or via autoreduction that opens a path for the formation of binuclear Fe(II) cationic complexes. However, these pure DFT results are not sufficient to conclude on the nature and structure of predominant species formed in Fe/ZSM-5.

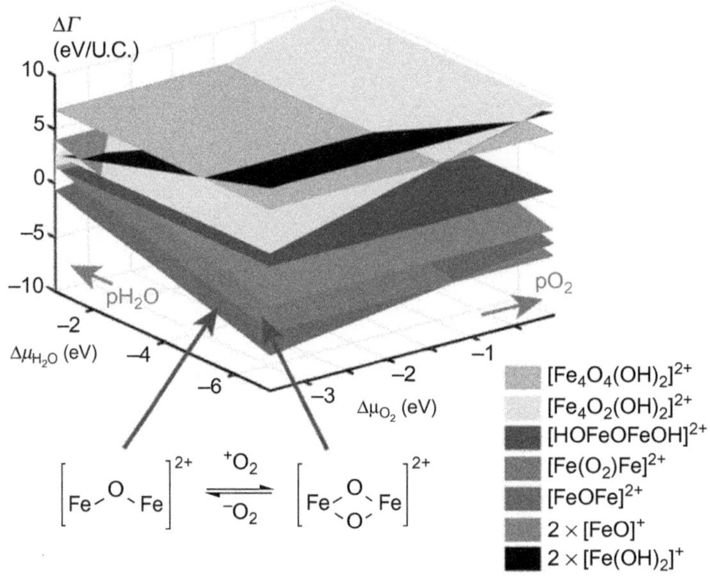

Fig. 7
Reaction free energy for the formation of various $Fe_xO_yH_z$ complexes in ZSM-5 ($\Delta\Gamma$) as a function of oxygen chemical potential ($\Delta\mu_O$) and water chemical potential ($\Delta\mu_{H_2O}$). *Figure from Ref. [45] with permission.*

To predict the chemical composition of a catalyst under the realistic conditions, one has to consider the stability of different extra-framework iron complexes at a finite temperature and also to take into account the presence of H_2O and O_2 during the catalyst preactivation. This information can be obtained via a statistical thermodynamic analysis based on the energetics predicted by DFT calculations. In the case of Fe/ZSM-5 zeolite, the following equilibrium reaction was considered to evaluate the thermodynamic stability of different species inside the zeolite channels:

$$\frac{2m-n-3x+2}{4}\mathrm{O_2}+\frac{n-2}{2}\mathrm{H_2O}+\frac{x}{2}\mathrm{Fe_2O_3}+2\mathrm{H-ZSM-5}\rightleftharpoons \mathrm{Fe}_x\mathrm{O}_m\mathrm{H}_n/\mathrm{ZSM-5} \tag{6}$$

Based on the DFT results, reaction energies for the formation of each of the potential extra-framework Fe complexes following this reaction can be calculated. The associated free energy change under particular conditions can be then estimated by correcting the DFT-computed energetics for the chemical potential of O_2 and H_2O in the gas feed:

$$\Delta\Gamma(T,P)=\frac{1}{A}\left[\Delta E-\frac{2m-n-3x+2}{4}\Delta\mu_{\mathrm{O}}-\frac{n-2}{2}\Delta\mu_{\mathrm{H_2O}}\right] \tag{7}$$

By plotting the resulting free energies of formation of different potential extra-framework complexes versus chemical potentials of O_2 and H_2O one can deduce which species are preferentially formed in a catalyst under chosen conditions (Fig. 7). It was found [45] that only two distinct extra-framework complexes are formed in Fe/ZSM-5, namely $[Fe(\mu\text{-}O)_2Fe]^{2+}$ and $[Fe(\mu\text{-}O)Fe]^{2+}$. Their relative distribution can be controlled by varying conditions of thermochemical activation of the zeolite. Whereas the former species are preferentially formed upon high-temperature calcination in O_2, the formation of the latter Fe(II)-containing complexes is favored when the catalyst is activated either in vacuum or in inert atmosphere.

5 Structure-Reactivity Relationships

Thus, DFT calculations complemented by statistical thermodynamic analysis showed that a realistic Fe/ZSM-5 catalyst should contain a small fraction of isolated Fe^{2+} species at specific positions inside the zeolite channels, while the predominant part of iron is present in the form of oxygenated cationic iron complexes. The questions about which of these different extra-framework complexes is actually responsible for the specific catalytic properties of Fe/ZSM-5 and what the role of other species were still open. They were addressed to a large extent only when the mechanisms of different catalytic reactions over different potential intrazeolite iron sites were thoroughly investigated by DFT calculations [43,46]. The influence of the nature and structural properties of Fe sites on two important catalytic processes promoted by Fe/ZSM-5, namely, the selective oxidation of benzene to phenol and the direct catalytic N_2O decomposition, was investigated [43,46].

It was demonstrated that in the case of benzene oxidation, the intrinsic reactivities of all potential extra-framework complexes for benzene activation, the initial dissociation of N_2O, and the final phenol formation were very similar. The energetics of elementary reactions steps were influenced only slightly by the nature and the structural properties of the catalytic site. Representative reaction energy diagrams for the mononuclear Fe^{2+} and Fe^{3+} sites are shown in Fig. 8.

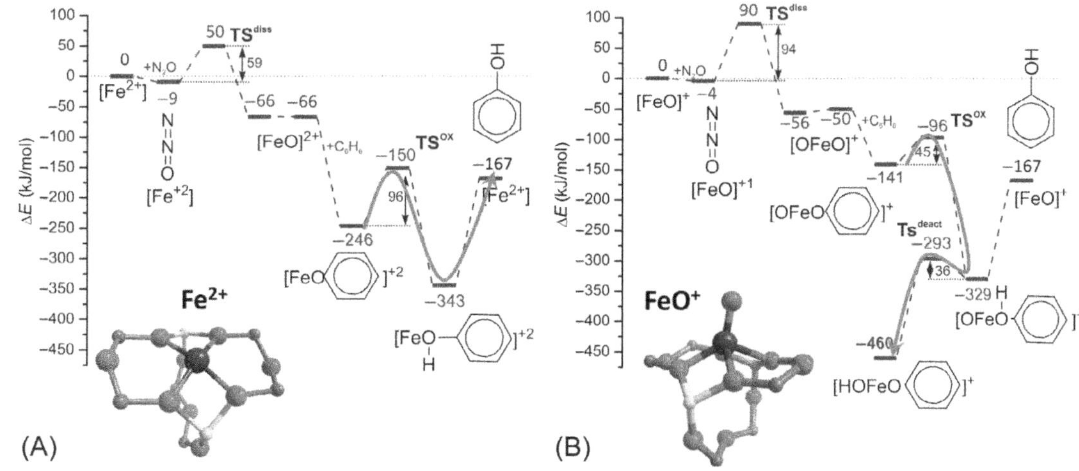

Fig. 8
Reaction energy diagrams for benzene oxidation over mononuclear (A) Fe(II) and (B) Fe(III) species in ZSM-5 zeolite *(adapted from Ref. [1])*.

The main difference between the isolated Fe^{2+} cations and all alternative oxygen-containing extra-framework complexes is observed only at the stage of the active site regeneration [43,44]. At the end of the catalytic reaction, phenol molecule has to be desorbed from the reaction site. However, in the presence of a basic extra-framework oxygen ligand an alternative reaction pathway is opened that leads to a deactivation of the active site. Such ligand acts as a proton-acceptor that facilitates the dissociation of an acidic OH group of the adsorbed phenol leading to the formation of stable grafted phenolate species (Fig. 8B). The resulting iron-phenolate complexes are inactive in the benzene oxidation reaction. This deactivation path is preferred over the path involving the phenol desorption and the regeneration of the active complex for all iron-containing sites except the isolated exchangeable Fe^{2+}. Thus, the results of DFT calculations show that the catalytic benzene to phenol oxidation can be promoted by only a very small fraction of intrazeolitic Fe sites. The remaining iron in the zeolite contributes to the long-term catalyst deactivation by generating bulky phenolate complexes inside the zeolite pores that not only decrease the accessibility of the active sites but also act as coke precursors.

An opposite situation is found for the catalytic N_2O decomposition over Fe/ZSM-5 [46]. Although mononuclear Fe sites are highly active for the dissociation of the first N_2O molecule, further catalytic reaction is hampered in this case by a very high barrier for the activation of the second N_2O molecule necessary for the formation O_2 and completion of the catalytic cycle. The binuclear $[Fe(\mu\text{-}O)_2Fe]^{2+}$ and $[Fe(\mu\text{-}O)Fe]^{2+}$ complexes are much more catalytically active. Each iron center in such complexes acts as an independent N_2O dissociation center. This allows the formation of two proximate labile extra-framework O species, which easily recombine to form molecular O_2. The reaction mechanism in this case is highly complex (Fig. 9). It involves multiple changes in the spin-state of the iron complexes as well as the interconversion between the $[Fe(\mu\text{-}O)_2Fe]^{2+}$ and $[Fe(\mu\text{-}O)Fe]^{2+}$ active complexes in the course of the catalytic process. The identification of the preferred reaction channel based only on the results of DFT calculations is not possible.

The complexity of the reaction network underlying the catalytic decomposition of N_2O over the binuclear Fe sites can be strongly reduced by translating the DFT-computed reaction energy diagrams into a MKM. In MKM, the kinetics of the reaction is described by the mean field differential equations based on the rate constants of elementary reaction steps [3]. The output of such a simulation is the overall reaction rate and the concentrations of different reaction intermediates under the catalytically relevant conditions. In agreement with the results of pure DFT calculations, the rates of O_2 formations over mononuclear sites were much lower than those predicted for the binuclear iron complexes. These findings led to conclusion that the N_2O decomposition activity of Fe/ZSM-5 is mainly due to the presence of the predominant binuclear cationic iron complexes. The MKM results were crucial for the description of the mechanism of catalytic N_2O decomposition by Fe/ZSM-5 [46]. More important is that they allowed one to rationalize previous experimental observations and suggest potential routes for stirring the catalytic reactivity of Fe/ZSM-5 for selected chemical transformations.

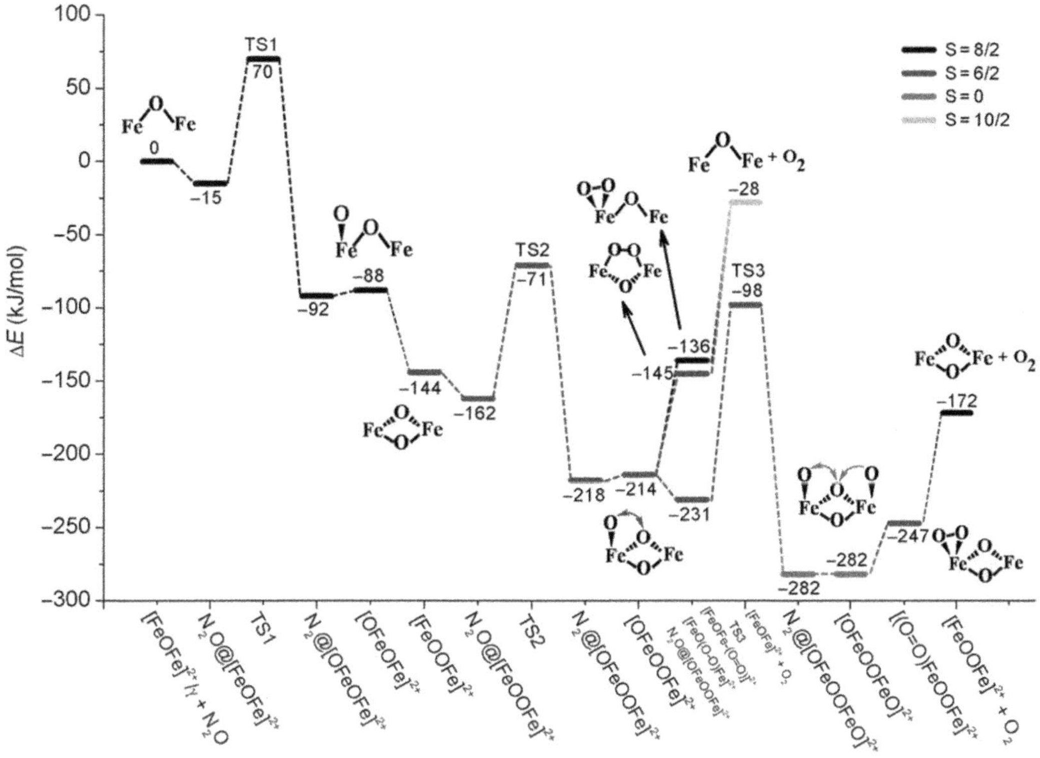

Fig. 9
Minimum energy path of N_2O decomposition over binuclear $[FeOFe]^{2+}$ active site including the spin state-transitions of the iron complexes in the course of the catalytic reaction. *Figure from Ref. [46] with permission.*

6 Conclusion

In this chapter we illustrated basic aspects and main capabilities and limitations of computational catalysis approaches. A comprehensive review of all of the available theoretical chemistry methodologies was not the aim of this review. Instead, we limited ourselves to highlighting the important principles and concepts underlying different quantum chemical methods, their capabilities and limitations as applied to chemistry of zeolites in order to orient the reader in the computational zeolite chemistry. The capabilities of the state-of-the-art computational methodologies for the realistic molecular-level description of highly complex catalytic systems are illustrated with relevant examples from zeolite chemistry. The power of computational catalysis methodologies for unraveling the structural complexity of zeolite catalysis as well as for creating well-defined molecular-level description of their reactivity is demonstrated with the example of recent comprehensive theoretical study of the structural and chemical properties of high-silica zeolites modified with iron.

Acknowledgment

This work was supported by the Netherlands Center for Multiscale Catalytic Energy Conversion (MCEC), an NWO Gravitation programme funded by the Ministry of Education, Culture and Science of the government of the Netherlands.

References

[1] van Santen RA, Pidko EA, Hensen EJM. Computational heterogeneous catalysis. In: Catlow R, Sokol A, Walsh A, editors. Computational approaches to energy materials. Weinheim: Wiley-VCH; 2013.
[2] Woodley SM, Catlow R. Nat Mater 2008;7:937.
[3] Dumesic JA, Rudd DF, Aparicio LM, Rekoske JE, Trevino AA. The microkinetics of heterogeneous catalysis. Washington, DC: American Chemical Society; 1993.
[4] Jansen APJ. Kinetic Monte Carlo. In: van Santen RA, Sautet P, editors. Computational methods in catalysis and materials science. Weinheim: Wiley-VCH Verlag GmbH & Co; 2009.
[5] Lewars EG. Computational chemistry. 2nd ed. New York: Springer, Springer Science+Business Media B.V.; 2011.
[6] Smit B, Maesen TLM. Chem Rev 2008;108:4125. Smit B, Maesen TLM. Nature 2008;451:671.
[7] Wilmer CE, Leaf M, Lee CY, Farha OK, Hauser BG, Hupp JT, et al. Nat Chem 2012;4:83.
[8] Young DC. Computational chemistry: a practical guide for applying techniques to real-world problems. New York: Wiley-Interscience; 2001.
[9] Born M, Oppenheimer JR. Ann Phys 1927;84:457.
[10] Hohenberg P, Kohn W. Phys Rev 1964;136:864.
[11] Kohn W, Sham L. Phys Rev 1965;140:1133.
[12] Perdew JP, Chevary JA, Vosko SH, Jackson KA, Pederson MR, Singh DJ, et al. Phys Rev B 1992;46:6671.
[13] Perdew JP, Burke K, Ernzerhof M. Phys Rev Lett 1996;77:3865.
[14] Becke AD. Phys Rev A 1988;38:3098.
[15] Sautet P. Quantum chemistry methods. In: Che M, Védrine JC, editors. Characterization of solid materials and heterogeneous catalysts: from structure to surface reactivity. New York: Wiley-Interscience; 2012.
[16] Becke AD. J Chem Phys 1993;98:5648.
[17] Adamo C, Barone V. J Chem Phys 1999;110:6158.
[18] Riley KE, Pitoňák M, Jurečka P, Hobza P. Chem Rev 2010;110:5023.
[19] Grimme S. WIREs Comput Mol Sci 2011;1:211.
[20] Frisch MJ, Trucks GW, Schlegel HB, Scuseria GE, Robb MA, Cheeseman JR, et al. Gaussian 09. Wallingford, CT: Gaussian, Inc.; 2009.
[21] Ahlrichs R, Bär M, Häser M, Horn H, Kölmel C. Chem Phys Lett 1989;162:165.
[22] Valiev M, Bylaska EJ, Govind N, Kowalski K, Straatsma TP, van Dam HJJ, et al. Comput Phys Commun 2010;181:1477.
[23] Segall MD, Lindan PJD, Probert MJ, Pickard CJ, Hasnip PJ, Clark SJ, et al. J Phys Condens Matter 2002;14:2717.
[24] Marx D, Hutter J. In: Modern methods and algorithms of quantum chemistry. Jülich: NIC, FZ Jülich; 2000. p. 301. Andreoni W, Curioni A. Parallel Comput 2000;26:819.
[25] Kresse G, Hafner J. Phys Rev B 1994;49:14251. Kresse G, Furthmüller J. Comput Mater Sci 1996;6:15. Kresse G, Furthmüller J. Phys Rev B 1996;54:11169.
[26] Dovesi R, Saunders VR, Roetti C, Orlando R, Zicovich-Wilson CM, Pascale F, et al. CRYSTAL2006 user's manual. Torino: Universita di Torino; 2006. http://www.crystal.unito.it/.
[27] Tosoni S, Tuma C, Sauer J, Civalleri B, Ugliengo P. J Chem Phys 2007;127:154102.
[28] Pidko EA, van Santen RA. Theoretical chemistry of zeolite reactivity. In: Cejka J, Corma A, Zones S, editors. Zeolites and catalysis: synthesis, reactions and applications. Weinheim: Wiley-VCH; 2010.
[29] Tuma C, Sauer J. Chem Phys Lett 2004;387:388. Tuma C, Sauer J. Phys Chem Chem Phys 2006;8:3955.
[30] Svelle S, Tuma C, Rozanska X, Kerber T, Sauer J. J Am Chem Soc 2009;131:816.

[31] Svelle S, Rønning PO, Kolboe S. J Catal 2004;224:115. Svelle S, Rønning PO, Olsbye U, Kolboe S. J Catal 2005;234:385.
[32] Svelle S, Arstad B, Kolboe S, Swang O. J Phys Chem B 2003;107:9281.
[33] Göltl F, Grüneis A, Bučko T, Hafner J. J Chem Phys 2012;137:114111.
[34] Kerber T, Sierka M, Sauer J. J Comput Chem 2008;29:2088.
[35] Zhao Y, Schultz NE, Truhlar DG. J Chem Phys 2005;123:161103.
[36] Zhao Y, Truhlar DG. Theor Chem Acc 2008;120:215.
[37] Panov GI, Sobolev VI, Kharitonov AS. J Mol Catal 1990;61:85. Panov GI, Sheveleva GA, Kharitonov AS, Romannikov VN, Vostrikova LA. Appl Catal A 1992;82:31. Sobolev VI, Kharitonov AS, Paukshtis YA, Panov GI. J Mol Catal 1993;84:117. Dubkov KA, Ovanesyan NS, Shteinman AA, Starokon EV, Panov GI. J Catal 2002;207:341.
[38] Sobolev VI, Panov GI, Kharitonov AS, Romannikov VN, Volodin AM, Ione KG. J Catal 1993;139:435. Perez-Ramirez J, Kapteijn F, Mul G, Moulijn JA. Appl Catal B 2002;35:227. El-Malki EM, van Santen RA, Sachtler WMH. Microporous Mesoporous Mater 2000;35:235.
[39] Sun Q, Gao ZX, Chen HY, Sachtler WMH. J Catal 2001;201:89. Wichterlova B. Top Catal 2004;28:131.
[40] Panov GI. CATTECH 2000;4:18. Ribera A, Arends I, de Vries S, Perez-Ramirez J, Sheldon RA. J Catal 2000;195:287.
[41] Sun K, Xia H, Feng Z, van Santen R, Hensen E, Li C. J Catal 2008;254:383. Chernyavsky VS, Pirutko LV, Uriarte AK, Kharitonov AS, Panov GI. J Catal 2007;245:466.Pérez-Ramírez J, Santhosh Kumar M, Brückner A. J Catal 2004;223:13.
[42] El-Malki EM, van Santen RA, Sachtler WMH. J Catal 2000;196:212. Chen H-Y, Sachtler WMH. Catal Today 1998;42:73.
[43] Li G, Pidko EA, van Santen RA, Feng Z, Li C, Hensen EJM. J Catal 2011;284:194.
[44] Fellah MF, Pidko EA, van Santen RA, Onal I. J Phys Chem C 2011;115:9668.
[45] Li G, Pidko EA, van Santen RA, Li C, Hensen EJM. J Phys Chem C 2013;117:413.
[46] Li G, Pidko EA, Filot IAW, van Santen RA, Li C, Hensen EJM. J Catal 2013;308:386.
[47] Pidko EA, Hensen EJM, van Santen RA. Proc R Soc A 2012;468:2070.

CHAPTER 4

Basics of Solid-State NMR for Application in Zeolite Science: Material and Reaction Characterization

A.G. Stepanov
Boreskov Institute of Catalysis SB RAS, Novosibirsk, Russia

1 Introduction

Intense fundamental studies of zeolite and other zeotype materials as well as their broad practical applications require detailed characterization of the chemical composition, structure, and morphology, active sites, presence of functional groups, porosity, ability to sorb and retain the molecules and chemically convert them on these materials, and so on. This important information can be provided by using the number of physico-chemical methods and spectroscopic techniques, such as X-ray diffraction analysis (XRD), electron microscopy, infrared (IR), nuclear magnetic resonance (NMR) spectroscopy, and the like. Among numerous spectroscopic methods, solid-state NMR spectroscopy could be considered as one of the most powerful and routine techniques allowing characterization of both the material itself and adsorbate species being involved in the chemical reaction proceeding on zeolites as catalysts.

In the 1960s to 1970s a conventional NMR technique was straightforwardly applied to solids. The spectra were very broad, the resonance lines were featureless, making the structural characterization obscured due to impossibility to distinguish the chemical shifts of the nuclei in different local environments. A new impetus for application of NMR spectroscopy to study the solid materials was received at the end of 1970s when commercial solid-state NMR spectrometers appeared with superconducting magnets, sophisticated electronics, and NMR probes allowing the line narrowing by the sample rotation with a speed of a few kHz at the angle of 54.7° between the direction of magnetic field B_0 and the direction of the rotation axis. The situation was referred to as "magic angle spinning" (MAS) and the angle of 54.7° was called the "magic angle." A procedure of registration of NMR spectra of solids with rotation under the angle of 54.7° was termed MAS NMR.

Zeolites and Zeolite-like Materials. http://dx.doi.org/10.1016/B978-0-444-63506-8.00004-5

The development of the novel MAS NMR technique opened the possibility to measure NMR spectra of solids comparable in spectral resolution to those of liquids. Subsequently, MAS NMR spectroscopy has developed rapidly into an effective and widely used tool of elucidating the structural properties of solid materials. The first applications of MAS NMR to zeolites were reported in 1981 and related to alumosilicate framework characterization of various zeolites in hydrated states with ^{29}Si MAS NMR [1,2] and ^{27}Al MAS NMR [3]. Further, the researchers have learned to record MAS NMR spectra of anhydrous zeolites. This made it possible to identify surface hydroxyl groups in zeolites [4,5], characterize Brønstead acidity [6], porosity, and adsorption sites with adsorbed probe molecules [7–13]. Later on, the adsorbed reaction products were characterized with ^{13}C MAS NMR [14], which provided a stimulus of application of MAS NMR to characterize the chemical reaction occurrence on zeolites [15] with the aim of unraveling the mechanisms of the reactions. Note, however, that researchers sometimes do not have to narrow the NMR signals with MAS to characterize the solids. For example, analysis of the line shape evolution with temperature for ^{2}H wide-line NMR spectra of adsorbates in zeolites affords a valuable information on the adsorbate molecular dynamics [16,17].

There are excellent reviews and a book on zeolite characterization with solid-state MAS NMR [18–21] that go in scope and depth far beyond the current presentation. The aim of this tutorial is to offer some benchmarks for students and young researchers who are trying applications of the solid-state NMR in zeolite science.

Here we focus on the outline of the phenomenon of NMR and basic approaches of both the use of solid-state NMR to derive the information on zeolite characteristics, dynamics of adsorbates, and reaction occurrence, and the performance of the solid-state NMR experiment itself to derive any particular useful for researcher information.

2 Phenomenon of NMR

The phenomenon of NMR is based on the property of nucleus to possess intrinsic angular momentum. It is often called spin angular momentum or simply "spin." Spin is a form of the angular momentum that is not due to the rotation of the particle, but is an intrinsic property of the particle itself. The concept of spin is difficult to visualize, because it is a quantum mechanical property with no analogue in the macroscopic world.

Spin angular momentum is usually represented by a vector **S**. Quantum mechanics tells us that all forms of the angular momentum and spin S as well are quantized, $\vec{S} = I \cdot \hbar$. The z-component of the spin angular momentum vector I_z takes values: $I_z = I, I-1, \ldots, -I$. Spin angular moments give rise to magnetic moments. The magnetic moment $\boldsymbol{\mu}$ is directly proportional to the spin $\vec{\mu} = \gamma \cdot \vec{S}$, where γ is the magnetogyric ratio measured in rad s^{-1}T^{-1}, which is a property of the nucleus.

As spin I is quantized, it produces $2I+1$ discrete values of the angular momentum m_I and corresponds to $2I+1$ levels in the magnetic field B_0. The energy of a nuclear spin in a magnetic field is determined by the so-called Zeeman interaction. It is a vector product of the magnetic moment and the magnetic field $E = -\vec{\mu} \cdot \vec{B}_0$. The energy of the nuclear spin in the magnetic field depends on γ and the magnitude of the B_0 field. For $S = 1/2$ there are two energy levels corresponding to two orientations given by $m_I = \pm 1/2$ (Fig. 1). If we apply a weakly oscillating magnetic field $h\nu$ to the system with the energy of this field being equal to the distance between two energy levels in the magnetic field, then we will observe the signal from the resonating nucleus, as Bloch and Purcell observed such a signal in 1945 (phenomenon of NMR) [22,23].

In the modern NMR spectrometers, the spectra are registered by the following manner. All the nuclei are excited by a short radio frequency pulse (rf pulse) of a few microseconds in length and then a response to this excitation or free induction decay (FID) is registered. Further Fourier transformation (FT) of the FID gives rise to the NMR spectrum, which represents a dependence of the intensity of the signals with different surrounding on the frequency.

NMR is inherently an insensitive method, as compared to the other methods, such as UV spectroscopy or even IR spectroscopy. It does not detect directly the transitions between the

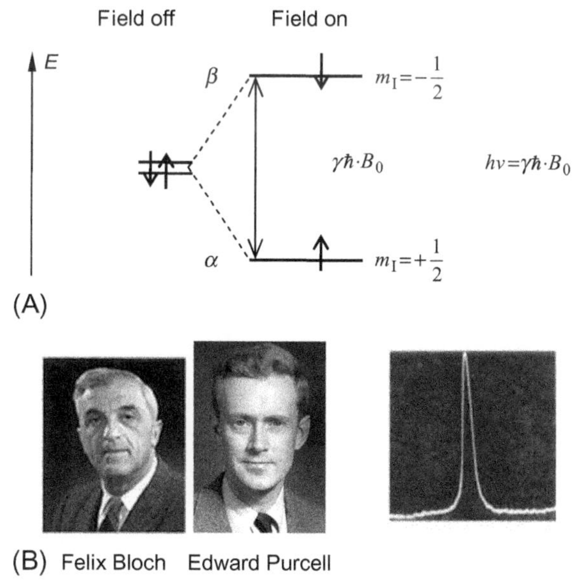

Fig. 1
(A) Energy levels in magnetic field B_0 for spin $S = 1/2$; (B) discoverers of NMR phenomenon and proton NMR signal of paraffin wax observed by Ed Purcell in 1945.

energy levels, it detects the total magnetic moment or magnetization M_0 is proportional to the population difference of spins with different energies $n_\alpha - n_\beta = \Delta n_0$. The difference in population is defined by the Boltzmann equation $n_\beta / n_\alpha = e^{-(E_\beta - E\alpha)/kT}$. As far as the energy difference between two energy levels is extremely low $E_\beta - E_\alpha = \gamma \hbar B_0 \approx 10^{-3}$ cm^{-1} compared to the thermal energy $kT \approx 200\text{cm}^{-1}$, then $n_\alpha - n_\beta = \Delta n_0 = \frac{N\gamma\hbar B_0}{2kT}$ and $M_0 = \frac{1}{2}\gamma\hbar\Delta n_0$. The difference in the population for protons is about $\Delta n / N \approx 1\times 10^{-5}$. This is the reason for poor sensitivity of the NMR method. A bulk magnetic moment or magnetization is proportional to the square of the magnetogyric ratio, the strength of external magnetic field B_0, and a total amount of nuclear spins in the sample:

$$M_0 = \frac{N\gamma^2\hbar^2 B_0}{4kT}$$

The proton has a large magnetogyric ratio γ, so the Zeeman energy splitting is large. This results in a large population excess in the lower Zeeman level and a large bulk magnetization M_0 (Fig. 2). Therefore, the sensitivity of proton NMR is good compared to other nuclei. The carbon-13 nucleus has a smaller magnetogyric ratio, so the Zeeman energy splitting is smaller. This results in a smaller bulk magnetization, and therefore the sensitivity of ^{13}C NMR is poor.

NMR spectroscopy of different nuclei is used for solid material characterization. These nuclei, in dependence on their magnetogyric ratio and the natural content of the magnetic isotope, can be considered as easily detected in NMR, or not easily detectable. The following nuclei could be considered as easily detectable: ^{1}H, ^{7}Li, ^{11}B, ^{14}N, ^{19}F, ^{23}Na, ^{27}Al, ^{29}Si, ^{31}P, ^{51}V, ^{129}Xe, ^{133}Cs, ^{195}Pt. The detection of such nuclei as ^{2}H, ^{13}C, ^{15}N, ^{17}O, ^{65}Cu, ^{67}Zn, ^{69}Ga, ^{71}Ga, ^{95}Mo, ^{119}Sn, ^{183}W is not easily to be performed. Of course, protons, with the highest value of the magnetogyric ratio, exhibit the highest sensitivity.

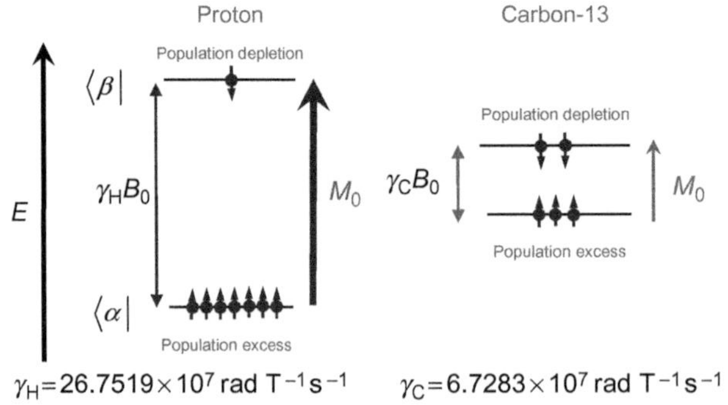

Fig. 2
Scheme demonstrating the difference in energy splitting and populations excess for protons and ^{13}C nuclei, providing the difference in the sensitivity of these two nuclei.

3 Basic Types of Nuclear Interactions in Solids

The peculiarity of the interactions among nuclei in the solids is that they are anisotropic; that is, they are orientationally dependent due to the ordered position (with fixed orientations) of the molecules in the solid materials. This results in essential broadening of NMR signals of solids compared to the NMR signals in liquids, where fast molecular motion results in the signal narrowing due to this motion. Three main interactions that may contribute to line broadening in the solid-state NMR spectra are the following: (1) direct dipole-dipole interactions; (2) magnetic shielding by the surrounding electrons (chemical shift anisotropy); (3) quadrupolar interactions. In the particular solid-state systems, one or two of the contributions usually dominate and hence determine the spectrum.

Dipole-dipole interaction is the interaction through a space between the observed nucleus and the neighboring nuclei. It depends on the magnitude of the magnetic moments of the interacting nuclei, sharply decreases with the distance between the nuclei ($1/r^3$), and is independent of the applied magnetic field B_0. Dipole-dipole interaction with protons, such as ^{1}H–(^{1}H), ^{13}C–(^{1}H), ^{29}Si–(^{1}H) is the dominant line-broadening interaction in ^{1}H, ^{13}C, and ^{29}Si NMR spectra in proton-containing solids. The value of this interaction reaches 20–100 kHz.

The *chemical shift anisotropy* (*CSA*) is due to the spatial dependence of the nuclear shielding, which is determined by the symmetry of the electron charge distribution around the resonance nucleus. The interaction is linear with the applied magnetic field and will be proportionately larger and more important at higher magnetic field strengths. Because it involves the surrounding electrons, it is the most sensitive interaction to the geometry and identity of the other atoms surrounding a particular nucleus. Typical line broadening from CSA is about 100 ppm but may be considerably larger in special cases.

The *quadrupole interaction* occurs only when the nuclear spin number $I > 1/2$ (eg, ^{2}H, ^{11}B, ^{17}O, ^{23}Na, and ^{27}Al) and arises from the interaction of the nuclear electric quadrupole moment eQ with the electric field gradient eq produced at the nuclear site by a nonspherical charge distribution around the nucleus. The quadrupole interaction is usually characterized by the quadrupole constant e^2qQ/h and asymmetry parameter η_Q. The magnitude of the quadrupole interaction is such (up to several MHz) that it completely dominates the spectrum for most quadrupolar nuclei, even in the case of ^{2}H NMR, which has the smallest quadrupole moment.

Although these aforementioned interactions obscure the important "high-resolution" characteristics of solid-state NMR spectra of powders such as chemical shift, these interactions could offer the information on the structure and bonding. Dipole-dipole interaction can yield information on internuclear distances and molecular geometries. CSA gives information on the coordination symmetry. Symmetry of the charge distribution around the nucleus may be derived from analysis of quadrupole interactions. Also, due to the sensitivity of the line shape of quadrupole nuclei (eg, ^{2}H NMR line shape) to the molecular motions, analysis of the ^2H NMR line shape gives information on the peculiarities of local intramolecular mobility in solids.

4 Main Procedures for Line Narrowing

Analysis of the structure of zeolites or organic adsorbates in zeolite pores is usually performed on the basis of the analysis of the chemical shifts of the structurally different nuclei constituting the analyzed material or adsorbate. As we just noted, the main type of interactions existing in solid materials make their NMR signals too broad, which makes determination of chemical shifts obscure. We also noted earlier that NMR is inherently a low sensitivity method. So we need to take some approaches to make broad NMR signals narrow to allow chemical shift identification and to enhance the sensitivity of this method. There are three main experimental techniques to achieve the line narrowing in solids:

Dipolar decoupling removes the heteronuclear interaction by irradiating the resonance frequency of the nucleus that produces the dipolar broadening (usually ^1H) while observing the nucleus under study (eg, ^{13}C, ^{29}Si).

Multiple pulse sequences are used to remove homonuclear ^1H–^1H dipolar interactions in ^1H NMR spectra by irradiation of precisely defined sequences of short, intense radio-frequency pulses, which average the dipolar interactions by reorienting the nuclear spins. Note, however, that the use of special pulse sequences to make NMR signals narrow is not widely applied due to the complexity of their application and sometime poor efficiency.

Magic angle spinning is a widely and routinely applied technique for line narrowing. This experimental technique was introduced by Andrew et al. in 1958 [24]. The idea of the experiment is the following. If some solid organic or inorganic material as a powder in a special container (rotor) is placed in the magnetic field and then one starts rotating this container with a high speed ν_{rot} (a few kHz) in the manner that the axis of rotation is inclined by the angle $\theta_m = 54.7°$ with respect to the direction of the magnetic field B_0 (Fig. 3), then the dipole-dipole interactions and CSA, containing the angle dependence of the form $(3\cos^2\theta - 1)$, are averaged to zero. In this case, the NMR spectrum of a solid powdered material resembles the spectrum of a liquid. The angle of 54.7° was designated as the magic angle. The procedure of registration of NMR spectra of solid materials with rotation under the angle of 54.7° was termed MAS NMR. MAS is the most important technique of high-resolution solid-state NMR spectroscopy for investigations of the structure of zeolites and analysis of adsorbates at chemical reaction studies on zeolites. Nowadays, the rapid spinning of samples is performed in special MAS NMR probes, provided with dry air (or nitrogen) bearing and driving systems, which allow the rotation of 2.5–7.0 mm in o.d. and 1 cm length zirconia rotors with the spinning rates of 5–40 kHz.

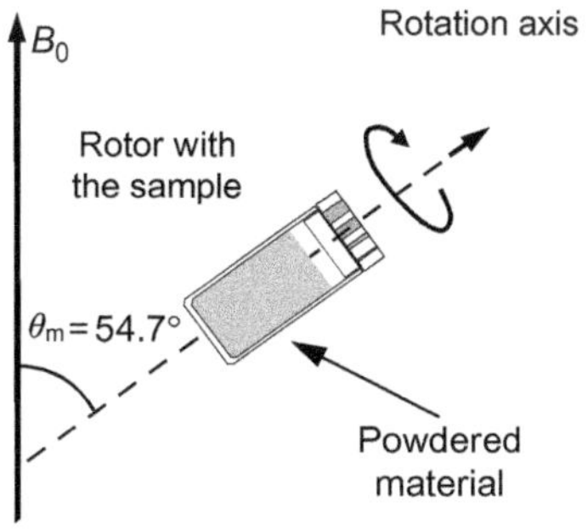

Fig. 3
Scheme of the MAS technique based on the rapid rotation of the sample around an axis tilted by the magic angle of $\theta_m = 54.7°$ with respect to the external magnetic field B_0.

4.1 Pulse Sequences Frequently Used in Solid-State NMR Experiments

Usually a standard procedure for registration of MAS NMR spectra is used, such as by all the nuclei excitation with a short radio frequency pulse (rf pulse) of a few microseconds length and then a response to this excitation or FID is registered (Fig. 4). The spectra recorded in this manner may exhibit undesired false signals that interfere with authentic desirable signals.

MAS makes the signals narrow but removes dipolar interactions in solid materials. However, the information about this interaction is sometimes needed, for example, when we want to determine the structure of solid materials. So a recoupling of dipolar interactions is required. Recoupling of dipolar interaction, which is removed by MAS, is a complicated procedure, but in principle it is possible to do this recoupling. Also, as we noted earlier, NMR spectroscopy, especially in the case of nonabundant nuclei such as ^{13}C and ^{15}N isotopes is the low sensitivity method. To overcome the problems of the solid-state NMR method limitation with respect to the sensitivity, observation of false signals, vanished dipolar interaction under MAS, and so on, the special type experiments are performed with the use of specific pulse sequences for spectra excitation.

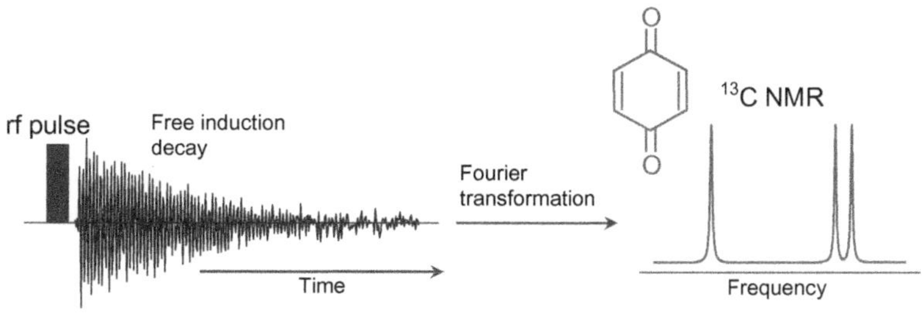

Fig. 4
Scheme of registration of NMR spectrum in modern spectrometers.

4.1.1 Hahn-echo sequence

^{1}H MAS spectra are usually registered with a rotor-synchronized Hahn-echo pulse sequence. The principle of Hahn-echo pulse sequence is demonstrated in Fig. 5A. With this sequence, the signal of echo is registered after evolution and refocusing periods at the time 2τ equal to an even number n of rotation periods τ_R ($\tau = n \cdot \tau_R$). This manner of the spectrum recording is used often to remove the background signal from residual protons.

4.1.2 REDOR

The procedure that is useful for intermolecular distance determination in the solid materials is termed REDOR (rotational echo double resonance) experiment [25,26]. It allows recovering dipolar interaction under MAS. REDOR is a spin-echo double-resonance experiment. A typical REDOR pulse sequence is shown in Fig. 5B. The rotor-synchronized echo sequence is applied to the spin I system and the echoes are detected after the time 2τ equal the even number n of rotation periods τ_R. For recoupling the dipolar interaction between the spins I and S, π-pulses are applied to the spin S system at every half rotation period τ_R. The dipolar coupling is determined by measuring the REDOR fraction $\Delta S/S_0$, which describes the decrease of the echo amplitude S as a function of the number of rotation periods n [25]:

$$\Delta S / S_0 = 1 - \frac{1}{2\pi} \int_0^{\pi} \int_0^{\pi} \cos\left[\frac{n\tau_R d}{\pi} \sqrt{2} \sin 2\beta \sin\alpha \right] \sin\beta \, d\alpha \, d\beta$$

Here S_0 and S describe the signals without recoupling π-pulse and with recoupling pulse, respectively, $\Delta S = S_0 - S$, τ_R is the dephasing period, n is the number of rotor periods, $d = \gamma_I \gamma_S / R_{I-S}^3$ is the dipolar coupling, R_{I-S} is the distance between I and S atoms, α and β are polar angles that describe the orientation of the rotor axis with respect to the I–S vector.

An example of how the REDOR technique was used was for determining the O–H distance of bridging hydroxyl groups in zeolites H–Y and H–ZSM-5 in high magnetic fields [27] and the P–B distances in SAPO-based mesoporous materials [28].

4.1.3 TRAPDOR

The next pulse sequence is termed TRAPDOR (transfer of polarization in double resonance) (Fig. 5C). It is used to study the dipolar coupling between spin I and spin S, involved in strong quadrupolar interactions (usually Al). In 1990 van Eck [29] has found that if at registration of echo signals of proton, arising from the pulse sequence consisting of $\pi/2$ and π-pulses, the system is additionally irradiated during the evolution period or at the evolution period and refocusing period at the frequency corresponding to the resonance of aluminum, then the dipolar interaction between the protons and aluminum-27, removed by MAS, is recoupled. This results in a decrease of the intensities of the signals from protons located in the vicinity of aluminum atoms. TRAPDOR is a qualitative experiment [29,30], because the decrease of

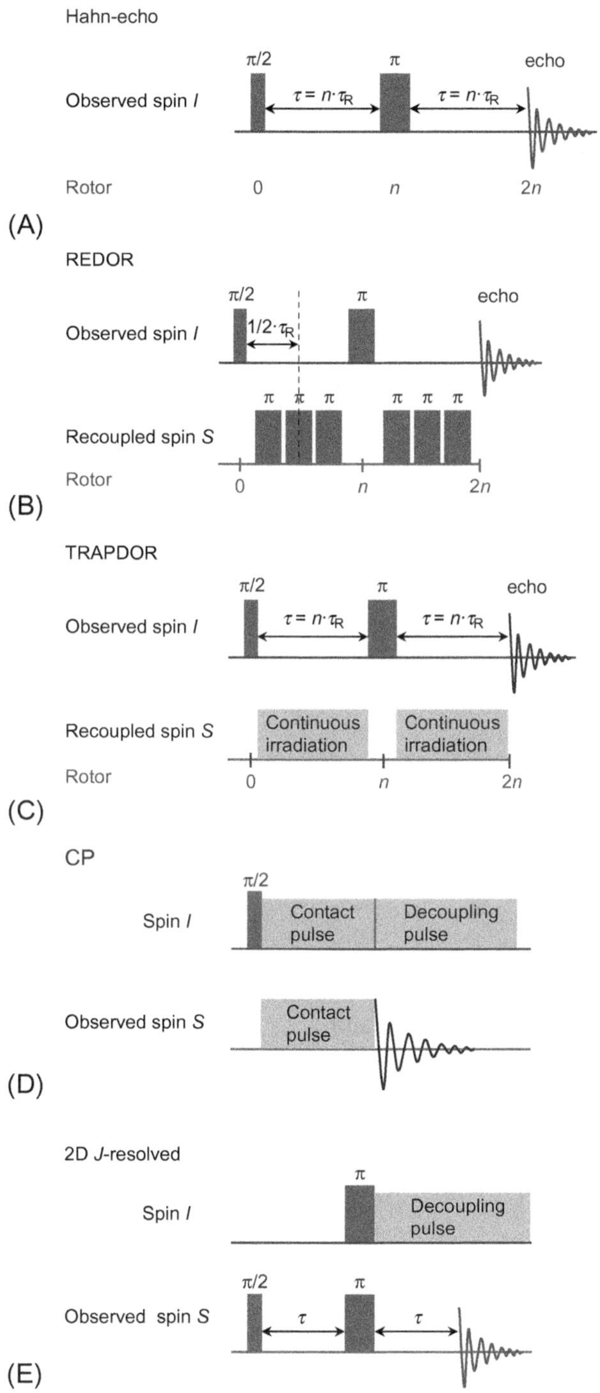

Fig. 5
Pulse sequences for Hahn-echo (A), REDOR (B), TRAPDOR (C), CP (D), 2D *J*-resolved (E) spectroscopy experiments.

the echo intensity is difficult to calculate. The comparison of spectra obtained by spin-echoes recorded with and without irradiating the spin *S* system allows distinguishing spins *I* with and without dipolar coupling to spins *S*.

In zeolite science, the TRAPDOR sequence is usually used to characterize the Brønsted acidity in zeolites, or more exactly it can be used to identify which OH groups of zeolite are attached to aluminum atoms and which OH groups are attached to silicon atoms. With TRAPDOR, the assignment of the ^{1}H MAS NMR signals of AlOH and bridging OH groups (SiOHAl) in dehydrated zeolite H-beta [31,32], HY and H–ZSM-5 [33] (*vide infra* Fig. 13), and of silanol groups in the vicinity of framework aluminum atoms in dehydrated SBA-15 [34] was performed.

4.1.4 Cross-polarization

To enhance the sensitivity of the NMR method the materials under study should be enriched with a magnetic isotope (eg, ^{13}C, ^{2}H, and ^{15}N) or a special procedure called cross-polarization (CP) is very often used for registration of ^{13}C MAS NMR spectra of hydrocarbons on the catalyst surfaces [35–37].

CP is an effective technique to significantly increase the sensitivity of nonabundant spins *S* dipolarly coupled with abundant spins *I*. The principle of the CP experiment is demonstrated in Fig. 5D. In this technique, an indirect excitation of dilute spins *S* (which are observed) by polarization transfer from abundant spins *I* (mostly ^{1}H) is employed. The CP experiment consists of three steps: (1) excitation of the *I* spins by $\pi/2$ pulse, (2) polarization transfer from *I* to *S* spins by simultaneous irradiation of the rf fields B_{1I} and B_{1S} of the *I* and *S* spins according to the Hartmann-Hahn condition [38] $\gamma_I B_{1I} = \gamma_S B_{1S}$ (where γ_I and γ_S are the magnetogyric ratios of the nuclei *I* and *S*), and (3) acquisition of the FID of the *S* spins with continued irradiation of the ^{1}H field for heteronuclear dipolar decoupling. The gain in sensitivity is given by the ratio γ_I/γ_S and amounts to a maximum factor of about four for ^{1}H–^{13}C and about five for ^{1}H–^{29}Si CP experiments. In addition, shorter pulse repetition times can be applied in CP experiments than in measurements with direct excitation of the *S* spins, because the optimum pulse delay in CP depends on the ^{1}H relaxation time $T_I(^{1}H)$, which is usually shorter than T_1 of the spins ^{13}C or ^{29}Si. The CP experiment may thus be repeated with much shorter intervals, giving a further increase in the signal-to-noise ratio of the *S* nucleus spectrum in a given period of time. The CP efficiency depends on the strength of the *I–S* dipolar interaction, that is, on the distance between the *I* and *S* nuclei. The CP technique may thus also be used to detect close coordinations between protons and Si or ^{13}C atoms, for example.

4.1.5 Two-dimensional J-resolved ^{13}C MAS spectroscopy

Fig. 5E shows a pulse sequence that can be used for detecting multiplicity of ^{13}C signals arising from scalar carbon-hydrogen interaction for hydrocarbons adsorbed in zeolites. This sequence was applied for the first time by Anderson and Klinowski in 1990 [39] to analyze with ^{13}C MAS NMR the products of methanol-to-gasoline (MTG) conversion process on zeolite ZSM-5. For registration of ^{13}C spectra they used the echo sequence consisting of subsequent $\pi/2$ and π pulses. If they apply simultaneously a π pulse in the ^{1}H channel with π pulse in the ^{13}C channel, then the ^{13}C signals of adsorbed alkanes become modulated by scalar J-coupling with attached protons as it occurs in the spectra of organic molecules in solutions [40]. The intensity of ^{13}C carbon signals from different CH_n groups changes the evolution time τ depending on the number of protons attached to the carbon atoms [40]. Thus they were able to record 2D J-resolved spectra for hydrocarbons in the adsorbed state [39]. In 2D J-resolved spectrum, the chemical shift of the ^{13}C signals is detected in the F2 direction and multiplicity of the signals arising from scalar J-coupling is registered in the F1 direction (*vide infra*, Fig. 15). This technique allows the researcher to perform the analysis of adsorbed hydrocarbons more reliably on the basis of both the ^{13}C chemical shifts and multiplicities of signals arising from J_{13C-1H} coupling.

4.2 Requirement for the Samples to Perform Zeolite and Reaction Characterization. Arrangement of NMR Experiment

For studying zeolites with MAS NMR, a powder of a zeolite sample (40–100 mg) is usually placed in 4 or 7 mm ZrO_2 rotor (Fig. 6) and is closed with a Kel-F cap. MAS with the rate of 5–15 kHz makes the signals from the powdered material narrow and analysis is performed on the basis of specific chemical shifts. This approach is suitable for the study of hydrated or moisture insensitive materials with ^{27}Al and ^{29}Si MAS NMR. The MAS NMR study of surface sites (eg, hydroxyl groups) or adsorbates requires the dehydrated zeolites. Dehydration of zeolite samples is performed at high temperature (400–500°C) under a vacuum and further dehydrated material is transferred into the gas-tight MAS NMR rotor using a glove box purged with an inert and dry gas. Another way of recording dehydrated zeolite samples is to seal the dehydrated zeolite powder in small glass ampoules [41,42]. Dehydration of a zeolite sample occurs in a simple glass device, which could be connected to the vacuum line (Fig. 6). After dehydration and loading the zeolite sample with reactants at a vacuum line, the glass insert is sealed at the waist (Fig. 6A). To prevent heating the sample, the glass ampoule is placed in liquid nitrogen while sealing with the flame. The sealed glass ampoule is then fit precisely into the MAS NMR rotor (Fig. 6B and C). The sealed glass ampoules containing the zeolite catalyst and the organic adsorbate represent a model of a batch reactor. It can be used for in situ studies of catalytic reactions at variable temperatures under controlled atmospheres [43–46]. The increase of the temperature of the sealed ampoule in the rotor located in the NMR probe, placed preliminarily in the magnet of NMR spectrometer, initiates the chemical reaction and allows monitoring the reaction proceeding with NMR. The NMR spectra are measured in situ while the reaction is in progress at a suitably high temperature (currently up to 300°C). This allows analysis of the

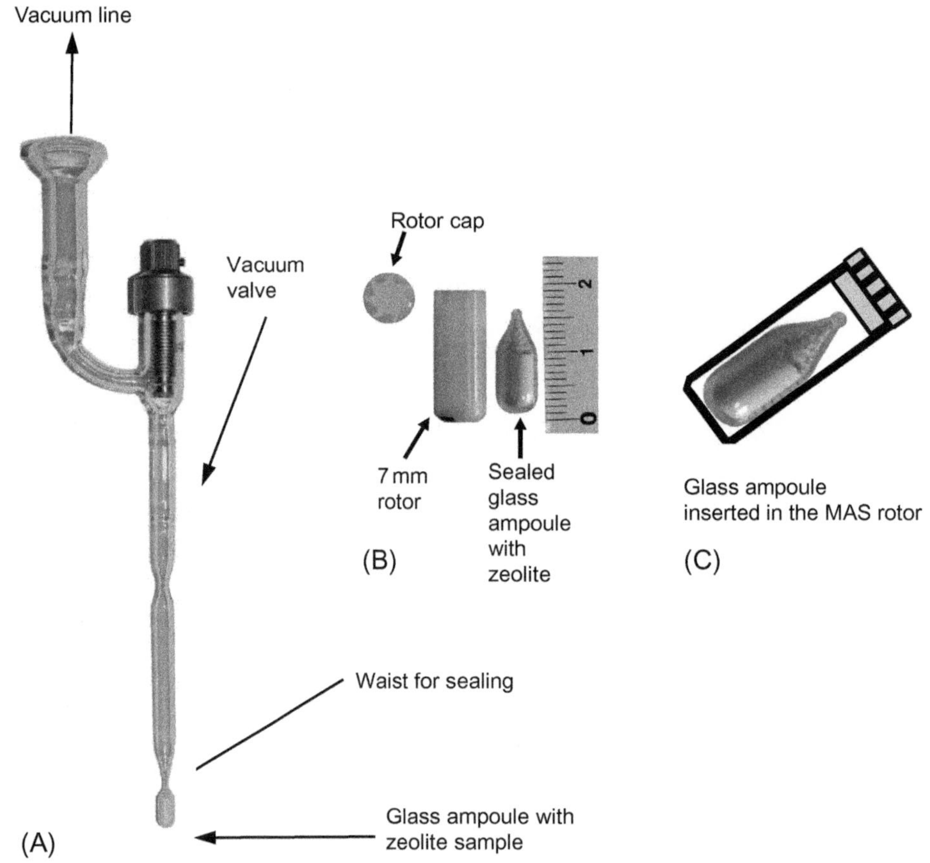

Fig. 6
(A) Equipment for preparation of zeolite samples under vacuum in a glass ampoule; (B) 7 mm ZrO_2 MAS rotor with zirconia cap and the sealed glass ampoule; (C) glass ampoule inserted in a MAS rotor.

intermediates and products in the course of the reaction and to follow the kinetics of the reaction with ^{13}C and ^{1}H MAS NMR [47–49].

Another approach to the study of chemical reactions with MAS NMR applies continuous-flow conditions, which simulate to some extent the situation in a catalytic flow-reactor. In this method, a continuous stream of a carrier gas and the gaseous reactant flow through the zeolite catalyst during the NMR experiment [50–52]. The gas is introduced through a narrow tube placed in an axial hole of the cylindrical catalyst bed within the MAS rotor and leaves the rotor through a hole in the rotor cap (Fig. 7). It is obvious that NMR studies of catalytic reactions using the reagent flow better to approximate the actual conditions in a catalytic flow-reactor than those carried out in sealed ampoules. However, certain tricks are required to operate with this system; therefore, not so many applications of the flow system exist so far [51,52].

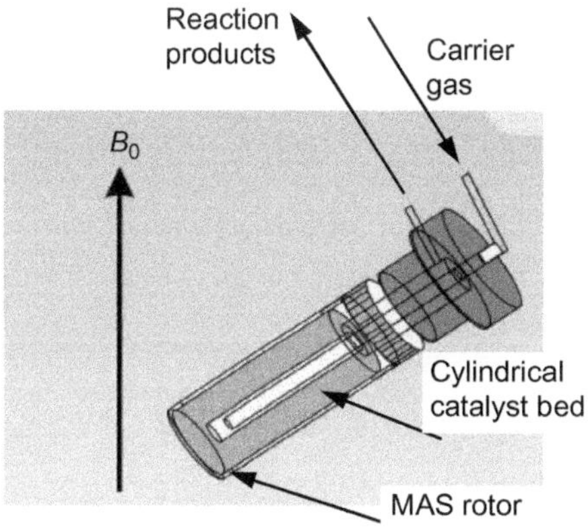

Fig. 7
Flow system for in situ studies of reactions on zeolites developed by Hunger [50].

5 Zeolite Characterization

5.1 ^{29}Si MAS NMR for Structure Characterization and Si/Al Ratio Definition

Although zeolites are highly crystalline they exhibit several characteristics limiting the structural information available by diffraction methods. They are usually microcrystalline, so the diffraction studies are limited to powders. Silicon and aluminum have very similar atomic masses and, therefore, almost identical structure factors. The distribution of silicon and aluminum atoms in the lattice is usually disordered. Therefore, ^{29}Si MAS NMR is often used for zeolite structure characterization.

The basic units of the zeolite framework are TO_4 tetrahedra with silicon atoms at the central T-positions. Some of these T-positions may be occupied by aluminum or other atoms such as boron, gallium, iron, or titanium, among others. ^{29}Si MAS NMR spectra of simple zeolites show a maximum of five well-resolved peaks, which correspond to the five possible distributions of silicon and aluminum atoms around the central silicon, the SiO_4 units (Q_4): Si(4Al), Si(3Al), Si(2Al), Si(1Al), and Si(0Al).

Each type of Si(nAl) (n=0, 1, 2, 3, 4) species gives a characteristic ^{29}Si MAS NMR signal in a well-defined range of chemical shifts (Fig. 8) [18,20,53]. The composition of the zeolite framework influences the relative intensities of the ^{29}Si MAS NMR signals of the above-mentioned Si(nAl) species. The relative signal intensities are directly related to the relative concentrations of the various Si(nAl) units present in the zeolite structure. Consequently, from a careful analysis of the chemical shifts and line intensities, the specific types and relative population of the distinct Si(nAl) units present in a zeolite can be determined. The relative intensity of the lower frequency peaks increase with an increase of the zeolite Si/Al ratio (Fig. 9). Hence, the Si/Al ratio of the lattice can be calculated from the spectral intensities.

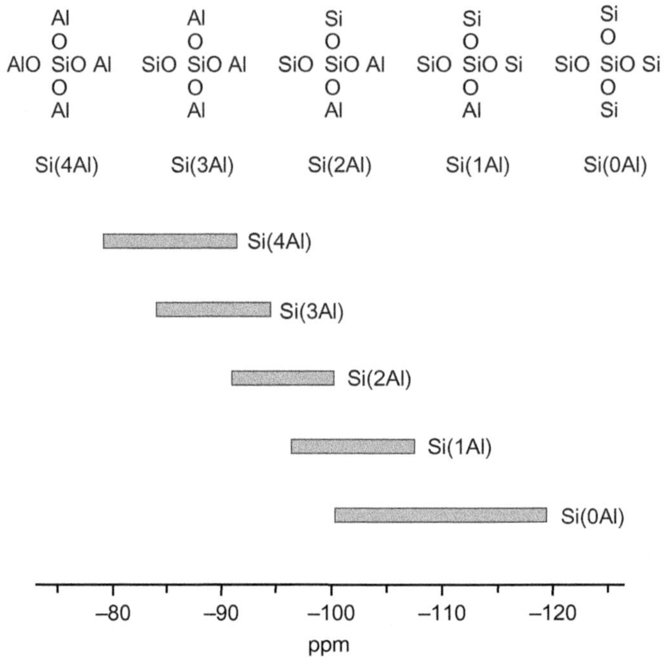

Fig. 8
^{29}Si chemical shifts of Si(*n*Al) units in zeolites.

One way to do this is the Loewenstein's rule, which postulates that Al–O–Al linkages do not occur. Each aluminum atom is surrounded by four O–Si linkages and therefore each Si–O–Al linkage in a Si(*n*Al) unit incorporates 1/4 Al atom and the whole unit has *n*/4 Al atoms. These considerations lead to an expression for the Si/Al ratio:

$$\frac{\text{Si}}{\text{Al}} = \frac{\sum_{n=0}^{4} I_{\text{Si}(n\text{Al})}}{\sum_{n=0}^{4} \frac{nI_{\text{Si}(n\text{Al})}}{4}}$$

Deconvolution of the spectrum into the separate lines and analysis of their relative intensities allows one to estimate the Si/Al ratio, for example, for NaX from the spectrum in Fig. 9:

$$\frac{\text{Si}}{\text{Al}} = \frac{400}{\sum_{n=0}^{4} nI_{\text{Si}(n\text{Al})}} = \frac{400}{4\times 23.37 + 3\times 38.49 + 2\times 26.46 + 1\times 9.61} = 1.47$$

Silicon atoms bearing OH groups, for example, $\text{Si(OH)(OSi)}_{3-n}\text{(OAl)}_n$, located in the framework defects exhibit peaks, shifted to the low field from the corresponding Si(*n*Al)

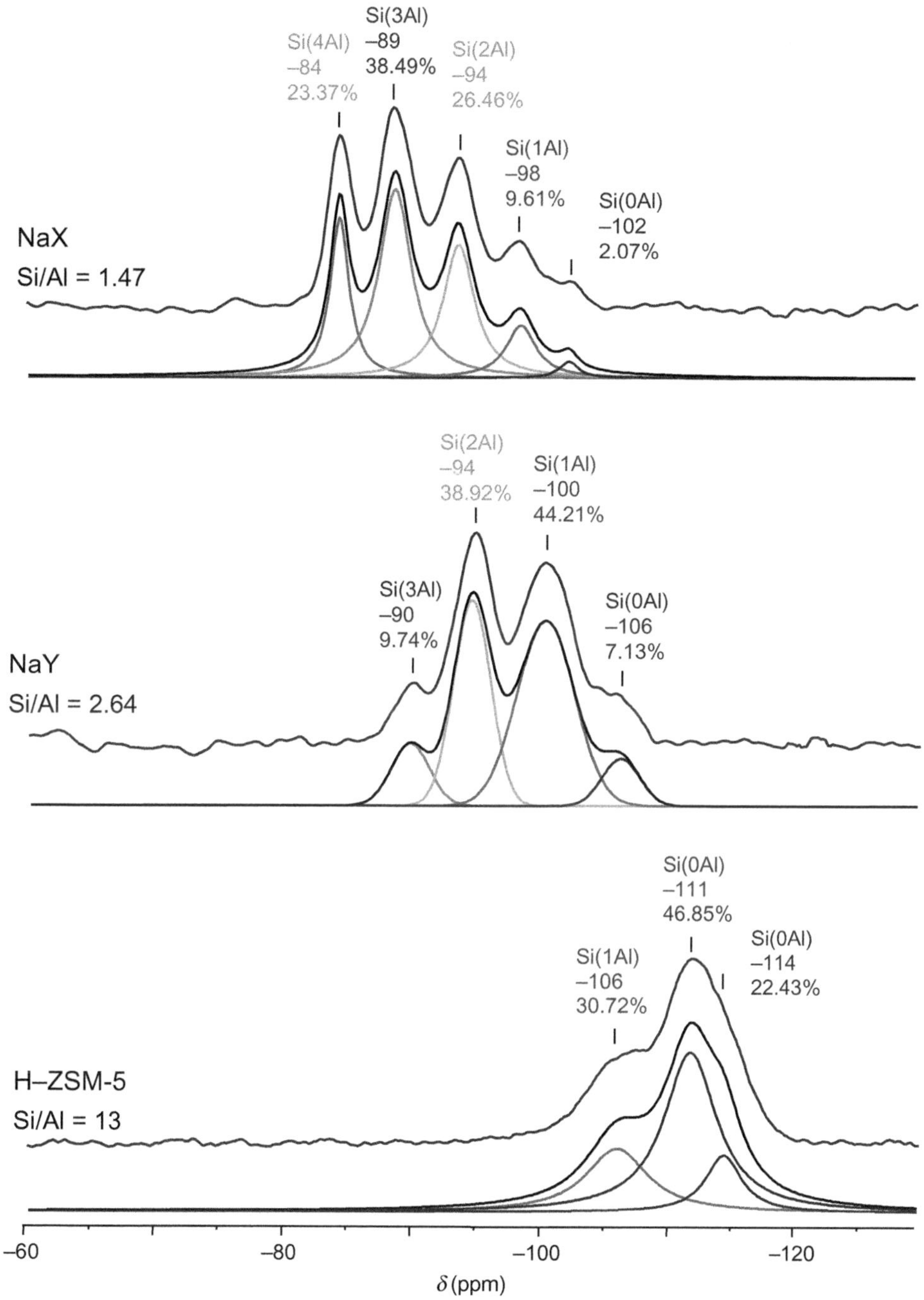

Fig. 9

^{29}Si MAS NNR spectra of some zeolites: NaX (Si/Al = 1.47), NaY (Si/Al = 2.64), H-ZSM-5 (Si/Al = 13).

units and may overlap with Si[(*n* + 1)Al] signals. The SiOH peaks can, however, be clearly identified by their intensity enhancement in the CP spectra [54,55].

^{29}Si MAS NMR allows identification of crystallographically nonequivalent Si sites in zeolites as these sites in chemically equivalent environments may have different chemical shifts [56]. It has been found that the ^{29}Si MAS NMR spectrum of aluminum-free zeolite ZSM-5 (silicalite) [57] exhibits at 295 K 20 well resolved lines (two with a threefold intensity), representing the 24 crystallographically inequivalent Si sites of the monoclinic form (Fig. 10). At 393 K only 10 lines exist in the spectrum (two with a double intensity), which correspond

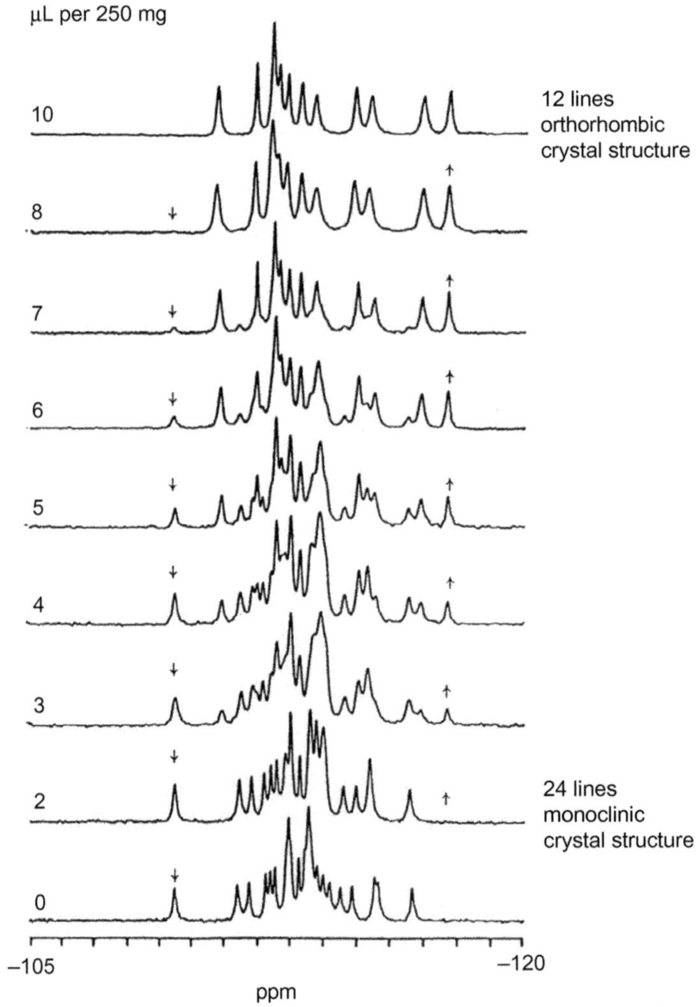

Fig. 10
^{29}Si MAS NMR spectra of aluminum-free ZSM-5 at 295 K with different quantities of adsorbed *p*-xylene [57].

to 12 inequivalent Si sites in the orthorhombic form [57]. Fully reversible changes in the spectra are observed also upon loading the ZSM-5 zeolite with organic sorbates, such as p-xylene, pyridine, acetylacetone (Fig. 10). This shows that ^{29}Si MAS NMR can follow phase transitions in zeolites and the influence of adsorbates on the zeolite structure.

The differences in the ^{29}Si chemical shifts of Si atoms located in the same chemical environment species but at crystallographically nonequivalent T-positions are mainly due to changes in the bonding geometry of the respective Si(*n*Al) structural unit. Experimental studies and theoretical considerations indicated that the ^{29}Si NMR shifts values, $\delta_{^{29}\text{si}}$, are linearly related to the average value of the four Si–O–T bond angles, α, at the silicon atom in the center of the tetrahedron. The following expression for the correlation between the shift values, $\delta_{^{29}\text{si}}$, and functions of the mean angle, α, was found for Si(*n*Al) units with $n=0$–4 [18,58–61]:

$$\delta_{^{29}\text{si}} = -223.9\cos\alpha / (\cos\alpha - 1) + 5n - 7.2$$

The quantitative relationship between δ and α can be used to estimate mean SiOT bond angles from ^{29}Si chemical shifts or, vice versa, if the bond angles are known from X-ray studies, to calculate the shifts, which may be helpful in the correct assignment of the spectra.

^{29}Si MAS NMR spectroscopy allows identification of bond connectivities between silicon atoms at crystallographically nonequivalent T-positions [62–64]. These investigations are performed, for example, by two-dimensional COSY (correlation spectroscopy) and INADEQUATE (incredible natural abundance double quantum transfer experiment) experiments [65].

5.2 ^{27}Al MAS NMR. Framework and Extra-Framework Aluminum Characterization

The ^{27}Al MAS NMR spectra of zeolites are in general, much simpler than their ^{29}Si NMR counterparts since, according to the Lowenstein's rule (which forbids Al–O–Al pairings), only a single tetrahedral Al environment, namely, $Al(OSi)_4$, exists in the zeolite framework. The AlO_4 units in hydrated zeolites show only small deviations from the tetrahedral symmetry and thus only weak quadrupole interactions (quadrupole constant of about 1–2 MHz). Consequently, a single narrow line is usually observed in the ^{27}Al MAS NMR spectra of hydrated zeolites (Fig. 11) with a typical chemical shift of about 60 ppm. The isotropic ^{27}Al chemical shifts of tetrahedral framework aluminum in hydrated zeolites of different structure type cover a rather small range from about 55 to 68 ppm [18].

In dehydrated zeolites, the symmetry of the AlO_4 tetrahedra is heavily distorted, which causes large quadrupole interactions (quadrupole constant of about 5–16 MHz) and thus much broader resonances often not observable under MAS.

Nonframework aluminum in zeolites, which has typically an octahedral AlO_6 coordination, gives rise to signals at about 0 ppm; this means it is well separated from the tetrahedral

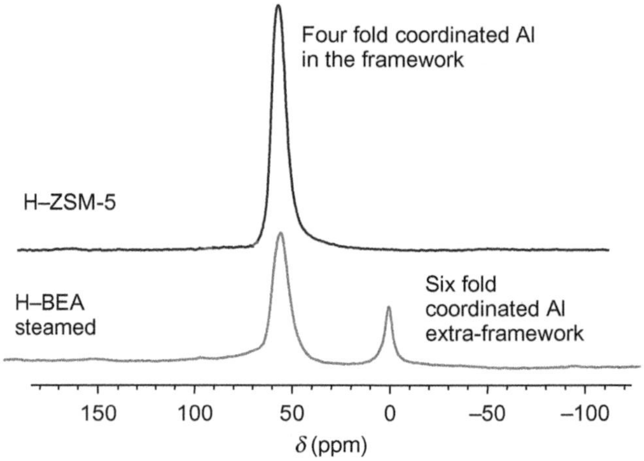

Fig. 11
^{27}Al MAS NMR spectra of hydrated zeolites H–ZSM-5 (Si/Al = 13) and H–BEA (Si/Al = 18).

Al sites of the framework (Fig. 11). However, if the nonframework Al exists as polymeric aluminum oxides or oxide hydrates in the zeolite pores, strong quadrupole line broadening may be observed owing to heavy distortions of the octahedral symmetry of the AlO_6 sites.

An important application of ^{27}Al MAS NMR is the detection and characterization of nonframework aluminum species formed by various thermal or hydrothermal treatments of zeolites, as is usually applied in the dealumination processes or the preparation of the acidic H-forms of zeolites. Provided that all aluminum is visible in the spectrum, such that no signal intensity is lost as a result of very strong quadrupole line broadening, the relative proportions of framework and nonframework Al in zeolites can be directly determined from the intensities of the signals at about 60 and 0 ppm.

Further complication can arise in the application of ^{27}Al MAS NMR if broad signals of Al sites subject to different quadrupolar interactions overlap in the spectrum. For assigning the nature of aluminum species responsible for the broad signals ^{27}Al DOR NMR [66] (DOR = double rotation) or ^{27}Al MQ MAS NMR (MQ stands for multiple quantum) [67] spectroscopy is suitable.

^{27}Al quadrupole coupling constant C_Q is highly sensitive to the symmetry distortion at tetrahedral AlO_4 or octahedral AlO_6 coordination sites. This property can be used to follow the structural changes not only in zeolites, but also in other solid crystalline Al-containing materials, such as Al-based metal-organic frameworks (MOFs). MIL-53(Al) possesses a flexible framework which forms one-dimensional rhombic-shaped channels. In the absence of adsorbed molecules, the channel dimension is 8.5 × 8.5 Å. The ^{27}Al NMR spectrum of this so-called open framework exhibits a well-defined quadrupolar pattern with a chemical shift of $\delta = 3.3$ ppm, $C_Q = 8.36$ MHz, and $\eta_Q = 0$ (Fig. 12A). Adsorption of some molecules like water

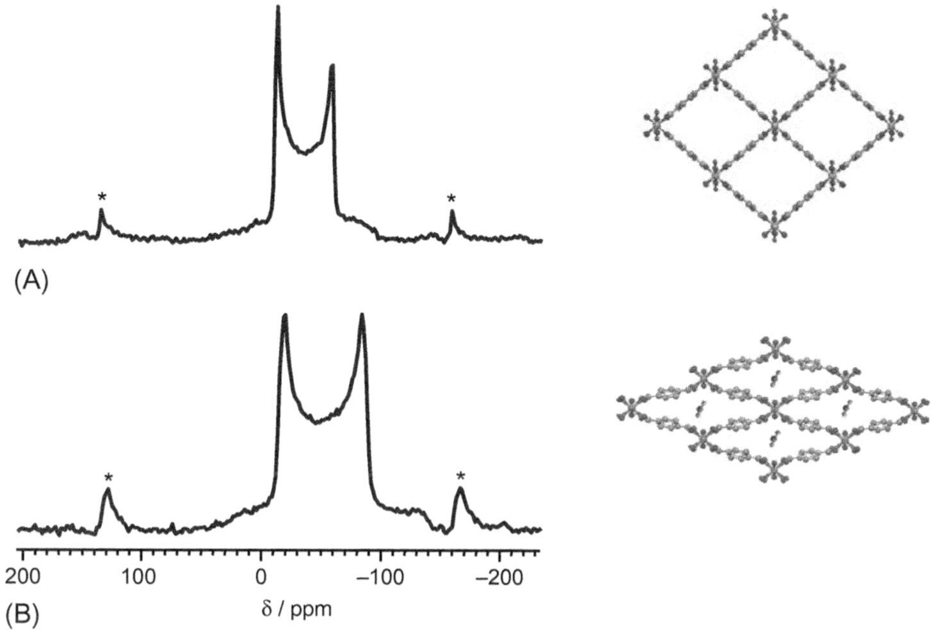

Fig. 12
^{27}Al MAS NMR spectra of MIL-53(Al) (A) for open-framework with empty pores and (B) for closed framework with adsorbed water. The corresponding structures for open and closed forms of MIL-53 are shown on the right-hand side.

or acetonitrile, ammonia, and pyridine [68] induces a significant shrinkage of the framework system resulting in the channel dimension of 2.6×13.6 Å (closed framework). The ^{27}Al NMR spectrum is substituted for quadrupolar pattern with a chemical shift of $\delta = 3.4$ ppm, C_Q of 10.67 MHz, and η_Q of 0.15 (Fig. 12B). This finding indicates the high sensitivity of ^{27}Al solid-state NMR spectroscopy on framework aluminum atoms in MIL-53(Al) to their interactions with adsorbate molecules and to structural transformations. So, the breathing effect (open-closed framework interconversion) of the framework of MIL-53 and influence on it of different adsorbates can be investigated with ^{27}Al MAS NMR.

5.3 Brønsted Acidity Characterization With ^1H MAS NMR

Zeolites are often used as acid type catalysts for heterogeneously catalyzed reactions of hydrocarbons. The active centers of the zeolite catalysts are represented by the surface hydroxyl groups acting as Brønsted acid sites. The hydroxyl groups could be detected and characterized with ^1H MAS NMR provided that adsorbed water is removed from zeolite pores. For this purpose, a zeolite sample is dehydrated at elevated temperatures in a vacuum in a special glass tube (vide supra, Fig. 6). This tube is further sealed off and transferred further into a zirconia rotor for spectrum registration.

^{1}H MAS NMR signals of hydroxyl groups in dehydrated zeolite catalysts cover a range of chemical shifts of δ_{1H}=0–7 ppm [6,69,70]. The lowest chemical shifts of −0.5 to 1.0 ppm have been observed for metal or cation OH groups, for example, AlOH [71,72]. The signals of terminal silanol SiOH groups at the external surface of the zeolite crystallites and framework defects occur at 1.2–2.2 ppm. The silanol groups involved in strong hydrogen bonding can show the signals at chemical shifts of up to 16 ppm [73,74]. Dealuminated zeolites exhibit often the signals at δ_{1H}=2.4–3.6 ppm due to hydroxyl protons bound to extra-framework aluminum species (AlOH) [69,75]. Isolated bridged Al–OH–Si groups are displayed at 3.6–5.5 ppm, while these groups involved in the hydrogen bonding with neighbor framework oxygens are found at 5–7 ppm [6].

Fig. 13 shows ^{1}H MAS NMR typical spectra of dehydrated zeolites H–ZSM-5 and H–BEA. The spectra of both zeolites exhibit the signals at 1.8, 4.1, and 5.4 ppm from the silanol groups (SiOH) formed at framework defects and external surface of zeolite crystallites, free bridged OH groups (SiOHAl), and SiOHAl groups interacting with neighboring oxygen atoms by hydrogen bonding, respectively. For H–BEA zeolite, the signal of SiOH groups is rather intense with respect to the signals from SiOHAl groups [31,76–78]. The high intensity of SiOH could result from the faulted structure of the zeolite BEA [79,80].

5.3.1 Extra-Framework Aluminum Species Displayed With TRAPDOR

$^{1}H\{^{27}Al\}$ spin-echo double resonance (TRAPDOR [29,30]) experiments (pulse sequence used is shown in Fig. 5D) performed with zeolite H–BEA show the suppression of signals from protons, which are in a close vicinity to aluminum atoms, that is, in the SiOHAl or AlOH groups, whereas the signals of protons, which are far from aluminum atoms (SiOH) are unaffected (cf. Fig. 13B and C). The difference spectrum (Fig. 13D) shows three plainly distinguishable resonances of OH groups connected to aluminum: at 5.4, 4.1, 2.6, and 1.0 ppm. The two first signals are clearly identified in the usual ^{1}H MAS spectrum (Fig. 13B). The other two signals at 2.6 and 1.0 ppm, which belong to the different hydroxyl groups bound to extra-framework aluminum species (AlOH), are identified only in the difference spectrum [81]. Thus, TRAPDOR experiments allow identification of the signals of hydroxyl groups, which are not evident in the usual ^{1}H Hahn-echo MAS NMR spectrum.

5.3.2 Determination of the quantity of hydroxyl groups

As the intensity of the signals in the NMR spectrum is proportional to the concentration of corresponding species, the ^{1}H MAS NMR signal intensities can be used to determine the concentration of hydroxyl groups in dehydrated zeolites by comparison of the intensity of different hydroxyl groups with the intensity of some internal reference, for example, certain quantity of adsorbed methane or TMS (tetramethylsilane).

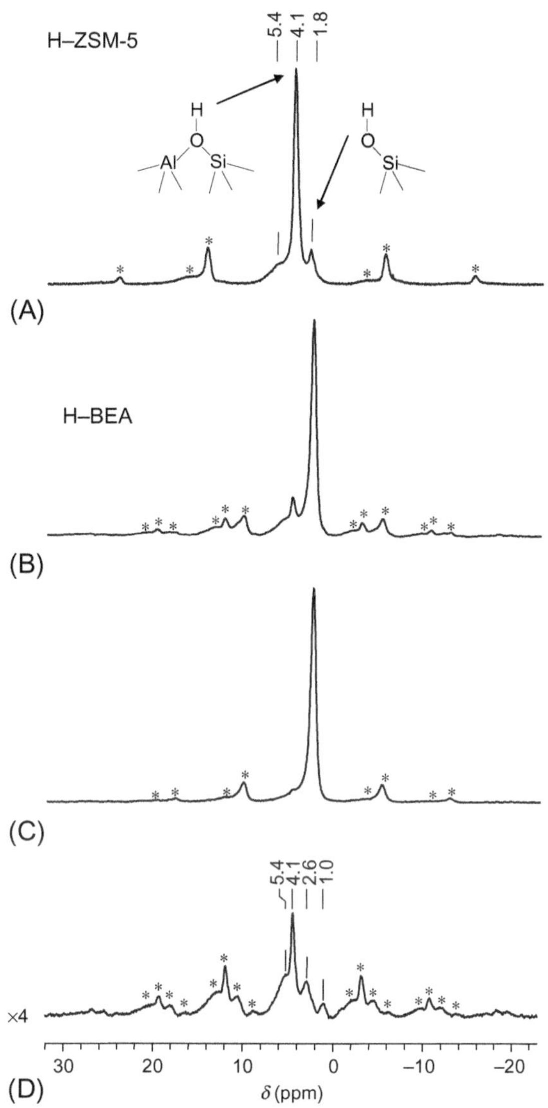

Fig. 13
^{1}H MAS NMR spectra of dehydrated zeolites H–ZSM-5 and H–BEA. (A) H–ZSM-5; (B)-(D) H–BEA; (B) without ^{27}Al irradiation and (C) with ^{27}Al irradiation (on-resonance). The difference spectrum of (B) and (C) is shown in (D). Asterisks (*) belong to the spinning sidebands.

Comparison of the intensity of the signal of adsorbed methane at 0 ppm with intensities of signals of zeolite H–ZSM-5 (Si/Al = 13) at 4.1 and 5.4 ppm from isolated and hydrogen-bonded AlOHSi groups (including the signals of the corresponding spinning sidebands) (Fig. 14) allowed determination of the quantity of AlOHSi groups as 1200 μmol/g, which is in a good agreement with the expected concentration of Brønsted acid sites in H–ZSM-5 with Si/Al = 13.

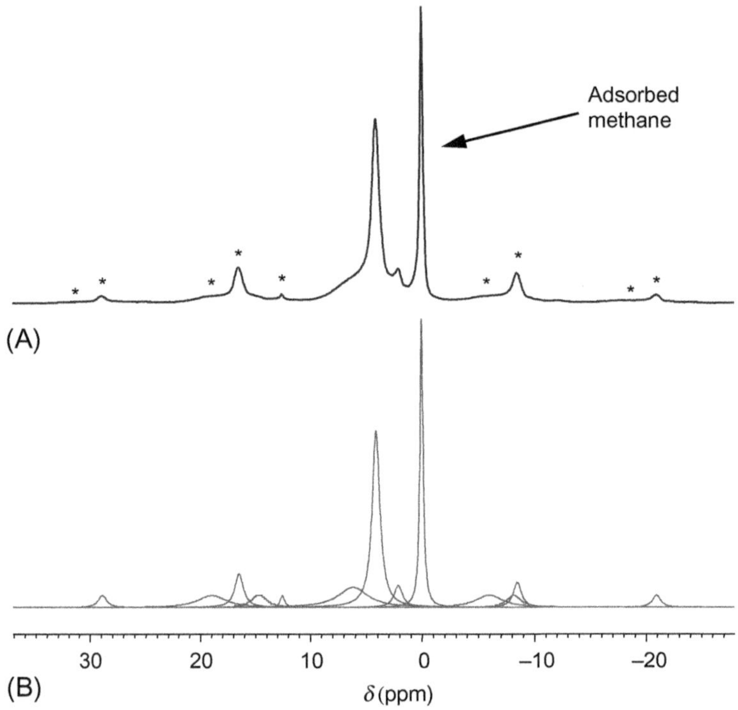

Fig. 14
Determination of the quantity of the Brønsted acid sites for zeolite H-ZSM-5 (Si/Al = 13) from the ^1H MAS NMR spectrum with the use of adsorbed methane as an internal reference. 200 μmol/g of methane was adsorbed in the sample. (A) ^1H MAS NMR spectrum of H-ZSM-5 zeolite with adsorbed methane. (B) Deconvolution of the spectrum into separate lines for the calculation of relative signal intensities. The spectrum was recorded using the Hahn-echo pulse sequence (Fig. 5A) and repetition time $t = 60$ s. Asterisks (*) belong to the spinning sidebands.

For quantitative studies, the repetition time t of the pulse experiments should be essentially longer in comparison with the spin-lattice relaxation times T_1 of different OH species, usually $t \geq 5 \times T_1$. In most cases, T_1 is about 1–10 s for different hydroxyl groups [82].

5.3.3 Characterization of acidity with adsorbed probe molecules

The strength of Brønsted acid sites in zeolites can be characterized by adsorption of probe molecules, such as pyridine, acetone, acetonitrile, trimethylphospine, deuterated alkanes, and so on. Among probe molecules used, acetone is applied more frequently [83–86]. It has been found that the carbonyl group of adsorbed acetone exhibits different chemical shifts $\delta_{^{13}C}$ for different zeolites [83]: 216.8 ppm for H-SAPO-5, 219.6 ppm for H-Y, 221.8 ppm for H-MOR, 223.4 ppm for H-ZSM-12, 223.6 ppm for H-ZSM-5, and 225.4 ppm for H-ZSM-22. These chemical shifts were compared with the shifts of 205 and 245 ppm for the carbonyl group of acetone dissolved in $CDCl_3$ and in 100% sulphuric acid [87]. A scale of the Brønsted acid strength was then introduced, which is based on the experimentally determined dependence of the chemical shift for the carbonyl group of acetone dissolved in aqueous sulphuric acid of different concentrations [88]. According to this scale, the acid strength of bridging OH groups in zeolite H–ZSM-5 corresponds to that of 80% H_2SO_4 in water [88].

5.4 Characterization of Adsorbates

Besides characterization of the zeolite structure and surface hydroxyl groups active in catalytic transformations of hydrocarbons, solid-state NMR represents an indispensable tool for characterization of adsorbates in regard to their identification or establishing the peculiarities of intramolecular dynamics for molecules confined inside the pores of zeolites or other nanoporous materials. Identification of adsorbates, such as hydrocarbons formed in the course of the reaction and remained occluded in the zeolite pores due to impossibility to desorb for some reasons can be performed with ^{13}C or ^{1}H MAS NMR. Establishing the intramolecular dynamics of the occluded molecules is performed by the analysis of the lineshape of the ^{2}H NMR spectrum of deuterated analogues of these adsorbed molecules.

5.4.1 Identification of hydrocarbons in the adsorbed state

MAS techniques make the signals of hydrocarbons adsorbed in zeolite pores sufficiently narrow to identify the signals of different fragments of hydrocarbon molecules [14,15,39,43,89,90]. Analysis of hydrocarbons in the adsorbed state is performed on the basis of the specific ^{13}C chemical shifts $\delta_{^{13}C}$ (more rarely, on the basis of δ_{1H}) of the separate CH_n ($n=0$–4) fragments of hydrocarbon molecules, because $\delta_{^{13}C}$ of the CH_n groups of a hydrocarbon molecule in the adsorbed states are, as a rule, identical to the corresponding chemical shifts in solutions.

Fig. 15A shows the ^{13}C MAS NMR spectrum of the adsorbed products of tert-butyl alcohol transformation on zeolite H–ZSM-5 at 448 K [91]. About 20 separate and overlapped signals can be distinguished in the spectrum, which belong to a mixture of linear and branched alkanes formed in the reaction. All the singlet lines can be assigned to the CH_n fragments of certain alkanes (propane, butanes, and pentanes) only tentatively, because the positions of signals from the CH, CH_2, and CH_3 groups of different alkanes in the spectrum are overlapping. More reliable assignment of the observed signals to the CH_n fragments of various alkanes can be made with the help of two-dimensional *J*-resolved ^{13}C NMR spectroscopy, allowing simultaneous identification of both the chemical shift of the signal (in the F2 direction) and its multiplicity (F1 direction), arising from scalar *J*-coupling with protons attached to the carbon atoms in CH_n fragments (Fig. 15B). For example, a singlet signal at 25.5 ppm in the F2 direction is in fact a quartet (four peaks along the direction F1 with relative intensities 1:3:3:1), which belongs to the methyl group. A badly resolved signal

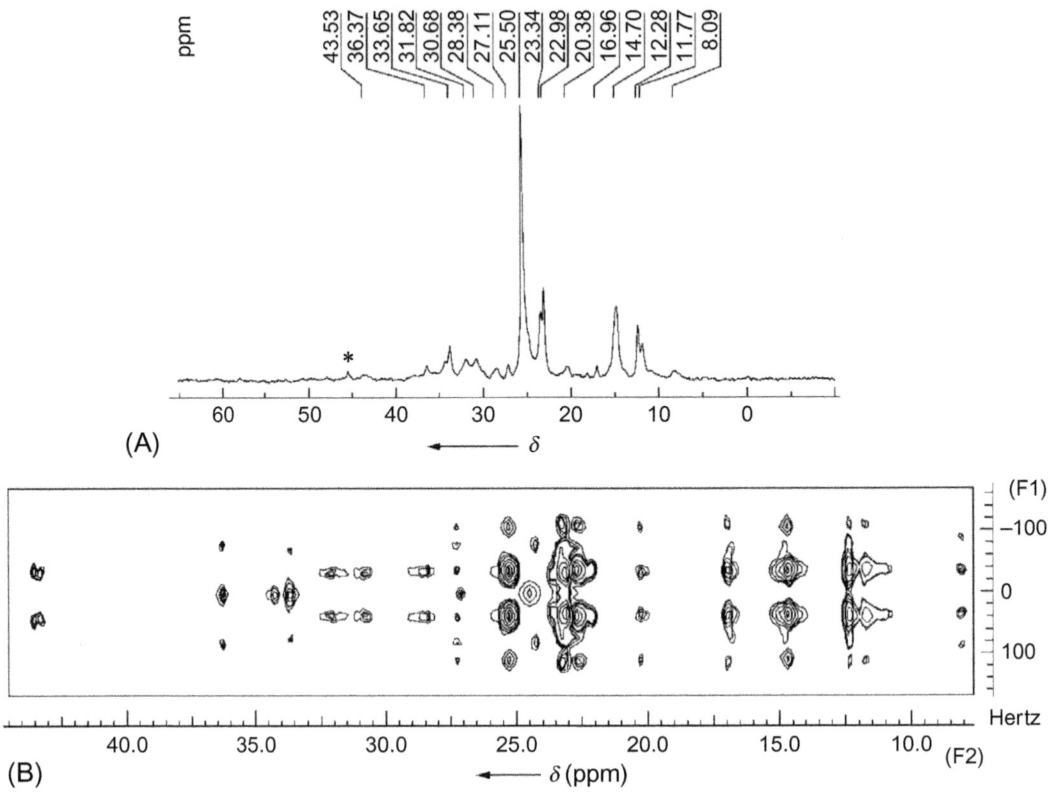

Fig. 15
(A) One-dimensional ^{13}C MAS spin-echo NMR spectrum and (B) contour plot of 2D *J*-resolved ^{13}C solid-state MAS NMR spectrum of the products of *tert*-BuOH dehydration on H-ZSM-5 zeolite at 448 K. The observed value of scalar $J(^{13}C–^{1}H)$ coupling is equal to half of the real $J(^{13}C–^{1}H)$ value, because of proton high-power decoupling during the second half of the evolution time τ (pulse sequence in Fig. 5E) [39]. Asterisk denotes a spinning sideband.

at 24.5 ppm is the methylene CH_2 group (a triplet, three peaks along the direction F1 with relative intensities 1:2:1). This approach allows a more reliable assignment of the signals to the certain CH_n groups of hydrocarbons with some particular structure and facilitates the analysis of the mixture of hydrocarbons, formed in the course of the reaction and remained on zeolite in the adsorbed state [39,91,92].

5.4.2 Mobily characterization with ^2H NMR

Translational diffusion and intramolecular mobility of hydrocarbon molecules inside the pores of zeotype materials plays a critical role in the performances of these materials as catalysts and adsorbents. Therefore, it is no wonder that much work has been done to measure and characterize the mobility of molecules in zeotype materials [93].

The translational mobility (diffusion) of hydrocarbons sorbed within zeolites are usually probed with pulsed-field gradient NMR (PFG NMR) [94], while intramolecular mobility can be characterized with deuterium solid-state (^2H) NMR [16,95,96].

The line shape for ^2H NMR is defined completely by intramolecular quadrupole interaction [97,98]. An important point is that the line shape of the ^2H NMR spectrum is especially sensitive to the type of molecular motion and its rate [98]. Therefore, by analyzing ^2H NMR line shapes as well as relaxation rates, one can evaluate the nature of the molecular motion and its rate.

^2H NMR spectra of deuterated immobile molecules or their fragments (eg, methyl CD_3-groups in hydrocarbons) represent Pake-type powder patterns [97,98]. The dominant features of these line shapes are two strong peaks separated by the splitting $(3/4)Q_0$ and two shoulders separated by $(3/2)Q_0$ (see Fig. 16A).

If a CD_3 group of an organic molecule undergoes fast diffusional rotation about the C–CD_3 bond with an angle $\alpha = 109.47°$ between the rotation axis C_3 and the direction

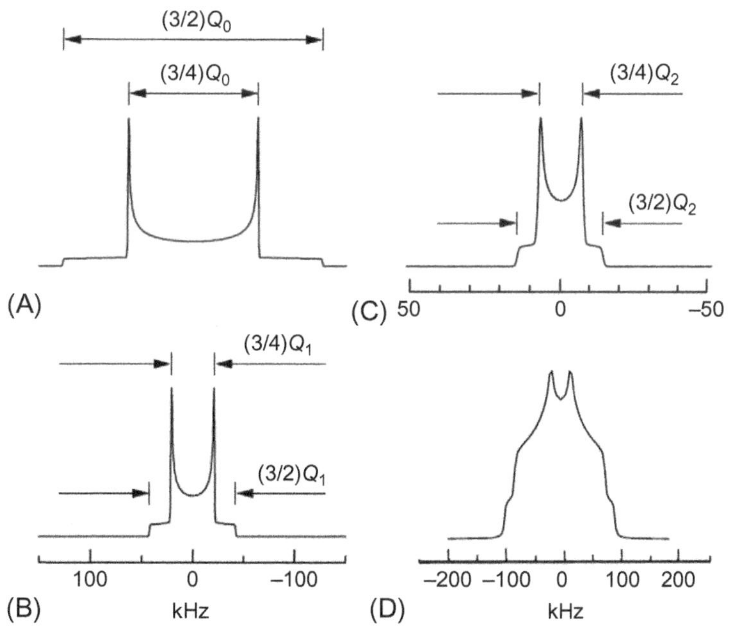

Fig. 16
Theoretical ^2H solid-state NMR spectra. (A) Polycrystalline sample (immobile methyl CD_3-group in an adsorbed hydrocarbon, $C_Q = e^2qQ/h = Q_0 = 169.5$ kHz, $\eta = 0$; (B) Rapid rotation of the CD_3-group about one C_3 axis, angle $\alpha = 109.47°$ between the rotation axis C_3 and the direction of the C–D bond, $Q_1 = (1/3)Q_0$; (C) Rapid rotation of the CD_3-group about two successive C_3 axes, rotation angles $\alpha = \beta = 109.47°$, $Q_2 = (1/9)Q_0$; (D) Reorientation of the benzene phenyl ring by π-flipping about its para axis.

of C–D bond, then the line shape of the spectrum for CD_3 groups will be similar to that of rigid molecules but with the quadrupole splitting reduced to a value of $(3/4)Q_1$, and $Q_1 = (1/3)\,Q_0$ (Fig. 16B). If a CD_3 group is involved in additional rotation about a second axis tilted at the angle β with respect to the first axis α, then, again, its ^{2}H NMR spectrum has a quadrupole splitting reduced to $(3/4)Q_2$ (Fig. 16C) with an effective quadrupole constant $Q_2 = (1/9)\,Q_0$. For the phenyl ring of the benzene molecule reorienting by π-flipping about its para axis, the spectrum has a line shape shown in Fig. 16D with $Q_{eff} \sim (1/2)Q_0$ and $\eta \sim 1$.

High sensitivity of the line shape of the ^{2}H NMR spectrum to the type and rate of the molecular motion allows characterization of the peculiarities of the intramolecular motion of alkanes [99–102], olefins [17,103], and alcohols [104,105] in zeolite frameworks of different types. Also the dynamics of zeolite Brønsted acid sites and nonacidic hydroxyl groups has been unraveled [106].

Dimensions of alkane molecules sorbed by zeolites are comparable to the dimensions of zeolite pores. So the diffusion and intramolecular dynamics of alkanes confined in the narrow pores of zeolite is defined by the size and geometry of the zeolite pores as well as by dimensions and the structure of the adsorbed alkane. This means that alkanes with different structures and dimensions can exhibit a unique ^{2}H NMR line shape in the zeolite structure of the same type.

Fig. 17 shows ^{2}H NMR line shapes for some of C_4–C_{10} alkanes adsorbed in zeolite ZSM-5. Due to the difference in the intramolecular motion of these alkanes, the line shapes are different. This property can be used as a fingerprint test for identification of adsorbed alkanes in zeolites of similar and different structures.

^{2}H NMR can be also used to characterize the organic framework mobility in MOFs. Organic linkers that bridge the metal clusters in MOFs and constitute the three-dimensional porous structure of these materials could be rather mobile and are involved in a certain type of rotational motion. This rotational motion can be probed with ^{2}H NMR, if the studied material is deuterated. Thus, valuable information on the mechanism of intramolecular motion of MOFs' organic fragments and its kinetic parameters can be derived.

Fig. 18 shows an evolution with the temperature of the ^{2}H NMR line shape of the deuterated phenyl group of a terephthalic acid linker in MIL-53(V). Simulation of the line shape with regard to the correspondence of the observed line shape to some certain mechanism of phenyl ring reorientation allows the researcher to establish that reorientation of the phenyl ring occurs by π-flipping with the rate of 1–10^3 kHz at 300–483 K and activation barrier of 45 kJ/mol [107]. Similar mechanism of phenyl ring reorientation is detected in other MOFs' structures [108–110].

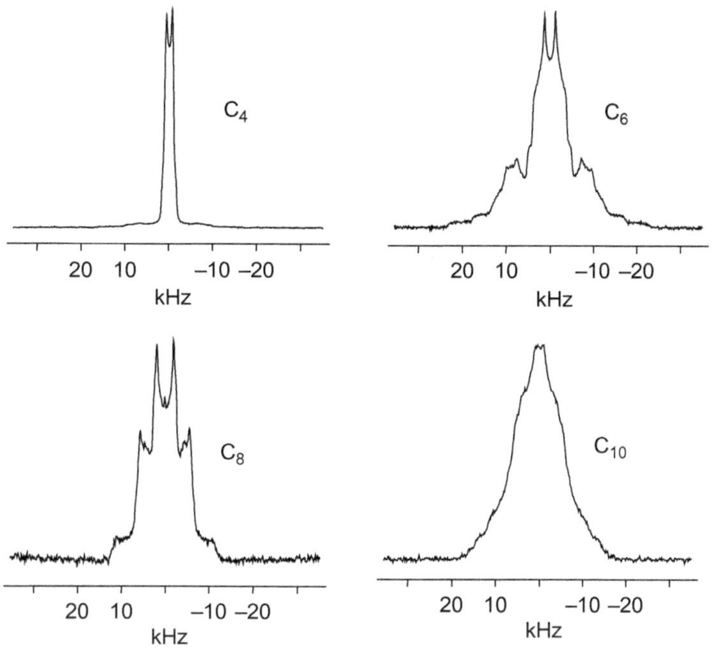

Fig. 17
^{2}H NMR line shapes for alkane adsorbed in zeolite H-ZSM-5: C_4, isobutane-d_{10}; C_6, *n*-hexane-d_{14}; C_8, *n*-octane-d_{18}; C_{10}, *n*-decane-d_{22}.

5.4.3 *Probing intracrystalline voids with ^{129}Xe NMR*

NMR studies using ^{129}Xe to characterize zeolites were pioneered in the early 1980s by the groups of Fraissard [7] and Ripmeester [8]. The use of ^{129}Xe as a probe molecule is now well established. ^{129}Xe has a half spin and 26% natural abundance, and the range of Xe chemical shifts is over 5000 ppm. It is chemically inert, nonpolar, spherical, and easily polarizable, with a large electron cloud sensitive to the pore structure. All of these features make it a convenient probe for nanoporous materials such as zeolites [111] or MOFs [112]. This technique provides information on the dimensions of cavities and channels [10], the short-range crystallinity, the nature of structural defects [13], and the effect of cations in zeolites [11], as well as structural modifications during any dealumination process, steaming, or acid leaching [113]. It can also be used to probe the nature of the binding sites, to determine the kinetics of molecular exchange between different sites, and to determine the zeolite surface area. All of these parameters derived by ^{129}Xe NMR have a pronounced effect in catalysis by zeolites.

Fig. 19 shows the example of application of ^{129}Xe NMR for probing the binding sites in zeolite mordenite (MOR) [114]. MOR contains two distinct environments for adsorption: a 12-ring main channels with smaller 8-ring side pockets off the main channel. At 190 K, two relatively broad lines are evident [8,9]. The low field line is assigned to xenon in the side

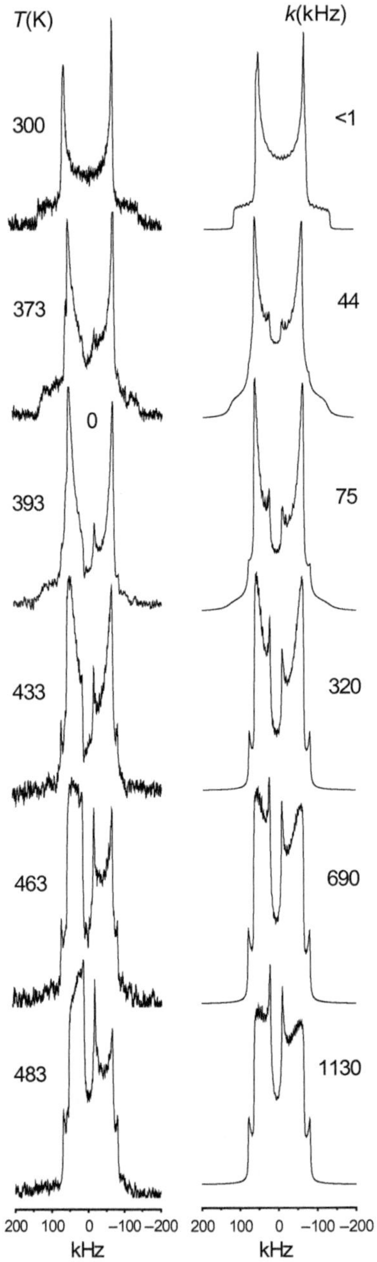

Fig. 18
Experimental (*left column*) and simulated (*right column*) ^{2}H NMR spectra *line shape* temperature dependence of the aromatic rings in MIL-47(V), k is the π-flip rate constant.

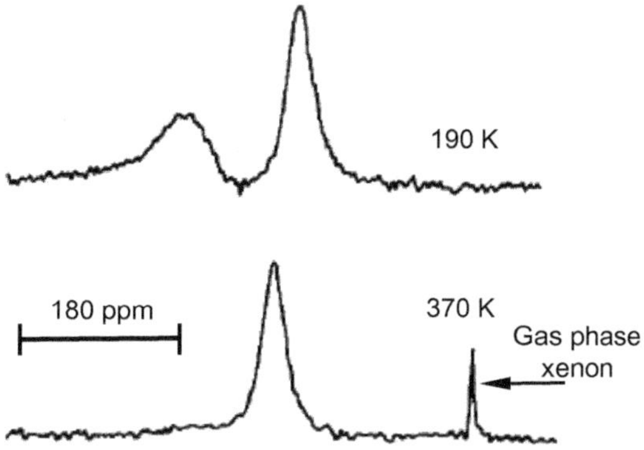

Fig. 19
^{129}Xe NMR spectra of xenon adsorbed on Na-MOR [8,9].

pockets of the MOR and the high field line to xenon in the main channel. Evidently at 190 K there is no exchange of the two types of xenon. The relatively large line widths suggest that within the two types of sites there is a further distribution. At 370 K only a single line is observable so that xenon species in the main channel and the side pockets are in a state of a rapid exchange. Thus, two signals of xenon in the spectrum are in a good accordance with the expected spatially different adsorption sites of MOR. The relative intensities of two signals are proportional to the relative intracrystalline voids at these different adsorption sites. The merger of two signals into one line at 370 K indicates that the molecules with the dimensions like a xenon atom can easily migrate between different adsorption sites.

6 Reaction Characterization on Zeolites

The possibility of solid-state NMR spectroscopy to perform the analysis of hydrocarbons in the adsorbed states on zeolites at room and higher temperatures and directly in the course of the reaction is the basis for application of this method for reaction characterization occurring on zeolites and other solid catalysts. Solid-state NMR provides in situ monitoring of the reaction proceeding: identification of the reaction intermediates, analysis of the reaction kinetics. This allows establishing the mechanisms of the reaction as well as the pathways of hydrocarbon activation. Here we provide some examples to demonstrate the approaches that are used in NMR studies to identify the nature of the intermediate on zeolite surface, to follow the kinetics of the reaction and establishing the reaction mechanisms.

6.1 The Nature of the Intermediates

6.1.1 In situ NMR identification of alkoxide intermediates

The nature of intermediates in heterogeneous catalysis by acid form zeolites attracted considerable interest since the first experiments by Derouane et al. on characterization of adsorbed hydrocarbons with ^{13}C CP/MAS NMR [14]. Alkyl carbenium ions, which were expected to be the intermediates in hydrocarbon conversions on solid acid catalysts, were not detected on the catalyst surface by any spectroscopic methods, even though they were identified in superacid solutions [115]. Already earlier quantum chemical studies of the stability of hydrocarbon structures on the surface of aluminosilicates [116,117] demonstrated that, among the possible intermediates of olefin conversion, the alkoxide structures are energetically the most stable and the bridging carbonium ions are the least stable. It was hypothesized that alkyl carbenium ions can form on solid acids only as transition states. ^{13}C MAS NMR can clearly distinguish whether alkoxy species or alkyl carbenium ions are formed as intermediates due to dramatic difference of the ^{13}C chemical expected for species with different nature. Carbenium ions should exhibit a characteristic chemical shift about 330 ppm (tert-butyl cation) [118,119], whereas alkoxy species should yield the signals at 50–80 ppm for the carbon atom attached to the surface oxygen atom (similar to the signals in corresponding aliphatic alcohols) [120]. Note that NMR is a low-sensitivity method and the surface intermediates are formed in low concentrations. Therefore, the methods of sensitivity enhancement should be applied to detect the intermediate species. Two approaches are used in case of detection of the hydrocarbon intermediates in zeolites: the application of the CP technique (*vide supra*) and the use of reactant hydrocarbons enriched with the ^{13}C isotope (up to 99%) (natural abundance of the magnetic ^{13}C isotope is only 1.07%). Sometimes the reactants with the selective ^{13}C or ^{1}H labels (in the latter case, deuterated hydrocarbons) are used for reliable identification of the intermediates (products) or for unraveling the mechanisms of the label transfer, if inter- or intramolecular label migration ^{13}C (^{1}H) occurs in the course of the reaction.

With the use of these sensitivity-enhancement approaches the stable alkoxide intermediates were indeed detected by ^{13}C CP/MAS NMR spectroscopy. Isopropoxide was the first alkoxy intermediate reliably identified in propylene (labeled with ^{13}C in the CH= group) conversion on zeolite HY [15]. Isopropoxide exhibits the signal at 87 ppm from the labeled ^{13}C atom, which is characteristic of the $(CH_3)_2CH$ fragment bonded to an oxygen atom of the zeolite framework (Fig. 20). Later, other alkoxide intermediates were detected and characterized. It was demonstrated that methoxides [121,122] and ethoxides [122,123] formed from methyl and ethyl iodides and also from methanol and ethanol on H–ZSM-5 and CsX zeolites. Isobutoxy [124] and *tert*-butoxy [90] intermediates resulted from the dehydration of isobutanol and *tert*-butanol on HZSM-5. Alkoxides are highly reactive species. For example, surface methoxides are effective methylating agents in their reactions with methanol, water, ammonia, alkyl halides, HCl, CO, acetonitrile, and aromatic compounds [125].

6.1.2 Monitoring the ^{13}C label scrambling in adsorbed olefins on acid forms of zeolites. Evidence for the formation of carbenium ion intermediates

The possibility of ^{13}C NMR to follow the intramolecular migration of the ^{13}C selective label in adsorbed hydrocarbons gave the researches a chance to provide an evidence for the formation of alkyl carbenium ions as intermediates in reactions on solid acid catalysts, including zeolites. Carbenium ions in superacids exhibit a unique property to scramble the selective

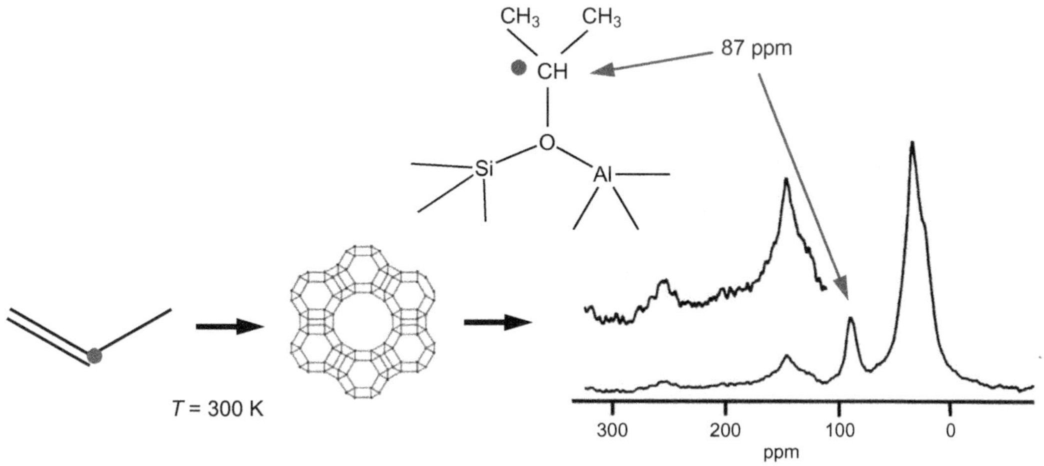

Fig. 20
^{13}C CP/MAS NMR identification of isopropoxide as an intermediate in propylene-2-^{13}C conversion on zeolite HY. The signals at 150 and 250 ppm are due to the cyclopentenyl cations also formed under reaction conditions [15]. Here • denotes the ^{13}C-label in propylene.

^{13}C label, which occurs with involvement of protonated cyclopropane [126–128]. So the observation of intramolecular ^{13}C-label scrambling in a hydrocarbon adsorbed on zeolite could be considered as the evidence that carbenium ions are formed as intermediates in the hydrocarbon conversion on zeolite catalysts.

For but-2-ene adsorbed on zeolite H-ferrierite (H–FERR) the scrambling of the ^{13}C -label between the 1- and 2-positions of the carbon backbone was observed [129,130] (Fig. 21).

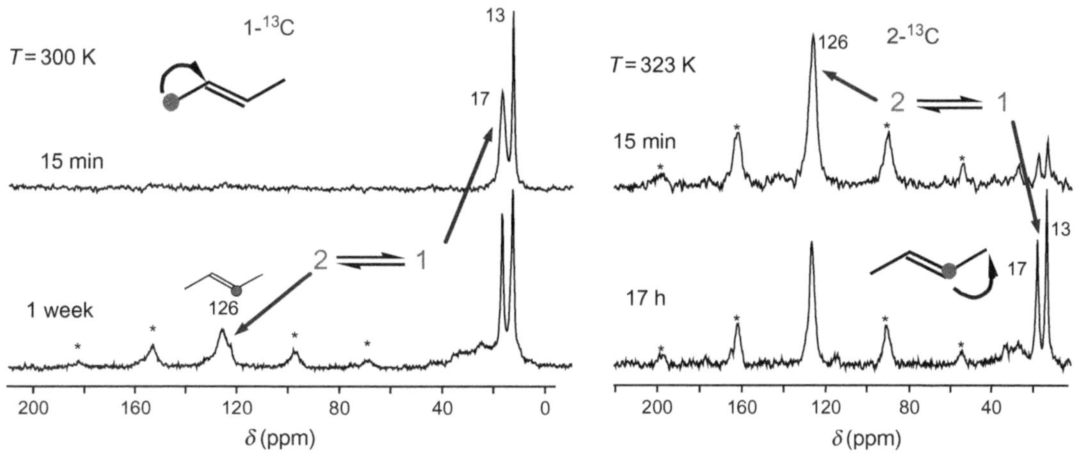

Fig. 21
Scrambling of the selective ^{13}C label (*dark circles*) between the 1- and 2-positions in but-2-ene adsorbed on H-ferrierite, according to in situ ^{13}C CP/MAS NMR data [129,130].

But-2-ene-1-^{13}C selectively labeled with ^{13}C in the methyl group (labeled in the 1-position) exhibits two signals at 13 and 17 ppm from methyl groups of *cis*- and *trans*-isomers of but-2-ene. After 1 week of keeping the sample at room temperature, the signal at 126 ppm from the inner CH= group of but-2-ene appears in the spectrum. This shows that the ^{13}C selective label migrates from the CH_3 group to the neighboring CH= group. If but-2-ene-2-^{13}C with the ^{13}C label in the 2-position was taken for but-2-ene adsorption, then the signal at 126 ppm from the label decreases, while two signals at 13 and 17 ppm from the methyl groups of but-2-ene increase with time in the spectrum (Fig. 21). So the scrambling of the ^{13}C label between the 1- and 2-positions in bute-2-ene occurs.

Under mild conditions, the ^{13}C label scrambling was also observed in oct-1-ene on H–ZSM-5 [131] and in propane on H–ZSM-5 at 523–573 K [132]. Because the scrambling of the ^{13}C label between the 1- and 2-positions is impossible without the formation of protonated methylcyclopropane, which is in the equilibrium with the butyl cation [127] (Scheme 1), one can conclude that carbenium ions as transient intermediates are formed on solid acid catalysts. Recent theoretical estimates of the stability of some cations in zeolites proved this possibility [133–135].

6.2 Monitoring the Reaction Kinetics In Situ With Solid-State NMR. Application to the Study of the Pathways of Alkane Activation

The study of the kinetics of the reaction on zeolite catalysts with solid-state NMR is usually performed in the following manner. The zeolite sample with adsorbed reactants in a sealed glass tube (batch reactor) is inserted into a zirconia rotor (*vide supra*, Fig. 6). A quick increase of the temperature of the spinning rotor directly in the MAS probe located in the NMR magnet initiates the chemical reaction. Further, NMR spectra are periodically registered in the course of the reaction. Reaction kinetics is analyzed on the basis of either an increase of the intensities

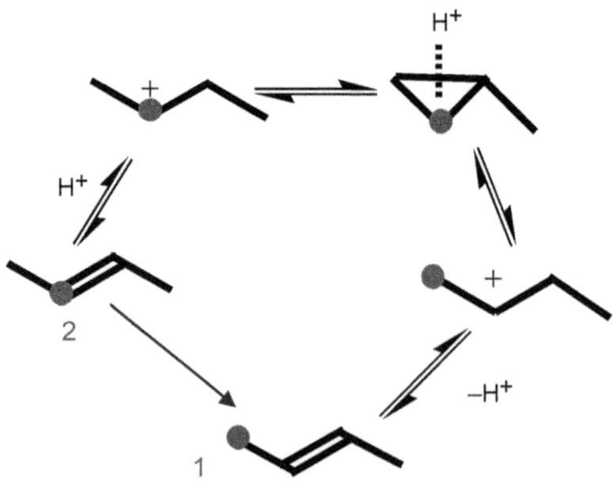

Scheme 1
Scrambling of a selective ^{13}C label (•) in but-2-ene on H-FERR via the formation of an intermediate butyl cation in the equilibrium with a protonated cyclopropane intermediate.

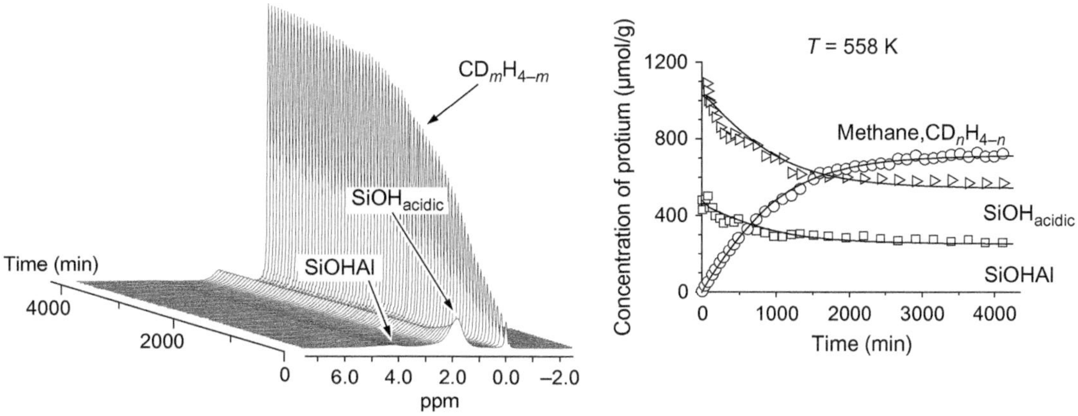

Fig. 22

Stack plot of the ^{1}H MAS NMR spectra at 558 K of methane-d_4 adsorbed on H-BEA (*left*) and kinetic curves (*right*) for H/D exchange of methane-d_4 with zeolite OH groups. The first spectrum (*bottom*) was recorded 5 min after the temperature was raised and equilibrated at 558 K and the last spectrum (*top*) after 4110 min of the reaction duration. The time between subsequent spectra recording was 60 min.

of the signals of the products or a decrease of the signal intensity of the reactants. Sometimes the intensity of the signals of both reactants and the products are analyzed.

Analysis of the reaction kinetics on zeolites with MAS NMR in situ has been applied to study the H/D exchange reactions [32,48,136–146], ^{13}C-label scrambling reactions [49,130], light alkane aromatization [47], etc. Some examples of the kinetics studies are offered here.

6.2.1 Mechanism of H/D exchange for methane on zeolite H–BEA

The simplest example of the application of solid-state NMR for in situ monitoring of the kinetics of H/D exchange concerns the study of hydrogen exchange between acidic OH groups of zeolites and deuterated methane-d_4. Fig. 22 shows the change with time of the ^{1}H MAS NMR spectrum of deuterated methane adsorbed on H–BEA zeolite. The intensity of the signal of methane at 0.0 ppm increases while the intensities of acidic bridged SiOHAl groups (4.0 and 5.4 ppm) and SiOH groups (1.8 ppm) (which are also acidic in H–BEA [81]) decrease in the course of the reaction. This occurs due to protium transfer from zeolite OH groups onto the alkane molecule and a reverse transfer of deuterium from the alkane to zeolite hydroxyl groups.

The temperature dependence of the rate constants k obtained from the analysis of the kinetics of the H/D exchange gives the activation energy for this reaction ($E_a = 138$ kJ/mol) [32,146], which is in agreement with theoretically estimated E_a, provided that the exchange occurs by a synchronous mechanism with the formation of a penta-coordinated carbon atom in a transition state [147] (Scheme 2). So, the kinetic study of H/D exchange with solid-state NMR confirms the theoretically suggested mechanism of hydrogen exchange between methane and the acid form of zeolite.

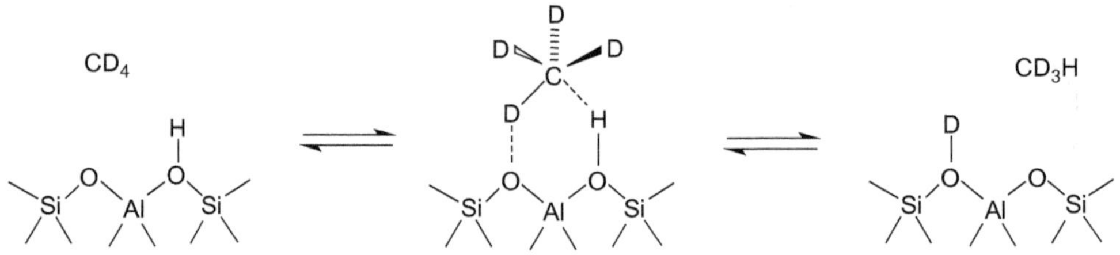

Scheme 2

Mechanism of H/D exchange between methane and acidic OH groups of zeolite that was theoretically justified [147] and confirmed with ^{1}H MAS NMR in situ kinetic measurements [32,146].

6.2.2 *Kinetics of H/D exchange for propane and n-butane on the acid form and Zn-modified zeolites*

Due to the expected noticeable difference in the ^{1}H chemical shifts of the methyl and methylene groups (eg, 1.0 and 1.5 ppm for the CH_3 and CH_2 groups of propane), ^{1}H MAS NMR allows identification of these separate fragments in adsorbed alkanes [136]). This gives us a chance to get into details of the mechanism of H/D exchange for C_3,C_4 alkanes and therefore to clarify the pathways of these alkanes activation on zeolite catalysts.

Fig. 23 illustrates the use of in situ ^{1}H MAS NMR kinetic study of H/D exchange between C_3,C_4 alkanes and acidic OH groups of zeolites H–ZSM-5 and Zn/H–ZSM-5 (Zn/H–BEA). In the case of propane-d_8 and *n*-butane-d_{10} on zeolite H–ZSM-5, both methyl and methylene groups are involved in the exchange (Fig. 23 A,C). This is indicated by the increase of the signals at 1.0 ppm (CH_3) and 1.5 ppm (CH_2) with time, which suggests that protium is transferred from OH groups to both methyl and methylene groups of deuterated alkane. For the zinc-modified zeolite Zn/HZSM-5 or Zn/H–BEA, the regioselectivity of the hydrogen exchange is observed: only the methyl groups of propane and *n*-butane are involved in the exchange (Fig. 23 B,D,E,F). The fact that zeolites differ in the regiospecificity of the exchange implies different mechanisms of exchange and different pathways of alkane activation by acid form and Zn-modified zeolites.

6.2.3 *Mechanism of H/D exchange for propane on zeolite H–ZSM-5*

For H/D exchange in propane on zeolite H–ZSM-5, the apparent rate constants for the methyl and methylene groups are equal [136]. This can be explained in terms of two alternative kinetic schemes: parallel and consecutive. In the parallel scheme, both the methyl group and the methylene group exchange hydrogen with acidic OH groups of the zeolite.

In the consecutive scheme, only the methyl group is directly involved in H/D exchange with the OH groups. There is also intramolecular hydrogen transfer from the methyl group to the methylene group. The latter process is much more rapid. It turned out that the exchange mechanism can be ascertained by examining the H/D exchange kinetics for three differently

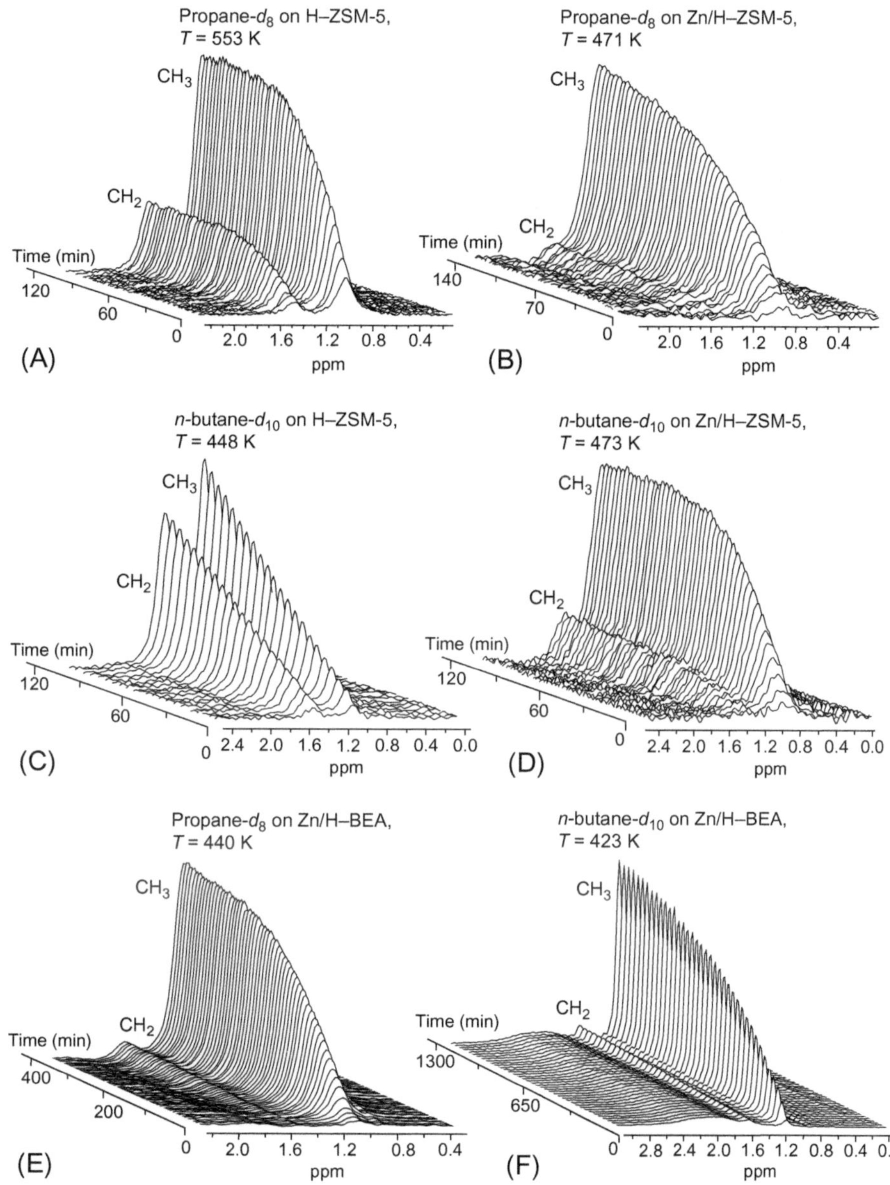

Fig. 23
Stack plots of ^{1}H MAS NMR spectra for propane-d_8 and n-butane-d_{10} adsorbed on acidic and Zn-modified zeolites ZSM-5 and BEA. The increase of the intensity of only methyl groups with the reaction time for Zn/H-ZSM-5 and Zn/H-BEA demonstrates regioselectivity of the H/D exchange [48,136,139,141,143,145].

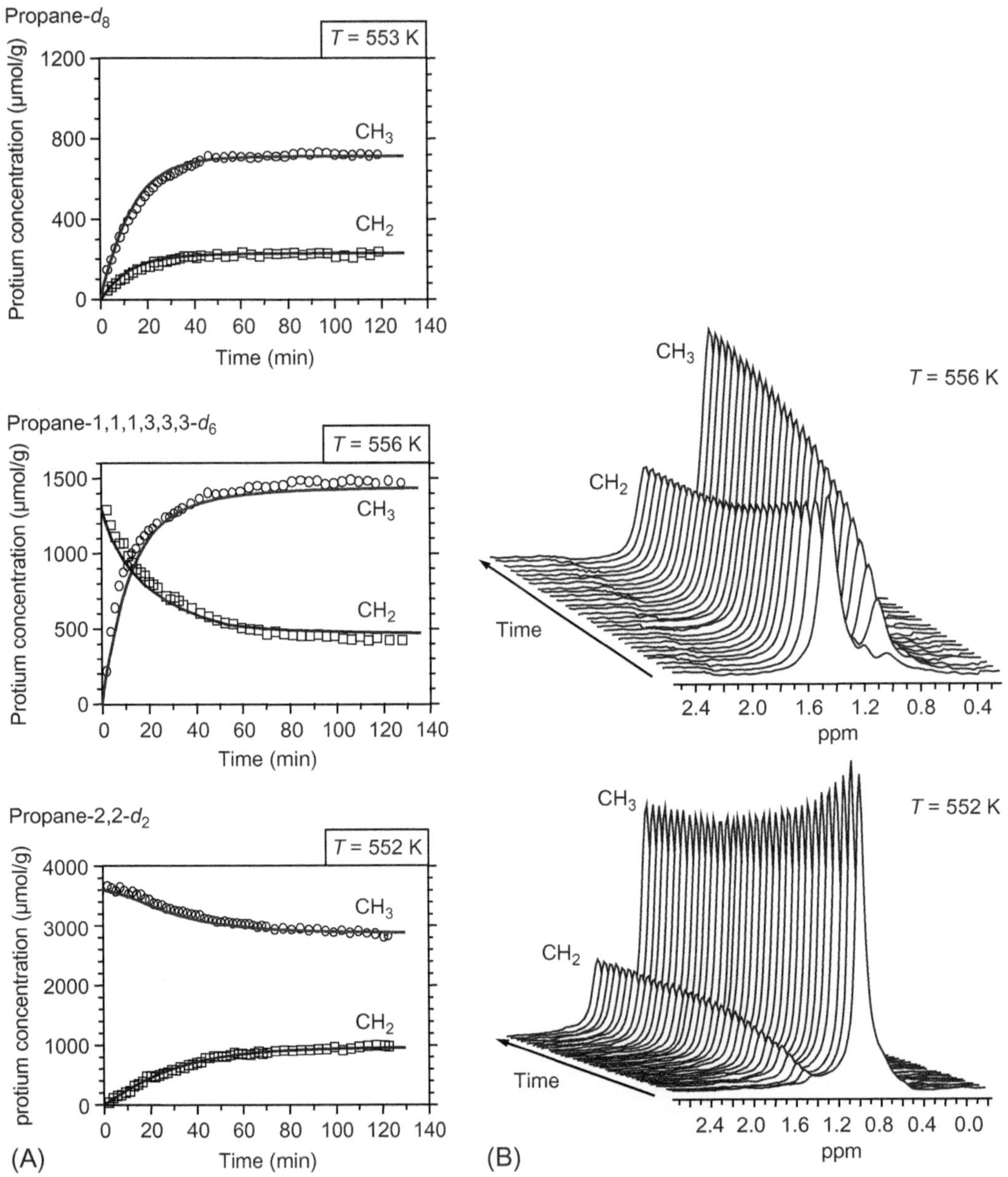

Fig. 24

(A) Kinetics of H/D exchange in propane-d_8, propane-1,1,1,3,3,3-d_6, and propane-2,2-d_2 on zeolite H-ZSM-5; (B) corresponding time evolution of the ^{1}H MAS NMR spectra of propane-1,1,1,3,3,3-d_6 and propane-2,2-d_2 on zeolite H-ZSM-5 [48,136,139]. For propane-d_8, the evolution of the NMR spectrum during the reaction is shown in Fig. 23A.

deuterated propanes: propane-d_8, propane-1,1,1,3,3,3-d_6, and propane-2,2-d_2 (Fig. 24). Simulation of experimental kinetics for these differently deuterated propanes demonstrates that H/D exchange in propane proceeds via a cyclic mechanism (Scheme 3). The criterion of the mechanism validity was the equality of the constants k_{CH_3}, k_{CH_2}, and k_{H_SCR} for the three propanes [48]. It was found for the cyclic mechanism that the exchange rate constants for the methyl group is 1.5 times larger than that for the methylene group, $k_{CH_3} \approx 1.5k_{CH_2}$, but intramolecular hydrogen scrambling takes place five times more slowly: $k_{CH_3} \approx 5\ k_{H\_SCR}$. It has been concluded that both the methyl group and the methylene group are independently involved in hydrogen exchange with Brønsted acid sites of the zeolite via a concerted mechanism, similar to the exchange of methane (Scheme 2). This inference was confirmed by the exchange activation energy data (107–125 kJ/mol) [48,136], which coincide with the quantum chemical estimates of the activation energy for H/D exchange via the intermediate formation of a pentacoordinated carbonium-ion transition state [147–151].

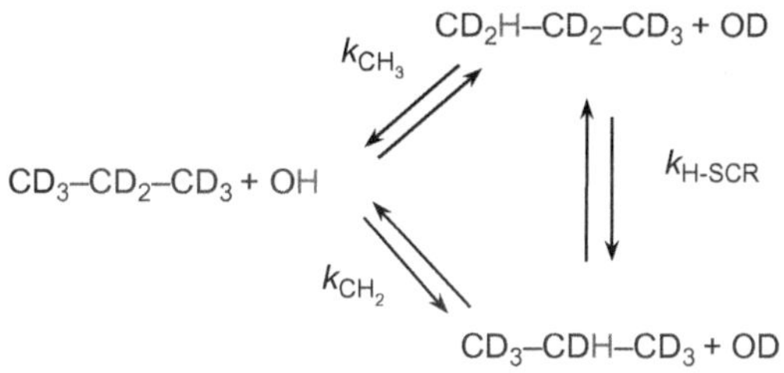

Scheme 3
Cyclic scheme of H/D exchange between propane and zeolite HZSM-5.

6.2.4 Analysis of ^{13}C-Label Scrambling in Adsorbed Propane

Kinetic analysis of intramolecular ^{13}C scrambling (discovered by Derouane et al. [132]) from the methylene group of propane to its methyl groups using in situ ^{1}H MAS NMR [49] and ^{13}C MAS NMR was also performed [48,139] (Fig. 25). This analysis demonstrated that the rate of intramolecular carbon scrambling is two times lower that the rate of intramolecular hydrogen scrambling: $k_{H\_SCR} \approx 2k_{C\_SCR}$. However, the activation energies of these processes are roughly equal: $E_{H\_SCR} \approx E_{C\_SCR}$ (Fig. 26). Similar kinetics of intramolecular carbon and hydrogen scrambling were observed in 1992 by Saunders et al. [152] for the isopropyl cation in a superacid solution: the activation energies of carbon scrambling and intramolecular hydrogen scrambling were equal, and the hydrogen scrambling rate was approximately three times higher, $k_{H\_SCR} \approx 3k_{C\_SCR}$. The kinetic similarity of the carbon and proton scrambling observed in the superacid to those observed on the zeolite suggests that the carbon and proton scrambling in propane on H–ZSM-5 can take place via the intermediate formation of the isopropyl cation.

Thus, the in situ monitoring by solid-state NMR of the kinetics of H/D exchange and ^{13}C label scrambling for propane on zeolite H–ZSM-5 affords information on the mechanism

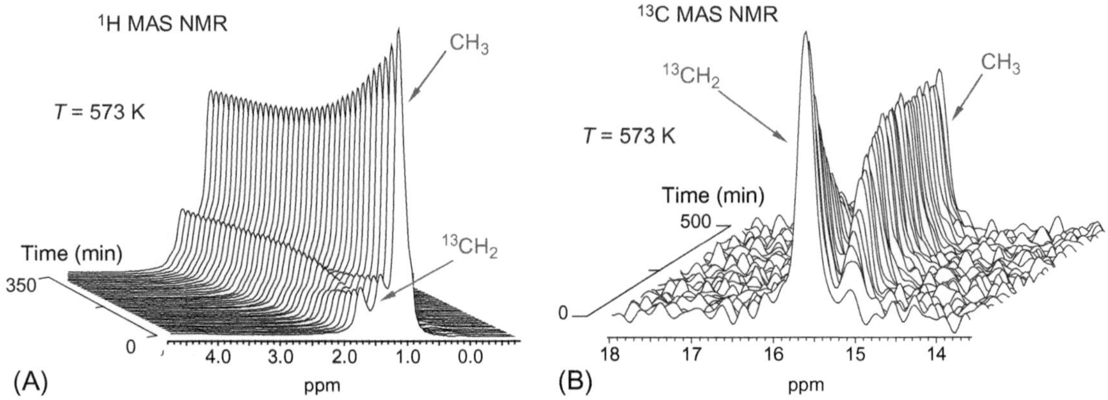

Fig. 25
In situ (A) ^1H and (B) ^{13}C MAS NMR monitoring of the kinetics of intramolecular ^{13}C scrambling in propane-2-^{13}C on zeolite H-ZSM-5 at 573 K [48,49].

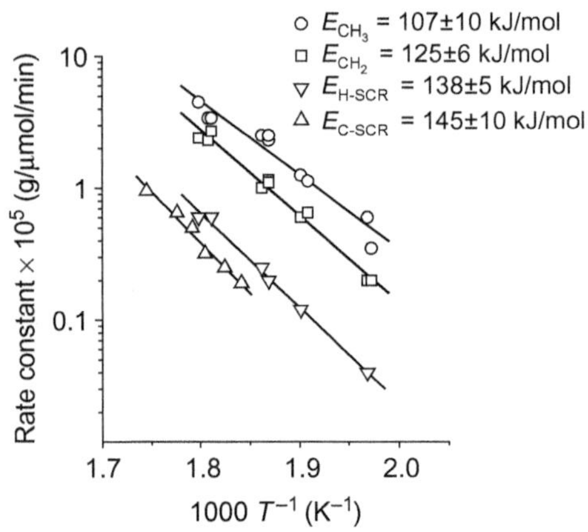

Fig. 26
Arrhenius plots of the rate constants (○) k_{CH_3}, (□) k_{CH_2}, (∇) k_{H_SCR}, and (Δ) k_{C_SCR} for the cyclic scheme of H/D exchange in deuterated propane and for ^{13}C scrambling in propane-2-^{13}C on zeolite H-ZSM-5 and the corresponding activation energies [48,49,139].

of H/D exchange of propane hydrogens with acidic OH groups, demonstrates the existence of intramolecular hydrogen and carbon scrambling in propane, provides the arguments for involvement of both carbonium and carbenium ion intermediates in these processes.

6.2.5 The mechanism of H/D hydrogen exchange for C_1–C_4 alkanes on Zn-modifed zeolites ZSM-5 and BEA

As we noted earlier, it is possible to detect the H/D exchange into the different fragments of an alkane molecule with ^1H MAS NMR. This allows establishment of peculiarities of the mechanism of hydrogen exchange and alkane activation pathway on Zn-modified zeolites, active catalysts of alkane aromatization. The activation pathway is established on the basis of comparative analysis of the rate and regioselectivity of hydrogen exchange for H-form and Zn-modified zeolites.

A kinetic analysis of hydrogen exchange between C_1–C_4 alkanes and Brønsted acid sites of the zeolites has demonstrated that the introduction of zinc into a zeolite resulted in an increase of the rate of hydrogen exchange and a decrease of the activation energy. For methane and ethane, the rate increases by three orders of magnitude and the activation energy decreases from 138 to 88 kJ/mol [32] (Fig. 27A). For propane and *n*-butane, the rate of H/D exchange is higher one order of magnitude and the activation energy is lower for the Zn-modified zeolites (Fig. 27B). The regioselectivity of H/D exchange is observed for the Zn-modified zeolites in the case of exchange of propane or *n*-butane: only the methyl groups of these two alkanes are involved in H/D exchange [141,145] (Fig. 23).

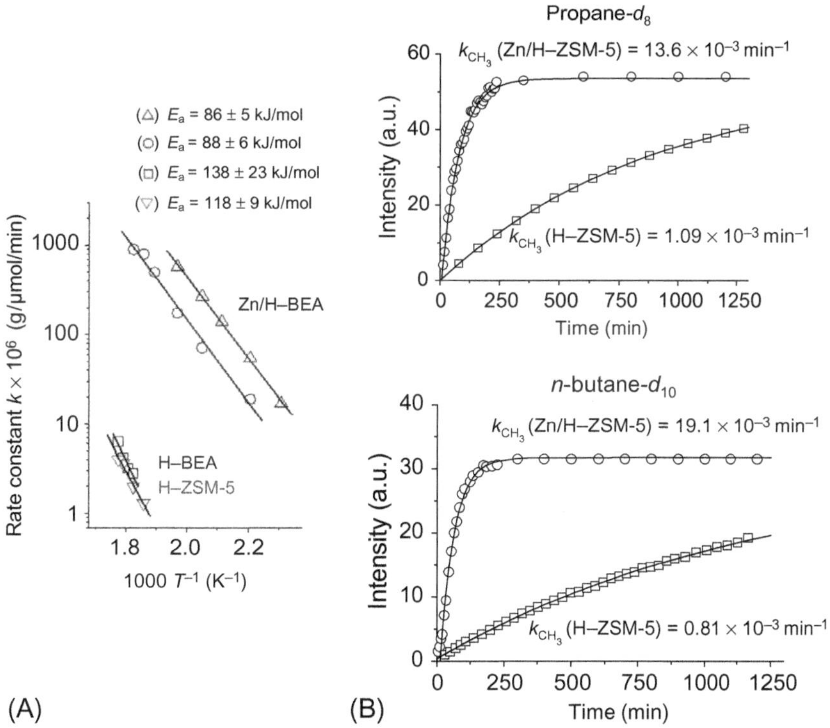

Fig. 27
(A) Arrhenius plots for the H/D exchange reactions: methane-d_4 on H-BEA (□); methane-d_4 on H-ZSM-5 (∇); methane-d_4 (Δ) and ethane-d_6 (o) on Zn/H-BEA. (B) Kinetics of proton transfer from the acidic SiOHAl groups of zeolites: (o) Zn/H-ZSM-5 (Si/Al = 35) and (□) H-ZSM-5 (Si/Al = 35) to the methyl groups of adsorbed propane-d_8 (T = 453 K, $E_a^{Zn/H\ ZSM\text{-}5} = 80 kJ/mol$, $E_a^{H\text{-}ZSM\text{-}5} = 107 kJ/mol$) and *n*-butane-$d_{10}$ (T = 433 K, $E_a^{Zn/H\ ZSM\text{-}5} = 99 kJ/mol$, $E_a^{H\text{-}ZSM\text{-}5} = 108 kJ/mol$) [32,48,136,141,145].

A dramatic increase of the reaction rates with lowering the apparent activation energies evidently demonstrates a promoting effect of Zn on the reaction of H/D hydrogen exchange for C_1–*n*–C_4 alkanes. Also, the regioselectivity of H/D exchange for propane and *n*-butane is provided evidently by an assistance of the Zn species loaded in zeolite.

Altogether, a dramatic change of the kinetic parameters of H/D exchange upon modification of the zeolites with Zn toward acceleration of the reaction of the H/D exchange for C_1–*n*–C_4 alkanes and the involvement in the exchange of only structurally distinct sites of propane and *n*-butane (methyl groups) points to the assistance of the reaction of H/D exchange of alkanes with Brønsted acid sites by Zn species. Acceleration of the reaction of H/D exchange with its decreased activation energy for Zn-modified zeolites demonstrates that the C–H bonds in alkanes become more easily broken and reversibly formed with intermediate involvement of Zn-species.

A promoting effect of Zn on H/D exchange is accounted for by involvement of Zn-alkyl species formed by the alkane dissociative adsorption on Zn-modified zeolite [142,153–156] in this reaction. The formation of Zn-alkyl species presumably occurs on small ZnO clusters existing inside the zeolite pores [153].

The pathway of H/D exchange for C_1–*n*–C_4 alkanes with involvement of Zn-alkyl species is outlined in Scheme 4 (pathway 1). Dissociative adsorption of the deuterated alkane on zinc oxide sites, which precedes the reaction of H/D exchange, leads to Zn-alkyl species and Zn–OD groups. Protonation of the zinc-alkyl species by the vicinal acidic OH group of the zeolite with synchronous reverse transfer of the deuterium from the Zn–OD to the

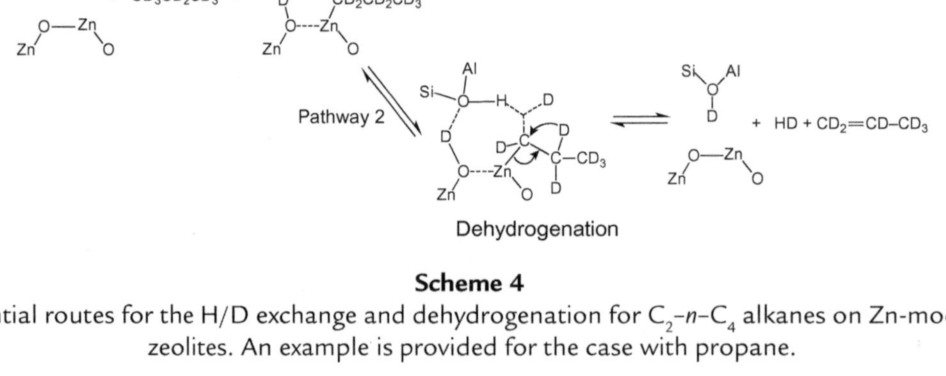

Scheme 4
Potential routes for the H/D exchange and dehydrogenation for C_2-*n*-C_4 alkanes on Zn-modified zeolites. An example is provided for the case with propane.

zeolite acid site, provides finally a transfer of protium from the SiOHAl group of the zeolite to the deuterated alkane molecule. This mechanism could provide the kinetic parameters of the exchange significantly different from those for pure acidic zeolite. The mechanism also rationalizes involvement of only methyl groups of C_3 and n-C_4 alkanes in the exchange. Only the hydrogens of the CH_2 group attached directly to Zn can be involved in the exchange. So, the data are the experimental demonstration of the earlier suggested [157] synergetic action of Zn-species and acid sites in breaking and formation of the C–H bonds in alkane activation.

It becomes clear from our data that the activation of the C–H bond in light alkanes by the zeolite Brønsted acid site does depend on Zn-species, which forms an intermediate with the alkane. One can assume that the interaction of Zn-alkyl species with zeolite acid sites can provide not only the H/D exchange reaction, but also a dehydrogenation step (Scheme 4, pathway 2) offering olefins for their subsequent aromatization steps. Pathway 2 rationalizes the synergy of the acid and Zn-sites (ZnO clusters) for effective aromatization of light alkanes by Zn-modified zeolites. The synergetic action of Zn and Brønsted sites in enhancing the aromatization ability could be achieved by involvement in the activation of the alkane C–H bond of both the zeolite acid OH groups and Zn-alkyl species formed on Zn–sites. The intermediate or transition state, which involves the zeolite acid sites, the alkane and Zn-sites provide an alkane dehydrogenation presumably by the energetically most favorable way. Thus, the enhancement of the alkane conversion toward aromatic molecules formation on Zn-modified zeolites could be achieved. So, the effect of Zn, which was earlier attributed either to the dehydrogenation ability of the loaded metal [158] or its ability to afford a recombinative desorption of H-atoms [159], is displayed as an influence of zinc directly on activation of alkane C–H bonds by the catalyst Brønsted acid sites.

6.3 Elucidation of Reaction Mechanisms by Identification of the Reaction Intermediates

A classical approach for establishing the mechanism of the reaction by any spectroscopic technique consists of identifying reaction intermediates during in situ monitoring of the reaction occurrence. The signals from the intermediate species should disappear while those of the final products increase in the course of the reaction. For the reaction on the acid forms of zeolites, it is vital to establish the nature of intermediates: whether the intermediate represents carbenium ion or alkoxy species.

A glowing example of establishing the reaction mechanism with identification of the nature and structure of the intermediate concerns the Ritter reaction [160] occurring on acid zeolites. This reaction represents the interaction between an olefin or alcohol and acetonitrile (or another nitrile) yielding *N*-alkylacetamide. Although some hypotheses as to the mechanism of the Ritter reaction were made as early as this reaction was discovered, only the use of ^{13}C CP/MAS NMR spectroscopy made it possible to identify the intermediate and to establish the reaction mechanism conclusively [161,162].

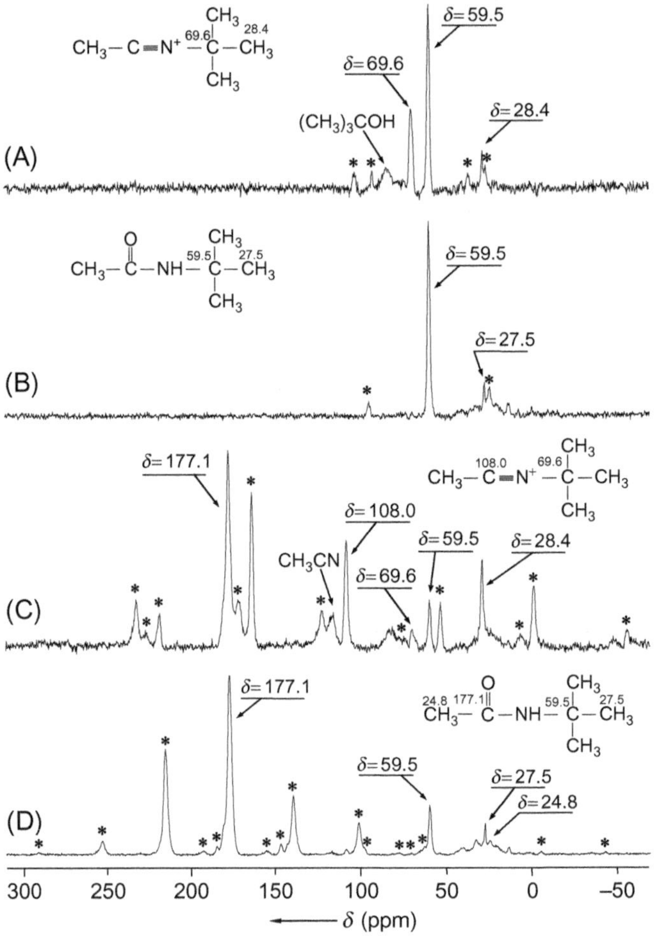

Fig. 28
^{13}C CP/MAS NMR spectra of the products of the reaction between *tert*-butyl alcohol and acetonitrile on zeolite H-ZSM-5 at 296 K: (A, B) reaction between *tert*-BuOH-1-^{13}C (82% ^{13}C) and unlabeled CH_3CN; (C, D) reaction between *tert*-BuOH-1-^{13}C (10% ^{13}C) and CH_3CN-1-^{13}C (82% ^{13}C). The spectra were recorded (A, C) 4 h and (B, D) 4 days after the beginning of the reaction. Asterisks (*) denote spinning side-bands.

In situ NMR monitoring of the reaction between *tert*-butyl alcohol and acetonitrile on zeolite H–ZSM-5 revealed signals characteristic of the *N-tert*-butylacetonitrilium cation at 69.6 and 108 ppm (Fig. 28), whose intensity decreased during the reaction, while the signals from the final product, *N-tert*-butylacetamide, at 59.5 and 177 ppm increased as the reaction proceeded. Thus, using ^{13}C CP/MAS NMR spectroscopy, it was demonstrated for the first time that the *N*-alkylnitrilium cation was the key intermediate in the Ritter reaction and this reaction proceeded according to Scheme 5.

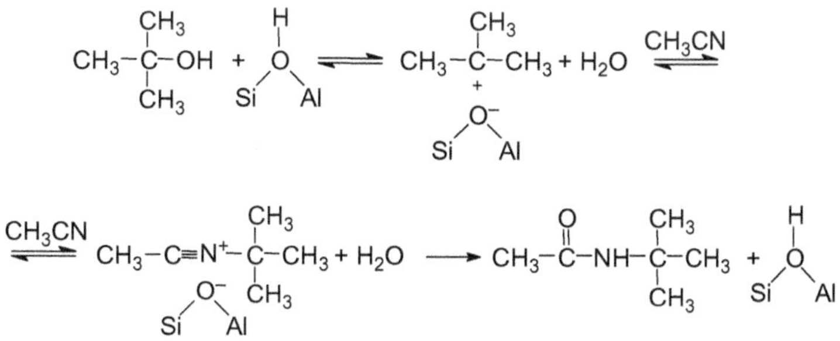

Scheme 5
Mechanism of the Ritter reaction on an acidic zeolite [161,162].

6.4 Direct Observation of Shape Selectivity in Zeolite ZSM-5 by ^{13}C MAS NMR

The phenomenon of shape selectivity in the reactions performed on zeolites arises from the fact that the probabilities of forming various products in the narrow intracrystalline cavities and channels are largely determined by the dimension and configuration of the zeolite pores [163]. Although there was extensive indirect evidence of the shape selectivity of catalysis by zeolites [164,165], there was no direct observation of this effect based on analysis of the reaction products inside the pores. The ability of ^{13}C MAS NMR to identify the structure of organic molecules directly in the intracrystalline voids of zeolites allowed direct observation of the shape selectivity effect. Anderson and Klinowski analyzed the products of methanol conversion into gasoline on zeolite ZSM-5 by in situ ^{13}C MAS NMR and reliably established the formation in zeolite pores of a number of methyl-substituted benzenes that had never been observed at the reactor outlet [166]. The origin of this effect was obvious (Scheme 6): different trimethylbenzenes are formed in the zeolite pores, however, the bulkiest isomers cannot leave the pore interior because of the limited pore exit size in the zeolite structure. Thus, the

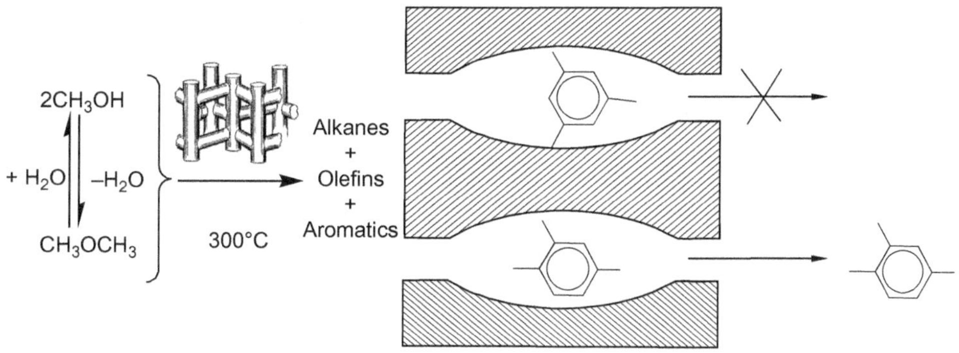

Scheme 6
Direct observation of the shape selectivity of zeolite ZSM-5 by in situ ^{13}C MAS NMR spectroscopy [166].

phenomenon of the shape selectivity was proved experimentally by applying in situ NMR spectroscopy to the comparative analysis of the adsorbed products inside and outside the zeolite.

6.5 Elucidation of the Mechanism of Methane and Propane Co-Aromatization on Zn-Modified High Silica Zeolite

Methane was claimed to be converted nonoxidatively into aromatic compounds over Ga or Zn modified high-silica zeolites in the presence of heavier alkanes or olefins [167]. The reported conversion of methane reached 40%. Later on, the experiments in which ^{13}C-labeled methane was used did not confirm the presence of the ^{13}C-labeled atoms from methane in the aromatization products [168]. This result gave rise to skepticism as to whether methane-involved aromatization occurred at all [168].

The capability of solid-state NMR to trace the ^{13}C label transfer from the initial reactants into the reaction products in the course of the reaction made it possible to demonstrate the possibility of the reaction of methane and propane co-aromatization on zeolite catalysts and to establish the mechanism of this reaction [155,169].

The ^{13}C MAS NMR spectrum of the products of the reaction between ^{13}C-labeled methane and unlabeled propane indicate the transfer of the ^{13}C-label from methane to the aromatic products. The two signals from methyl-substituted benzenes at 20 and 130 ppm, which were observed only when $^{13}CH_4$ was used, provide the evidence that methane is involved in the aromatization reaction (Fig. 29, spectra A,B). In addition, it was demonstrated that

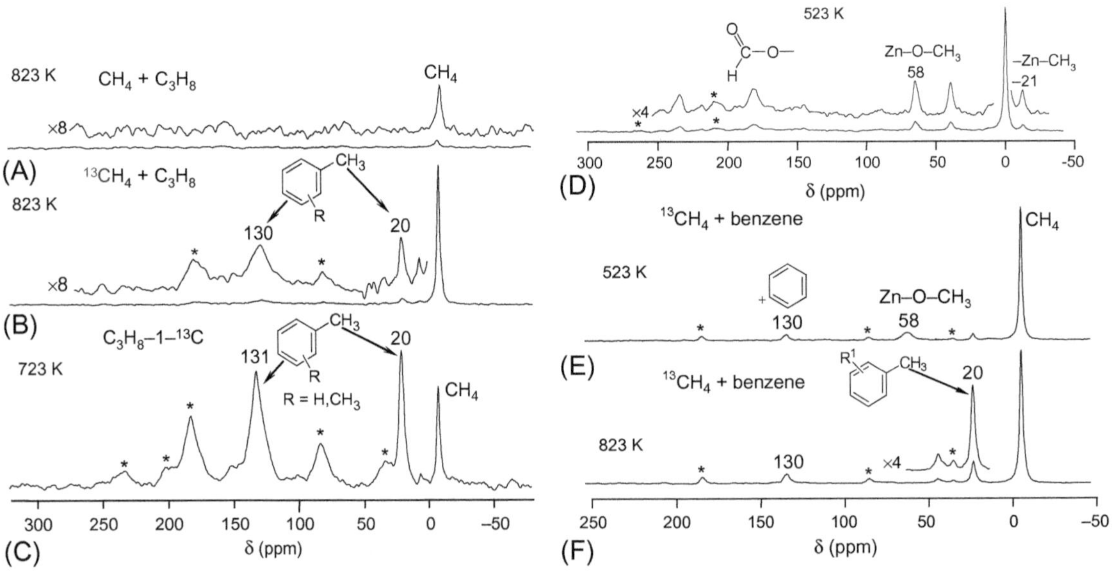

Fig. 29
^{13}C CP/MAS NMR spectra of the products of methane-^{13}C and propane conversion on zeolite Zn/HBEA. Asterisks (*) denote spinning side bands.

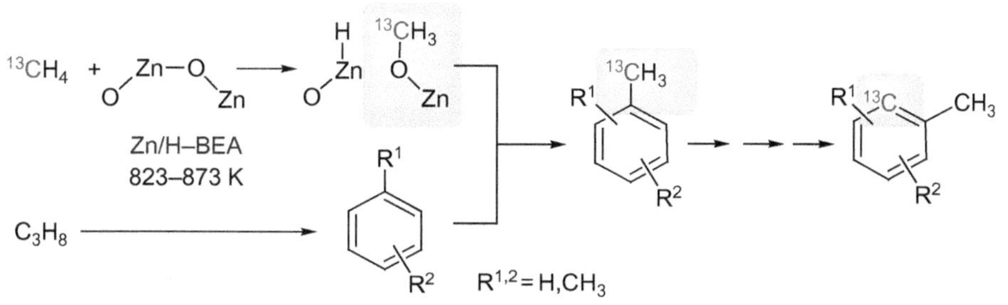

Scheme 7

Mechanism of methane-propane co-aromatization on zeolite Zn/HBEA.

propane in the absence of methane is converted into aromatic compounds well below the temperature at which methane is involved in aromatization (Fig. 29, spectra B,C). In the absence of propane, the interaction between methane and zeolite Zn/HBEA yields methylzinc ($ZnCH_3$) and methoxide ($ZnOCH_3$) species and formate fragments, which undergo further conversion into acetaldehyde and acetic acid (Fig. 29D). In the presence of benzene, only the formation of the methoxide $ZnOCH_3$ is observed, which is apparently not oxidized by oxygen of the defected ZnO structure (Fig. 29E). At 823 K, benzene is methylated by $ZnOCH_3$, yielding methyl-substituted aromatics, namely, toluene and xylenes (Fig. 29F). It was thereby found that methane participated in the methane–propane co-aromatization reaction by alkylating the aromatic compounds that resulted from propane, as is illustrated by Scheme 7.

References

[1] Lippmaa E, Mägi M, Samoson A, Tarmak M, Engelhardt G. Investigation of the structure of zeolites by solid-state high-resolution ^{29}Si NMR spectroscopy. J Am Chem Soc 1981;103:4992–6.

[2] Ramdas S, Thomas JM, Klinowski J, Fyfe CA, Hartman JS. Ordering of aluminum and silicon in synthetic faujasites. Nature 1981;292:228–30.

[3] Freude D, Behrens JH. Investigation of ^{27}Al NMR chemical-shifts in zeolites of the faujasite type. Cryst Res Technol 1981;15:K36-K8.

[4] Freude D, Hunger M, Pfeifer H. Study of bronsted acidity of zeolites using high-resolution proton magnetic-resonance with magic-angle spinning. Chem Phys Lett 1982;91:307–10.

[5] Zamaraev KI, Mastikhin VM. New possibilities of NMR-spectroscopy in studies of adsorption and catalysis. Colloids Surf 1984;12:401–27.

[6] Hunger M. Bronsted acid sites in zeolites characterized by multinuclear solid-state NMR spectroscopy. Cat Rev Sci Eng 1997;39:345–93.

[7] Ito T, Fraissard J. ^{129}Xe NMR study of xenon adsorbed on Y zeolites. J Chem Phys 1982;76:5225–9.

[8] Ripmeester JA. Nuclear shielding of traped xenon obtained by proton-enhanced, magic-angle spinning ^{129}Xe NMR-spectroscopy. J Am Chem Soc 1982;104:289–90.

[9] Ripmeester JA. Characterization of void spaces in mordenites by means of ^{129}Xe NMR. J Magn Reson 1984;56:247–53.

[10] Benslama R, Fraissard J, Albizane A, Fajula F, Figueras F. An example of the technique of studing adsorbed xenon by ^{129}Xe NMR - approximate determination of the internal void space of zeolite beta. Zeolites 1988;8:196–8.

[11] Fraissard J, Ito T. ^{129}Xe NMR study of adsorbed xenon. A new method for studing zeolites and metal zeolites. Zeolites 1988;8:350–61.
[12] Ito T, Springuel-Huet MA, Fraissard J. ^{129}Xe NMR study of xenon adsorbed on Na-OMEGA and Na and K-ferrierite. Characterization of void space, defects, and intergrowth of ferrierite-mordenite structures. Zeolites 1989;9:68–73.
[13] Springuel-Huet MA, Bonardet JL, Gedeon A, Fraissard J. ^{129}Xe MMR for studying surface heterogeneity: well-known facts and new findings. Langmuir 1997;13:1229–36.
[14] Derouane EG, Gilson J-P, Nagy JB. In situ characterizatin of carbonaceous residues from zeolite-catalysed reactions using high resolutin solid state ^{13}C NMR spetroscopy. Zeolites 1982;2:42–6.
[15] Haw JF, Richardson BR, Oshio IS, Lazo ND, Speed JA. Reactions of propene on zeolite HY catalyst studied by in situ variable-temperature solid-state nuclear magnetic resonance spectroscopy. J Am Chem Soc 1989;111:2052–8.
[16] Eckman RR, Vega AJ. Deuterium solid-state NMR study of the dynamics of molecules sorbed by zeolites. J Phys Chem 1986;90:4679–83.
[17] Boddenberg B, Burmeister R. ^{2}H NMR studies on the dynamics of ethene molecules in silver-exchanged X-type zeolites. Zeolites 1988;8:480–7.
[18] Engelhardt G, Michel D. High-resolution solid-state NMR of silicates and zeolites. Chichester, UK: John Wiley and Sons; 1987.
[19] Hunger M. Solid-state NMR spectroscopy. In: Chester AW, Derouane E, editors. Zeolite characterization and catalysis. New York: Springer; 2009. p. 65–105.
[20] Engelhardt G. Solid-state NMR spectroscopy applied to zeolites. In: van Bekkum H, Flanigen EM, Jacobs PA, Jansen JC, editors. Studies in surface science and catalysis. New York: Elsevier; 2001. p. 387–418.
[21] Jentys A, Lercher JA. Techniques for zeolite characterization. In: van Bekkum H, Flanigen EM, Jacobs PA, Jansen JC, editors. Studies in surface science and catalysis. New York: Elsevier; 2001. p. 345–86.
[22] Bloch F, Hansen WW, Packard M. Nuclear induction. Phys Rev 1946;69:127.
[23] Purcell EM, Torrey HC, Pound RV. Resonance adsorption by nuclear magnetic moments in a solid. Phys Rev 1946;69:37–8.
[24] Andrew ER, Bradbury A, Eades RG. Nuclear magnetic resonance spectra from a crystal rotated at high speed. Nature 1958;182:1659.
[25] Gullion T, Schaefer J. Rotational-echo double-resonance NMR. J Magn Reson 1989;81:196–200.
[26] van Eck ERH, Veeman WS. Spin-density description of rotational-echo double resonance, transferred-echo double-resonance and 2-dimensional transferred-echo double resonance solid-state nuclear-magnetic-resonance. Solid State Nucl Magn Reson 1993;2:307–15.
[27] Peng L, Huo H, Liu Y, Grey CP. ^{17}O magic angle spinning NMR studies of Broensted acid sites in zeolites HY and HZSM-5. J Am Chem Soc 2007;129:335–46.
[28] Huang Y, Yan Z, Richer R. Characterization of BAPO- and SAPO-based mesoporous materials by solid-state NMR spectroscopy. Chem Mater 2005;17:6545–54.
[29] van Eck ERH, Janssen R, Maas WEJR, Veeman WS. A novel application of nuclear spin-echo double-resonance to aluminophosphates and aluminosilicates. Chem Phys Lett 1990;174:428–32.
[30] Grey CP, Vega AJ. Determination of the quadrupole coupling constant of the invisible aluminum spins in zeolite Hy with ^{1}H/ ^{27}Al TRAPDOR NMR. J Am Chem Soc 1995;117:8232–42.
[31] Deng F, Yue Y, Ye CH. Observation of nonframework Al species in zeolite beta by solid-state NMR spectroscopy. J Phys Chem B 1998;102:5252–6.
[32] Stepanov AG, Arzumanov SS, Gabrienko AA, Toktarev AV, Parmon VN, Freude D. Zn-promoted hydrogen exchange for methane and ethane on Zn/H-BEA zeolite. In situ ^{1}H MAS NMR kinetic study. J Catal 2008;253:11–21.
[33] Deng F, Yue Y, Ye CH. H-1/Al-27 TRAPDOR NMR studies on aluminum species in dealuminated zeolites. Solid State Nucl Magn Reson 1998;10:151–60.

[34] Hu W, Luo Q, Su Y, Chen L, Yue Y, Ye C, et al. Acid sites in mesoporous Al-SBA-15 materials as revealed by solid-stateNMR spectroscopy. Microporous Mesoporous Mater 2006;92:22–30.
[35] Pines A, Gibby MG, Waugh JS. Proton-enhanced nuclear induction spectroscopy. ^{13}C chemical shielding anisotropy in some organic solids. Chem Phys Lett 1972;15:373–6.
[36] Pines A, Gibby MG, Waugh JS. Proton-enhanced nuclear induction spectroscopy. A method for high resolution NMR of dilute spins in solids. J Chem Phys 1972;56:1776–7.
[37] Pines A, Gibby MG, Waugh JS. Proton-enhanced NMR of dilute spins in solids. J Chem Phys 1973;59:569–90.
[38] Hartmann SR, Hahn EL. Nuclear double resonance in the rotating frame. Phys Rev 1962;128:2042–53.
[39] Anderson MW, Klinowski J. Monitoring organic products of catalytic reactions on zeolites by two-dimensional J-resolved solid-state NMR. Chem Phys Lett 1990;172:275–8.
[40] Brown DW, Nakashima TT, Rabenstein DL. Simplification and assignment of carbon-13 NMR spectra with spin-echo Fourier transform techniques. J Magn Reson 1981;45:302–14.
[41] Carpenter TA, Klinowski J, Tennakoon DTB, Smith CJ, Edwards DC. Sealed capsules for convenient acquasition of variable temperature controlled-atmosphere magic-angle-spinning NMR spectra of solids. J Magn Reson 1986;68:561–3.
[42] Beck LW, White JL, Haw JF. Reliable method for sealing glass ampuls for MAS-NMR experiments on air-sensitive samples. J Magn Reson 1992;99:182–3.
[43] Ivanova II, Derouane EG. Advances in the insitu ^{13}C MAS NMR characterization of zeolite catalyzed hydrocarbonreactions. Stud Surf Sci Catal 1994;85:357–90.
[44] Ernst H, Freude D, Mildner T, Wolf I. Laser-supported high-temperature MAS NMR for time-resolved in situ studies of reaction steps in heterogeneous catalysis. Solid State Nucl Magn Reson 1996;6:147–56.
[45] Haw JF. In situ spectroscopy in heterogeneous catalysis. Weinheim: Wiley VCH Verlag GmbH; 2002. p. 276.
[46] Hunger M, Wang W. Characterization of solid catalysts in the functioning state by nuclear magnetic resonance spectroscopy. Adv Catal 2006;50:149–225.
[47] Arzumanov SS, Gabrienko AA, Freude D, Stepanov AG. In situ high temperature MAS NMR study of the mechanisms of catalysis. Ethane aromatization on Zn-modified zeolite BEA. Solid State Nucl Magn Reson 2009;35:113–9.
[48] Arzumanov SS, Reshetnikov SI, Stepanov AG, Parmon VN, Freude D. In situ ^{1}H and ^{13}C MAS NMR kinetic study of the mechanism of H/D exchange for propane on zeolite HZSM-5. J Phys Chem B 2005;109:19748–57.
[49] Stepanov AG, Arzumanov SS, Ernst H, Freude D. ^{1}H MAS NMR monitoring of the ^{13}C-labeled carbon scrambling for propane in zeolite H-ZSM-5. Chem Phys Lett 2006;420:574–6.
[50] Hunger M, Horvath T. A new MAS NMR probe for in situ investigations of hydrocarbon conversion on solid catalysts under continuous-flow conditions. J Chem Soc Chem Commun 1995;1423–4.
[51] Hunger M. In situ flow MAS NMR spectroscopy: state of the art and applications in heterogeneous catalysis. Prog Nucl Magn Reson Spectrosc 2008;53:105–27.
[52] Sundaramurthy V, Cognec JP, Thomas K, Knott B, Engelke F, Fernandez C. A continuous gas flow MAS NMR probe for operando studies of hydrocarbon conversion on heterogeneous catalysts. CR Chim 2006;9:459–65.
[53] Klinowski J. Solid-state NMR-studies of molecular-sieve catalysts. Chem Rev 1991;91:1459–79.
[54] Engelhardt G, Lohse U, Samoson A, Magi M, Tarmak M, Lippmaa E. High-resolution ^{29}Si NMR of dealuminated and ultrastable Y zeolites. Zeolites 1982;2:59–62.
[55] Rakiewicz EF, Mueller KT, Jarvie TP, Sutovich KJ, Roberie TG, Peters AW. Solid-state NMR studies of silanol groups in mildly and highly dealuminated faujasites. Microporous Mater 1996;7:81–8.
[56] Fyfe CA, Gobbi GC, Klinowski J, Thomas JM, Ramdas S. Resolving crystallographically distinct tetrahedral sites in silicalite and ZSM-5 by solid-state NMR. Nature 1982;296:530–3.
[57] Fyfe CA, Strobl H, Kokotailo GT, Kennedy GJ, Barlow GE. Ultra-high-resolution ^{29}Si solid-state MAS NMR investigation of sorbate and temperature-induced changes in the lattice structure of zeolite ZSM-5. J Am Chem Soc 1988;110:3373–80.

[58] Thomas JM, Klinowski J, Ramdas S, Hunter BK, Tennakoon DTB. The evaluation of non-equivalent tetrahedral sites in zeolites and related aluminosilicates. Chem Phys Lett 1983;102:158–62.
[59] Radeglia R, Engelhardt G. Correlation of Si-O-Si or Si-O-Al angles and ^{29}Si NMR chemical shifts in silicates and aluminosilicates - interpretation by semi-empirical quantum chemical considerations. Chem Phys Lett 1985;114:28–30.
[60] Engelhardt G, Luger S, Buhl JC, Felsche J. ^{29}Si MAS NMR of aluminosilicate sodalites – correlations between chemical shifts and structure parameters. Zeolites 1989;9:182–6.
[61] Fyfe CA, Feng Y, Grondey H. Evaluation of chemical shift-structure correlations from a combination of X-ray diffraction and 2D MAS NMR data for highly siliceous frameworks. Microporous Mater 1993;1:393–400.
[62] Fyfe CA, Feng Y, Gies H, Grondey H, Kokotailo GT. Natural abundance two-dimensional solid-state ^{29}Si NMR investigations of three-dimensional lattice connectivities in zeolite structures. J Am Chem Soc 1990;112:3264–70.
[63] Fyfe CA, Grondey H, Feng Y, Kokotailo GT. Natural abundance two-dimensional ^{29}Si MAS NMR investigation of the three-dimensional bonding connectivities in the zeolite catalyst ZSM-5. J Am Chem Soc 1990;112:8812–20.
[64] Fyfe CA, Grondey H, Feng Y, Kokotailo GT, Ernst S, Weitkamp J. Two-dimensional solid-state ^{29}Si NMR investigation of the three-dimensional bonding connectivities and structure of zeolite ZSM-23. Zeolites 1992;12:50–3.
[65] Ernst RR, Bodenhausen G, Wokaun A. Principles of nuclear magnetic resonance in one and two dimensions. Oxford: Clarendon Press; 1994.
[66] Ray GJ, Samoson A. Double-rotation and variable field ^{27}Al NMR study of dealuminated Y zeolites. Zeolites 1993;13:410–3.
[67] Jiao J, Kanellopoulos J, Wang W, Ray SS, Foerster H, Freude D, et al. Characterization of framework and extra-framework aluminum species in non-hydrated zeolites Y by ^{27}Al spin-echo, high-speed MAS, and MQMAS NMR spectroscopy at B_0=9.4 to 17.6T. Phys Chem Chem Phys 2005;7:3221–6.
[68] Loiseau T, Serre C, Huguenard C, Fink G, Taulelle F, Henry M, et al. A rationale for the large breathing of the porous aluminum terephthalate (MIL-53) upon hydration. Chem Eur J 2004;10:1373–82.
[69] Freude D, Hunger M, Pfeifer H. Investigation of acidic properties of zeolites by MAS NMR. Z Phys Chem (NF) 1987;152:171–82.
[70] Hunger M. Multinuclear solid-state NMR studies of acidic and non-acidic hydroxyl protons in zeolites. Solid State Nucl Magn Reson 1996;6:1–29.
[71] Hunger M, Freude D, Pfeifer H, Prager D, Reschetilowski W. Proton MAS NMR studies of hydroxyl groups in alkaline-earth cation-exchanged zeolite Y. Chem Phys Lett 1989;163:221–4.
[72] Mastikhin VM, Mudrakovskii IL, Nosov AV. ^{1}H-NMR magic angle spinning (MAS) studies of heterogeneous catalysis. Prog Nucl Magn Reson Spectrosc 1991;23:259–99.
[73] Koller H, Lobo RF, Burkett SL, Davis ME. Si…OSi hydrogen-bonds in As-synthesized high-silica zeolites. J Phys Chem 1995;99:12588–96.
[74] Wolf I, Gies H, Fyfe CA. Ordering of silicate layers by hydrogen-bonding networks: solid state NMR investigation of the perfect three-dimensional registration in the layer silicate RUB-18. J Phys Chem B 1999;103:5933–8.
[75] Huang J, Jiang Y, Reddy MVR, Thomas B, Romanova E, Hunger M. Characterization and acidic properties of aluminum-exchanged zeolites X and Y. J Phys Chem C 2008;112:3811–8.
[76] Beck LW, Haw JF. Multinuclear NMR studies reveal a complex acid function for zeolite beta. J Phys Chem 1995;99:1076–9.
[77] Hunger M, Ernst S, Steuernagel S, Weitkamp J. High-field H-1 MAS NNR investigations of acidic and non-acidic hydroxyl groups in zeolites H-beta, H-ZSM-5, HZSM-58 and H-MCM-22. Microporous Mater 1996;6:349–53.
[78] Paze C, Zecchina A, Spera S, Cosma A, Merlo E, Spano G, et al. Comparative IR and ^{1}H MAS NMR study of adsorption of CD_3CN on zeolite H-beta: evidence of the presence of two families of bridged bronsted sites. Phys Chem Chem Phys 1999;1:2627–9.

[79] Higgins JB, LaPierre RB, Schlenker JL, Rohrman AC, Wood JD, Kerr GT, et al. The framework topology of zeolite beta. Zeolites 1988;8:446–52.
[80] Treacy MMJ, Newsam JM, Deem MW. A general recursion method for calculating diffracted intensities from crystals containing planar faults. Proc R Soc Lond A 1991;433:499–520.
[81] Gabrienko AA, Danilova IG, Arzumanov SS, Toktarev AV, Freude D, Stepanov AG. Strong acidity of silanol groups of zeolite beta: evidence from the studies by IR spectroscopy of adsorbed CO and ^{1}H MAS NMR. Microporous Mesoporous Mater 2010;131:210–6.
[82] Kennedy GJ, Afeworki M, Calabro DC, Chase CE, Smiley RJJ. ^{1}H MAS NMR (magic-angle spinning nuclear magnetic resonance) techniques for the quantitative determination of hydrogen types in solid catalysts and supports. Appl Spectrosc 2004;58:698–704.
[83] Biaglow AI, Gorte RJ, Kokotailo GT, White D. A probe of Bronsted site acidity in zeolites - ^{13}C chemical shift of acetone. J Catal 1994;148:779–86.
[84] Biaglow AI, Gorte RJ, White D. ^{13}C NMR studies of acetone in dealuminated faujasites - a probe for nonframework alumina. J Catal 1994;150:221–4.
[85] Xu T, Munson EJ, Haw JF. Toward a systematic chemistry of organic reactions in zeolites - in situ NMR studies of ketones. J Am Chem Soc 1994;116:1962–72.
[86] Xu MC, Arnold A, Buchholz A, Wang W, Hunger M. Low-temperature modification of mesoporous MCM-41 material with sublimated aluminum chloride in vacuum. J Phys Chem B 2002;106:12140–3.
[87] Haw JF. Zeolite acid strength and reaction mechanisms in catalysis. Phys Chem Chem Phys 2002;4:5431–41.
[88] Haw JF, Nicholas JB, Xu T, Beck LW, Ferguson DB. Physical organic chemistry of solid acids – lessons from in situ NMR and theoretical chemistry. Acc Chem Res 1996;29:259–67.
[89] Anderson MW, Klinowski J. Solid-state NMR studies of the shape-selective catalytic conversion of methanol into gasoline on zeolite ZSM-5. J Am Chem Soc 1990;112:10–6.
[90] Stepanov AG, Zamaraev KI, Thomas JM. ^{13}C CP/MAS and ^{2}H NMR study of tert-butyl alcohol dehydration on H-ZSM-5 zeolite. Evidence for the formation of tert-butyl cation and tert-butyl silyl ether intermediates. Catal Lett 1992;13:407–22.
[91] Stepanov AG, Sidelnikov VN, Zamaraev KI. In situ ^{13}C-solid-state NMR and ex situ GC-MS analysis of the products of tert-butyl alcohol dehydration on H-ZSM-5 zeolite catalyst. Chem Eur J 1996;2:157–67.
[92] Stepanov AG, Zudin VN, Zamaraev KI. Two dimensional J-resolved ^{13}C solid-state NMR analysis of the products of ethylene conversion on zeolite H-ZSM-5. Solid State Nucl Magn Reson 1993;2:89–93.
[93] Karger J, Ruthven DM. Diffusion in zeolites and other microporous solids. New York: Wiley-Interscience; 1992.
[94] Heink W, Karger J, Pfeifer H, Stallmach F. Measurement of the intracrystalline self-diffusion of xenon in zeolites by the NMR pulsed field gradient technique. J Am Chem Soc 1990;112:2175–8.
[95] Vega AJ, Luz Z. Characterization of NH4-rho and vaccum-calcined H-rho zeolites by multinuclear NMR spectroscopy. J Phys Chem 1987;91:365–73.
[96] Kustanovich I, Vieth HM, Luz Z, Vega S. NMR-studies of the sorption of p-xylene, toluene, and benzene on ZSM5 zeolite. J Phys Chem 1989;93:7427–31.
[97] Abragam A. The principles of nuclear magnetism. Oxford: Oxford University Press; 1961.
[98] Spiess HW. Rotation of molecules and nuclear spin relaxation. In: Diehl P, Fluck E, Kosfeld R, editors. NMR basic principles and progress. New York: Springer-Verlag; 1978. p. 55.
[99] Stepanov AG, Shubin AA, Luzgin MV, Jobic H, Tuel A. Molecular dynamics of n-octane inside zeolite ZSM-5 as studied by deuterium solid-state NMR and quasi-elastic neutron scattering. J Phys Chem B 1998;102:10860–70.
[100] Stepanov AG, Alkaev MM, Shubin AA, Luzgin MV, Shegai TO, Jobic H. Dynamics of isobutane inside zeolite ZSM-5. A study with deuterium solid-state NMR. J Phys Chem B 2002;106:10114–20.
[101] Stepanov AG, Shubin AA, Luzgin MV, Shegai TO, Jobic H. Dynamics of n-hexane inside silicalite, as studied by ^{2}H NMR. J Phys Chem B 2003;107:7095–101.
[102] Kolokolov DI, Jobic H, Stepanov AG. Mobility of n-butane in ZSM-5 zeolite studied by ^{2}H NMR. J Phys Chem C 2009;114:2958–66.

[103] Boddenberg B, Burmeister R. ^{2}H NMR study on the rotation and diffusion kinetics of propene and benzene in NaX and Ag NaX zeolites. Zeolites 1988;8:488–94.
[104] Stepanov AG, Alkaev MM, Shubin AA. Molecular dynamics of isobutyl alcohol inside zeolite H-ZSM-5 as studied by deuterium solid-state NMR spectroscopy. J Phys Chem B 2000;104:7677–85.
[105] Nishchenko AM, Kolokolov DI, Gabrienko AA, Stepanov AG. Mobility of tert-butyl alcohol in MFI framework type studied by deuterium NMR. J Phys Chem C 2012;116:8956–63.
[106] Kobe JM, Gluszak TJ, Dumesic JA, Root TW. Deuterium NMR characterization of Bronsted acid sites and silanol species in zeolites. J Phys Chem 1995;99:5485–91.
[107] Kolokolov DI, Jobic H, Stepanov AG, Guillerm V, Devic T, Serre C, et al. Dynamics of benzene rings in MIL-53(Cr) and MIL-47(V) frameworks studied by ^{2}H NMR spectroscopy. Angew Chem Int Ed 2010;49:4791–4.
[108] Gould SL, Tranchemontagne D, Yaghi OM, Garcia-Garibay MA. Amphidynamic character of crystalline MOF-5: rotational dynamics of terephthalate phenylenes in a free-volume, sterically unhindered environment. J Am Chem Soc 2008;130:3246–7.
[109] Gonzalez J, Devi RN, Tunstall DP, Cox PA, Wright PA. Deuterium NMR studies of framework and guest mobility in the metal–organic framework compound MOF-5, $Zn_4O(O_2CC_6H_4CO_2)_3$. Microporous Mesoporous Mater 2005;84:97–104.
[110] Kolokolov DI, Stepanov AG, Guillerm V, Serre C, Frick B, Jobic H. Probing the dynamics of the porous Zr terephthalate UiO-66 framework using ^{2}H NMR and neutron scattering. J Phys Chem C 2012;116:12131–6.
[111] Chen QJ, Springuelhuet MA, Fraissard J. ^{129}Xe NMR of xenon adsorbed on the molecular sieves AlPO4-5, SAPO-5, MAPO-5, and SAPO-37. Chem Phys Lett 1989;159:117–21.
[112] Hoffmann HC, Debowski M, Muller P, Paasch S, Senkovska I, Kaskel S, et al. Solid-state NMR spectroscopy of metal-organic framework compounds (MOFs). Materials 2012;5:2537–72.
[113] Springuel-Huet MA, Fraissard JP. A ^{129}Xe NMR study of dealuminated mordenites. Zeolites 1992;12:841–5.
[114] Baerlocher C, Meier WM, Olson DH. Atlas of zeolite framework types. 5th revised ed. Amsterdam, The Netherlands: Elsevier; 2001.
[115] Olah GA, Baker EB, Evans JC, Tolgyesi WS, McIntyre JS, Bastien IJ. Stable carbonium ions. V. Alkylcarbonium hexafluoroantimonates. J Am Chem Soc 1964;86:1360–73.
[116] Zamaraev KI, Zhidomirov GM. Active sites and the role of the medium in homogeneous, heterogeneous and enzymatic catalysis: similarities and differences. In: Yermakov Y, Likholobov V, editors. Proc. Fifth Int. Symp. Relat. Homogen. Heterogen. Catal; Novosibirsk: VNU Science Press; 1986. p. 23–74.
[117] Kazansky VB, Senchenya IN. Quantum chemical study of the electronic structure and geometry of surface alkoxy groups as probable active interemediates of heterogeneous acidic catalysts: what are the adsorbed carbenium ions? J Catal 1989;119:108–20.
[118] Olah GA, White AM. Stable carbonium ions. XCI. Carbon-13 nuclear magnetic resonance spectroscopic study of carbonium ions. J Am Chem Soc 1969;91:5801–10.
[119] Olah GA, Donovan DJ. ^{13}C Nuclear magnetic resonance spectroscopic study of alkyl cations. The constancy of ^{13}C nuclear magnetic resonance methyl substituent effects and thier application in the study of equilibrating carbocations and the mechanism of some rearrangements. J Am Chem Soc 1977;99:5026–39.
[120] Breitmaier E, Voelter W. ^{13}C NMR spectroscopy, methods and applications in organic chemistry. Weinheim: VCH; 1978.
[121] Bosacek V. Formation of surface-bonded methoxy groups in the sorption of methanol and methyl iodide on zeolite studied by ^{13}C MAS NMR spectroscopy. J Phys Chem 1993;97:10732–7.
[122] Murray DK, Chang JW, Haw JF. Conversion of methyl halides to hydrocarbons on basic zeolites. A discovery by in situ NMR. J Am Chem Soc 1993;115:4732–41.
[123] Wang W, Jiao J, Jiang YJ, Ray SS, Hunger M. Formation and decomposition of surface ethoxy species on acidic zeolite Y. ChemPhysChem 2005;6:1467–9.
[124] Stepanov AG, Romannikov VN, Zamaraev KI. ^{13}C CP/MAS NMR study of isobutyl alcohol dehydration on H-ZSM-5 zeolite. Evidence for the formation of stable isobutyl silyl ether intermediate. Catal Lett 1992;13:395–405.

[125] Wang W, Hunger M. Reactivity of surface alkoxy species on acidic zeolite catalysts. Acc Chem Res 2008;41:895–904.

[126] Saunders M, Hagen EL. Rearrangement reactions of secondary carbonium ions. Isopropyl cation. J Am Chem Soc 1968;90:6881–2.

[127] Saunders M, Hagen EL, Rosenfeld J. Rearrangement reactions of secondary carbonium ions. Protonated cyclopropane intermediates formed from sec-butyl cation. J Am Chem Soc 1968;90:6882–4.

[128] Walker GE, Kronja O, Saunders M. Isotopic scrambling in Di-C-13-labeled 2-butyl cation: evidence for a protonated cyclopropane intermedia. J Org Chem 2004;69:3598–601.

[129] Stepanov AG, Luzgin MV, Arzumanov SS, Ernst H, Freude D. n-Butene conversion on H-ferrierite studied by ^{13}C MAS NMR. J Catal 2002;211:165–72.

[130] Stepanov AG, Arzumanov SS, Luzgin MV, Ernst H, Freude D. In situ monitoring of n-butene conversion on H-ferrierite by ^{1}H, ^{2}H, and ^{13}C MAS NMR: kinetics of a double-bond-shift reaction, hydrogen exchange, and the ^{13}C-label scrambling. J Catal 2005;229:243–51.

[131] Stepanov AG, Luzgin MV, Romannikov VN, Zamaraev KI. Carbenium ion properties of octene-1 adsorbed on zeolite H-ZSM-5. Catal Lett 1994;24:271–84.

[132] Derouane EG, Hamid SBA, Ivanova II, Blom N, Hojlundnielsen PE. Thermodynamic and mechanistic studies of initial stages in propane aromatisation over Ga-modified H-ZSM-5 catalysts. J Mol Catal 1994;86:371–400.

[133] Teraishi K. Computational study on the product selectivity of FCC zeolitic catalyst. J Mol Catal A 1998;132:73–85.

[134] Boronat M, Viruela PM, Corma A. Reaction intermediates in acid catalysis by zeolites: prediction of the relative tendency to form alkoxides or carbocations as a function of hydrocarbon nature and active site structure. J Am Chem Soc 2004;126:3300–9.

[135] Tuma C, Sauer J. Protonated isobutene in zeolites: tert-butyl cation or alkoxide? Angew Chem Int Ed 2005;44:4769–71.

[136] Stepanov AG, Ernst H, Freude D. In situ ^{1}H MAS NMR studies of the H/D exchange of deuterated propane adsorbed on zeolite H-ZSM-5. Catal Lett 1998;54:1–4.

[137] Sommer J, Habermacher D, Jost R, Sassi A, Stepanov AG, Luzgin MV, et al. Activation of small alkanes on solid acids. An H/D exchange study by liquid and solid-state NMR: the activation energy and the inhibiting effect of carbon monoxide. J Catal 1999;181:265–70.

[138] Haouas M, Walspurger S, Taulelle F, Sommer J. The initial stages of solid acid-catalyzed reactions of adsorbed propane. A mechanistic study by in situ MAS NMR. J Am Chem Soc 2004;126:599–606.

[139] Stepanov AG, Arzumanov SS, Luzgin MV, Ernst H, Freude D, Parmon VN. In situ ^{1}H and ^{13}C MAS NMR study of the mechanism of H/D exchange for deuterated propane adsorbed on HZSM-5. J Catal 2005;235:221–8.

[140] Truitt MJ, Toporek SS, Rovira-Truitt R, White JL. Alkane C-H bond activation in zeolites: evidence for direct protium exchange. J Am Chem Soc 2006;128:1847–52.

[141] Stepanov AG, Arzumanov SS, Parmon VN, Kolyagin YG, Ivanova II, Freude D. Regioselective H/D exchange of propane on Zn/H-MFI zeolite. Catal Lett 2007;114:85–90.

[142] Stepanov AG, Parmon VN, Freude D. In situ NMR spectroscopy in heterogeneous catalysis: kinetic study of hydrocarbon conversion mechanisms. Kinet Catal 2007;48:521–34.

[143] Arzumanov SS, Stepanov AG, Freude D. Kinetics of H/D exchange for n-butane on zeolite H-ZSM-5 studied with ^{1}H MAS NMR in situ. J Phys Chem C 2008;112:11869–74.

[144] Huang J, Jiang YJ, Marthala VRR, Ooi YS, Hunger M. Regioselective H/D exchange at the side-chain of ethylbenzene on dealuminated zeolite H-Y studied by in situ MAS NMRUV/Vis spectroscopy. ChemPhysChem 2008;9:1107–9.

[145] Stepanov AG, Arzumanov SS, Gabrienko AA, Parmon VN, Ivanova II, Freude D. Significant influence of Zn on activation of the C–H bonds of small alkanes by Bronsted acid sites of zeolite. ChemPhysChem 2008;9:2559–63.

[146] Gabrienko AA, Arzumanov SS, Toktarev AV, Freude D, Haase J, Stepanov AG. Hydrogen H/D exchange and activation of C_1 - n-C_4 alkanes on Ga-modified zeolite BEA studied with ^{1}H magic angle spinning nuclear magnetic resonance in situ. J Phys Chem C 2011;115:13877–86.

[147] Kramer GJ, van Santen RA, Emeis CA, Nowak AK. Understanding the acid behavior of zeolites from theory and experiment. Nature 1993;363:529–31.
[148] Blaszkowski SR, Jansen APJ, Nascimento MAC, van Santen RA. Density functional theory calculations of the transition states for hydrogen exchange and dehydrogenation of methane by a Bronsted zeolitic proton. J Phys Chem 1994;98:12938–44.
[149] Blaszkowski SR, Nascimento MAC, van Santen RA. Activation of C–H and C–C bonds by an acidic zeolite: a density functional study. J Phys Chem 1996;100:3463–72.
[150] Zheng X, Blowers P. A computational study of alkane hydrogen-exchange reactions on zeolites. J Mol Catal A Chem 2005;242:18–25.
[151] Zheng X, Blowers P. A computational study of methane catalytic reactions on zeolites. J Mol Catal A Chem 2006;246:1–10.
[152] Saunders M, Hewett AP, Kronja O. Deuterium and ^{13}C scrambling processes in isopropyl cation. Croat Chem Acta 1992;65:673–8.
[153] Kolyagin YG, Ordomsky VV, Khimyak YZ, Rebrov AI, Fajula F, Ivanova II. Initial stages of propane activation over Zn/MFI catalyst studied by in situ NMR and IR spectroscopic techniques. J Catal 2006;238:122–33.
[154] Kolyagin YG, Ivanova II, Ordomsky VV, Gedeon A, Pirogov YA. Methane activation over Zn-modified MFI zeolite: NMR evidence for Zn-methyl surface species formation. J Phys Chem C 2008;112:20065–9.
[155] Luzgin MV, Rogov VA, Arzumanov SS, Toktarev AV, Stepanov AG, Parmon VN. Understanding methane aromatization on a Zn-modified high-silica zeolite. Angew Chem Int Ed 2008;47:4559–62.
[156] Kolyagin YG, Ivanova II, Pirogov YA. ^{1}H and ^{13}C MAS NMR studies of light alkanes activation over MFI zeolite modified by Zn vapour. Solid State Nucl Magn Reson 2009;35:104–12.
[157] Buckles G, Hutchings GJ, Williams CD. Aromatization of propane over Ga/HZSM-5: an explanation of the synergy observed between Ga^{3+} and H^{+}. Catal Lett 1991;11:89–94.
[158] Hagen A, Roessner F. Ethane to aromatic hydrocarbons: past, present, future. Cat Rev Sci Eng 2000;42:403–37.
[159] Biscardi JA, Iglesia E. Structure and function of metal cations in light alkane reactions catalyzed by modified H-ZSM5. Catal Today 1996;31:207–31.
[160] Ritter JJ, Minieri PP. A new reaction of nitriles. I. Amides from alkenes and mononitriles. J Am Chem Soc 1948;70:4045–8.
[161] Luzgin MV, Stepanov AG. The Ritter reaction in zeolite H-ZSM-5. NMR observation of the intermediate N-alkylnitrilium cation formed on interaction between t-BuOH and MeCN. Mendeleev Commun 1996;238–9.
[162] Stepanov AG, Luzgin MV. Interaction of acetonitrile with olefins and alcohols in zeolite H-ZSM-5. In situ solid-state NMR characterization of the reaction products. Chem Eur J 1997;3:47–56.
[163] Dwyer J. Shaping up for selective catalysis. Nature 1989;339:174–5.
[164] Weisz PB, Frilette VJ. Intracrystalline and molecular shape-selective catalysis by zeolite salts. J Phys Chem 1960;64:382.
[165] Chen NY, Garwood WE. Industrial application of shape-selective catalysis. Catal Rev Sci Eng 1986;28:185–264.
[166] Anderson MW, Klinowski J. Direct observation of shape selectivity in zeolite ZSM-5 by magic-angle-spinning NMR. Nature 1989;339:200–3.
[167] Choudhary VR, Kinage AK, Choudhary TV. Low-temperature nonoxidative activation of methane over H-galloaluminosilicate (MFI) zeolite. Science 1997;275:1286–8.
[168] Naccache CM, Meriaudeau P, Sapaly G, Tiep LV, Taarit YB. Assessment of the low-temperature nonoxidative activation of methane over H-galloaluminosilicate (MFI) zeolite: a C-13 labelling investigation. J Catal 2002;205:217–20.
[169] Luzgin MV, Rogov VA, Arzumanov SS, Toktarev AV, Stepanov AG, Parmon VN. Methane aromatization on Zn-modified zeolite in the presence of a Co-reactant higher alkane: how does it occur? Catal Today 2009;144:265–72.

CHAPTER 5

Methanol to Olefins: An Insight Into Reaction Pathways and Products Formation

N. Nesterenko, J. Aguilhon, Ph. Bodart, D. Minoux, J.-P. Dath
Refining & Base Chemicals, Total Research & Technology Feluy, Seneffe, Belgium

1 Introduction and Scope

The continuous rise in petroleum prices in addition to the increase in world demand for light olefins and, in particular, for propylene, led to innovation by the chemical industry and to the development of production routes other than those involving oil. Thus, the further economic and environmental benefits arising from the use of natural gas encouraged an alternative route for olefins production by using relatively inexpensive methanol, which is deemed to be a readily stored and manageable intermediate product for the use of natural gas. Methanol to olefins (MTO) represents a real paradigm shift as the whole process is new and the feedstock, methanol, is different from the conventional feedstocks directly available from fossil resources, namely naphtha, liquid pressurized gas (LPG), or ethane [1].

While all fossil hydrocarbons are finite resources and increasingly difficult to explore, they are still far from being limited and will provide the vast majority of fuels and chemicals for the next decades. In addition, many proven "conventional" and "unconventional" fossil reserves may not find a market in the near future because of their remote locations. The prospect of transporting the gas and coal is not always economical. To achieve economies of scale, these resources have to be monetized by using another production option. Among the potential solutions, the methanol route is poised as the most profitable, simple, and, in many cases, the preferred one. In addition, the methanol-based emerging technologies may significantly contribute to a sustainable and less oil-dependent future. Last, but not least, methanol as a building block will contribute to the commitments to limit the greenhouse gas emissions and to better manage a product life cycle. However, all of these solutions need to be prepared and managed properly [1].

Zeolites and Zeolite-like Materials. http://dx.doi.org/10.1016/B978-0-444-63506-8.00006-9

1.1 Market Environment

The light olefins (ethylene and propylene) are important intermediates in the production of plastics, fibers, and other organic chemicals, which are ultimately consumed in the packaging, transportation, and construction industries and in a myriad of industrial and consumer markets. Demand for these light olefins is thereby driven primarily by growth in the production of polyolefins.

In 2011 world ethylene production was about 127 million metric tons. In the next 20 years, world demand for ethylene is projected to grow by 3.5%/year [2,3]. For the most part, ethylene is mainly produced by steam crackers and from refinery off-gases. On the other hand, world propylene production in 2011 was about 70 million metric tons. Between 2011 and 2021, world demand for propylene is expected to grow at an average annual rate of 4.4% [2,3]. Currently propylene is mainly produced as a by-product of ethylene plants and gasoline machine/fluid catalytic cracking (FCC) units. With the startup of additional ethane-based ethylene capacities (lighter fractions such as ethane or LPG are considered as very advantageous feedstocks), the production of propylene via the steam cracker has been decreasing [4] (Fig. 1). Further, the demand for propylene is actually growing faster than that of ethylene because it is driven by the increasing consumption of derivatives such as poly-propylene. Finally, the price ratio of propylene to ethylene has been gradually increased since 1990 [5].

To some extent the propylene production can be optimized by running the steam cracker at low severity or by transforming an FCC unit into a petrochemical FCC after a major revamp. However, the gasoline surplus as well as the changing motor gasoline regulations and specifications are expected to result in a gradual slowdown in FCC production growth. Therefore, there is an increased interest in developing secondary propylene production

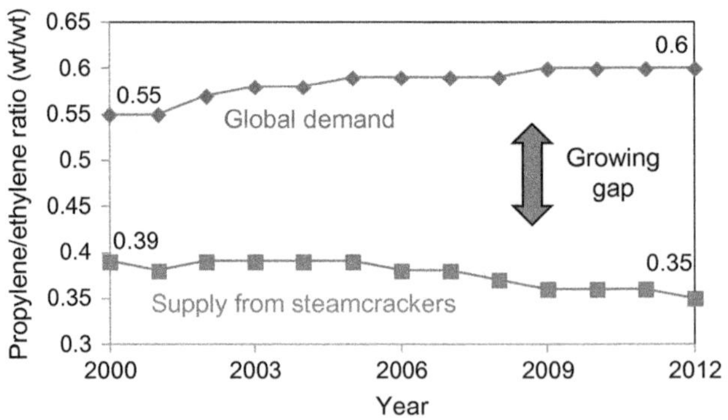

Fig. 1
Supply of propylene from steam cracker versus global demand. *Adapted from Ding J, Hua W. Game changers of the C_3 value chain: gas, coal, and biotechnologies. Chem Eng Technol 2013;36:83–90.*

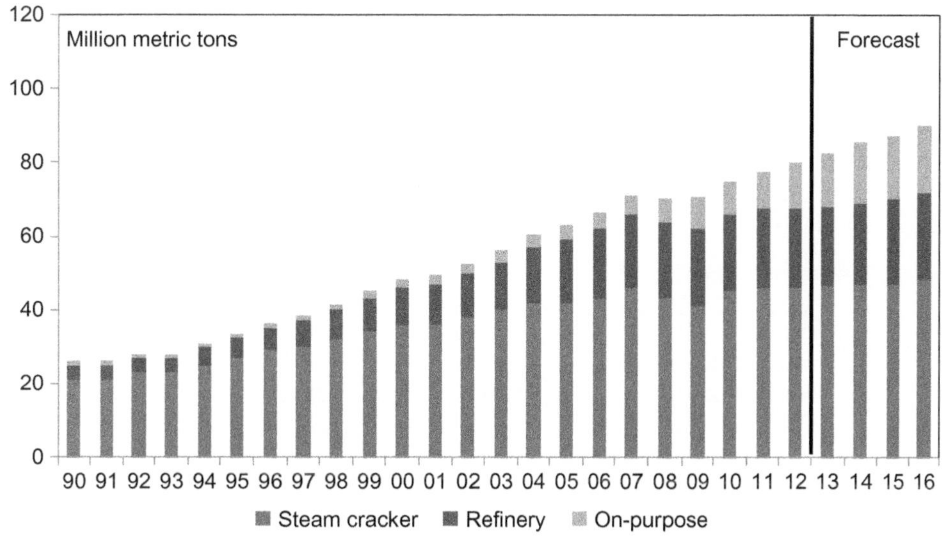

Fig. 2

Propylene production by technology (PG: polymer grade, CG: chemical grade). *Adapted from World light olefins analysis. Chemical Market Associates, INC.; 2012.*

technologies (Fig. 2). Production through on-purpose propylene processes is expected to almost double within the next 5 years (Fig. 2).

New technologies have evolved to the extent that on-purpose processes are capable of making competitive supplies of propylene under different circumstances. The share in propylene production between the different on-purpose technologies is illustrated in Fig. 3. New feedstocks to existing products (*X*-to-*Y*) technology is increasingly becoming an opportunity to introduce the nonconventional resources into the energy and chemical markets.

1.2 Methanol Production Sources

Unlike conventional feedstocks, methanol can be produced economically with a selectivity close to 100% from a variety of sources via synthesis gas (syngas) derived from coal, natural gas, petroleum liquids, petroleum coke, recycled plastics, and municipal waste. A synthesis of methanol is a very well-established technology with a huge industrial experience [6]. A number of other alternatives to syngas route sources of methanol from new feedstocks that are under extensive development [6]. Currently, most of world methanol capacity is produced from syngas via natural gas steam reforming or coal gasification (Fig. 4) [6]. The vast variety of sources and the versatility of the production routes position methanol as a main nonpetroleum source for light olefins production. In China, the recent implementation of numerous coal-to-chemicals projects in various regions has increased the supply of low-cost

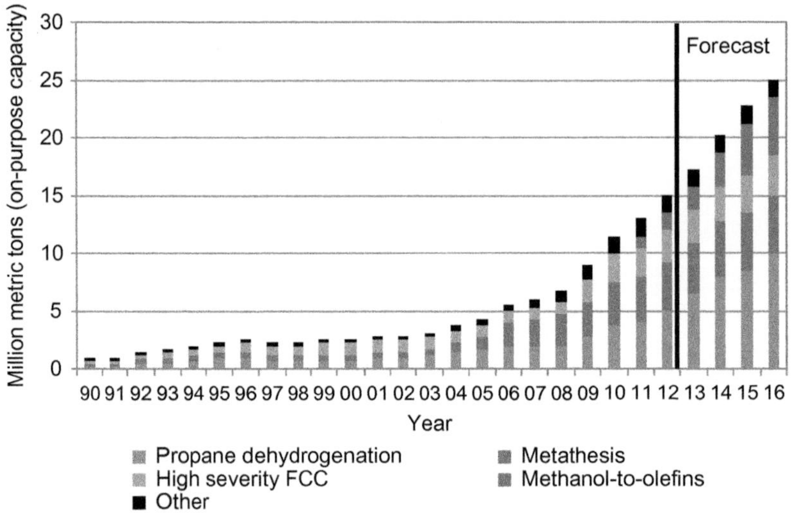

Fig. 3

Propylene capacity from on-purpose units (PG: polymer grade, CG: chemical grade). *Adapted from World light olefins analysis. Chemical Market Associates, INC.; 2012.*

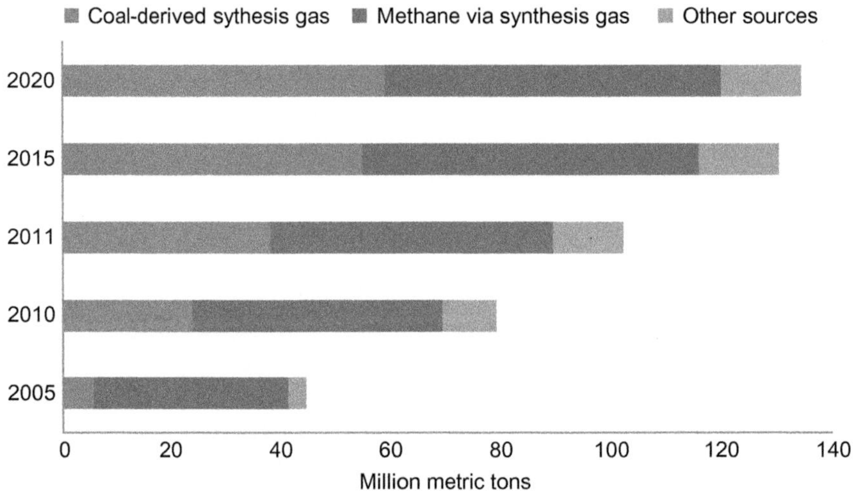

Fig. 4

Methanol production by origin. *Adapted from Coal to olefins. SRI Consulting; 2011.*

methanol, making the country the first to re-adopt coal as a major feedstock for chemicals production. Other coal-rich countries such as the United States are also turning to coal and shale gas to reduce dependence on imported oil.

So far, the best feedstock has been used with minimal energy difference between *X* and *Y*. New feedstock utilization will require new catalysis and technologies to adjust energy-level content of new feedstock to our conventional products. Among the emerging feedstocks and platform molecules, methanol is of particular interest because it is chemically close to the light olefins (carbon in methanol has the same oxidation state), is easy to transport and handle, and can be transformed through various routes to existing products with high selectivity [7].

1.3 Beyond the Status Quo

In the aftermath of the 1973 oil crisis, independently from market environment, MTO route has always attracted a lot of attention and was poised as a potential way for on-purpose catalytic conversion of non-oil-based feedstocks or intermediates. One could hardly call methanol transformation to olefins a new technology. Since the early 1980s, MTO technology has been considered to be the second best among the emerging solutions. Nobody knows whether the oil had been too accessible or the lack of industrial experience has delayed a sustainable implementation of the process for several decades.

In pursuit of the pioneers of MTO development, Mobil Oil in 1984–88 [8,9], a number of companies successfully demonstrated MTO/MTP technologies. The major MTO technologies include the UOP/Norsk Hydro (now INEOS) MTO process [10,11], the Lurgi MTP process [12,13], the DMTO (DICP MTO) process [with methanol or dimethyl ether (DME) as the feedstock] developed by the Dalian Institute of Chemical Physics in China [14,15], and the SMTO (SRIPT MTO; Shanghai Research Institute of Petrochemical Technology) process developed by SINOPEC [16,17]. Recently, Total Petrochemicals and UOP (now Honeywell) have successfully demonstrated in a fully integrated semicommercial demonstration plant (10 ton/day), the conversion of methanol into light olefins by means of MTO and OCP process and subsequent polymerization into polyethylene and polypropylene at industrial scale [18,19].

Since the beginning of the 20th century, highly efficient and robust catalysts have been developed for the selective conversion of methanol into ethylene and propylene. SAPO-34 is regarded as the most carbon-efficient catalyst today to produce light olefins [11,20]. The ZSM-5 modified samples, on the other hand, provide the highest propylene yield, the most versatile product distribution, and a possibility to recycle back heavier olefins to boost the yield to light olefins [8].

The great potential of MTO technologies regularly attracted a lot of attention in both patent and public literature (Fig. 5) [21–25]. Throughout the history of MTO, there have been two big highlights in patent literature: one was related to a methanol to gasoline (MTG)/MTO project of Mobil in New Zealand (~1984) and the other began on the eve of the "anniversary of SAPO-34," which coincided with the historic oil price growth just before the financial crisis of 2007–08. Nevertheless, a rocketing progress in the MTO development in the recent decade still requires better understanding. Today, 20 years after the formulation of the

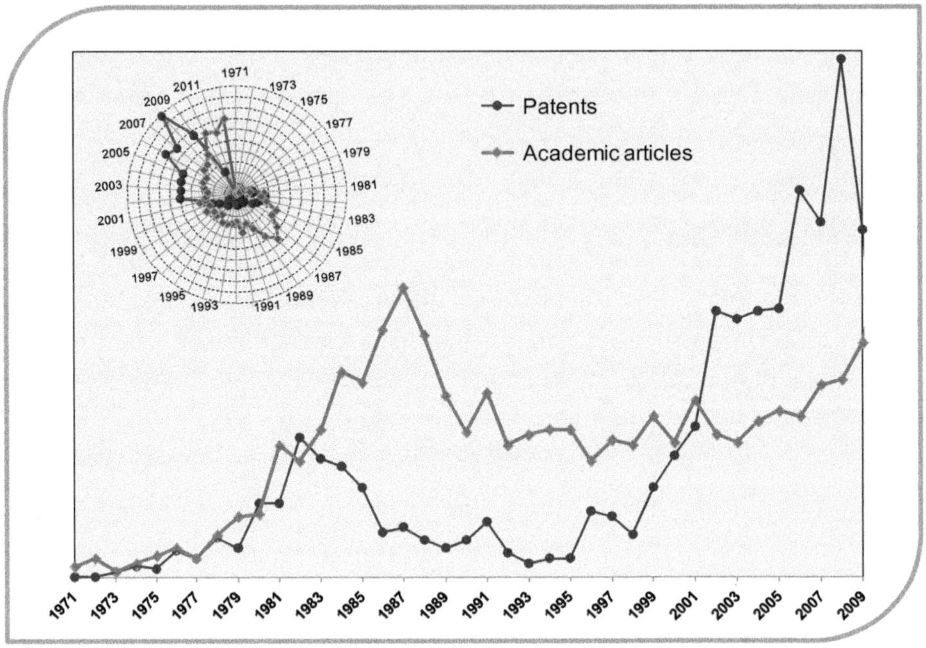

Fig. 5
Statistics on patent and academic research. *From Derwent and Chemical abstract databases.*

hydrocarbon pool mechanism for methanol conversion by Dahl and Kolbe, we are still looking for more precision, better explanations, and the tools to use it in everyday life [26]. While the introduction of the hydrocarbon pool brought a step change for the MTO development in the beginning of 1990s, the "black box" spirit of the concept with a number of blind spots has remained until the present time.

Whereas some evidence of the transformation of MTO were present in the prior arts, it is truly the Mobil discovery of the early 1970s that launched the transformation of the methanol to hydrocarbons (MTH), which has led to a new discipline in catalysis and petrochemical or organic chemistry. Since then this discipline has pursued at a continuous pace in other works. Fig. 6 shows the relative number of zeolite-related publications, between 1970 and 2012. It demonstrates that the rate of publications is always increasing, highlighting a continuous scientific interest in these materials.

The aim of this contribution is not another updated compilation of the literature on the MTO mechanism. Many reviews and some very recent ones already exist and appeal to more detailed studies [21–25,27–29]. In particular, some of them could be recommended for compulsory studies due to a fundamental breadth and depth of coverage [21–25].

In this review we aim to provide the reader with the approach on how to use the accumulated knowledge on mechanisms and where we should be critical relative to the prior arts. A number

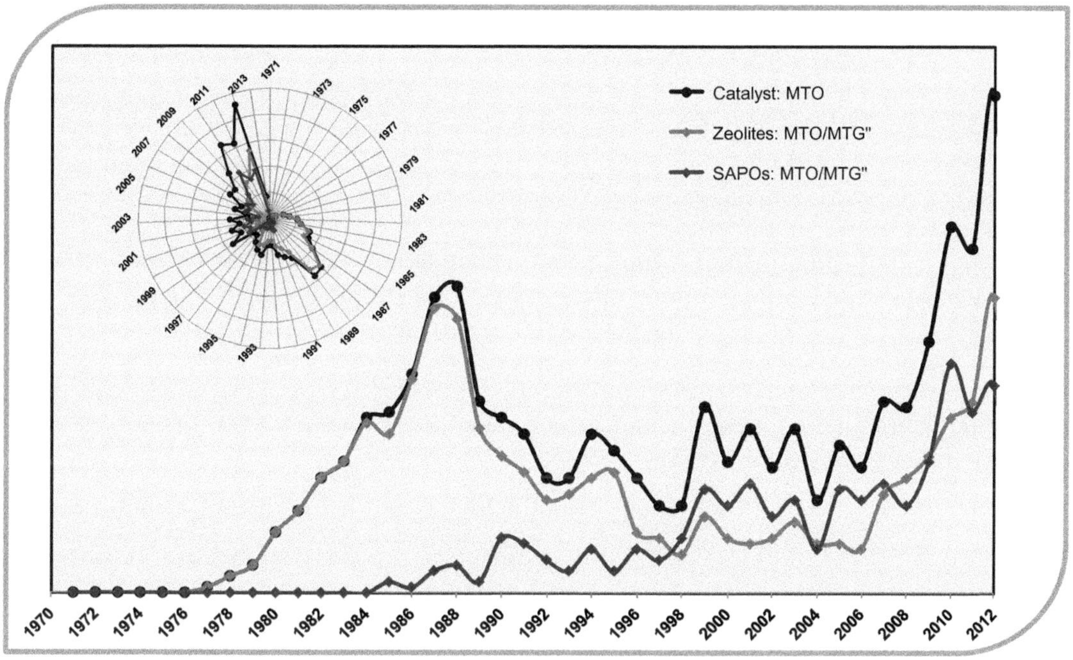

Fig. 6
Public research reports on a use of zeolites and SAPOs on methanol to olefins (MTO)/methanol to gasoline (MTG) technologies.

of MTO-related hypotheses and paradigms have been victimized by past successes. It is hard for newcomers to the world of MTO to build their own and independent point of view. The paradigms are not necessarily incorrect, but the imposed generalization and the blind cloning of them may limit the attention to relevant experimental details. The primary learning outcome from this contribution is to separate the facts from the concepts and to provide the reader with a background in the field of MTO.

In a first part, we present a brief overview of the historical background of MTO, and then we describe the mechanisms existing in the literature. In a second phase, we will try to be critical of those results, understand the formation and the maximization of ethylene and propylene selectivities, as well as refine the existing concepts.

2 MTO History

2.1 Before Mobil Discovery

The first reports of the formation of olefins from MeOH date back to 1880 [21]. In 1914, Sernagiotto [30] reported the decomposition of methanol by P_2O_5. The reaction products were propylene, butene, hexene, and hexane. Adkins and Perkins [31] studied the reaction of methanol over Al_2O_3 at 300–350°C with the main product being DME, while above 350°C, the main products were CH_4 and C_2H_4 as well as CO, CO_2, and H_2. Cullinane et al. [32] found that methanol contacted with γ-Al_2O_3 at 450°C yielded small quantities of hexamethylbenzene (HMB). Topchieva and Ballod [33] compared the activity of silica gel, Al_2O_3 and silica-alumina (Al_2O_3 30%, SiO_2 70%) for methanol conversion. The products obtained were CO_2, C_2H_4, CO, and C_2H_5 in addition to DME, and unreacted methanol.

Gorin reacted a mixture of DME (18.2 mol%) and isobutane over amorphous silica-alumina at 370°C, 10 atm, and 6.8 h^{-1} weight hourly space velocity (WHSV). The conversion of DME to hydrocarbons was 94.5%, while no net change in isobutane occurred (Table 1) [34].

In 1977 a Process Economic Program (PEP) report was prepared by Stanford Research Institute (SRI) and first discussed methanol as a chemical raw material [35]. The report deals mainly with the manufacture of ethylene and acetic acid from methanol. In addition, the routes toward many other chemicals produced from methanol were reviewed briefly.

According to the (PEP) report, the ethylene manufacturing process from methanol involves two chemical steps, each of which occurs at close to 100% selectivity. In the first step, using Mobil Oil technology, methanol vapor was converted at 75% per pass to DME. The reaction takes place over a zeolite catalyst at temperatures close to 204°C and 1 atm pressure.

In the second step, DME vapor was converted, by an approach patented in the Soviet Union, at 85% conversion per pass to ethylene [36]. A rather slow but highly selective reaction takes place over Al_2O_3-TiO_2 catalyst at about 135°C and 1 atm pressure. The report shows 99.9% of selectivity to ethylene from methanol [36]. The exothermic reaction was carried out at a low vapor space velocity for DME of 16 h^{-1}. Even a slightly higher temperature or pressure caused a fall-off in selectivity [36]. Regeneration had to be performed every 15–20 h by circulating hot ethylene through the catalyst to remove the adsorbed water [35]. One can understand that it was probably the first example of the unconscious use of the pre-coked (with ethylene) catalyst in DME transformation.

Table 1: DME conversion in presence of isobutane over SiO_2-Al_2O_3 (370°C, 1 atm, GHSV 6.8 h^{-1})

Products	C-Basis (wt%)
Methane	5.7
Ethylene	10.2
Propane + propylene	8.1
n-Butane + butylene	5.2
i-C_5	26.4
C_6^+	35.6
Coke	8.6

Adapted from Gorin E. Conversion of dimethyl ether; 1948.

2.2 Development of Methanol to Hydrocarbons by Mobil

It was not until the early 1970s that the industrial potential of the reaction emerged due to the first oil crisis. The discovery of the MTG process on zeolite happened by accident [8]. Two teams of Mobil scientists working on unrelated projects discovered by accident the formation of hydrocarbons from methanol over the synthetic zeolite ZSM-5. The group at Mobil Chemical in Edison, New Jersey, had been trying to convert methanol to ethylene oxide, while workers at Mobil Oil's Central Research Laboratory in Princeton were attempting to methylate isobutene with methanol in the presence of ZSM-5. Neither reaction proceeded according to expectations. Instead, aromatic hydrocarbons were found. Mobil Oil's Central Research team tried to find out whether methanol could serve as a precursor in homologation of isobutene, to form, presumably, neopentane. An equimolar mixture of methanol and isobutene was passed over H-ZSM-5. Methanol was quantitatively converted, whereas only about 27% of the isobutene reacted. An experiment carried out with pure methanol also showed a complete conversion of methanol [21,27]. While the group of Kaeding tried to turn the project to the chemicals, the Mobil Oil management was strategically oriented more toward fuel technology [8]. The PEP report on methanol as a chemical raw material issued in 1977 does not contain any traces about Mobil's methanol to ethylene technology. This confirms again that the MTO option was not considered at that time [35].

The MTG developments were successfully pursued by Mobil [37]. However, according to Ref. [8], at one point in the MTG reaction, the product mix showed about 40% C_2–C_5 olefins. This was a starting point to reconsider the MTO option. Eventually, by modifying the catalyst to "lower activity" and by adjusting the reaction conditions, the production of olefins could have been roughly doubled [9]. This discovery led to the development of the first ZSM-5-based MTO process [38,39]. The MTG fluidized bed pilot completed in 1984 was extended to demonstrate MTO [37]. A 100 barrels/day MTO demonstration unit was successfully operated in Germany [40]. Because the product distribution of the ZSM-5-based methanol-to-hydrocarbon reaction can be controlled by changes in zeolite structure and/or reaction conditions, Mobil developed three principal modes of operation: the production of an ethylene-rich olefins mixture (as in the MTE or methanol-to-ethylene process), the production of a mainly C_3^+ lower olefins-rich product (as in the MTO or methanol to olefins process), and the production of an aromatics-rich gasoline mixture (as in the MTG or methanol-to-gasoline process). The MTO technology was developed by Mobil as an extension of MTG and primary to make ethylene from methanol. Probably, the C_3^+ lower olefins-rich products option was mostly useful at that time as a part of MOGD technology.

ZSM-5 catalyst was not very selective to ethylene production. The catalyst was stabilized hydrothermally but generated mostly propylene and butylenes, with high gasoline yield as a by-product. Mobil modified the ZSM-5 crystal to be more selective to ethylene, but even after that the selectivity to ethylene was still limited. Probably for this reason, many small-pore zeolites such as chabazite or etionite have received attention as potential catalysts for methanol conversion to ethylene and propylene from the researchers at Mobil Oil. Methanol conversion over a variety of small-pore zeolites was reported by Chang et al. in 1977 [41]. These materials showed significantly higher ethylene selectivity relative to ZSM-5. Fast deactivation of small-pore zeolites by coking could not be a big problem for the leading company in fluidized bed engineering. So it is most likely an issue with hydrothermal stability that was one of the main constraints along with focus on MTG that stopped Mobil from commercializing the small-pore zeolites for MTO. So it comes as no surprise then that when a company obtained access to a hydrothermally stable small-pore molecular sieve, it stepped immediately into the market.

2.3 UOP/Hydro (INEOS) MTO Process

In 1984, scientists in the Molecular Sieve Division of the Union Carbide Corporation reported the synthesis of a new family of hydrothermally stable SAPO molecular sieve materials with a broad range of structure types and compositions [20,42,43]. In 1988, the Molecular Sieve Division became a part of UOP. In 1992, UOP established a relationship with Norsk Hydro in Norway to accelerate the development and commercialization of the MTO technology. Eventually, in November 1995, UOP and Norsk Hydro jointly announced the development of the technology and its availability for license. Today, this technology is known in the literature as the UOP/Hydro (INEOS) MTO process [10,11]. An integral part of this process is the catalytic conversion of methanol to light olefins in fluidized bed reactor over the UOP proprietary MTO-100™ SAPO-34 based catalyst. A 0.9 ton/day demonstration unit integrating the process and catalyst regeneration has been in operation at the Hydro Research Center in Porsgrunn, Norway, since 1995. Over the next few years, details on this technology have been disclosed in numerous publications [10,44].

Since that time, a major milestone for advanced MTO commercialization was the startup of a semicommercial, fully integrated MTO demonstration unit in Feluy, Belgium, on Total Petrochemicals premises in 2009. The plant has a capacity exceeding 10 tons methanol/day; more than 13 times the capacity of the first smaller demo unit built at Hydro in Norway. The process demonstration unit (PDU) includes a complete MTO process with product recovery and purification as well as integration with ATOFINA/UOP olefin cracking process (OCP) to maximize the yields of ethylene and propylene. Following some initial work by Total Petrochemicals (formerly Atofina) in the mid-1990s, Total Petrochemicals and UOP formed a joint-development alliance in late 2000 [3,45].

MTO and OCP are fully integrated with C_4^+ produced via MTO being converted in the olefin cracking process, and olefin cracking light olefins products being purified and fractionated along those produced in MTO. The PDU demonstrates all unit operations from feed processing through purified polymer-grade ethylene and polymer-grade propylene. A semicommercial polyolefins demonstration unit on the same Total Petrochemicals site is also included downstream to demonstrate the direct industrial production of polyethylene and polypropylene from propylene and ethylene ex MTO-OCP using various polymer catalysts [18,19].

Methanol conversion exposes SAPO-34 to both thermal and hydrothermal cycling at elevated temperatures. The SAPO-34 catalyst requires frequent carbon-burn regenerations to remove nonvolatile coke deposits formed within the micropores of the molecular sieve. Accelerated hydrothermal aging of the SAPO-34 powder was studied by Barger and Lesch to understand the structural changes that might occur during the multiple reaction–regeneration cycles [46]. They observed no degradation of the structure after either low-temperature hydrothermal aging (450°C) or 205 cycles of reaction and regeneration in the MTO process. Partial SAPO-34 structural collapse did occur, but only under more severe hydrothermal conditions (above 650°C). These findings indicated that the SAPO-34 structure collapse does not appear to be a major concern for the long-term stability of the MTO catalyst. In later studies, UOP reported 450 cycles of process-regeneration of their MTO-100™ catalyst in order to evaluate its long-term stability [46]. According to this contribution the observed level of deactivation could be overcome by the catalyst makeup required to replace material losses due to physical attrition in the fluidized bed reactors [46].

2.4 Lurgi MTP® Methanol to Propylene Process

This section reviews the Lurgi MTP® technology for production of propylene from methanol on a proprietary ZSM-5 type catalyst MTPROP® supplied from Süd Chemie [12,13]. Lurgi started the development of methanol to propylene (MTP®) in 1993. The proprietary zeolite-based MTP® catalyst is exclusively supplied by Süd Chemie [47]. The process begins with a vapor phase dehydration of methanol to DME to produce an equilibrium mixture of DME, methanol, and steam. This mixture is then converted to propylene in a fixed-bed MTP® reactor at 400–500°C in the presence of steam and recycled C_4–C_5 olefins. By-products of the Lurgi MTP® process are gasoline, LPG, and fuel gas. The ethylene could be recycled back or used as a copolymer.

For a better approach to isothermal conditions, the MTP® reactor is designed with several catalyst beds in the series. Steam is added to the feed of the first MTP® reaction stage to reduce partial pressure and coking. A fraction of cold mixture exiting the DME reactor is divided into several streams (number of catalyst beds—1), each feeding one of the fixed catalyst beds in the MTP® reactor. Aiming at optimal catalyst performances, the inlet temperature is kept at 400–450°C in the MTP® reactor by adjusting the interstage cooling to maintain similar catalyst/reaction conditions in each catalyst bed. In order to maximize overall propylene yield, several olefin-containing streams are recycled to the first MTP reaction stage [48]. All olefins other than propylene are recycled to extinction, purged as fuel gas, or end up as gasoline range hydrocarbon product. The recycled hydrocarbons and steam also serve as a heat sink.

A decrease of the partial pressure also tends to enhance the olefins formation at low temperature by suppressing the aromatization reaction. Thus, olefins yields are maximized by operating at partial methanol conversions and by using steam as inert diluents. Even at reduced methanol conversions, however, the production of aromatics is still significant.

The viability of the Lurgi MTP® process was first demonstrated at a unit operated jointly with Statoil at Statoil's methanol plant in Norway [49] from 2001 to 2004. The skid-mounted MTP unit comprised of three reactors used 360 kg/day methanol feed per reactor. In 2003, to confirm the quality of the propylene obtained in the MTP demonstration unit, samples were sent to Borealis's Innovation center in Rønningen, Norway, where it was polymerized with Borealis's Borstar process to a polypropylene that met all specifications, and converted into thermoformed cups. The results from the demonstration and further tests proved that the MTP® catalyst life exceeds 1 year of operation. The results also showed that the zeolite-based catalyst could be easily regenerated more than a dozen times.

2.5 JGC/Mitsubishi DTP® Process

Since 2007, JGC Corporation and Mitsubishi Chemicals have been jointly developing a new propylene production process that uses DME or methanol as feedstock. In some configurations, it is also possible to use olefin products from other sources. The process is called the Dominant Technology for the Propylene Production (DTP®) [50–52].

The JGC/Mitsubishi DTP process is similar to the MTP process licensed by Lurgi, except for important changes in the catalyst and, consequently, on the reaction temperature range and regeneration steps configurations. Although the DTP process was originally designed for producing propylene from DME, it works better with a DME/methanol mixture. JGC also commercializes a DME synthesis reactor that partially converts methanol in DME [50–52].

In the DTP process, the DME/methanol mixture is thought to react in the presence of steam on an alkaline earth metal-containing zeolite catalyst from the ZSM-5 type. The preferable alkaline earth metal is said to be calcium [53]. The reaction proceeds at a temperature in the 500°C range under near atmospheric pressure through multiple fixed-bed adiabatic reactors connected in series.

JGC claims to have developed a stable catalyst due to its modification with an alkaline earth metal. JGC argues that the DTP® catalyst does not deactivate with steam, which can be used as dilution gas during the regenerating step.

3 Mechanistic Concepts

The conversion of MTH over acidic zeolite catalysts has been the subject of widespread research activity since the reaction was discovered by Mobil researchers in the aftermath of the 1973 oil crisis [21]. Originally intended to produce gasoline (MTG) [8], it was, in

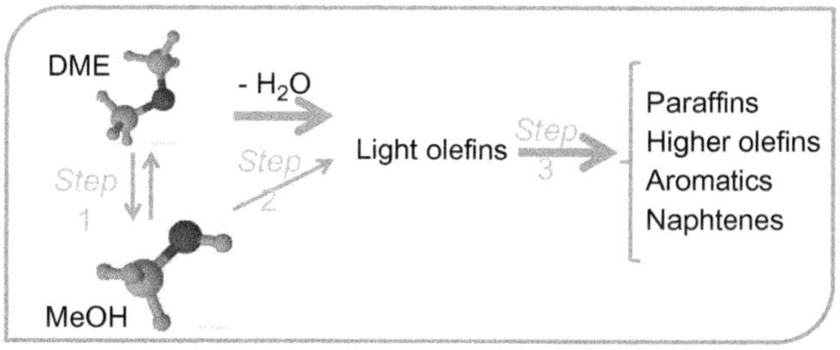

Fig. 7
Reaction steps of methanol (MeOH) conversion into hydrocarbons. *Adapted from Chang CD, Silvestri AJ. ChemTech 1987;10:624.*

a second phase of research, focused first on the formation of light olefins (MTO) [54] and subsequently for obvious economic reasons on propylene (MTP) [55].

According to Refs. [21,56], over an acid catalyst, methanol (MeOH) is first dehydrated to DME (Fig. 7). Next, the equilibrium mixture of MeOH, DME, and H_2O is further converted into light olefins (step 2). The following step (step 3) is the formation of *n*/*iso* paraffins, higher olefins, aromatics, and naphthenes by hydrogen transfer, alkylation, and polycondensation.

The dehydration of MeOH to DME, the first step in the reaction mechanism, is widely described in the literature [57–59]. As long as only DME and water are observed as the products in the effluent, no detectable coke amount is formed on the catalyst. The reaction proceeds via surface methoxy groups produced by dehydration of MeOH, which further undergo a nucleophilic attack by another MeOH molecule in order to form DME (Fig. 8) [57–59].

The reaction mechanism for the conversion of MTH over molecular sieve catalysts has been extensively reviewed [21–24,27–29,60–62]. The jump of activity was observed by increasing the contact time [63] or by increasing the reaction temperature [64]. The conversion increased by more than 80% in a temperature range of less than 20°C [64]. These sudden increases in the conversion levels might be interpreted in both cases as proof of the autocatalytic nature of the reaction. However, the jump is observed at different temperatures on catalysts of different activity and nature.

Associated with the problem of the first C–C bond formation is the question of the nature of the "first" olefin formation. The question was extensively addressed in the early 1980s by various research groups [65–68] and has also generated a lot of controversy.

The experimental evidence demonstrates that the formation of the first C–C bond should be mechanistically separated from the further pathways, which contribute to formation of olefins. The formation of the first C–C bond is the limiting step in the kinetic of MeOH/DME

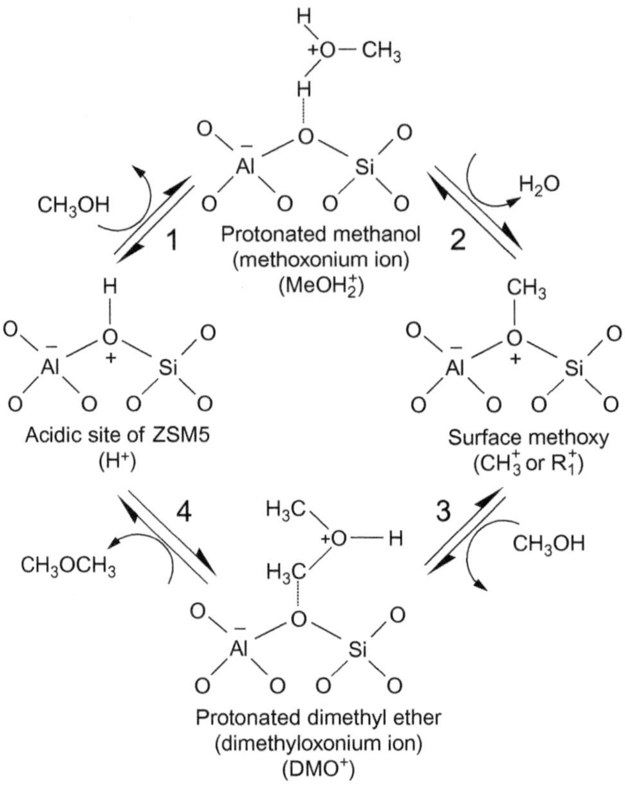

Fig. 8
Reaction mechanism for the formation of DME by dehydration of methanol over H-ZSM-5 catalyst. (H^+) acidic site, $MeOH_2(+)$ protonated methanol (methoxonium ion), CH_3^+ / R_1^+) surface methoxy species, DMO^+) protonated DME (dimethyloxonium ion). *Adapted from Park T-Y, Froment GF. Analysis of fundamental reaction rates in the methanol-to-olefins process on ZSM-5 as a basis for reactor design and operation. Ind Eng Chem Res 2004;43:682–9; Park T-Y, Froment GF. Kinetic modeling of the methanol to olefins process. 2. Experimental results, model discrimination, and parameter estimation. Ind Eng Chem Res 2001;40:4187–96; Park T-Y, Froment GF. Kinetic modeling of the methanol to olefins process. 1. Model formulation. Ind Eng Chem Res 2001;40:4172–86.*

transformation to hydrocarbons. As soon as the first hydrocarbon has been formed, the jump in activity is observed. This shows that other transformation routes dominate in further transformation of MeOH.

3.1 Formation of the First C–C Bond

Much work on the mechanistic understanding of the MTO reaction pathways had already been done by the early 1990s. Hutchings and Hunter [57–59] have summarized the mechanisms proposed at an early stage of MTO development for the formation of the primary hydrocarbon products in four types. Generally speaking, all of the major mechanisms received roughly

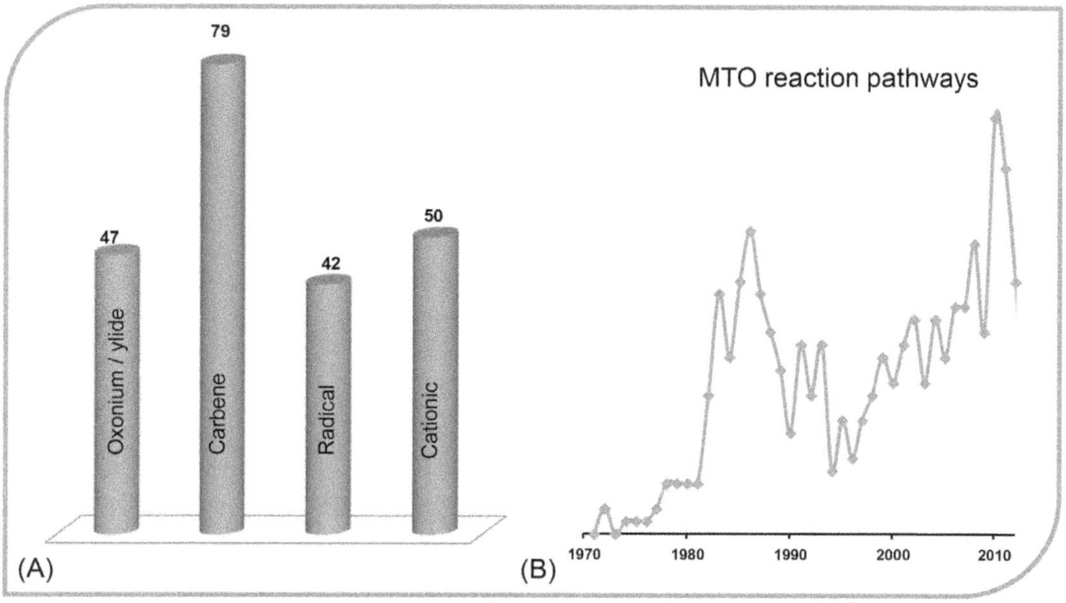

Fig. 9
(A) Citations scores for the different mechanisms of the first C–C bond formation. (B) Publication of research reports on methanol to olefins (MTO) reaction pathways over the years.

equal attention in the literature (Fig. 9). However, due to the pioneer publications of Mobil researchers, the carbene mechanism shows a slightly higher citations score.

3.1.1 The carbene mechanism

The carbene mechanism proposes an elimination of water from methanol resulting in the formation of carbenes (:CH_2) [66,69]. This occurred via the cooperative action of acid and basic sites of the zeolite (Fig. 10). For the next step, it was first stated that carbene formation was followed by polymerization of the resultant carbenes to alkenes. Later it was concluded that carbene polymerization was less likely than the insertion of the carbene into the C–O bond of methanol or DME to form the first C–C bond [57–59].

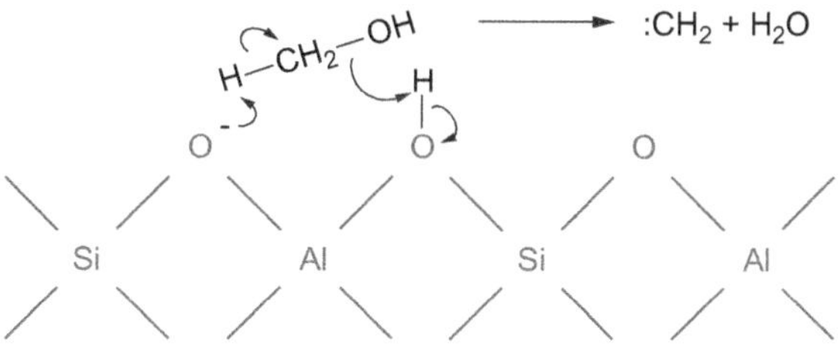

Fig. 10
Possible mechanism for carbene formation on the zeolite surface.

Blaszkowski and van Santen reported that the free carbene route is very unlikely, as it involves very high activation barriers [70].

3.1.2 The oxonium ylide mechanism

The oxonium ylide mechanism involves the trimethyloxonium (TMO) ion as an intermediate [54,67,71,72]. This ion is formed through methylation of a DME molecule via a Brønsted acid site of the zeolite (1). The TMO ion is subsequently deprotonated (2) by the conjugate base site (O lattice atom) of the zeolite creating the dimethyloxonium methylide. The next two possible steps were proposed in (Fig. 11). Either the dimethyloxonium methylide undergoes a Stevens rearrangement (3) resulting in methylethylether which then leads to ethylene (5), or the dimethyloxonium methylide reacts via an intermolecular methylation (4) to form an ethyl dimethyloxonium ion. The latter can either lead to ethylene via an elimination (6) or to propylene through a second Stevens rearrangement (2′, 3′, and 5′). According to this mechanism, ethylene and propylene are the primary products in the reaction.

3.1.3 The free radical mechanism

The free radical mechanism has been proposed partly to eliminate the requirement of the strong basic sites in zeolite for proton removal from C–H, a major drawback of the oxonium ylide mechanism [23,73–75]. The involvement of free radicals in the conversion of MTO had already been proposed by Zatorski et al. toward the end of the 1970s [23]. Clarke et al. [74] reported on the detection of free radicals during decomposition of DME over H-ZSM-5 at 240°C. Afterward, it was confirmed that DME can be a source of methylradicals over H-ZSM-5. Eventually, Hutching et al. reported the inhibition of MeOH conversion to hydrocarbons over H-ZSM-5 by NO [76–78]. However, it was concluded that the inhibition effect of NO does not act as a radical scavenger. The direct coupling of radicals leading to a C–C bond formation mechanism was considered unlikely given the probability of radical interaction with the zeolite surface.

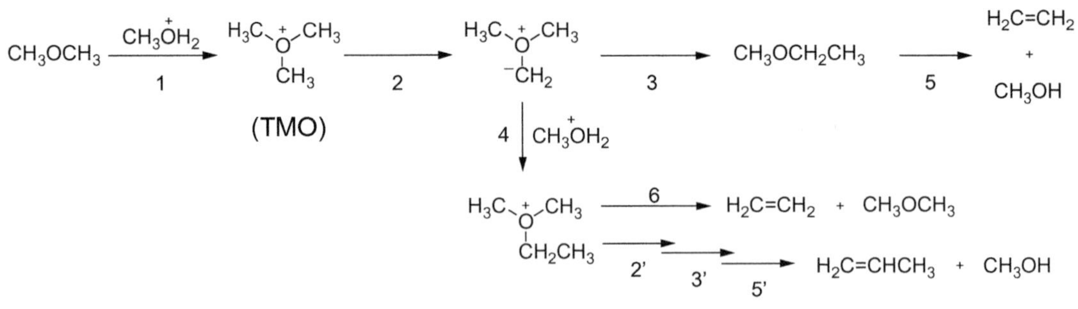

Fig. 11
Oxonium ylide mechanism. *Adapted from Spivey JJ, Froment GF, Dehertog WJH, Marchi AJ. In: Spivey JJ, editor. Catalysis, vol. 9; 1992. p. 1–64.*

3.1.4 The surface methoxy mechanism (carbocationic mechanism)

According to this mechanism [65,70,79–82], methoxy or surface methyloxonium groups operate as crucial intermediates. The MTH reaction could proceed via a *rake mechanism* as first proposed by Cormerais et al. [82] (Fig. 12). In this case, the formation of olefins proceeds via ethoxy, propoxy, butoxy groups, or the like, formed by alkylation with a C_1 group (MeOH or DME), allowing the growth of the chain (Fig. 13). When the olefins reach six carbon atoms, they can crack to again form light olefins. In other words, the adsorbed species play a key role in MeOH transformation via continually adding and splitting off reactants and products.

Two more recent alternative pathways are worth mentioning: the methane-formaldehyde route and the impurities-driven route.

3.1.5 Methane-formaldehyde route (CH_4/HCHO)

In the proposal for the methane-formaldehyde mechanism, methane and formaldehyde are formed from the surface methoxy intermediates and react to form ethanol [83–86]. Tajima et al. [83] used theoretical calculations to indicate that CH_4/HCHO are also potential reaction intermediates. van Santen with Blaszkowski [86] have theoretically found a transition state for hydride transfer between methanol and surface methoxy to form methane and potentially formaldehyde. Because methane is very slow to react, almost all conventional experimental and theoretical studies have supposed methane to be independent of the first C–C bond

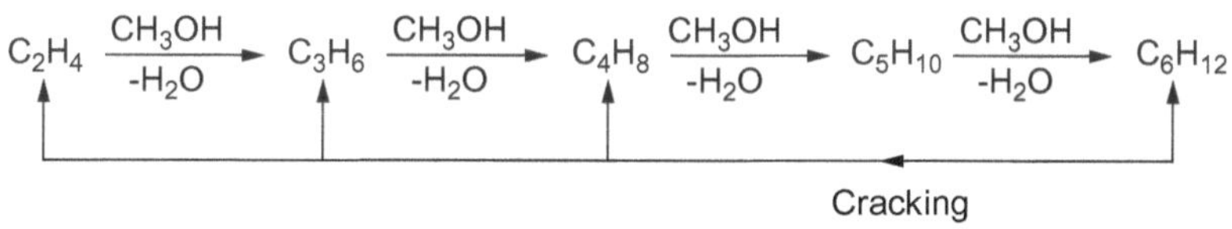

Fig. 12
Scheme of rake mechanism. *Adapted from Stöcker M. Methanol-to-hydrocarbons: catalytic materials and their behavior. Microporous Mesoporous Mater 1999;29:3–48; Cormerais FX, Perot G, Chevalier F, Guisnet M. J Chem Res Synop 1980:362.*

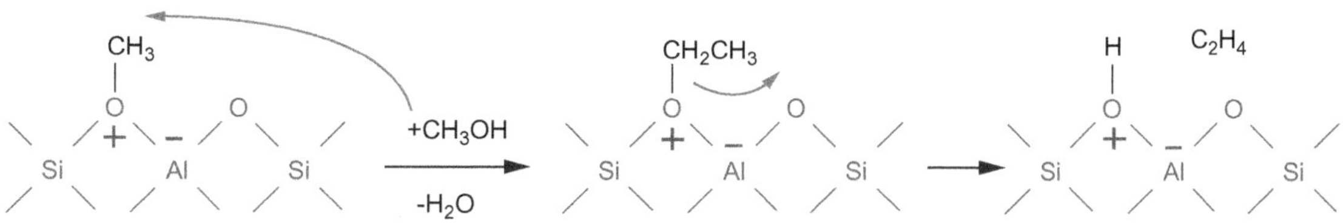

Fig. 13
Surface methoxy mechanism. *Adapted from Stöcker M. Methanol-to-hydrocarbons: catalytic materials and their behavior. Microporous Mesoporous Mater 1999;29:3–48.*

formation [84]. On the contrary, formaldehyde could be one of the potential active species during the induction period. Dewaele et al. [85] studied the conversion of methanol to light olefins by means of temporal analysis of products and found that methane and formaldehyde could be detected in equal amounts at low conversion. Less formaldehyde was detected at higher conversion and higher temperatures.

3.1.6 Impurities-driven route

The idea of an impurities-driven route had been speculated in prior arts. A mechanism for the initiation of the MTO process by organic impurities instead of any direct route from pure methanol and DME was also suggested [28,87]. Moreover, it was stated that the rate of formation of the initial reaction centers and, therefore, the duration of the kinetic induction period for the first C–C bond formation are governed by impurities [87]. According to Ref. [28], these first hydrocarbons could be the result of impurities present in the MeOH feed or could originate from the incomplete combustion of organic templates from the catalysts, or formed in situ due to, for example, a presence of some transition metal cations on the catalyst, or being formed with the help of the metallic surface of the reactor walls.

According to Ref. [88], the activation of methanol via a trimethyl oxonium rearrangement for "first carbon/carbon bond formation" is relevant only during an incubation period at low temperature. They suggested that the impurity-driven mechanism may gain more importance at higher temperatures [88].

3.1.7 Conclusions

Formation of the first C–C bond is governed at least by catalyst nature, reaction conditions, and nature of the feed. The only mechanistic concepts that take into consideration all of these parameters are insufficiently developed. Among the six mechanisms proposed for the formation of the first C–C bond, a few of them need a catalyst (acid or basic sites) to occur. The carbene and the free radical types of mechanisms are considered to be highly unlikely today. Depending on the acidity of the catalyst and/or the operating conditions, one mechanism will be more favored than the others, but they still could take place in parallel. The surface methoxy mechanism appears quite probable since the activation barriers are not so high compared to the other routes.

A large number of proposals have been made concerning the mechanism of the formation of the primary hydrocarbon products. Despite the number of proposed mechanisms, none of them has any monosemantic supporting experimental evidence. The common peculiarity of all routes resides in a high energetic barrier and in a difficult predictability. So we should be ready to "unlearn" the things we learned because all unexpected details matter in MTO.

3.2 Reaction Pathways to Produce Light Olefins From MeOH

The formation of the first C–C bond is the kinetically limited step of MeOH transformation. As soon as the first C–C bond has been formed, the subsequent conversion of light olefins to paraffins, aromatics, naphthenes, and higher olefins proceeds via classical carbenium ion mechanisms with concurrent hydrogen transfer taking place [54]. This chemistry is well known from hydrocarbon chemistry in the presence of acid sites [54]. Hutchings also suggested that the aromatics are most likely formed from primary olefins [22].

All these reaction pathways are kinetically much faster than the formation of the first hydrocarbon from MeOH. The formed olefins may be further cracked to light olefins or undergo a hydrogen transfer reaction to form aromatics and paraffins [58,59]. So the aromatics are coproduced at this stage with ethylene, propylene, and some heavier olefins. According to the prior arts, two main routes can lead to the reaction products [26,89,90]:

- A hydrocarbon pool mechanism via alkylation/dealkylation of hydrocarbon scaffolds.
- Olefins interconversion via methylation, oligomerization, and cracking.

The "dual-cycle" mechanism, recently introduced by Svelle et al. after confirmation by theoretical calculations [91], combines the features of both mechanistical approaches (Fig. 14) [29,92,93]. This mechanism represents the most recent state on olefins formation from methanol and allows the best fitting with the experimental data.

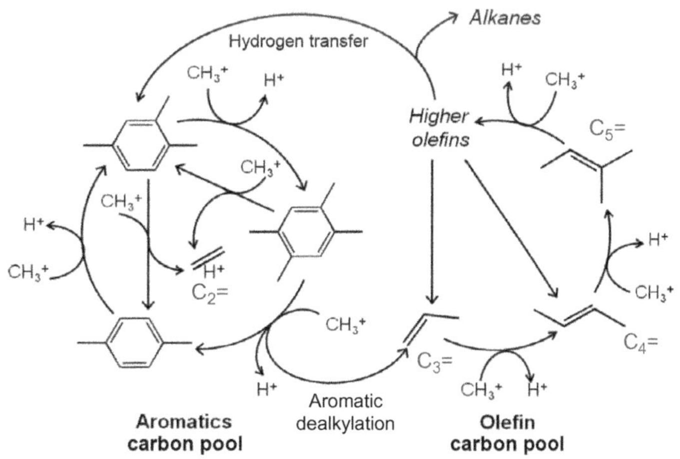

Fig. 14

Dual olefin and aromatic methylation catalytic cycle for methanol to hydrocarbons. *Adapted from Ilias S, Bhan A. Mechanism of the catalytic conversion of methanol to hydrocarbons. ACS Catal 2013;3:18–31.*

Originally, the idea was to mechanistically separate ethylene formation from the formation of higher alkenes [93]. It was found that ethylene was formed from the (lower) methylbenzenes (ie, from the aromatics carbon pool) whereas propylene and higher alkenes were to a considerable extent formed from alkene methylations and interconversions (eg, cracking reactions).

3.2.1 Hydrocarbon pool: "aromatics carbon pool"

The first brick to the hydrocarbon pool concept was brought in the early 1980s. First, Mole and coworkers [94,95] reported that deliberately introduced toluene acted as a "co-catalyst" for methanol conversion over H-ZSM-5. They proposed a mechanism by which methyl substituents on benzene rings underwent side-chain alkylation followed by olefin elimination (Fig. 15).

In the same year, Chang published a scheme diagrammed in (Fig. 16) [96] for the methanol transformation with propylene, aromatics, and naphthenic intermediates in methanol transformation to olefins [21]. The only difference between it and Mole's concept was the role of propylene as an intermediate in the formation of the secondary aromatics intermediates and heavier olefins.

In the beginning of the 1990s, Dahl and Kolbe [26,90] formalized a "hydrocarbon pool" mechanism. The publication initiated an immediate growth of the academic investigations in the area of methanol transformation (Fig. 9).

Initially, the experimental evidence for the hydrocarbon pool mechanism was formulated by showing through labeling experiments that ^{13}C-methanol and ethylene fed together to an H-SAPO-34 catalyst failed to form propylene by methylation of ethylene [26,90]. In the proposed mechanism, methanol is added in a continuous manner to a pool of adsorbed hydrocarbons $(CH_2)_n$, which may have many characteristics in common with

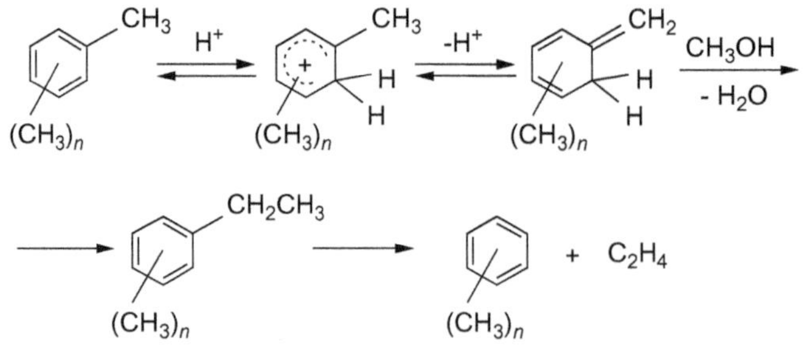

Fig. 15

Mole's reaction path for olefin formation via toluene. *Adapted from Mole T, Bett G, Seddon D. Conversion of methanol to hydrocarbons over ZSM-5 zeolite: an examination of the role of aromatic hydrocarbons using 13carbon- and deuterium-labeled feeds. J Catal 1983;84:435–45; Mole T, Whiteside JA, Seddon D. Aromatic co-catalysis of methanol conversion over zeolite catalysts. J Catal 1983;82:261–6.*

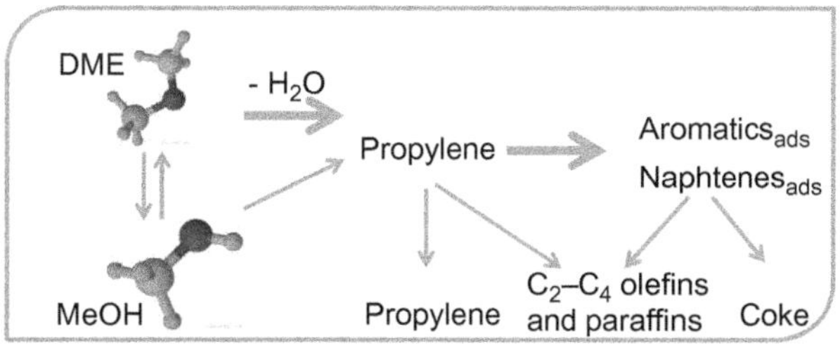

Fig. 16
Reaction steps of MeOH conversion into hydrocarbons. *Adapted from Chang CD. Hydrocarbons from methanol. Catal Rev Sci Eng 1983;25:1–118; Ceckiewicz S. Methanol conversion to hydrocarbons and dimethyl ether on decationized zeolite-T. Kinet Catal Lett React 1981;16:11.*

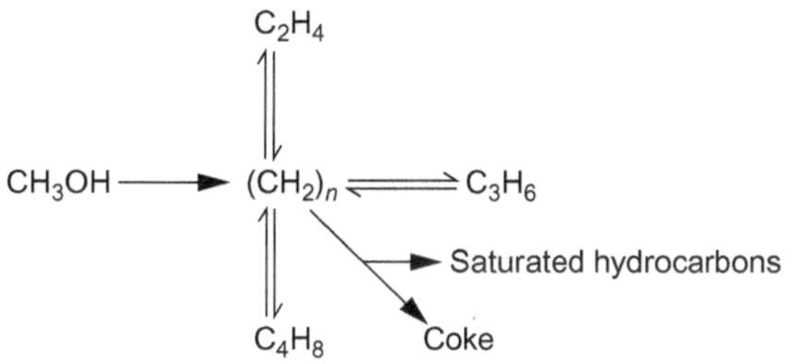

Fig. 17
Original scheme of hydrocarbon pool mechanism. *Adapted from Dahl IM, Kolboe S. On the reaction mechanism for propene formation in the MTO reaction over SAPO-34. Catal Lett 1993;20:329–36; Dahl IM, Kolboe S. On the reaction mechanism for hydrocarbon formation from methanol over SAPO-34: I. Isotopic labeling studies of the coreaction of ethene and methanol. J Catal 1994;149:458–64.*

ordinary coke (Fig. 17). The initial buildup of this "hydrocarbon scaffold," essential for this mechanism, was assigned to a short, unspecified induction period. Most probably the formation and further constant supply of the scaffolds occurs via olefins methylation/oligomerization optionally followed by dehydrocyclization.

Afterward, the notion of "unspecified carbon deposition" with an olefin-like composition $(CH_2)_n$ has been gradually transformed to Polymethylbenzenes (PMBs) by many research groups [87,97]. Those PMBs serve as scaffolds/cocatalysts, where methanol is added and olefins are eliminated in a closed catalytic cycle [87,98]. It is therefore indicated that the interplay between the inorganic framework and the organic reaction centers dictates the activity and selectivity. However, according to Ref. [97], the role of PMBs as the major hydrocarbon pool species appears to be independent of the zeotype catalyst chosen. Haw et al. [87] provided both experimental and theoretical evidence in favor of PMBs as the driving force for the hydrocarbon pool mechanism. In 1998, by means of pulse-quench reactions on an H-ZSM-5 catalyst and GC-MS and ^{13}C MAS NMR analysis, it was reported that after a double-pulse ^{12}C–^{13}C methanol experiment, propylene contained carbon atoms from both reagent pulses. This was considered to be a strong enough support for the hydrocarbon pool mechanism. Further, analysis of the NMR spectrum pointed out that it was dominated by alkyl-substituted cyclopentyl carbenium ions [87]. Reaction of the aromatic product with methanol or DME under MTO conditions would regenerate the original methylbenzene, completing a catalytic cycle.

At the same time, Tsai reported [99] that even in the absence of methanol, PMBs may reach equilibrium with ethyl PMBs through a five-membered ring carbocation reaction (Fig. 18). The hypothesis was supported by the formation of ethane and butane in transalkylation products.

In the hydrocarbon pool mechanism (ie, aromatics carbon pool), two parallel routes, namely the side-chain route or "*exocyclic methylation*" route and the "*paring*" route, have been suggested for the formation of olefins [97].

The paring mechanism, which was first suggested by Sullivan et al. in 1961 [100] for hydrocracking reaction on Ni-containing silica-alumina mesoporous support, was well known for transalkylation reactions [99] and was also suggested for an MTO case in [97]. According to Ref. [97], the paring mechanism should mostly explain a formation of propylene and butenes by the carbon atom interchange between the rings of PMBs. According to the proposed mechanism, heptamethylbenzene (HPMB) splits off olefins with low-energy barriers [101,90] after appropriate rearrangements (ring contraction followed by expansion, provide the means to extend an alkyl chain and hence eliminate an olefin from methylbenzene by dealkylation) leading to the formation of lower methylbenzenes that are re-methylated back by methanol (Fig. 19) [97].

The "*side-chain*" route involves the stepwise growth of alkyl side chains on PMBs and the subsequent elimination of the side chain into light olefins by dealkylation (Fig. 20).

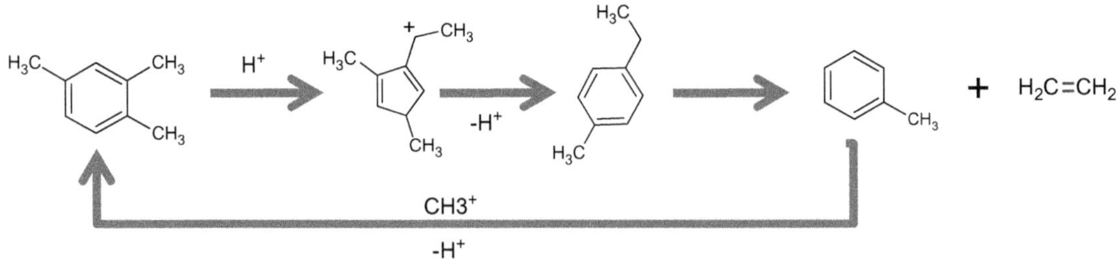

Fig. 18
Formation of ethylene via alkyl cyclopentyl carbenium ion. *Adapted from Tsai T-C, Liu S-B, Wang I. Disproportionation and transalkylation of alkylbenzenes over zeolite catalysts. Appl Catal A 1999;181:355–98.*

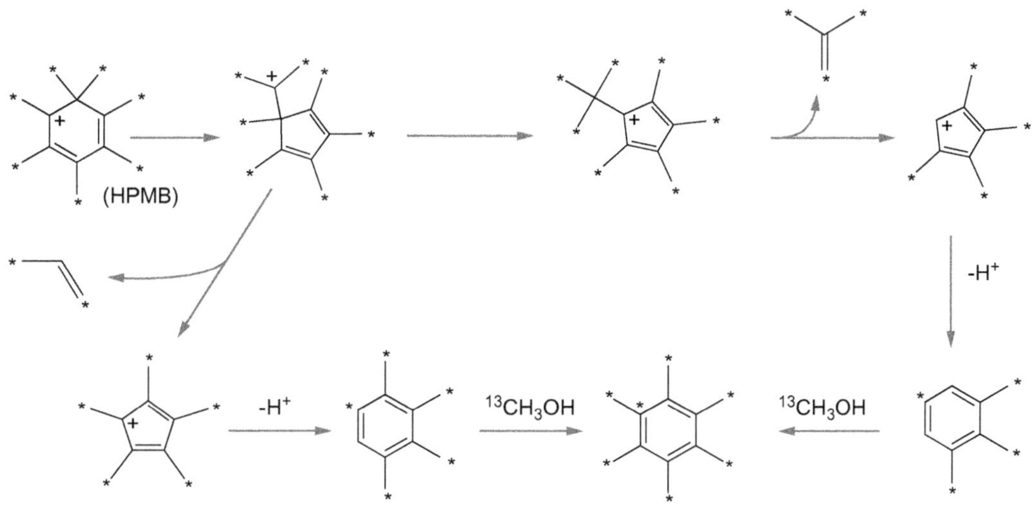

Fig. 19
The paring mechanism (* stands for ^{13}C). *Adapted from Olsbye U, Bjørgen M, Svelle S, Lillerud K-P, Kolboe S. Mechanistic insight into the methanol-to-hydrocarbons reaction. Catal Today 2005;106:108–11.*

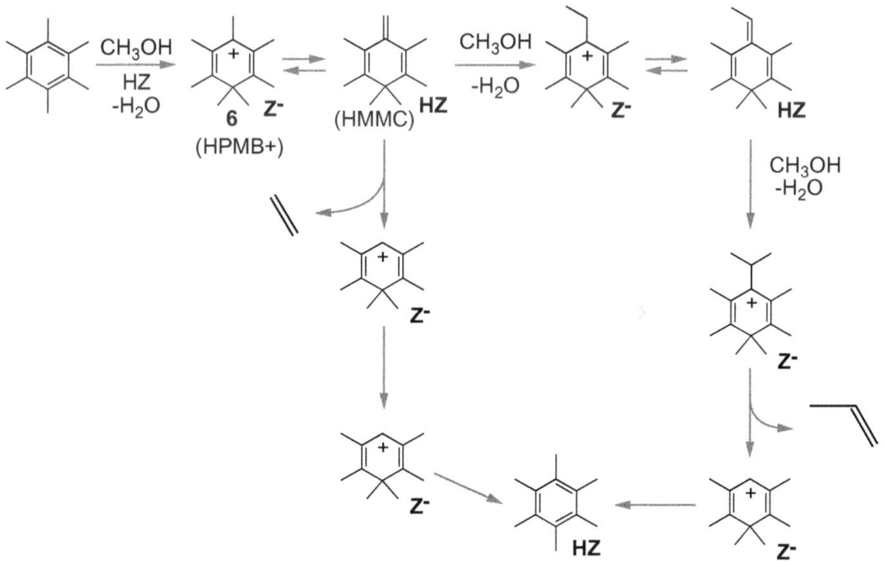

Fig. 20
The side-chain methylation mechanism (or exocyclic methylation mechanism).
Source: *Olsbye, Bjorgen, et al., 2005.*

In the exocyclic methylation route, first proposed by Mole et al. [94,95] and later refined by Haw et al. [87], HPMB+ is deprotonated into HMMC (1,2,3,3,4,5-hexamethyl-6-methylene-1, 4-cyclohexadiene). The exocyclic double bond then undergoes reaction with an incoming methanol molecule, resulting in an ethyl group on the benzene ring, which is subsequently dealkylated as ethylene (Olsbye, Bjorgen, et al. 2005). Propylene formation is also possible from this reaction scheme, while butene formation is not, due to potential steric hindrance [102]. Theoretical calculations indicated that the rate-determining step in the exocyclic methylation route is deprotonation of the HPMB+ ion, and that the essential intermediate in the MTH cycle is two gem-methyl groups attached to the aromatic ring [102].

The kinetics of the zeolite catalysis is strongly affected by both the topology of the inorganic framework and the geometry of the organic reaction center. Wang et al. suggested that the reaction pathways involving organic hydrocarbon pool species other than MBs such as simple olefins may well be present in parallel with the MBs pathways [61,103]. Further theoretical insight performed by the same group showed a minor role of paring mechanism in the methanol-to-olefins conversion. The overall energy barriers of MBs to produce propylene and isobutene through a paring route are much larger than those to produce propylene and ethylene through the side-chain route (Fig. 21). That shows that the paring mechanism may not be operative in the MTO. So the alkene methylation and the cracking mechanism should deserve more attention. This is in line with the previous experimental findings that the side-chain methylation is the predominant pathway and the paring route is just a possible minor pathway in HBeta zeolite [104] and on mesoporous silica-alumina supports [100].

Song et al. suggested that both the MTO activity and propylene selectivity increase with the average number of methyl groups per benzene ring [105]. According to Ref. [102], the ethylene-to-propylene ratio could be increased by lowering the number of methyl groups on the aromatic ring.

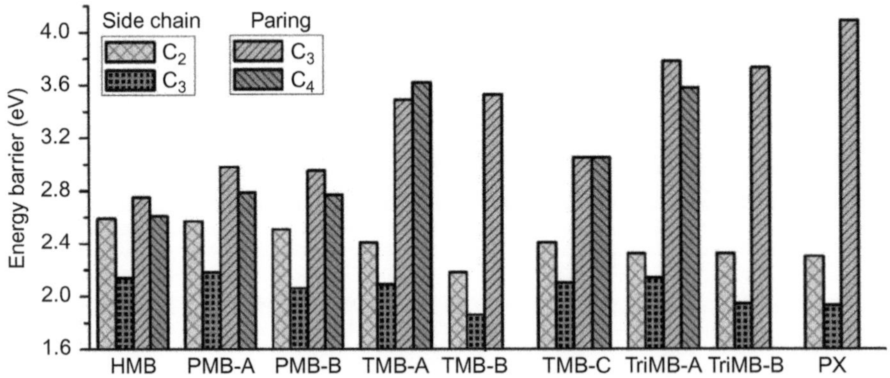

Fig. 21
Overall energy barriers of the MTO conversion catalyzed by different PMBs in SAPO-34: side chain versus paring route [61,103].

Some very recent first-principle calculations together with kinetic Monte Carlo simulations have shown that the MBs with five or six methyl groups are not more active than those with fewer methyl groups [103]. Propylene is intrinsically more favorable than ethylene when the reaction is not diffusion limited based on a side-chain hydrocarbon pool mechanism. The theoretical results are consistent with some experimental observations and can be rationalized based on the shape selectivity of key reaction intermediates and transition states in the pore of catalyst [61,103].

The complexity of the MeOH to hydrocarbons reaction as a result of rapid secondary reactions as well as limitations of suitable techniques to follow this reaction has always been a problem to get conclusive evidence for the chemistry of the methanol conversion. Besides the topic of formation of the first C–C bond, the number and nature of intermediates have been a matter of large interest. One debate is the effect of MBs structure on catalytic activity and selectivity. Furthermore, the overall energy barriers in the side-chain route is calculated to be high as well [103]. The study focused only on MBs as the hydrocarbon pool scaffold limits the incomplete understanding of the MTO transformation. The pursuit of other types of hydrocarbon pool species is therefore required.

3.2.2 Olefins interconversion cycle: "olefin carbon pool"

In the olefins interconversion cycle or a consecutive mechanism, the olefins are successively methylated, oligomerized, and partially cracked to smaller ends, such as ethylene and propylene [62]. In fact, the olefins higher than C_6, which have been formed by a combination of methylation/oligomerization could easily be recracked into light olefins. The olefins C_2–C_5 are more prone to methylation, alkylation, and oligomerization (than cracking directly) to form higher olefins, which will be further cracked. Oligomerized products could also be transformed to polyolefinic coke precursors or to aromatics and paraffins via a hydrogen transfer reaction.

3.2.2.1 Olefin methylation

Olefin methylation is the most tangible route by which methyls from MeOH/DME are incorporated into hydrocarbon products, thus resulting in a following chain growth. Regarding chain growth, Espinoza [106,107] presented the experimental evidence that at the normal operating temperature, the oligomerization reaction of C_3^+ alkenes is dominant over alkene methylation; the methylation of ethylene was proved to be relatively negligible [108].

Recently, Wang et al. [103] suggested that alkene methylation, firstly proposed by Dessau, should receive more attention in the MTO conversion even on SAPO-34 [109]. The overall energy barriers for the production of ethylene and propylene are much lower than those in side chain and paring hydrocarbon pool mechanisms. That is to say, hydrocarbon pool mechanism, where alkenes themselves are the organic active centers, may be operative in the MTO conversion [61,103].

There are two proposed mechanisms for olefins methylation [62]: (1) a co-adsorbed mechanism in which methanol and an olefin are adsorbed on a single acid site and react in a single, concerted step; and (2) a surface methoxide mechanism in which methanol or DME dehydrates to form a methoxide that desorbs upon reaction with an olefin.

The kinetics of C_2–C_4 olefins methylation have been studied on H-ZSM-5 using ^{13}C-methanol by Svelle et al. [110] and using DME on H-ZSM-5, H-BEA, H-FER, H-MOR by Hill et al. [111,112]. The following ratios were obtained by Svelle with coworkers for the methylation rates at 350°C: $k_{\text{ethylene}}:k_{\text{propylene}}:k_{n\text{-butene}} = 1:17:50$ [110]. Despite these reactions taking place over a large range of experimental conditions (47–440°C), both studies show that olefin methylation has a first-order dependence on the olefin pressure and a zero-order dependence on the methylating agent (methanol or DME) and that this kinetic behavior is consistent across the zeolite frameworks studied. The coreaction of propylene or *n*-butene and methanol over an H-ZSM-5 acidic zeolite catalyst showed [110] that the methylation products are dominating, and the isotopic composition is in agreement with the methylation formation mechanism. It has been found that alkene interconversion (oligomerization/cracking) reactions are strongly suppressed by the presence of methanol. Ethylene was always detected in trace amounts in all the coreaction experiments. The ethylene was fairly rich in ^{13}C, and its isotopic distribution was close to random.

Olefin methylation has also been investigated on H-SAPO-34 at 400°C by Dahl and Kolboe [26,90,113]. They showed that olefin methylation is not the dominant route for propene and butene formation on H-SAPO-34; rather, aromatic dealkylation reactions are responsible for light olefins formation.

Hill et al. also measured olefin methylation rates for all four butene isomers in the absence of C_4 isomerization reactions on H-ZSM-5 and H-BEA, observing that the rate constant of isobutene methylation was an order of magnitude greater than that of the other butene isomers [112]. These observations are consistent with the conclusion that the reactivity of olefins is dependent on the degree of substitution about the double bond, which stabilizes intermediate carbocations through inductive electron donation. Similarly, Iglesia and coworkers [114] showed for DME conversion over H-BEA that under conditions when skeletal isomerization is suppressed (<227°C), highly branched alkenes and alkanes, particularly triptane and triptene (Fig. 22), are the dominant product as a result of methylation reactions favoring the formation of the most substituted carbenium ion as a reaction intermediate. Computational chemistry studies, using DFT on 30–46T cluster sizes to account for dispersion effects, have also investigated C_2–C_4 olefin methylation via the co-adsorption mechanism and, in agreement with experimental work, show that activation energy barriers decrease while rate constants increase systematically with olefin size [114].

An important conclusion is that the olefins methylation reactions are propagated to different extents, depending on the identity of the zeolite framework [112].

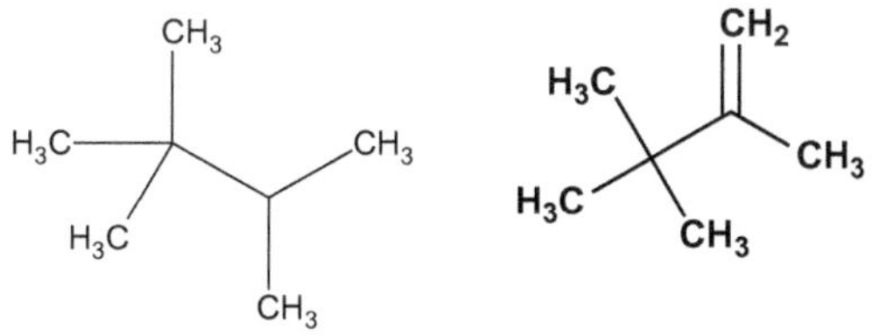

Fig. 22
Triptane and triptene.

3.2.2.2 Olefin cracking

Olefin cracking as a route to light olefins production in MTO was first proposed by Dessau and LaPierre [115]. The mechanism of olefin cracking requires protonation of an olefin to form an alkoxide intermediate, followed by β-scission of the alkoxide to form a smaller olefin and a smaller alkoxide. The smaller alkoxide subsequently desorbs to form another olefin and leaves behind a proton to regenerate the acid site.

Weitkamp et al. [116] first developed nomenclature to describe the various modes of cracking based on carbenium ion types for products and reactants, a useful method for understanding the difference in cracking rates as a function of olefin size and skeletal structure (Fig. 23).

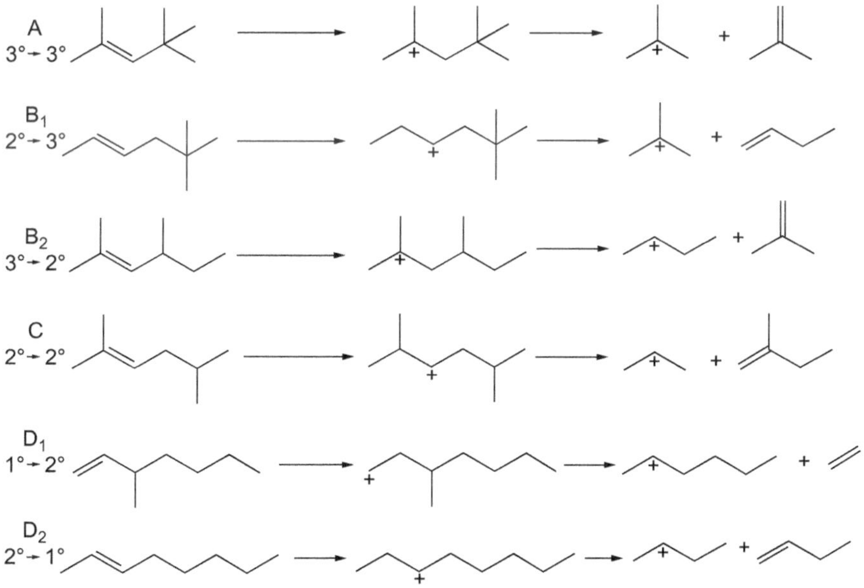

Fig. 23
Different modes of β-scission for C_8 isomers with nomenclature. Developed by Weitkamp et al. *Adapted from Ilias S, Bhan A. Mechanism of the catalytic conversion of methanol to hydrocarbons. ACS Catal 2013;3:18–31; Buchanan JS, Santiesteban JG, Haag WO. Mechanistic considerations in acid-catalyzed cracking of olefins. J Catal 1996;158:279–87.*

Buchanan et al. [117] extended this nomenclature in a study on the relative rates of monomolecular alkene cracking for C_5–C_8 olefins on H-ZSM-5 at 510°C (Fig. 23). First-order rate constants for cracking were reported, with relative rates of **C_5:C_6:C_7:C_8** alkene cracking to be **1:24:192:603**. A near-equilibrium distribution of hexene isomers was achieved at 26% conversion of 1-hexene feed, and the reaction of three different skeletal isomers of heptene resulted in identical product distributions, suggesting that numerous adsorption, isomerization, and desorption events take place prior to and are significantly faster than β-scission. In contrast to 1-octene, 2,4,4-trimethyl-2-pentene cracked 50 times more slowly, even though this branched isomer cracked through a more energetically favored 3° → 3° cracking mode (Fig. 23). The slower reaction rate of 2,4,4-trimethyl-2-pentene compared with 1-octene is mostly likely due to diffusion limitations of this molecule in H-ZSM-5.

Speciation of olefin isomers is critical for determining the product selectivity for MTO. For example, if olefin cracking is considerably faster than olefin methylation, then the product distribution should be rich in light olefins. In contrast, if olefin cracking is slower than olefin methylation, then the product distribution may be rich in larger olefins, which may cyclize to eventually form aromatics. Simonetti et al. [118] measured the rates of β-scission for a variety of C_5–C_8 alkene isomers while cofeeding ^{13}C-DME on H-BEA at 200°C and found that in general, the rate of β-scission is at least 40 times slower than olefin methylation and is of the same order of magnitude as skeletal isomerization.

Elucidating rates of β-scission on various zeolites will be key to determining what fraction of light olefins, particularly propylene and butenes, are products of the aromatic- versus olefin-based carbon pool; however, the prevalence of secondary reactions, such as olefin oligomerization and dehydrocyclization, may hinder the measurement of kinetic parameters of β-scission.

An important conclusion is that the cracking of the short C_4–C_6 olefins is a relatively slow reaction. If the molecular sieve imposes diffusion limitations and steric constraints for chain growth by methylation and oligomerization, the formation of the coke precursors dominates over the direct cracking.

3.2.2.3 Aromatic formation and alkylation

There are two possible generalized routes to olefin cyclization and aromatization [62]. One involves dehydrogenation of olefins to form dienes and trienes that undergo cyclization to aromatics. In the second route, olefins first form cycloalkanes and are subsequently dehydrogenated to form aromatics. In both of these routes, dehydrogenation occurs through hydrogen transfer reactions in which olefins or cycloalkanes donate hydrogen to other hydrocarbons that act as hydrogen acceptors.

Hydrogen transfer reaction is a bimolecular reaction in which a hydrogen atom is transferred between an adsorbed surface alkoxide and a cyclic or acyclic alkene. This reaction leads to a stoichiometric formation of aromatic and paraffins.

Isotopic labeling studies of methanol co-fed with an aromatic show that aromatic methylation reactions involve sequential methylation steps; however, varying incorporation of ^{13}C atoms into methylbenzenes also shows that aromatic methylation is not the only route by which methylbenzenes are formed [62].

3.2.3 $C_2^=/C_3^=$ as primary products

Two possible carbocation reaction schemes have been published. There is presently much debate over the nature of the intermediate from which the primary olefins—ethylene, propylene, and the butenes—are formed. According to Hutchings and Hunter [22] this would be an oxonium methylylide, produced from the methoxy carbenium ion, CH_3^+ generated out of methanol by dehydration (Fig. 24). According to Arstad et al. [102], Wei et al. [28], and Dahl and Kolboe [90] the intermediate(s) they call "carbon pool" with which methanol reacts would be the hexamethylbenzenium ion or a similar compound. They mainly deduce this from the analysis of the catalyst after conversion of the methanol in a batch mode.

Froment, for example, even in his most recent works, reconciled the hydrocarbon pool mechanism even for SAPO-34 [119]. They claim that the so-called intermediates in the "carbon pool" with which methanol reacts are mainly deduced from the analysis of the catalyst after conversion of the methanol in a batch mode, very diluted with inert gas conditions (not steam), and under the conditions far beyond the MTO/MTG modes. PMBs are the stable dead-ends and not the true intermediates. Coke plays an important role, but

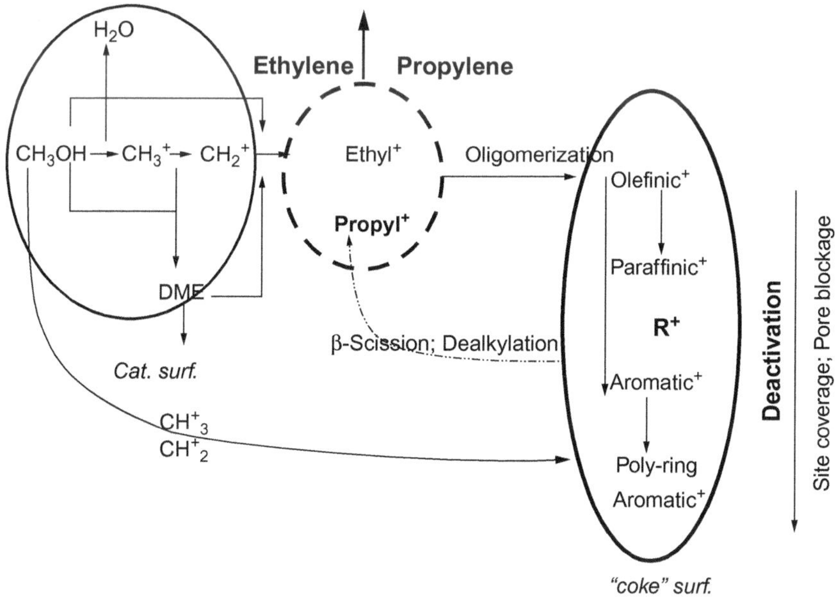

Fig. 24
Formation of ethylene and propylene via oxonium methyl ylide mechanism.

the hydrocarbon pool species are not necessarily the alkylbenzenes. The acid sites of the "coked" surface generate and stabilize the carbenium ions. While they do acknowledge that the large aromatics are not completely inert, the PMBs are not the main intermediates. They may undergo alkylation/dealkylation with MeOH or olefins followed by β-scission to yield ethylene and propylene. Organic hydrocarbon pool species other than MBs such as simple olefins are present in parallel with the MBs. Active coke precursors could be also olefinic.

Detailed kinetic studies performed by his group showed an extrapolation of the propylene yield to initial times [101]. This potentially proves that propylene is at least one of the primary products in MTO reaction after the initiation step, which is consistent with [61,103].

According to Froment [119], the propylene formation is very fast, but this olefin is very reactive on the nondeactivated catalyst.

The induction period predicted here and observed in the olefins yield in MTO on SAPO-34 depends as much on selectivation by deactivation as on the reaction scheme itself [119]. We should emphasize that the behavior described here, at the time, was based on the reaction scheme proposed by Hutchings and Hunter [22] and used by Alwahabi [101], not upon the carbon pool mechanism, often referred to for its explanation.

According to Froment [119], the so-called intermediates in "carbon pool" with which methanol reacts are mainly deduced from the analysis of the catalyst after conversion of the methanol in a batch mode, at very diluted conditions, and under the conditions far beyond the MTO/MTG modes [28,29,102,90].

On the other hand, the primary carbenium ions ethyl$^+$, propyl$^+$, and butyl$^+$, whose formation involves the oxonium methyl ylide, can be deprotonated into the corresponding olefins, but they can also undergo methylation and oligomerization into higher carbenium ions, leading to paraffins formation via H-transfer reactions and to aromatic via cyclization [119]. The latter transformation leads to polyring aromatics, which are too bulky to leave the catalyst because of the structure of SAPO-34. They cover sites and block pores and thus deactivate the catalyst. However, the catalyst should be partially deactivated to a primary formed propylene which could escape the pore system of the molecular sieve.

However, Froment [119] acknowledges that the large aromatics are not completely inert. They may react with the methoxy ion CH_3^+ and the methyloxonium ylide, CH_2^+. They can also undergo dealkylation or β-scission to generate ethylene and propylene. In such a generalized scheme there are two types of active sites upon which the elementary steps take place: the first type consists of acid sites of the catalyst surface itself, on which the carbenium ions are generated. The second type of the acid sites is located on the "coked" surface and consists of carbenium ions, which have been produced. It is on the latter sites that methylation, oligomerization, and β-scission occur, but also reactions with CH_2^+, CH_3^+, and with protonated methanol, as suggested by the carbon pool mechanism. This general scheme and

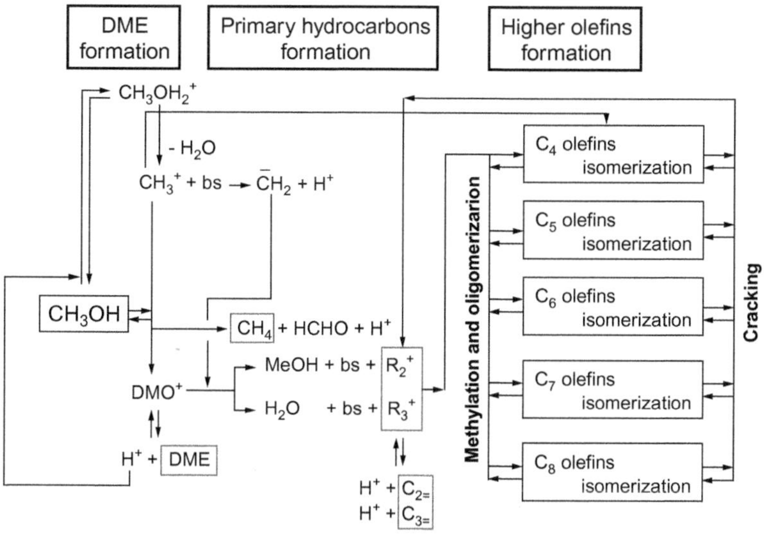

Fig. 25

Reaction scheme (bs: basic site of the catalyst, R_2: carbenium ion ethyl$^+$, R_3: carbenium ion propyl$^+$). *Adapted from Alwahabi SM, Froment GF. Conceptual reactor design for the methanol-to-olefins process on SAPO-34. Ind Eng Chem Res 2004;43:5112–22; Alwahabi SM, Froment GF. Single event kinetic modeling of the methanol-to-olefins process on SAPO-34. Ind Eng Chem Res 2004;43:5098–111.*

the kinetic modeling by Park and Froment [57–59] and Alwahabi [101,120] show an example of the interpretation of a number of features of the MTO process. The finally retained model corresponds to a mechanism that proceeds over oxonium methylide formed from a methoxy ion interacting with catalyst (Fig. 25). The ylide subsequently reacts with dimethyloxonium ions to generate in parallel the primary products ethylene and propylene. Through steps of carbenium ion chemistry, the latter lead to higher olefins but also, to a lesser extent, to paraffins and aromatics. The only weak point of this reaction scheme is a requirement for the presence of the basic sites on the zeolites.

3.2.4 Conclusion

The gradual acceptance of the dual aromatic and olefin interconversion catalytic cycles in methanol-to-hydrocarbons catalysis on acid zeolites has given a new context for rationalizing structure-function relationships for this complex chemistry. While there are still some conflicting interpretations, at least six major chemistries are involved in the formation of hydrocarbons from methanol: olefin methylation, olefin cracking, hydrogen transfer, dehydrocyclization, aromatic alkylation, and aromatic dealkylation. The current mechanistic understanding of MTO challenges the concepts of structure-performance relationships. The effect of the zeolite framework on the identity of the reaction pathways and the hydrocarbon pool still has to be better understood.

4 Individual Product Formation

4.1 Ethylene From Methanol

A number of authors considered ethylene to be an active intermediate in the reaction sequence [66–68]. According to Kaeding and Butter [121], ethylene is the initial olefin at low methanol conversion over phosphorus-modified ZSM-5.

Dessau and LaPierre [115] have studied methanol conversion over H-ZSM-5 using ^{13}C-labeled methanol in the presence of various olefins: they found that the resulting ethylene contained less ^{13}C than would be expected if the product were derived entirely from methanol. The amount of doubly labeled ethylene was further reduced when hexene-1 was used as the trapping olefin. They concluded that most of the ethylene was produced, not as a primary product (except during a short initiation phase) but as a product of cracking of built-up alkenes. A few years later, Dessau [109] presented convincing evidence for the involvement of propylene and butenes in the production of ethylene by the route involving long contact times on the catalyst and involving secondary olefins interconversion of the primary kinetic olefinic products.

More recently, in a theoretic study, Lesthaeghe et al. claimed that no ethylene can be formed through oxonium ylide mechanism, and their findings correspond to the fact that direct mechanisms fail in explaining the ethylene formation in MTO [122]. They found that ethylene cannot be eliminated from the oxonium intermediates and should be produced via hydrocarbon pool mechanism.

Experiments with conversion of MeOH in the presence of benzene [109,123], alkylation of ethylene with toluene [124], and alkylation of ethylene with ethylbenzene [125], over zeolite catalysts, show high yields and conversions to the corresponding ethylated analogs. This suggests that ethylene oligomerization and methylation (alkylation of ethylene by methanol) reactions are not very important under MTO conditions. So, the ethylene is relatively inert compared to other olefins under MTO conditions.

According to Svelle et al. [110], three routes can be envisaged for ethylene formation: (1) direct formation from methanol/DME, (2) alkene cracking, and (3) the hydrocarbon pool mechanism (aromatics cycle). Direct formation from methanol/DME might seem reasonable considering the high share of the $^{13}C_2$ isotopomer detected; nevertheless this possibility was discarded because no detectable amount of ethylene was formed when methanol was reacted alone. According to Ref. [110], a direct formation mechanism will not depend on the presence of products in the catalyst pores. Alkene cracking also may be ruled out, because the discrepancy between the total content of ^{13}C in ethylene and the higher alkenes (exemplified by hexene) is too large. This leaves the hydrocarbon pool mechanism as the only possible route for ethylene formation. The total label content and the isotopic distribution of ethylene matched those observed for *p*/*m*-xylene, which may be taken to represent the isotopic composition of the pool species reasonably well.

Svelle et al. proposed to separate the C_3^+ alkene cycle and the aromatics/ethylene cycle. However, the authors say that it is probably not the case for H-ZSM-5, as the constant production of aromatics during methanol conversion means that some of the C_3^+ alkenes continuously form new aromatics [93]. This means that for H-ZSM-5 the aromatics/ethylene cycle cannot run without the C_3^+ alkene cycle. Inversely, on the basis of the low reactivity of ethylene toward methanol relative to that of propylene and butenes, the contribution to the C_3^+ alkenes involving ethylene is very small and might not be required for the C_3^+ alkene cycle to occur [93].

According to very recent reports performed with participation of the same researchers [73], a reaction path analysis established that ethylene originates exclusively from the aromatic hydrocarbon pool. The same group compared the methanol conversion over H-beta and H-ZSM-5 [92]. It was found that the ethylene yield was seven times higher on H-ZSM-5 relative to H-beta. The authors suggested that on H-beta, penta- and hexa- MBs were mainly involved in a hydrocarbon pool-type mechanism, leading predominantly to propylene and the higher alkenes, whereas for H-ZSM-5, the lower MBs were the intermediates and represented the only route for ethylene formation [92]. However, it is not clear from the report why the cracking of the oligomerized product does not result in ethylene formation. This type of chemistry is well established for olefins interconversion [19,45,58,59].

Ni-SAPO-34 exhibits an extraordinarily high selectivity to ethylene as high as 85% at total methanol conversion conditions. The formation of methane was also more important and as high as 6.8% [126]. The cause of methane formation is considered to be that methanol is decomposed into hydrogen and carbon monoxide on the Ni part, which existed outside the framework of the catalysts and/or in an amorphous phase, and these gases react to form methane on the same Ni part. At the same time, by means of Ni-incorporation into the framework of SAPO-34, its acid strength is reduced to an appropriate level to produce primarily ethylene and limit further ethylene transformation. It is also well known that Ni-containing catalysts are very active in olefins transformation to aromatics, dealkylation, and transalkylation of aromatics [99] as well as in ethylene oligomerization [127]. However, the coke level found on the Ni-SAPO-34 was similar to the unmodified SAPO-34 catalyst. So, most probably, lower strength of the acid sites on Ni-SAPO-34 rather promotes alkylation of aromatic rings relative to olefin methylation, oligomerization-cracking transformation. Recently, Khadzhiev with coworkers also found that a modification of SAPO-34 with Ni leads to a dramatic increase of the selectivity to ethylene from methanol [128]. They showed 88% of selectivity with $C_2^= / C_3^=$ ratio of 17.6 at 450°C. This observation indirectly confirms that the formation of ethylene is much more related with the alkylation of the coke precursors than for propylene. The presence of Ni favors the dealkylation and avoids the transformation of $C_2^=$

into other heavier products. It is interesting that the modification of SAPO-34 with cobalt, which is also an aromatization catalyst, have not much changed the selectivity compared to nonmodified SAPO-34. So, to selectively produce ethylene it is important to suppress olefins methylation and oligomerization/cracking reactions and dispose of a permanent supply of the coke precursors on the catalyst.

So, the ethylene production does correlate with coke presence, in particular with aromatics formation as far as the diffusion limitations are not significant. However, it seems that the majority of ethylene is not always formed directly from MeOH [115]. The aromatics and other coke species could be the products of the conversion of primary carbenium ions, which are mobile and could equilibrate each other [28]. This may explain the isotopic distribution in products and retained coke molecules and the coexistence of aromatics and carbenium ions [28]. In addition to the coproduction of ethylene with aromatics in olefins interconversion cycle, formation of ethylene via alkylation-dealkylation of methyl aromatics with heavy olefins or with the equivalent carbenium ions like ethyl$^+$, propyl$^+$, and butyl$^+$ could be an option. The alkyl aromatics with the side chain length of two carbons or longer are not stable in the pore and dealkylates on the acid sites due to too long residence time and steric hindrances. This may lead to formation of ethylene, other olefins, and alkylaromatics with different structure, namely PMBs [129]. In other words, the ethylene is formed via interaction of the carbenium ions like ethyl$^+$, propyl$^+$, and butyl$^+$ formed from MeOH or heavy olefins with aromatics and other coke precursors followed by cracking and in a less extent by a direct alkylation of PMBs with methanol. The speculation is based properly on analysis of the prior arts and is not contradictory with the concept of the "aromatic" cycle for ethylene formation.

According to some very recent results [130], ethylene is produced through the cracking of the short C_5^+ and C_6^+ ions with higher cracking barriers (the cracking of C_5^+ ion forms both ethylene and propylene). This cracking of the precursors with high energetic barrier could occur with the help of the interaction with coke precursors. This is consistent with the above hypothesis. This is an explanation for why the topology with high space confinement and high coke makeup favors ethylene formation at a high level of MTO conversion.

4.2 Propylene From Methanol

The view that propylene is the initial olefin was suggested by a number of research groups [79,131,132]. They found ethylene to be substantially less reactive than propylene over a variety of catalysts. The last was regarded as evidence that ethylene could not be a primary intermediate.

The kinetic constants of olefin formation at 350°C follow the order $kC_3 > kC_4^+ > C_2^-$. The activation energy for propylene formation is lower than it is for other olefins [133].

The Anderson group pursued the research and found out that at 400°C propylene was readily converted to aromatic products, while ethylene generally exhibited low reactivity [132]. However, it was observed that certain catalyst preparation showed considerable activity in promoting ethylene conversion [132]. It was speculated that these particular catalysts may possess a few highly active sites, which are easily poisoned by coke formation.

Ceckiewicz, in Fourier transform infrared spectroscopy (FTIR) studies of methanol sorption on zeolite at 25–450°C, observed the propylene band pattern but failed to detect ethylene [96]. Upon temperature-programmed desorption, however, C_2–C_4 olefins were identified. In the suggested MTO scheme (Fig. 16), the propylene was nominated as a primary product (ie, obtained directly from MeOH or DME), and the formation of the other products occurs via two parallel routes.

Detailed kinetic studies performed by Froment's group shows an extrapolation of the propylene yield to initial times [101]. This proves that propylene might be at least one of the primary products in MTO reaction after the initiation step, which is formed directly from MeOH/DME. This result is consistent with [61,103]. According to Froment [119], the propylene formation is very fast, but this olefin is very reactive on the catalyst. With progressive deactivation by "coke," the reactor zone in which the propylene yield reaches its maximum moves further downstream [101].

Some very recent first-principle calculations, applied to methanol-to-olefins conversion in H-SAPO-34, showed that propylene is intrinsically more favorable than ethylene when the reaction is not diffusion limited based on side-chain hydrocarbon pool mechanism [103].

Lesthaeghe et al. [134], based on the theoretical studies available in the literature, claimed that no ethylene can be formed through direct mechanisms in MTO. However, the trimethyl oxonium and ethyldimethyloxonium ions are distinctly stabilized by the surrounding ZSM-5 framework, which does not impose steric constraints on further intermolecular reactions. This means that the potential intermediates for propylene formation are more probable.

4.3 Propylene From Ethylene

Because ethylene and propylene are formed independently in MTO, the ability of the catalyst to convert ethylene to propylene becomes an important property of the catalyst.

Already in the beginning of 80th, the researchers from Union Carbide have tested SAPO-34 in the inter-conversion of small olefins [135]. Ethylene, propylene, or butene in the presence of SAPO-34 at 375°C was interconverted into mixtures of these three olefins.

Then, the reaction has been studied by several research groups with a purpose of better understanding the MTO reaction mechanism [110,136–140]. In the early 1980s,

Dessau et al. [109,115] proposed an indirect reaction mechanism for the methanol conversion based on alkenes as the key intermediates.

In the beginning of 90th, Tau and Davis [136] investigated the coreaction of ethylene and methanol over silicalite S-115 by the isotopic tracer method (^{14}C labeled or unlabeled methanol and unlabeled or ^{14}C labeled ethylene) and concluded that ethylene was converted by adding a C_1 specie derived from methanol. However, the relative ^{14}C in the hydrocarbon products revealed that the alkylation of C_3^+ alkenes is more rapid than the formation of $C_2^=$ and C_3^+ alkenes from methanol only. For the conversion of ethylene only (in the absence of MeOH), the dimerization to form butenes was the dominant reaction. Adding a flow of water in an amount equimolar to ethylene significantly decreased the conversion of ethylene. The addition of methanol to the feed stream, in an amount equal to that of ethylene, increased the total conversion and altered the product distribution so that propylene is formed in about twice the amount of butenes. The ^{13}C labels on the C_3–C_5 number products were similar to that of ethylene. The authors concluded therefore that the C_3–C_5 products were formed by the successive addition of an unlabeled C_1 species derived from methanol to labeled ethylene. The data clearly show that the formation of ethylene from methanol is a slow reaction compared to the addition of the C_1 species to the C_3^+ products. Thus, the formation of ethylene is an important issue only for the reaction initiation. In those processes, where a small amount of C_3^+ alkenes are added to the methanol feed, the formation of ethylene directly from methanol represents a small part of the hydrocarbons produced from methanol.

The data with propylene and methanol showed that once C_3 and higher carbon number alkenes approach a significant molar concentration, the methanol reacts predominantly by addition of C_1 species to the C_3^+ alkenes. A rapid series of oligomerization/cracking reactions serves to redistribute the initial products toward a thermodynamic equilibrium and a statistical distribution of the added isotope.

In another study on the cofeeding of ethylene and methanol with isotopic labeling over a ZSM-5 catalyst, Ronning et al. [138] drew the conclusion that ethylene is unreactive and its methylation could not be an important part in the reaction network. They determined that the ethylene conversion is increased in the presence of methanol but only at rather low levels of conversion.

Svelle and coworkers [137] studied the [^{12}C]ethylene-[^{13}C]methanol coreaction on H-ZSM-5 using the isotopic labeling technology as well. The authors found that the methylation of ethylene to form propylene was dominating at very low conversions, but it was eclipsed by other pathways once there was an appreciable amount of hydrocarbon products. At very high feed rates the dominant coreaction product is propylene (approaching 90%). At the highest feed rates the $^{12}C_2^{13}C_1$ isotopomer was constituted of about 85% of the propylene molecules.

Song et al. [141] studied olefin homologation reactions (ie, reactions between olefins and different alcohols from MeOH to butanol) on zeolite ZSM-22, for which its one-dimensional pore structure inhibited the hydrocarbon pool mechanism and secondary reactions of the MTO conversion. Homologation of ethylene, propylene, and styrene were directly observed between ^{13}C labeled methanol and olefins. Isotopic tracking showed high selectivity for the homologation reaction, that is, starting from ethylene, propylene showed one ^{13}C atom, butene contained two ^{13}C atoms, and pentene three. Homologation of styrene's side chain also showed a very specific isotopic pattern. The role of olefin homologation is speculated as a key reaction during the induction time of the MTO process. With no overwhelming secondary reactions on ZSM-22, the olefin homologation reaction could be directly observed.

A mixture of methanol and ethanol (for ethylene) or propanol (for propylene) was pulsed onto the ZSM-22 catalyst [141]. Primary olefins such as ethylene or propylene were generated in situ by the dehydration of the respective alcohols. Soon after the primary olefin was produced, it reacted via methylation with MeOH to extend its olefin chain. Small amounts of propylene, butene and pentene along with a large amount of ethylene were produced when a mixture of MeOH and EtOH was pulsed onto ZSM-22 at 350°C. Similarly, after a MeOH/iso-PrOH pulse, butene and pentene were produced next to propylene. The yield of products from homologations is quite low, especially from ethylene. To investigate the temperature influence on the homologation reaction over ZSM-22, the mixture of alcohols (MeOH-^{13}C/EtOH or MeOH-^{13}C/*iso*-PrOH) were pulsed onto the catalyst at a series of temperatures (250, 300, 350, and 400°C). In both sets of experiments, no homologation product was detected at 250°C. However, when the MeOH-^{13}C/EtOH was pulsed onto the ZSM-22 at higher temperatures, interesting isotopic distribution patterns were observed. At 400°C, a clear trend of adding ^{13}C step by step via homologation of the olefins with the labeled alcohol was observed; this trend became less apparent at 350°C. At 300°C, the ^{13}C distribution of olefin products did not follow the expected trend. Propylene had one ^{13}C as expected, but most butene had no ^{13}C and most pentene had only one ^{13}C. The reason for these odd ^{13}C distributions on butene and pentene are perhaps due to the differences in energy barriers and reaction rates for the different possible reactions of olefins on ZSM-22. It has been found that the relative rate of alkene interconversion increased with the decreasing reaction temperature. At lower temperatures (300°C), the dimerization reaction of ethylene dominated over the homologation reaction, leading to no ^{13}C distribution in butene.

Li et al. coreacted ethylene and methanol over H-ZSM-5, P-La modified H-ZSM-5 (PLaHZSM-5), and hydrothermally treated PLaHZSM-5 catalysts [142]. Hydrothermal treatment at high temperatures significantly reduced the acidity of the catalyst, which leads to an almost complete suppression (conversion lower than 3%) of the direct conversion of ethylene or methanol/DME. However, the cofeeding of ethylene and methanol over the said catalyst resulted in considerable higher conversion of both reactants (MeOH conversion of 60% and $C_2^=$ conversion of 11%). High propylene selectivity (c.80%) was obtained only at lower conversions when cofeeding MeOH and $C_2^=$ (conversion decreases with increase of TOS). At the highest conversion of MeOH (in the presence of $C_2^=$), a level of selectivity of about 50% could be achieved for propylene. In general, it was the methylation of ethylene by methanol that was responsible for the increase in conversion and propylene selectivity in the coreaction system. The competition of different types of reactions in the coreaction of ethylene and methanol is influenced by the acidity of the catalysts. Lowering the acidity and/or decreasing the acid sites number of the catalyst helped to suppress the direct conversion of methanol or ethylene. Over the steamed PLaHZSM-5, methylation of ethylene played a dominant role in producing propylene at high selectivity, especially when conversion was lower. However, another route for methanol transformation toward propylene over steamed PLaHZSM-5 in the presence of ethylene was the repeated methylation of propylene and the cracking of the formed higher olefins. The role of the latter route in the complicated reaction system became more dominant with increasing reaction temperature. The results of the coreaction of ethylene and methanol over steamed PLaHZSM-5 varied with temperature [141]. When the temperature increased from 450 to 550°C, the propylene selectivity increased from 51.3% to 53.6%, and the butenes selectivity also increased from 19.4% to 22% with decreased selectivities for C_5^+ . Meanwhile, C_1–C_4 alkanes and BTX selectivities presented little change. The plot of the molar ratios of butenes to propylene and that of pentenes to butenes at different temperatures against total conversions was consistent with the mechanism of consecutive methylation by methanol to produce propylene, butenes, and pentenes. On the other hand, at the same conversion level the ratios decreased with the reaction temperature. This tendency was more obvious for heavy olefins, which are more liable to cracking compared to light olefins like propylene and ethylene. This implies that the cracking of higher olefins to form propylene and ethylene also occurs under current reaction conditions and becomes more prominent at higher temperatures. The molar ratios of converted methanol to converted ethylene at different space velocities plotted against total conversion was found larger than 1 under all conditions, and the ratio was nearly 1.5 when extrapolated to zero conversion. This implies that the methylation of ethylene by methanol would not be the only dominant reaction in this context and there must be another route for methanol transformation. Because the direct MeOH conversion was completely suppressed over this catalyst (PLaHZSM-5-steamed), it was supposed that the further methylation of propylene and butenes by methanol plays a great role in methanol transformation and a considerable part of the propylene produced, came from the cracking of the higher olefins that originated from the consecutive methylation reactions.

Yamaguchi et al. [143] studied the methylation of ethylene with methanol to propylene over MFI zeolites with different heteroatoms. They found that an efficient catalyst based on MFI with weak acidity prevents the formation of ethylene oligomers. It was demonstrated that the MeOH could be completely converted to hydrocarbons at 500°C with a selectivity of about 50–60% to propylene over a weak acid catalyst in presence of ethylene. The ethylene conversion under such conditions was nearly zero. The authors suggested that the enhancement of propylene selectivity was caused by the methylation of ethylene with methanol.

Struelens studied the reaction of DME in the presence of ethylene on H-ZSM-5 and H-SAPO-34 [144]. The purpose was to investigate the possibility of selective propylene production from ethylene and DME [144]. Through isotopic labeling experiments with singly $^{13}C_1$ labeled ethylene, it is shown that propylene to a considerable extent can be formed by DME propagation as well as ethylene methylation on both topologies. In conditions of reduced DME conversion, propylene formation occurs nearly exclusively and via methylation of the added ethylene. As contact time and/or reaction temperature and consequently feed conversion rises, the contribution of the ethylene methylation diminishes, while the production of propylene increasingly becomes the result of DME propagation. It was also demonstrated that DME propagation does not proceed via ethylene, on both topologies. When $C_2^=$ is added, two mechanisms occur: $C_2^=$ methylation and DME propagation. 100% selectivity to propylene from DME could be achieved at low conversion level in a presence of ethylene by limiting the further transformation of the propylene (Table 2). The chosen temperature is quite low (200°C) to minimize secondary cracking reactions. It is quite obvious that the coke amount of the catalyst in these experiments was extremely low. While the role of ethylene is not completely clear, DME shows almost the same conversion level whether in the absence or the presence of ethylene. It might be that ethylene is important to deactivate the most acidic sites via the mechanism of competitive adsorption and thus limit the further transformation of propylene. DME inhibits alkene oligomerization and at least a part of $C_3^=$ is formed through $C_2^=$ methylation (Table 2).

Consequently, while the role of aromatics could not be completely excluded for higher conversion level and higher temperature, the hydrocarbon pool or the "aromatic" cycle mechanism plays a secondary role in the formation of propylene. The experiments confirm the previous results and show an existence of an alkene methylation cycle on MFI and CHA topology. This sheds further light on the MTO/propylene mechanism [144].

Most of the proposed mechanisms predict that at low conversion rates propylene could be formed by ethylene methylation. The relatively low reactivity of ethylene under conditions of

Table 2: Conversion, X, and selectivity, S, of DME (feed 1), ethylene (feed 2), and an equimolar mixture of DME and ethylene (feed 3) over H-ZSM-5 (Si/Al = 140) at 200°C after TOS of 900 min

Feed	X_{DME} (%)	$X_{C_2^=}$ (%)	$S_{C_2^=}$ (%)	$S_{C_3^=}$ (%)	S_{C_4} (%)	$S_{C_5^+}$ (%)
1	3.1	–	24.9	76.2	0.0	0.0
2	–	2.4	–	23.4	46.0	30.6
3	3.6	2.2	–	100.0	0.0	0.0

Adapted from Struelens P. Dimethylether transformation in presence of ethylene with acid zeolites. K.U. Leuven; 2012.

methanol conversion makes it unlikely that the methanol is converted to higher hydrocarbons via ethylene. However, the presence of ethylene helps to improve selectivity in MeOH conversion to propylene.

In the cofeeding reaction system of ethylene and methanol, it should be noted that both methylation, oligomerization/cracking and so-called MTO conversion via the hydrocarbon pool mechanism are acid-catalyzed parallel reactions and require an appropriate nature of acid sites. The above contradictory conclusions from different researchers may come from the different catalysts used and the different conditions studied.

4.4 DME Versus MeOH Conversion in Coreaction With Ethylene

Methanol and DME conversion in the presence and absence of ethylene have been studied on phosphorous-modified ZSM-5 in the temperature range between 200 and 450°C [145].

Fig. 26 shows that the MeOH is much less reactive than DME. It was observed indeed that the conversion of the DME to hydrocarbon starts at 200°C and reaches unity at 400°C.

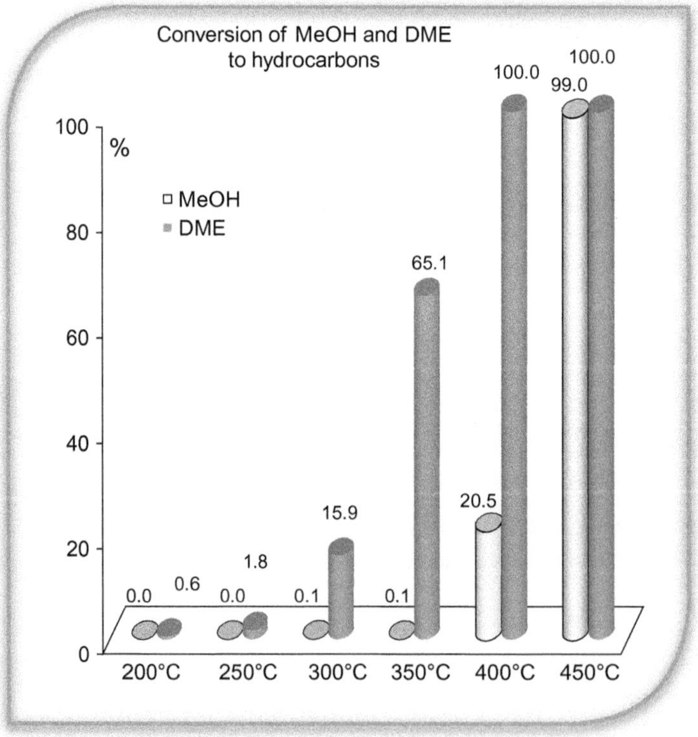

Fig. 26
Conversion of MeOH and DME to hydrocarbons on P-ZSM-5. *Adapted from Smits W. The catalytic conversion of ethylene to propylene. Unpublished results; 2010.*

Although the pure methanol experiment produces DME at low temperature, no hydrocarbons are produced before reaction temperature reached 350°C. It was hypothesized that the great discrepancy between the pure MeOH feed and the pure DME as feed is mainly caused by the fact that methanol produces water by formation of DME and water plays a role in adsorption equilibrium. It is interesting that the cofeeding of DME with the equivalent amount of water did not change the figures. Much higher concentration of the steam was required to decrease the reactivity of DME. This means that the water and the relative diffusion rate of DME and MeOH might also impact the reactivity of the MeOH [145].

However, the addition of ethylene to methanol drastically changes the breakeven temperature of the methanol conversion (Fig. 26). The conversion of MeOH in the presence of ethylene began already at 200°C, which is a strong contrast compared to the results obtained with the pure methanol feed experiment.

Second, the reaction does not consume ethylene (Table 3). On the contrary, ethylene is produced from MeOH. Selectivity to ethylene as high as 90% was observed at 21% conversion. This result was obtained in ethylene transformation, with a low methylation activity agent MeOH and in the presence of ethylene to initiate the transformation. It is important that the selective synthesis of ethylene from MeOH is possible for a higher level of conversion than for propylene because of low ethylene reactivity. In the case of propylene, the selectivity close to 100% could be obtained only up to the conversion level of 5% [144]. After 24 hours-on-stream, the effluent composition for a reaction temperature of 300°C was measured again and the amount of ethylene produced had decreased by less than 1%. So, the catalyst produces ethylene from methanol without any visible deactivation (Fig. 27).

This means that the catalyst with limited capability to transform ethylene to propylene (limited olefins interconversion activity) offers an opportunity to a selective formation of ethylene from methanol in the presence of ethylene. The experiments with MeOH and ethylene cofeeding have been repeated several times and confirmed in [144]. Kaeding and Butter [121] also suggested ethylene as the initial olefin at low methanol conversion over phosphorus-modified ZSM-5.

The explanation for that might be the interaction of the carbenium ions like ethyl$^+$, which originated from MeOH adsorption with coke species, produced from ethylene oligomerization. Oligomerization of ethylene could be observed at temperatures as low as 37°C on H-ZSM-5 and HNaY zeolites by FTIR coupled with GC-MS [84]. Upon heating to 197°C, C_3–C_7 aliphatic hydrocarbons were evolved followed by alicyclic hydrocarbons, and eventually at 327–427°C, methyl-substituted aromatics appeared. With HNaY, saturated hydrocarbons predominated, while H-ZSM-5 gave mainly unsaturates. Water vapor did not affect the ethylene reaction with NaHY, however, with H-ZSM-5, water vapor facilitated desorption of the products. Van der Berg performed detailed studies on the sorption of light olefins in H-ZSM-5 [67]. His work confirmed the oligomerization of ethylene at a low

Table 3: Transformation of DME and methanol in the presence of ethylene at low temperature and atmospheric pressure on P-ZSM-5

FEED (CH_2-Basis, wt%)		DME (33%)+C_2H_4 (67%)				MeOH (50%)+C_2H_4 (50%)			
Temperature (°C)		250	275	300	350	250	275	300	350
Conv. (%)	CH_2 as DME+MeOH	1.3	8.0	18.8	84.7	3.1	21.5	37.0	61.3
	CH_2 as C_2H_4	14.5	39.7	39.7	25.4	−8.0	−24.9	−29.2	0.7
Selectivity to C_2H_4 (%)		-	-	-	-	88.1	90.3	79.0	-
Effluent composition (CH_2-basis, wt%)									
DME		31.5	27.0	21.1	1.4	20.9	25.1	21.7	9.4
MeOH		0.6	2.9	5.3	3.6	24.6	11.1	9.8	9.9
Ethylene		66.7	62.5	57.3	40.4	54.0	62.4	64.6	49.7
Aromatics		0.0	0.4	1.4	7.1	0.0	0.0	0.4	4.4
Propylene		0.5	3.3	5.9	10.3	0.1	0.4	1.6	8.3
Propane		0.2	0.2	0.3	1.1	0.0	0.2	0.0	0.6
Paraffins		0.3	1.1	2.4	11.9	0.1	0.1	0.7	6.5
C_4^+		0.2	2.7	6.3	24.3	0.4	0.7	1.2	11.2

Adapted from Smits W. The catalytic conversion of ethylene to propylene. Vrije Universiteit Brussel; 2010.

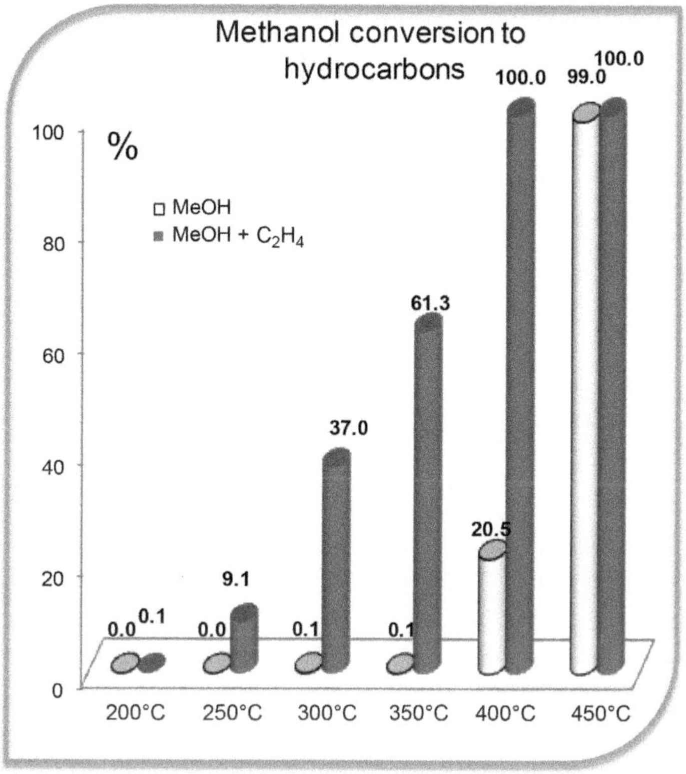

Fig. 27
Transformation of methanol to hydrocarbons on P-ZSM-5. *Adapted from Smits W. The catalytic conversion of ethylene to propylene. Unpublished results; 2010.*

temperature as well as a formation of the ethylene oligomers with an average chain length of about 25 carbon atoms. This work [67] also acknowledges the effect of water. Nevertheless, no visible deactivation proves that the coke amount is relatively low. Low-temperature region did not favor formation of aromatics. It seems that under these conditions, the second cycle of olefins interconversion in the methanol transformation, which follows the formation of the first C–C bond, is rendered negligible while the "aromatic/coke" cycle mainly producing ethylene from methanol remained active at some level.

Regarding DME/ethylene cofeeding, the mixture begins to react at 200°C and full conversion is reached at 400°C (Table 3). The results are comparable with those obtained from the pure DME feed. However, in the case of the DME/ethylene feed, the conversions were slightly higher in comparison with those for the DME feed at the same temperatures, for example, 20% conversion versus 16% at 300°C and 85% versus 65% at 350°C.

This could be explained by the fact that the first C–C bonds are already present (in the form of $C_2^{=}$), eliminating the most important kinetic barrier in the formation

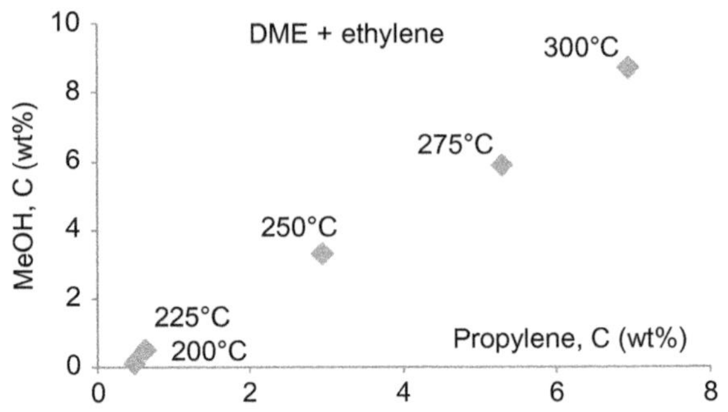

Fig. 28

Yield of MeOH as a function of propylene yield at low conversion level in DME+ethylene cofeeding. *Adapted from Smits W. The catalytic conversion of ethylene to propylene. Vrije Universiteit Brussel; 2010.*

of hydrocarbons from methanol or DME. The conversion of DME shows a correlation at low conversion levels with the conversion of ethylene but outgrows the ethylene conversion at higher temperatures and higher conversions. Fig. 28 shows that the methanol is produced simultaneously with propylene up to the temperature of 300°C [145]. At higher temperatures the correlation does not exist anymore (Fig. 28). One can understand that in contrast to methanol/ethylene mixture, the ethylene is consumed in the reaction with DME. It is evident from these experiments that the reaction pathways from DME and MeOH could be completely different and water plays an important role at the adsorption step. Because DME is more reactive than methanol it is preferably consumed while methanol is produced at rates that are—initially—higher than methanol consumption rates, thereby generating a MeOH/DME ratio much higher than thermodynamically possible at the given reaction temperatures.

5 Effect of Zeolite Topology on the Products Distribution

5.1 Small-Pore Zeolites

Molecular sieves with pore openings of about 0.45 nm show very interesting shape-selectivity properties for the conversion of MeOH to olefins. The most studied small-pore molecular sieves in the MTO process comprise zeolites: chabazite, erionite, zeolite T, ZK-5, ZSM-34, zeolite A, and zeotypes: SAPO-17 (ERI), SAPO-34 (CHA), SAPO-18 (AEI), SAPO-35 (LEV), and SAPO-44 (CHA). All of them can sorb only straight chain molecules but no branched isomers and aromatics because the pore opening is smaller than the kinetic diameter of branched and aromatic molecules, but large enough to permit the access of linear molecules. The pore-openings of small-pore molecular sieves are eight-membered oxygen rings. The cages of the three-dimensional channel system are normally much larger than the connecting windows and may admit aromatics.

The framework of chabazite possesses large, ellipsoidal cages composed of D6R (double-6-rings) at top and bottom, six 8-rings in rhombohedral positions and six pairs of adjacent 4-rings. The cavities are interconnected to six others by the puckered elliptical 8-rings.

The structure of **erionite** contains "supercages" that are supported by the columns formed by cancrinite units and the hexagonal prisms.

The result is a complex pore system interconnected by the 8-rings. The sorption cavity is the "supercage." Molecules have access to this cavity through the six elliptical openings formed by 8-rings.

Zeolite A has a large spherical cavity with twelve 4-rings, eight 6-rings, and six 8-rings. The three-dimensional porous structure is originated by the linkage of the large cavities through the 8-rings. ZK-5 structure is close to that of zeolite A.

Some zeolites possess sorption properties close to those of small-pore zeolites, even when they have pore openings exceeding 0.45 nm. This is due to blockage of pores by large cations or structural dislocations (eg, intergrowth of two crystallographic structures [146]). Offretite, zeolite T, ZSM-34, and clinoptiolite belong to this category [147].

Silicoaluminophosphates such as SAPO-17, SAPO-34, and SAPO-44 have pore openings of about 0.43 nm. SAPO-17 has an erionite-like structure, while SAPO-34 and SAPO-44 have a chabazite-like structure [148,43,149]. An interesting fact is that small-pore SAPO have porous and crystalline structures similar to those of small-pore zeolites but different acidic properties.

Chabazite, erionite, zeolite T, and ZK-5 have been used by Chang et al. [27,41] for the conversion of MeOH into olefins (Table 4). The C_2–C_4 olefins concentration in the hydrocarbon fraction was always less than 60 wt% at 100% MeOH conversion. However, when the conversion of MeOH decreases, the hydrocarbon fraction becomes richer in C_2–C_4 olefins.

Givens et al. [150] used erionite, TMA-offretite, zeolite T, and ZSM-34 as catalysts for MeOH conversion. The best yield in C_2–C_4 olefins at the highest conversion was obtained with ZSM-34 zeolite (Table 5).

Small-pore SAPOs were used as catalysts for MeOH conversion into light olefins [151,152,60]. These molecular sieves have shown a very high selectivity for light olefins at 100% MeOH conversion. SAPO-17 and SAPO-34 were the most active and selective for this reaction (Table 6) [151]. The SAPO-16 has a pore size too small (0.3 nm) to admit MeOH and produces essentially nonolefinic products. The reaction would take place principally on the external acid sites. SAPO-16 is capable of dehydrating MeOH to DME and, in small amounts, to C_2–C_3 olefins, but not of converting large amounts of oxygenated compounds to hydrocarbons. Although SAPO-34 and SAPO-44 both have the CHA framework topology, the two materials differ in their catalytic behavior. SAPO-44 gave actually a weight selectivity

Table 4: MTO performance of H-CHA, H-ERI, and ZK-5 zeolites in MTO (a: pulse experiment: 1 μL MeOH in He, 500 h^{-1} GHSV)

Catalyst	H-Chabazite		H-Erionite		ZK-5
T (°C)	538	370	538	370	538
WHSV (h^{-1})	a	1.3	a	1.3	a
Conversion (%)	100	3	100	5	100
Hydrocarbons (wt%)					
C_1	3.3	10.0	11.0	6.0	3.2
$C_2^=$	25.4	32.0	26.7	44.0	21.4
$C_3^=$	21.2	31.0	18.8	33.0	13.5
$C_4^=$	10.4	–	9.6	8.0	22.6
C_4		1.0		1.0	
C_3	33.3	2.0	27.0	1.0	31.8
C_2	4.4	14.0	5.0	5.0	0.0
C_5^+	2.0	9.0	1.9	2.0	7.5
C_2-C_4 olefins	57.0	63.0	55.1	85.0	57.5
$C_2^= / C_3^=$ molar ratio	1.80	1.55	2.13	2.00	2.38

Adapted from Chang CD. Methanol conversion to light olefins. Catal Rev Sci Eng 1984;26:323; Chang CD, Lang WH, Silvestri AJ. Manufacture of light olefins; 1977.

Table 5: Product distribution for MeOH conversion on erionite, TMA-offretite, ZSM-34, and zeolite T at 380°C

Catalyst	Offretite	Erionite	Zeolite T	ZSM-34
SiO_2/Al_2O_3	7.9	9.8	7.2	10.2
Conversion (%)	82.8	76.4	69.1	88.2
Product distribution (wt%)				
Water	32.1	35.8	34.4	40.3
DME	65.3	58.0	63.2	36.9
Hydrocarbons	2.4	6.1	2.0	22.4
Hydrocarbon distribution (wt%)				
C_1	8.6	3.6	2.1	2.1
$C_2^=$	20.2	45.7	40.5	42.5
$C_3^=$	20.2	30.0	22.1	26.1
$C_4^=$	16.8	10.0	7.4	6.7
C_2	0.0	0.7	0.0	0.4
C_3	0.0	0.0	0.0	1.8
C_4	20.9	6.5	22.9	3.8
C_5^+	12.6	3.1	4.6	16.5
C_2-C_4 olefins	57.2	85.7	70.0	75.3
$C_2^= / C_3^=$ molar ratio	1.50	2.29	2.75	2.44

Adapted from Gubish D, Bandermann F. Chem Eng Technol 1989;12:155.

Table 6: Comparison of several small-pore SAPO structures for the MTO reaction—Conditions: 375°C; 1.34 atm

	Molar Selectivity (%)				
SAPO	−17	−34	−44	−16	−35
Framework type	ERI	CHA	CHA	AST	LEV
$C_2^=$	36.5	35.0	17.7	0.5	42.8
C_2	0.5	0.6	6.3	Trace	0.4
$C_3^=$	29.3	43.0	13.3	0.6	31.2
C_3	Trace	0.4	9.5	Trace	1.3
$C_4^=$	12.2	15.8	7.4	Trace	8
C_5's	4.9	3.6	1.1	nd	2.9
C_6's	2	Trace	nd	nd	1.4
C_1	2.9	1.5	5.5	nd	11.5
CO_2	0.2	0.2	2.8	nd	0.6
DME	0	0	36.4	98.9	0
C_2–C_4 olefins	78.0	93.8	38.4	1.1	82.0
$C_2^=$ / $C_3^=$ molar ratio	1.25	0.81	1.33	0.83	1.37
Hours on stream	4.7	6.3	1	2	1
WHSV (h^{-1})					
MeOH	0.86	1.17	0.85	0.87	2.6
H_2O	2	2.73	1.99	2.03	2.43
Conversion (%)	100	100	45	53	100

Adapted from Kaiser SW. Production of lights olefins; 1985.

of 36.4% of DME after 1 h on stream, which may indicate rapid deactivation by coke and/or the presence of amorphous material. The more rapid deactivation in the case of SAPO-44 is due to the higher SiO_2 content compared to SAPO-34, which results in more acid sites.

Another interesting eight-membered ring zeolite is the MCM-35 zeolite with MTF topology [153]. The cages of MTF zeolites are interconnected to form a one-dimensional channel with pore size of 0.36×0.39 nm. It is noted that the pores of both zeolites are large enough for small organic chemicals such as methanol, ethanol, and ethylene to diffuse into. The CHA and MTF topologies having similar Al contents were used as catalysts for MeOH and EtOH transformation. It was observed that both alcohols were readily transformed into olefins on H/CHA whereas only dehydration reactions occurred on H/MTF. Moreover, ethylene is transformed into other hydrocarbons over H/CHA at elevated temperatures, whereas H/MTF shows poor catalytic performance in the ethylene transformation. The catalytic differences were attributed to the differences in the topology and inherent acidity between these two seemingly similar zeolites [154]. So, it seems that there is a sort of discrimination by a transition state to initiate the transformation of methanol even in presence of ethylene.

Görring [146] has determined that the diffusion coefficients of *n*-paraffins in T-zeolite at 340°C decrease as the number of chain C-atoms increase from 2 to 8. Also, for most of the small-pores zeolites and zeotypes, the length of the cavities is less or equal to the length of the n-heptane molecule. Thus, it may be assumed that the cavity length imposes a restriction on the formation of C_6^+ linear compounds. The combination of cavity dimensions and pore opening permits attaining high selectivities for C_2–C_4 linear hydrocarbons.

It follows from the above results that the MTO-effluent on the small-pore molecular sieves contains principally C_2–C_4 hydrocarbons, high amounts of ethylene, and very small amounts of C_6^+ compounds. At the same time, the greater part of this sieve suffers from deactivation due to a significant coking rate. This may be related to diffusion constraints and cavity dimensions.

5.2 Medium-Pore Zeolites

Medium-pore zeolites can generally be described as crystalline molecular sieves consisting of linked silica-and alumina-tetrahedra forming 10-membered oxygen ring channels. The dimensions of these medium-size pores are from 0.5 to 0.6 nm. With respect to the conversion of MeOH to olefins, only ZSM-5 or its isostructural analogs and, to a lesser extent, ZSM-11, ZSM-22, ZSM-23, and ZSM-48 have been studied.

The structure of ZSM-5 (0.54 × 0.56 nm, MFI) and ZSM-11 (0.51 × 0.55 nm, MEL) [155,156] contains two perpendicularly intersecting channel systems. Unlike ZSM-5 and ZSM-11, ZSM-48 (0.53 × 0.56 nm, MRE) [157] and ZSM-22 (0.55 × 0.45 nm, TON) [158] have no bidirectional intersecting channels. It consists of linear 10-membered ring channels [159,160].

The activity of ZSM-5 increases linearly with the Al-concentration varying from 10 to 10,000 ppm [161,162]. This correlation has also been observed with the conversion of MeOH and the dealkylation of ethylbenzene [163,63]. This one-to-one relationship between the activity of the zeolite and the amount of Al-atoms suggests that each active site contributes equally to the observed activity, whatever is the total concentration.

Jacobs et al. [164] compared the acidities of the pentasil zeolites and found that ZSM-11 was less acidic than the corresponding ZSM-5 for a similar Si/Al ratio. This was confirmed by Vedrine et al. [165].

Medium-pore zeolites have been shown to have excellent restricted transition state selectivity [164,166]. The high resistance toward coke formation on medium-pore zeolites has also been attributed to this type of shape selectivity [21,167–170]. Closely related to the shape selectivity on ZSM-5 is the effect of molecular traffic control [170]. This results from the existence of two types of intersecting channels: reactant molecules preferentially enter the catalyst through a given channel system while the products diffuse out through the other, so that counter diffusion limitations are avoided.

Studies on the conversion of MeOH as a function of contact time on H-ZSM-5 catalysts with varying Si/Al-ratios [171,172] revealed that the maximum in the yield of the intermediate

Table 7: Light olefin distribution in the conversion of methanol on H-ZSM-5 catalysts with varying Si/Al ratio at those conditions where the olefin yield was at a maximum ($T = 320°C$, $p = 1$ bar)

Si/Al (Atomic)	50	100	200	400
W/Fo (gcat h/mol) (contact time)	7.5	6	12.2	23.2
Conversion (%)	87.1	90.7	94.2	94.4
Yield (g formed/100 g fed)				
C_2H_4	7.3	8.1	8.7	7
C_3H_6	5.3	5.7	5	4.5
C_4H_8	2.9	4	4.5	3.9
$C_2^= - C_4^=$	15.5	17.8	18.2	16.1

Adapted from Katalytische Omzetting van Synthesegas in Lichte Olefins. Diensten voor Programmatie van het Wetenschapsbeleid; 1987.

light olefins ($C_2^= - C_4^=$) is shifted toward higher conversions of MeOH with increasing Si/Al-ratio. Table 7 shows [171] that ZSM-5 with Si/Al = 200 offers the highest yields of C_2–C_4 olefins (Table 7). This is in close agreement with the results of Chang et al. [39] who reported a maximum yield at Si/Al = 250. Prinz and Riekert [173], however, obtained maximal selectivities toward $C_2^= - C_3^=$ on H-ZSM-5 with Si/Al = 100. Liu et al. mentioned that, when Si/Al is higher than 100, the density of acid sites is lower and this results in higher propylene selectivity and negligible aromatics [174].

Because ZSM-5 normally may exhibit high selectivities to aromatics and paraffins, the zeolite and the process conditions have been modified to improve selectivity to light olefins. The effective shape selectivity was increased by the deposition of silica in the channel system [175]. Phosphorus modification [121] with trimethylphosphite ($P(OCH_3)_3$) followed by air calcinations improved olefin selectivity by lowering acid site strength and probably also by reducing the effective pore size. Ion exchange with Mn^{2+} [176], and higher SiO_2/Al_2O_3 ratios (lower site density) [39] were all reported to improve olefin selectivities.

Methanol conversion to olefins was investigated over H-ZSM-5 catalyst with an Si/Al ratio of 200 in a wide range of operating conditions [177]. Product distribution was affected by operation conditions remarkably only at low methanol conversions (80%). When methanol conversion approaches 100%, product distribution was hardly affected by operation conditions. Ethylene selectivity shows almost no dependence on methanol conversion. The formation of ethylene is increased significantly only at high contact time ranges via over-cracking of olefins. In addition, it was always found that propylene selectivity is enhanced at the expense of hexenes at high methanol conversions. According to Ref. [177], the dependence of product distribution on methanol conversion derives from the change of a dominant reaction pathway from olefin methylation and cracking at low conversion to oligomerization and cracking at high conversion.

5.3 Large-Pore Zeolites

Large-pore molecular sieves such as mordenite (MOR), Y (FAU) zeolite, SAPO-5 (AFI), have also been used as catalysts for MeOH conversion into olefins.

The channel system of mordenite is formed by parallel elliptical 12-rings of 0.67×0.70 nm and 8-rings of 0.29×0.57 nm. The parallel channels are interconnected via small side pockets of 0.29 nm [178]. The porous framework of mordenite behaves as a two-dimensional system for small molecules and as a one-dimensional system for large molecules.

Zeolites X and Y have faujasite-like structure. The faujasite framework is an array of truncated, octahedra linked by D6R prisms. The supercage that is formed in this way is larger than that in zeolite A. These large cavities are accessible through distorted 12-ring windows of 0.8–0.9 nm diameter. The supercages are 1.2–1.3 nm diameter and can contain 2.8 isooctane or 5.4 benzene molecules [179] at 25°C.

SAPO-5 possesses a one-dimensional pore structure and consists of noninterconnecting cylindrical 12-membered ring channels of 0.75 nm [149,179].

Mordenite is a highly active catalyst for MeOH conversion into hydrocarbons in the temperature range of 300–500°C [21]. The selectivity to heavy compounds, as C_5^+ aliphatics and A_{11}^+ aromatics, is very high. Besides, mordenite deactivates rapidly [180,181]. The selectivity to olefins and the resistance to deactivation of mordenite catalysts can be improved by dealumination [182,183], which reduces the acidic character, by barium exchange or by chemical deposition of $Si(OCH_3)_4$ [181].

Na-Y zeolite was found not active for MeOH conversion in the temperature range studied (20–350°C) [184]. Light hydrocarbons, predominantly butane and propylene, were detected at 250°C with H-Y. The higher activity of H-Y is due to the high Brønsted acid character of its surface. Proton-exchanged and cation-exchanged (Y-type zeolites with metal cations having hydrogenation activity [185], Ni^{2+}, Cr^{3+}, and Pb^{2+} exchanged CaY zeolites [186], rare-earth X-type zeolite) forms of large-pore zeolites are active for MeOH conversion into hydrocarbons. Sodesawa [187] investigated the effects of various degrees of proton-exchange on C–C formation with the aim of clarifying the strength of the active sites on HNaY zeolites. The temperature dependence suggests that the increasing proton exchange leads to stronger active sites on NaY zeolite, in agreement with the results found by Cormerais et al. [79]. Yields of paraffins and aromatics increase as the degree of proton exchange is raised. The conversion of olefins to paraffins and aromatics is favored by the strong acidity that can be found in large-pore zeolites. The high yield of heavy molecules, like penta- and hexamethylbenzenes, is favored by both the strong acidity and the large-pore size. The porous structure of large-pore zeolites does not impose any restrictions on the formation and diffusion of bulky molecules like HMBs. Consequently, hydrocarbon fractions with low-light olefins contents can be obtained.

SAPO-5 molecular sieves are also active catalysts for MeOH conversion into hydrocarbons. However, high concentrations of aromatics are obtained on this molecular sieve. In SAPO-5, the selectivity toward olefins can be improved by decreasing the Si/Al ratio [188]; therefore, the concentration of strong acid sites. Incorporating bivalent elements to the aluminophosphate framework also modifies the acid properties.

5.4 Conclusions on the Effect of Topologies on Product Selectivity

The small-pore molecular sieves give by far the highest selectivity to light olefins and to ethylene (Fig. 29). To increase the selectivity to light olefins, higher reaction temperature, low partial pressure, or lower contact time should be applied. These parameters will contribute to higher cracking conversion in the olefins interconversion cycle. However, the same effect could only hardly result in higher conversion on the large-pore zeolite due to their higher pore opening which favors high alkylation and aromatization activity. So the constraint to boost the yield of the light olefins over large-pore zeolites is more significant.

Both ZSM-5 and SAPO-34 use the shape selectivity of the material to guide the formation of a specific product composition [189]. ZSM-5, having a medium-pore opening size of about 0.55 nm, allows the larger molecules in the gasoline range to pass, such as C_5–C_8 paraffins and aromatics. On the other hand, SAPO-34, having a smaller pore opening, allows only

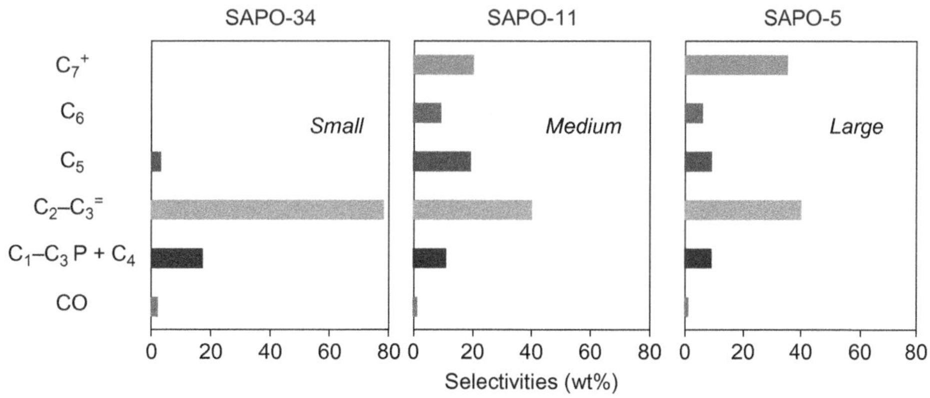

Fig. 29
Product distribution in MTO reaction over SAPO molecular sieves 350°C; WHSV = 0.3 h^{-1}; MeOH = 0.02 bar; N_2 = 0.98 bar. *Adapted from Yang SM, Wang SI, Huang CS, Holmen KJJA, Kolboe S, editors. Studies in surface science and catalysis, vol. 61. Elsevier; 1991. p. 429–35.*

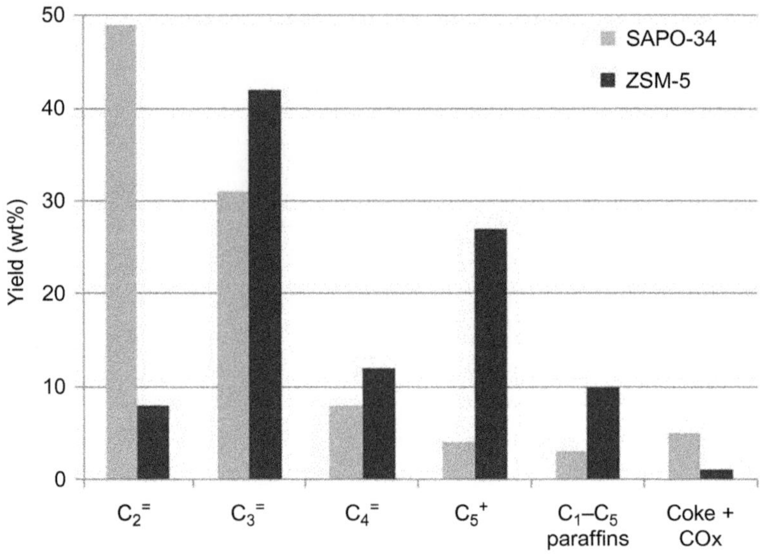

Fig. 30

Products yield from MeOH: SAPO-34 and ZSM-5 catalysts. *Adapted from Barger PT. In: Guisnet M, Gilson J-P, editors. Zeolites for cleaner technologies, vol. 3. Imperial College Press; 2001. p. 239–60.*

movement of linear hydrocarbons. This discrepancy in pore opening sizes leads to different product distribution in the MTO reaction, as shown in Fig. 30.

5.5 Refinement of the Dual-Cycle Mechanism

The emergence and gradual acceptance of the dual-cycle mechanism has contributed significantly to the general understanding of the methanol transformation to olefins over different zeolite topologies [29,92,93]. In the original version, the concept includes an aromatics/ethylene and an olefins cycles. However, a detailed analysis of the data in the prior art shows that a number of refinements should be implemented. In the most recent literature, more and more publications point out the role of olefins themselves as the dominating species involved in the hydrocarbon pool and a role of homologation [130].

First of all, ethylene could be formed in olefins cycle via cracking of oligomerized products and the contribution of this reaction is significant at high temperatures (>450°C). This is evident from a thermodynamic distribution of olefins as a function of temperature (Fig. 34).

The relative rates of C_5:C_6:C_7:C_8 alkene cracking is 1:24:192:603. This means that to crack C_5–C_6 olefins on the small-pore zeolite with significant diffusion constraints, the olefins should first interact with a coke precursor or aromatic precursor. The cracking and dealkylation of alkylaromatics is very fast and leads to ethylene and propylene. Taking into account that the methylation reactions are at least an order of magnitude faster than cracking, the reaction pathway via methylation coupled with alkylaromatic cracking looks very probable in the case of availability of organic "scaffolds." This is often the case of small-pore zeolites containing supercages. While the oligomerization is greatly suppressed by water presence at low temperature, the cracking of alkylaromatics could gain importance even on certain medium-pore zeolites due to the pore opening. This reaction could be even faster than the exocyclic methylation mechanism or the contraction of cycle via the cyclopentyl carbocation rearrangement.

Based on these arguments, a dual-cycle concept may be refined as a cracking cycle and coke catalyzing coupling cycle (Figs. 31 and 32). The coke catalyzing cycle or "aromatic/coke" cycle leads mainly to ethylene. While some propylene could be formed from coke, the probability of forming propylene should be lower. The propylene formation occurs mainly from the cracking cycle ("olefins cycle") (Fig. 31).

SAPO-34 and ZSM-22 are two frameworks that can essentially be viewed as two different extremes of the dual-catalytic cycle. Fig. 33 depicts that SAPO-34 could not produce any isobutene. In reality, the isobutene and aromatics could be formed on the SAPO-34 but stay inside of the pore structures. This means that the isobutene isomerizes first to *n*-butenes to escape the pore system. This leads to domination of the coke-catalyzing coupling cycle

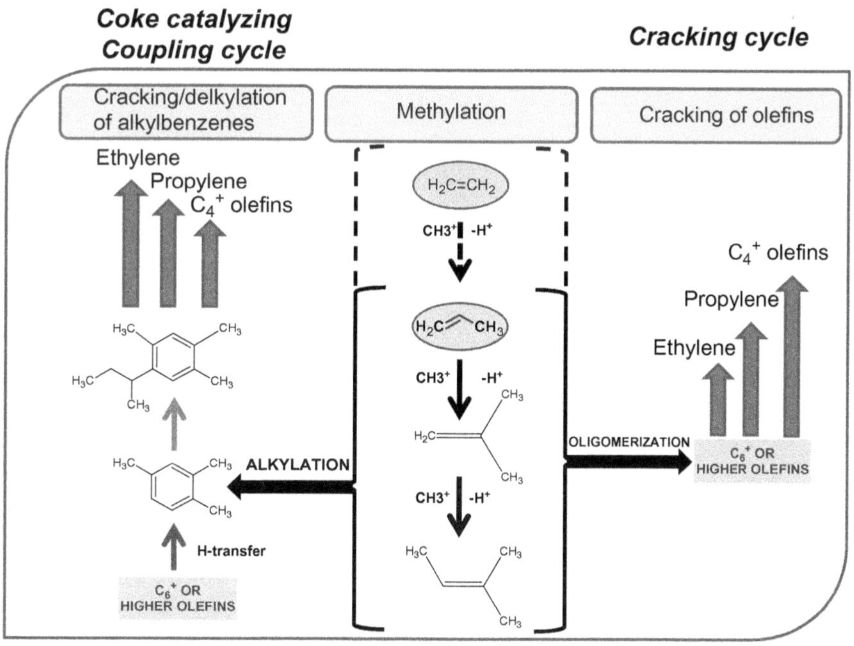

Fig. 31
Refined dual-cycle mechanism in methanol transformation to olefins.

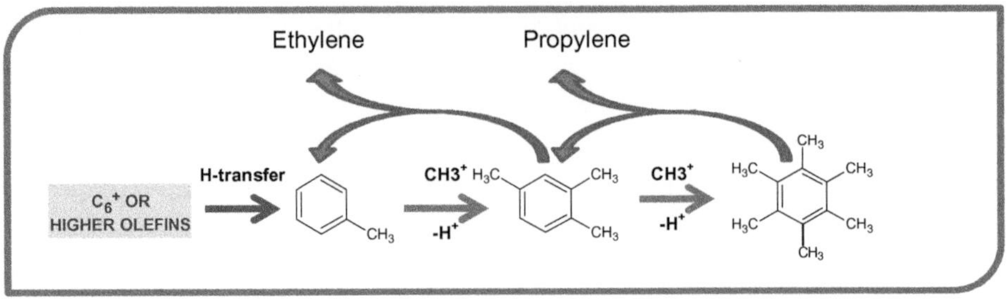

Fig. 32
Coke-catalyzing coupling cycle in methanol transformation to olefins.

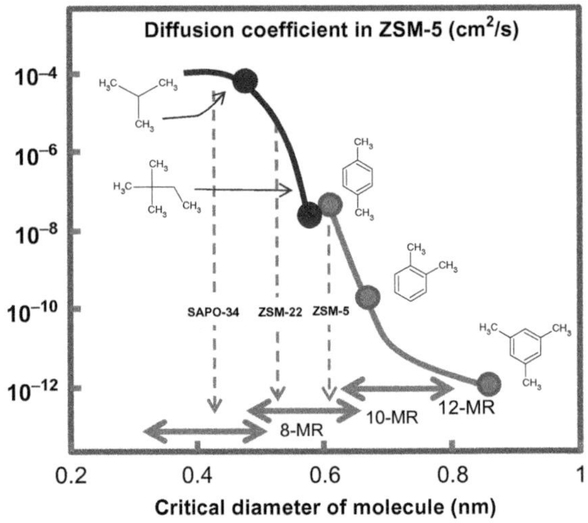

Fig. 33
Diffusion coefficient in ZSM-5 as a function of critical diameter of molecules. *Adapted from Tsai T-C, Liu S-B, Wang I. Disproportionation and transalkylation of alkylbenzenes over zeolite catalysts. Appl Catal A 1999;181:355–98.*

together with a left part of cracking cycle on SAPO-34. The small 8-MR windows hinder C_5^+ hydrocarbons from escaping the larger 12-MR cavities, resulting in a product selectivity rich in light olefins. On the contrary, the ZSM-22 could form and produce isobutene but does not allow formation of any aromatics products. The one-dimensional 10-MR pores of H-ZSM-22 are too small to bear with aromatics. On the other hand, the oligomerization/cracking requires higher temperature and good transportation properties. So on this zeolite, both cycles are suppressed and the only methylation reaction drives the selectivity, resulting in a formation of the fraction rich in C_5^+ aliphatics. Eventually, H-ZSM-5 and larger pore zeolites can propagate both the coke-catalyzing coupling and cracking cycles, and the contribution of these pathways is tunable by adjusting the operating conditions. It is interesting that at low conversion the propylene to ethylene ratio is close to 1 on ZSM-5, which is close to the ratio obtained on SAPO-34. That might indicate a dominance of the "aromatic/coke" cycle at low conversion on ZSM-5. On the contrary, higher conversion leads to significantly higher propylene to ethylene ratio and to the dominance of the cracking cycle.

Recently, Li reported the conversion of MTO over three zeolites with different topologies: SAPO-34, H-ZSM-5, and H-ZSM-22 [174]. The correlation between reaction mechanism and the zeolite topology was also investigated. SAPO-34 presented the highest selectivity for light olefins such as ethylene and propylene, and no aromatics were detected. H-ZSM-5 showed relatively high selectivity for ethylene and propylene, and large amounts of aromatics were detected in the effluent. An aromatic-based hydrocarbon pool mechanism leading to high ethylene yield was the most important reaction route of methanol conversion over SAPO-34 with enough spaces for the accommodation of large intermediates. Methanol conversion catalyzed by ZSM-5 both followed the hydrocarbon pool mechanism at low conversion and the olefin methylation-cracking route. Over H-ZSM-22, the selectivity for ethylene was very low, and a large amount of nonaromatic C_6^+ olefins was generated. Over the zeolite with one-dimensional 10-member ring channels, H-ZSM-22, contrary to the report by Cui et al. [141,190], the work by Li et al. suggested that methanol could be converted into olefins, and that owing to the absence of supercages or channel intersections for hydrocarbon pool formation, the olefin methylation was the main reaction route of methanol conversion. Hence, it was demonstrated that by carefully choosing catalyst topology, product formation via alkene cracking and methylation can be affected [191].

The consequences of zeolite structure in MTO products could be assessed by using experimentally measured rates, descriptors, or reaction coordinates.

This would bring the information about activation barriers for individual reaction steps and in terms of speciation preferences within the dual olefin- and aromatic-catalytic cycles to alter their relative propagation. A number of different concepts could be used for this purpose.

In the absence of detailed kinetic parameters for individual reaction steps, the relative propagation of the aromatic- to olefin-based catalytic cycles can be assessed by determining the rate and selectivity of termination products of these cycles that are predominantly unreactive in subsequent MTH reactions. One possibility is to use the ethylene/isobutane selectivity as a measure to describe the relative rates of propagation for the aromatic- and olefin-based cycles [62]. Kinetic studies show that ethylene methylation is at least an order of magnitude slower than other olefin methylation reactions, and isotopic switching studies show that the ^{13}C incorporation of ethylene matches that of aromatics on various zeolites; therefore, ethylene can be considered to be a termination product of the aromatics-based cycle. Isotopic switching studies on H-ZSM-5 also show that the incorporation of ^{13}C atoms in isobutene matches that of other C_3^+ olefins, showing that on H-ZSM-5, isobutene is predominantly formed via the olefin-based cycle. Isobutane can be formed from isobutene via hydrogen transfer reactions, and because alkanes are less reactive than olefins or aromatics on zeolites, isobutane can be considered a termination product of the olefin-based cycle. Although this specific ratio may not be an appropriate descriptor for large-pore zeolites, such as H-BEA, where prior work shows that isobutene incorporates carbons originating from aromatics [62] or small-pore zeolites, in which isobutane transport may be diffusion-limited, other similar ratios may be used to describe the relative propagation of the aromatic- and olefin-based cycles [62]. Descriptors such as ethylene and isobutane are still qualitative because it is not possible to assess the rate of individual steps, but they bring us closer to a quantitative understanding of the complex MTO chemistry.

Another possibility is to use the propane concentration, introduced by Mobil, as a measure of activity or hydrogen transfer reactions [192].

6 Effect of Reactions Conditions

The complexity of the reaction mechanism can give access to a wide range of products through several consecutive reactions. Hence, the influence of the operating parameters is extremely important, because they can drastically modify the intrinsic activity, the selectivity and also the stability of the catalyst. Moreover, modification of the reaction conditions might mask the important details and lead to the changes in reaction pathways resulting in incorrect catalyst screening.

One of the most important difficulties of catalyst screening is a high exothermic temperature rise due to MeOH transformation to hydrocarbon (at 450°C, $\Delta H^{\circ}_{\text{reaction}} = -858\,\text{kJ/kg}$ CH_3OH). This means that a comparison of the catalysts in a fixed-bed reactor at different conversion levels or at different methanol flow rates and equal conversion may potentially mean different temperatures and is relatively tricky. That is a reason why most of the publications use a highly diluted methanol feed [29,92,93]. The dilution media acts as a heat vector in this case to smooth over the adiabatic temperature increase. However, the presence of dilution media greatly decreases methanol partial pressure. The latter masks mechanistic details and may change reaction pathways. Some important studies also highlight the role of diffusion, adsorption equilibrium, role of water, and the role of surface species, which could hardly be seen with diluted feed [24,28,60,119].

6.1 Temperature

MeOH conversion to olefins, catalyzed by zeolites, occurs between 250 and 550°C. The product distribution over ZSM-5 is strongly dependent from the temperature [66]. Below 300°C, the main reaction is the MeOH dehydration into DME. The increase of the temperature favors first the olefins, then the formation of aromatics. Above 450°C, the quantity of light olefins, coming from secondary cracking, increases. Dissociation of MeOH into CO and H_2 is observed above 500°C.

Table 8: MeOH conversion to light olefins over SAPO-34—Conditions: 1.34 atm; WHSV (h^{-1}) 0.83–0.87 MeOH; 1.95–2.04 H_2O

	Molar Selectivity (%)			
	375°C	400°C	425°C	450°C
$C_2^=$	43	46.7	51.4	61.1
C_2	0.8	0.6	0.6	0.7
$C_3^=$	41.8	36.7	32.5	27.4
C_3	0.5	0.5	0.5	Trace
$C_4^=$	10.8	11.9	9.3	5.4
C_5's	1.7	1.6	1.4	0.6
C_6's	Trace	Trace	Trace	Trace
C_1	1.3	2	4.3	4.8
DME	Trace	–	–	–
C_2-C_4 olefins	95.6	95.3	93.2	93.9
$C_2^=/C_3^=$ molar ratio	1.03	1.27	1.58	2.23
Hours on stream	5.2	6.3	6.2	11
MeOH conversion (%)	100	100	100	100

Adapted from Kaiser SW. Production of lights olefins; 1985.

The effect of temperature on the distribution of olefin products over SAPO-34 is illustrated in Table 8. As the temperature increases, the molar selectivity to ethylene increases, selectivities to propylene and butene decrease, and the molar selectivity to CH_4 increases [151,193]. According to the equilibrium, propylene selectivity should also increase when temperature is raised (Fig. 34). However, propylene and butenes are more reactive than ethylene, so they can further undergoing transformation. Furthermore, diffusion of ethylene to outside cages of SAPO-34 is favored in comparison with the diffusion of propylene [194]. This highlights the fact that the thermodynamic equilibrium of light olefins may occur within SAPO-34 prior to diffusion out of the crystalline structure. But diffusion limitations and differences in reactivity between the light olefins influence the final composition outside the zeolyte crystal.

The influence of equilibrium on the ethylene/propylene product ratio obtained with SAPO-34 catalyst over a range of temperatures and pressures can be seen from (Fig. 35) [189]. The relationship is linear with a slope that is significantly greater than 1, indicating that ethylene is in excess of equilibrium on the small-pore zeolites. An explanation for this relationship is that ethylene and propylene are in equilibrium within the pore structure of the SAPO-34 molecular sieve. However, because the 0.43 nm pore diameter of the SAPO-34 is too small to allow propylene to take a straight-line path through the pore mouth, its diffusion is hindered relative to ethylene. This would imply that diffusion limitations are the key factor controlling the $C_2^=/C_3^=$ ratio [189].

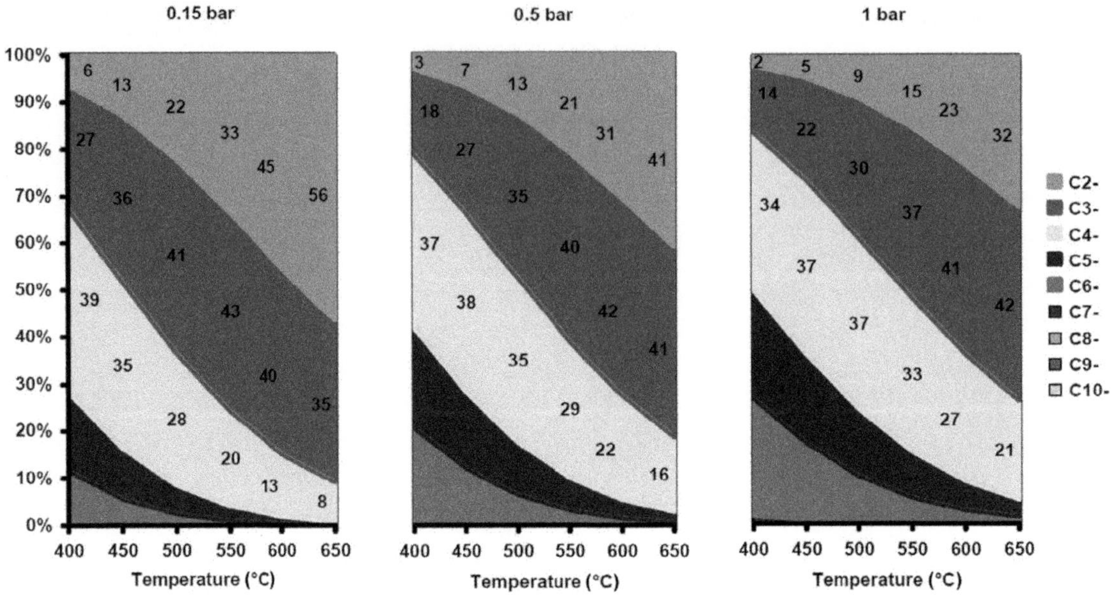

Fig. 34
Equilibrium distribution of C_2–C_5 olefins at 0 psig. *Adapted from Vermeiren W. Presented in part at the Natural gas conversion symposium 9, Lyon, Juin; 2010.*

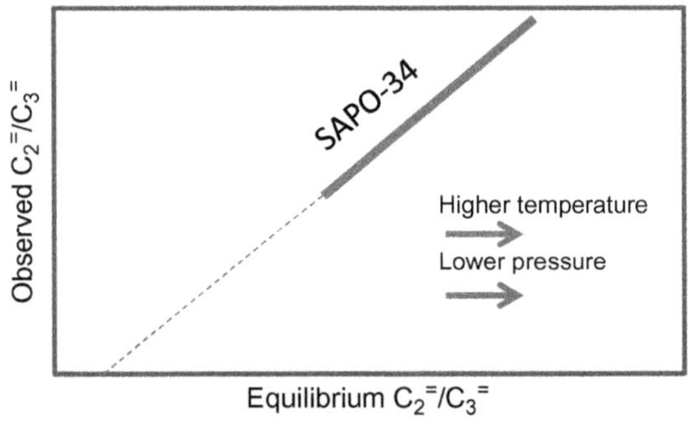

Fig. 35
Relationship of ethylene-to-propylene product ratio over SAPO-34 catalyst with thermodynamic equilibrium. *Adapted from Barger PT. In: Guisnet M, Gilson J-P, editors. Zeolites for cleaner technologies, vol. 3. Imperial College Press; 2001. p. 239–60.*

Several studies also evidenced a strong dependence between temperature and lifetime (conversion capacity) of the catalyst [195–197]. Generally there is an optimum depending on the topology and the acidity of the zeolite. For example, the maximum of conversion capacity of SAPO-34 was observed at 400°C in a fixed-bed reactor, whereas it is at 365°C for SSZ-13. The two catalysts have the same structure (CHA) and the same density of acid sites, but SSZ-13 presents stronger acid sites.

6.2 Partial Pressure

In the conversion of MeOH to hydrocarbons over ZSM-5 zeolite, the composition of the hydrocarbon products can be markedly influenced by changes in reactant partial pressure. Varying reactant partial pressure affects mainly the relative rates of the dehydration and aromatization steps in the reaction sequence [198,199]. It was also found that elevated pressures tended to enhance secondary alkylation reactions as reflected by an increase in the aromatic chain-to-ring ratio. This resulted in increased selectivity to tri- and tetramethylbenzenes in particular [200–202].

At subatmospheric partial pressures, high selectivity to olefins is possible at complete conversion as shown in Table 9. Reduction of the reactant partial pressure to 0.04 atm results in an altered reaction pathway: the main reaction products up to oxygenates extinction are seen to be light olefins. Only minor amounts of aromatics are formed while paraffins comprise less than 10% of the total product. The effect of lowered pressure is therefore to decouple the dehydration and aromatization steps of the reaction sequence. Ethylene increases with time albeit slowly and remains at relatively low levels. Propylene and butenes, on the other hand, rapidly become the predominant olefinic species after an initial induction period. Upon raising the reactant partial pressure to 50 atm, the overlap region of the dehydration and aromatization reactions is greatly expanded. This provides an explanation for the increase in PMBs with pressure. This increase is observed noticeably in durene (1,2,4,5-tetramethylbenzene) yields [198,203].

6.3 Contact Time

WHSV is defined as the ratio of mass flow rate of MeOH with mass of catalyst. It is inversely proportional to contact time. According to Ref. [66], the increase of contact time has the same effects on products distribution as the temperature does. Selectivity to light olefins increases

Table 9: MeOH conversion at low pressure over ZSM-5—Conditions: 427°C; LHSV 1 h^{-1}, 1 atm total pressure

MeOH partial pressure (atm) Conversion (%)	1 99^+	0.25 99^+	0.17 99^+	0.07 99^+
Hydrocarbon distribution (wt%)				
Ethylene	3.2	12.4	17.4	21
Propylene	4.8	18.2	26.5	38.7
Butenes	2.2	6.3	7.6	18.5
Pentenes	0.4	0.3	0.7	2.4
Total C_2–C_5 olefins	10.6	37.2	52.2	80.6
Methane	1.5	0.8	0.6	0.5
C_2–C_5 paraffins	43	39.4	24.2	15.6
C_6 nonaromatics	3.9	2.3	2.7	1.3
Aromatics	41	20.3	20.3	2

Adapted from Chang CD, Lang WH. Process for manufacturing olefins; 1977.

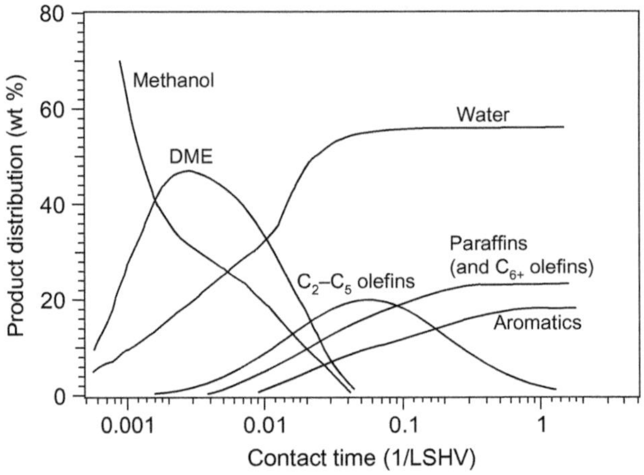

Fig. 36
Influence of contact time on MeOH conversion (H-ZSM-5, 371°C, 1 atm). *Adapted from Chang CD, Silvestri AJ. The conversion of methanol and other O-compounds to hydrocarbons over zeolite catalysts. J Catal 1977;47:249–59.*

with contact time until it reaches a maximum. This indicates that these olefins, as well as DME, are intermediate products in the reaction. Still increasing contact time, distribution shifts toward aromatics, even if the conversion is already complete.

The fact that the catalysts display considerable differences in selectivities after a certain reaction time might indicate that the species formed after the start of reaction influence the $C_2^= / C_3^=$ ratio, either by: (1) subtle changes in the reaction mechanism or (2) diffusion limitations. In support of the first explanation, Song et al. [105] have investigated the evolution of the $C_2^= / C_3^=$ ratio with time on stream over H-SAPO-34 and were able to correlate the increase in ethylene over propylene with a reduction in the number of methyl substituents on the hydrocarbon pool species (Fig. 36).

6.4 Effect of Water

Adding water to the feeding of MeOH allows researchers to strongly increase the selectivity to light olefins and particularly to ethylene [150]. Coke formation is also reduced, which enhances the catalyst stability [204]. When the feed contains 74–80 mol% water, the catalyst can process eight more times the amount of MeOH than a feed of pure MeOH before the conversion decreases below 100% [205]. Vedrine et al. [206] showed that the presence of water diminishes the number of acid sites and suppresses the strongest acid sites, limiting aromatization and hydrogen transfer reactions. Competitive adsorption between water and light olefins also favors desorption from the surface of the catalyst and thereby reduces the secondary reactions that lead to the formation of coke, hence increasing the lifetime of the catalyst.

For example, Marchi and Froment found that aromatics formation, chain growth, and light paraffin production were all suppressed by adding water to the feed [60]. They attributed this effect to competitive chemisorption between the water and olefin molecules on the strongest acid sites. When the partial pressure of methanol in the feed was reduced by nitrogen instead of water, the resulting effect was closer to that observed with pure methanol feed at a reduced partial pressure.

7 Role of Coke and Catalyst Deactivation

Coke deposition [207] is known to be the major cause of deactivation during the MTO reaction. On the other hand, the coke species play a role of the active species and a role of an activity regulator to moderate the hydrogen transfer activity of the zeolite and to block the most acidic sites. So, in the same time, the coke content of the catalyst is a factor governing the selectivity and activity of the catalyst [208]. Coke has both a promoting effect and a deactivating effect for the formation of olefins [209]. A comprehensive review dealing with coke formation and deactivation was recently published by Chen et al. [24].

7.1 Coke Formation and Nature of Coke

It was found that the coke formed from oxygenates on SAPO-34, referred to as active coke, promoted olefin formation while the coke formed from the cracking of the "external" olefins, referred to as inactive coke, had only a deactivating effect [210]. A possible explanation of the phenomenon is a different location of the coke species. The coke from oxygenates formation was mainly located close to the core of the crystal and acted as a "ship-in-a-bottle" scaffold, while the coke species from the cracking laid down close to the external surface of the zeolite blocking the porosity. Coke forms rapidly on SAPO-34 due to its large cages and narrow pores [211]. Deactivation by coke may be due to both covering of acid sites and blockage of pore structure [212–214]. The interconnecting three-dimensional network of pores with supercages existing in small-pore molecular sieves provides room for accommodating some coke without immediately blocking the pores [215]. It has been suggested that the deactivation by coke and its effect on the product distribution depend on the way the coke is deposited on the catalyst [60].

Catalyst deactivation depends not only on the coke content but also on the nature and location of coke molecules. SAPO-34 deactivates generally as a linear function of the coke content [204,208,216], but it deactivates faster at high temperatures, indicating that larger molecules are formed at higher temperatures at an identical coke content [216].

The yield of olefins during the MTO reaction was found to go through a maximum as a function of both time of stream and amount of coke. Coke from olefins cracking was found to reduce the DME diffusivity, which enhances the formation of olefins, particularly ethylene. The ethylene to propylene ratio on SAPO-34 increased with coke content, regardless of the nature of the coke [210].

The SAPO-34 crystal size and morphology contribute to determine the diffusion path and the residence time in the crystal. Large SAPO crystals deactivated faster than small crystals with similar coke contents, due to coke formation near the external surface of the catalyst particles, gradually blocking the diffusion path of oxygenates to the inner core of the catalysts [217]. The effective diffusivity of MeOH decreased by almost three orders of magnitude between the fresh catalyst and a catalyst with a coke content of 15 wt% [218]. It can therefore be concluded that small SAPO-34 crystals are preferred to reduce the diffusion limitation, thus reducing the deactivation rate or the deactivating effect of the coke formed [215,219,220].

Anderson et al. showed by in situ ^{13}C MAS NMR spectroscopy that the main species present in the intracrystalline space are branched C_4 and C_5 saturated hydrocarbons such as isobutane and isopentane [221]. The contribution acknowledges again a role of diffusion to the formation of the first olefins structure [221].

A lower space velocity resulted in a higher coking rate, as a result of high average concentration of olefins and low average concentration of oxygenates (MeOH/DME mixtures) in the reactor. It indicates that the rate of coke deposition is related to the concentration of olefins [222].

Higher temperatures favored the reactions leading to coke formation more than the reactions leading to olefins, hence a faster deactivation [223]. However, the coking rate was found to decrease with increasing temperature in the low-temperature range of 340–420°C on SAPO-34, which was explained by increased cracking of the coke precursors [207].

Temperature programmed oxidation (TPO) give indication on the nature of the coke: the coke corresponding to the low-temperature peak is referred to as soft coke with relatively high hydrogen to carbon ratio (ie, olefinic coke). The one corresponding to the high temperature peak is referred to as hard coke with a relatively low H/C ratio (ie, aromatic coke). The ratio of soft to hard coke decreased concurrently with increasing the coke content [224].

The most abundant and stable molecules in coke generated during MTO were found to be aromatics with multimethyl groups. This brought a basis to the development of the hydrocarbon pool mechanism [87,90,224]. The coke composition changed from monoaromatics to two rings, three rings, and four rings aromatics with increasing coke content [113]. Nevertheless, the post-extraction method has a drawback that the small molecules and reactive surface entities cannot be detected due to their volatile nature. Different techniques should be employed to provide more detailed information on all coke molecules.

It was shown recently, by operando RAMAN spectroscopy, that the primary coke species formed in/on SAPO-34 were mainly polyolefinic, some aromatic species were present on ZSM-5, and mainly polyaromatic or substituted aromatic species were predominant in USY at high temperatures [225].

7.2 Effects of Pore Structure on Catalyst Deactivation

Between four small-pore SAPOs, the lifetime decreased in the following order: SAPO-18 > SAPO-34 ~ SAPO-17 > SAPO-35 [226]. SAPO-34 has the smallest cage size and the lowest number of eight-membered ring openings in each cage. In addition, it should be mentioned that the number of acid sites is not the same, following an order of SAPO-34 > SAPO-17 > SAPO-18 > SAPO-35. The small cage appears to be the main cause for the faster deactivation: a small cage size provides a space constraint for the growth of coke molecules. The probability of molecular diffusion in and out of the cages is related to the number of ring openings in each cage and a small number of rings could cause more pronounced deactivation by a faster reduction in diffusion caused by blockage by coke molecules. A better stability of SAPO-18 could be explained by a different shape of the cage and a lower amount of acid sites [211].

The channels of large-pore molecular sieves can also accommodate carbonaceous residues that lead to loss of catalytic activity. Coke deposition can vary the selectivity of large-pore molecular sieves by coverage of acid sites and pore blockage. The effect of blockage is more pronounced for molecular sieves with one-dimensional noninterconnecting channels [180].

The rate of deactivation depends on the acidity: the higher the catalyst acidity, the faster the deactivation by coke formation. The H-form of chabazite, erionite, and erionite-offretite type zeolites deactivate faster than dealuminated H-chabazite or H-SAPO-34 and H-SAPO-17 [218]. Furthermore, acid site density and acidic strength are expected to have significant effects on reaction leading to coke formation, such as oligomerization, cyclization, and hydrogen transfer. These reactions require strong acidic sites. High acidic strength resulted in high reaction rates on SAPO-18 and SAPO-34 compared to SAPO-11 and Beta-zeolites [227]. Since SAPO-18 and SAPO-34 have a similar pore structure and a close acidic strength, a greater density of acid sites of SAPO-34 might be responsible for the faster deactivation compared to SAPO-18.

An intermediate acid site density has been found advantageous for the stability of H-SSZ-13 catalyst [228]. On SAPO-34 and SSZ-13 synthesized with the same crystal size and acid site density, it has been observed that acidic strength decreases the energy barrier and thus increases the reaction rates [229]. Ethylene increases in a shorter time on H-SSZ-13 than on H-SAPO-34, suggesting a faster coke formation on H-SSZ-13 due to its higher acidic strength.

7.3 Effects of Coke on Shape Selectivity

Olefin products were identified as the primary products of the MeOH-DME mixture, while the paraffins (C_2 and C_3) were identified as secondary products of the olefins. The methane was suggested as both the primary and secondary product of the large molecules such as coke [208,216]. Selectivity to ethylene increases with increasing coke content. The following decrease in selectivity of the olefins with coke content was found: $C_6 > C_5 > C_4$, C_3.

Both product-shape selectivity and transition-state shape selectivity have been reported to govern the product distribution in the MTO reaction. Product shape selectivity plays an obvious role in obtaining high selectivity to olefins in SAPO-34 compared to ZSM-5. The eight-membered ring openings limit aromatics diffusion out of the cages. However, there is still a debate about the role of shape selectivity with respect to the selectivity of C_2–C_4 olefins on SAPO-34.

According to Song et al. [105], the formation of olefins is governed by transition-state shape selectivity. Selectivity at the same conversion level is very different on coked SAPO-34, compared to fresh SAPO-34 [223,230]. The formation of olefins with different sizes would go through different-sized intermediates and the reaction with the larger intermediates deactivates faster. Coke deposition itself is also suppressed by coke molecules in the cages through transition-state shape selectivity, and the change in coke selectivity is rather similar as for the large olefins with increasing coke content.

With progressive deactivation by "coke," the reactor zone in which the propylene yield reaches its maximum moves further downstream. After some time, the remaining amount of nondeactivated catalyst became too small to permit the propylene yield to reach its maximum. As a consequence, the yield observed at the exit starts to decline. On the fresh and very active catalyst all the produced olefins are converted into C_5, C_6 aliphatic and into aromatics, which are simply too bulky to exit from the SAPO-34 catalyst and to be observed at the reactor exit. As the catalyst deactivates and the very narrow reaction zone migrates downstream in the fixed bed, the olefins are not completely converted into paraffins and aromatics anymore and break through at the exit. The active zone in which the olefins are consumed is shrinking with time and increasing deactivation so that their yield observed at the exit progressively increases, until the plateau is reached. The complete olefins yield versus time curve then shows a rising part, a plateau, and a decreasing part, substantiating the opinion of practitioners claiming that optimal MTO is possible only within a window of carbon content on the catalyst [119].

8 Conclusions

Despite the number of proposed mechanisms for the formation of the first C–C bond, none of them have any monosemantic supporting experimental evidence. The common peculiarity of all routes lies in a high energetic barrier, which results in a long induction period. Although many mechanisms do not account for the induction, neither do they reconcile it.

The experimental evidence demonstrates that the formation of the first C–C bond should be mechanistically separated from the further pathways, which contribute to the formation of olefins. Primary traces of any external impurities play an important role only at the initial step; as soon as the first hydrocarbon species are formed, a jump in activity is observed. Whatever is the mechanism behind the formation of the first C–C bond, the main practical conclusion of the investigations is a parallel formation of ethylene and propylene as secondary products at high conversion level of methanol.

Most probably, there are two types of active sites upon which the elementary steps take place: the first active site consists of acid sites of the catalyst surface itself, the second type of the acid sites is the "coked" surface.

In the most recent literature, more and more publications point out a role of olefins themselves as the dominating species involved in the hydrocarbon pool and a role of homologation. Based on these arguments, a dual-cycle concept may be refined as a cracking cycle and a coke-catalyzing coupling cycle. The homologation of olefins plays an important role in both cases.

Both ethylene and propylene could be formed via the cracking of the sufficiently long oligomeric species in cracking cycle at higher temperature.

As initially suggested by the carbon pool mechanism, the coke-catalyzing coupling cycle occurs via exocyclic methylation mechanism or the contraction of cycle via the cyclopentyl carbocation rearrangement. This transformation leads mainly to ethylene and to propylene to a lesser extent. However, the ethylene and propylene could be also formed in the coke-catalyzing cycle via interaction of the short olefinic species formed by oligomerization or homologation with the coke precursors followed by cracking.

The methylation reactions are at least an order of magnitude faster than cracking and are the most tangible ways to incorporate methyl's to hydrocarbon pool from methanol and DME in MTO.

The selectivity in MTO reaction is kinetically driven. The methanol conversion rate is greatly enhanced by the addition of ethylene or any other hydrocarbon to methanol. On the contrary, the conversion of ethylene is lower in presence of MeOH than when ethylene is converted alone. Methanol exhibits a very different reactivity and reaction pathways from DME in MTO.

The DME conversion rate is much less impacted by ethylene addition than is MeOH. The DME presence may increase ethylene conversion because of high activity in homologation of ethylene. DME cofeeding with water does not lead to the same products as MeOH.

Water plays an important role at the adsorption state, since it moderates acidity via an effect of competitive adsorption and limits the contribution of oligomerization at low level of conversion or on low-active catalysts.

Ethylene is not obviously produced directly from MeOH. So the ethylene is indeed mostly a product of aromatic cycle, which roles out by a transalkylation of the methylated products with coke species.

By a number of independent studies, it was found that propylene is at least one of the primary products in MTO reaction, which could be formed directly from DME/MeOH. At low temperature, propylene is mostly formed by DME propagation as well as ethylene methylation. In conditions of reduced DME conversion, propylene formation occurs nearly exclusively and via methylation of the added ethylene with selectivity close to 100% on SAPO-34 and H-ZSM-5. It was also demonstrated that DME propagation does not proceed via ethylene, on both topologies. The addition of ethylene to the DME feed switches the MTO mechanism toward a mechanism where propylene is a true primary product from DME. At higher temperature, the propylene is a product of the ethylene methylation, "aromatic/coke" and "olefins" cycles.

The fact that ZSM-22 does not produce a lot of ethylene and propylene is explained by a suppression of both "aromatic/coke" and olefins interconversion cycles on this catalyst. SAPO-34 favors "aromatic/coke" cycle and limits the "olefins" cycle. On the contrary, ZSM-5 demonstrates a contribution of the "aromatic/coke" cycle at low conversion level. However, higher conversions lead to significantly higher propylene-to-ethylene ratios and to the dominance of the "olefins" cycle.

The contribution of methylation together with the "aromatic/coke" or "olefins" cycles can be assessed by determining the rate and selectivity of termination products of each cycle. Ethylene, the sum of butanes, propane as well as the ratio between all of these compounds can be used as descriptors and activity coordinates for MTO performance assessment.

References

[1] Vermeiren W, Gilson JP. Impact of zeolites on the petroleum and petrochemical industry. Top Catal 2009;52:1131–61.
[2] Lacson J. Process Economics Program Report 279: coal to olefins. Menlo Park, CA: SRI Consulting, IHS; 2011.
[3] Vermeiren W, Andersenb J, Gregor J. Conversion of lower value by-product to higher value light olefins. In: ERTC petrochemical conference; 2008.
[4] Ding J, Hua W. Game changers of the C_3 value chain: gas, coal, and biotechnologies. Chem Eng Technol 2013;36:83–90.
[5] World light olefins analysis. Houston, TX: Chemical Market Associates, INC.; 2012.
[6] Gong T, Zhang X, Bai T, Zhang Q, Tao L, Qi M, et al. Coupling conversion of methanol and C_4 hydrocarbon to propylene on La-modified H-ZSM-5 zeolite catalysts. Ind Eng Chem Res 2013;51:13589–98.
[7] Olah GA, Goeppert A, Surya Prakash GK. Beyond oil and gas: the methanol economy. Weinheim: Wiley-VCH; 2006.
[8] Chang CD, Silvestri AJ. MTG: origin, evolution, operation. ChemTech 1987;624–31.
[9] Meisel SL. Catalysis research bears fruit. ChemTech 1988;32–7.
[10] Vora BV, Marker TL. Conversion of natural gas to ethylene and propylene: UOP/Hydro MTO process, New Delhi; 1997.

[11] Vora BV, Marker TL, Barger PT, Nilsen HR, Kvisle S, Fuglerud T, et al, editors. Studies in surface science and catalysis, vol. 107. Amsterdam: Elsevier; 1997. p. 87–98.
[12] Koempel H, Liebner W, Wagner M. In: Lurgi's gas to chemicals (GTC): advanced technologies for natural gas monetisation, Bilbao, Spain; 2005.
[13] Wurzel T. In: Presented in part at the 19th world petroleum congress, Madrid, Spain; 2008.
[14] Liu Z. Coal to chemicals, 23–24 May 2007, Beijing, http://scholar.google.fr/scholar_url?url=http%3A%2F%2Flive.belfercenter.org%2Ffiles%2F10-zhongmin_liu.pdf&hl=fr&sa=T&oi=ggp&ct=res&cd=0&ei=e8XeVsWyI8TSmAGy26iYDg&scisig=AAGBfm0G67cAHlRHu5_cchp9iyEO6xtU4A&nossl=1&ws=1920x955; 2007.
[15] Liu Z. In: Presented in part at the U.S.-China coal conversion and carbon management workshop, March 12; 2009.
[16] Teng J, Wang R, Xie Z, Gan Y. New olefins production technologies in SINOPEC SRIPT, Spain; 2008.
[17] Chen Q, Yang W, Teng J. Recent advances in coal to chemicals technology developed by SINOPEC. Cuihua Xuebao/Chinese J Catal 2013;34:217–24.
[18] Romers E, Senetar J. In: Presented in part at the AIChE, The 21st ethylene producers conference, March 14; 2011.
[19] Vermeiren W. In: Presented in part at the Natural gas conversion symposium 9, Lyon, Juin; 2010.
[20] Vora B, Chen JQ, Bozzano A, Glover B, Barger P. Various routes to methane utilization of SAPO-34 catalysis offers the best option. Catal Today 2009;141:77–83.
[21] Chang CD. Hydrocarbons from methanol. Catal Rev Sci Eng 1983;25:1–118.
[22] Hutchings GJ, Hunter R. Hydrocarbon formation from methanol and dimethyl ether: a review of the experimental observations concerning the mechanism of formation of the primary products. Catal Today 1990;6:279–306.
[23] Stöcker M. Methanol-to-hydrocarbons: catalytic materials and their behavior. Microporous Mesoporous Mater 1999;29:3–48.
[24] Chen D, Moljord K, Holmen A. A methanol to olefins review: diffusion, coke formation and deactivation on SAPO type catalysts. Microporous Mesoporous Mater 2012;164:239–50.
[25] Mokrania T, Scurrella M. Gas conversion to liquid fuels and chemicals: the methanol route-catalysis and processes development. Catal Rev 2009;51:1–145.
[26] Dahl IM, Kolboe S. On the reaction mechanism for propene formation in the MTO reaction over SAPO-34. Catal Lett 1993;20:329–36.
[27] Chang CD. Methanol conversion to light olefins. Catal Rev Sci Eng 1984;26:323.
[28] Wang W, Jiang Y, Hunger M. Mechanistic investigations of the methanol-to-olefin (MTO) process on acidic zeolite catalysts by in situ solid-state NMR spectroscopy. Catal Today 2006;113:102–14.
[29] Olsbye U, Svelle S, Bjørgen M, Beato P, Janssens T, Joensen VWF, et al. Conversion of methanol to hydrocarbons: how zeolite cavity and pore size controls product selectivity. Angew Chem Int Ed 2012;51:5810–8531.
[30] Sernagiotto E. On the chemistry of nascent methylene. Gazz Chim Ital 1914;44:587.
[31] Adkins H, Perkins PD. The behavior of methanol over aluminum and zinc oxides. J Phys Chem 1928;32:221.
[32] Cullinane NM, Chart SJ, Meatyard R. J Soc Chem Ind 1948;67:142.
[33] Topchieva KV, Ballod AV. Dokl Akad Nauk SSSR 1950;75:247.
[34] Gorin E. Conversion of dimethyl ether; 1948.
[35] Methanol as a chemical raw material # 111, California; 1977.
[36] Matushevskiy BV, Freydlin GN. Process to produce ethylene, SU186444; 1966.
[37] Socha RF, Chang CD, Gould RM, Kane SE, Avidan AA. Fluid-bed studies of olefin production from methanol; 1986.
[38] Socha RF, Chang CD, Gould RM, Kane SE, Avidan AA, editors. Industrial chemicals via C_1 processes, vol. 328. Washington: American Chemical Society; 1987. p. 34–41.
[39] Chang CD, Chu CTW, Socha RF. Methanol conversion to olefins over ZSM-5: I. Effect of temperature and zeolite $SiO_2Al_2O_3$. J Catal 1984;86:289–96.
[40] Chang CD, editor. Methanol production and uses. New York: Marcel Dekker Publishers; 1994.

[41] Chang CD, Lang WH, Silvestri AJ. Manufacture of light olefins, US4062905; 1977.
[42] Lok BM, Messina CA, Patton RL, Gajek RT, Cannan TR, Flanigen EM. Crystalline silicoaluminophosphates, US4440871; 1982.
[43] Lok BM, Messina CA, Patton RL, Gajek RT, Cannan TR, Flanigen EM. Silicoaluminophosphate molecular sieves: another new class of microporous crystalline inorganic solids. J Am Chem Soc 1984;106:6092–3.
[44] Foley T. In: Presented in part at the 9th international petrochemicals and gas conference, London, 12–13 January; 2006.
[45] Grootjans J, Vanrysselberghe V, Vermeiren W. Integration of the Total Petrochemicals-UOP olefins conversion process into a naphtha steam cracker facility. Catal Today 2005;106:57–61.
[46] Barger PT, Lesch DA. Hydrothermal stability of SAPO-34 in the methanol-to-olefins process. Arab J Sci Eng 1996;21:263–72.
[47] Burgfels G, Kochloefl K, Ladebeck J, Schmidt F, Schneider M, Wernicke HJ. Catalyst based on crystalline aluminosilicate. 1989.
[48] Hack M, Koss U, Kηnig P, Rothaemel M, Holtmann HD. Method for producing propylene from methanol, US7015369; 2002.
[49] Koss U. A brief glance on Lurgi's syngas-based activities in China, GTC conference, 1–4 October, Washington; 2006.
[50] Takahashi J, Okita A, Ohbayashi S, Honda K. In: Presented in part at the 110th catalyst society of Japan meeting, Fukuoka; 2012.
[51] Okita A, Honda K. Selective propylene production process using methanol/dimethyl ether and olefin as raw materials. PETROTECH 2012;35:581.
[52] Ohyama K, Takada H, Chikamatsu N, Honda K, Ohbayashi S. In: Presented in part at the 42nd meeting of the Japan Petroleum Institute, Akita, Japan; 2012.
[53] Murata C, Ito H, Honda K, Okita M. Alkaline earth metal compound-containing zeolite catalyst, its preparation method and manufacturing method of lower hydrocarbon; 2006.
[54] Froment G, Dehertog W, Marchi A. Zeolite catalysis in the conversion of methanol into olefins. In: Spivey JJ, editor. Catalysis, vol. 9. London: Royal Society of Chemistry; 1992. p. 1–64.
[55] Koempel H, Liebner W, Noronha F, Bellot MS, Falabella E, editors. Studies in surface science and catalysis, vol. 167. Amsterdam: Elsevier; 2007. p. 261–7.
[56] Chang CD, Silvestri AJ. ChemTech 1987;10:624.
[57] Park T-Y, Froment GF. Analysis of fundamental reaction rates in the methanol-to-olefins process on ZSM-5 as a basis for reactor design and operation. Ind Eng Chem Res 2004;43:682–9.
[58] Park T-Y, Froment GF. Kinetic modeling of the methanol to olefins process. 2. Experimental results, model discrimination, and parameter estimation. Ind Eng Chem Res 2001;40:4187–96.
[59] Park T-Y, Froment GF. Kinetic modeling of the methanol to olefins process. 1. Model formulation. Ind Eng Chem Res 2001;40:4172–86.
[60] Marchi AJ, Froment GF. Catalytic conversion of methanol to light alkenes on SAPO molecular sieves. Appl Catal 1991;71:139–52.
[61] Wang C-M, Wang Y-D, Liu H-X, Xie Z-K, Liu Z-P. Theoretical insight into the minor role of paring mechanism in the methanol-to-olefins conversion within HSAPO-34 catalyst. Microporous Mesoporous Mater 2012;158:264–71.
[62] Ilias S, Bhan A. Mechanism of the catalytic conversion of methanol to hydrocarbons. ACS Catal 2013;3:18–31.
[63] Chen NY, Reagan WJ. Evidence of autocatalysis in methanol to hydrocarbon reactions over zeolite catalysts. J Catal 1979;59:123–9.
[64] Ono Y, Imai E, Mori T. Autocatalytic nature of methanol conversion over ZSM-5 zeolites. Zeits Phys Chem 1979;115:99.
[65] Ono Y, Mori T. Mechanism of methanol conversion into hydrocarbons over ZSM-5 zeolite. J Chem Soc, Faraday Trans I 1981;77:2209–21.
[66] Chang CD, Silvestri AJ. The conversion of methanol and other O-compounds to hydrocarbons over zeolite catalysts. J Catal 1977;47:249–59.

[67] van der Berg JP, Wolthnizen JP, van Hoof JHC. The conversion of dimethylether to hydrocarbons on zeolite H-ZSM-5. The reaction mechanism for formation of primary olefins. In: Rees LV, editor. Proceedings 5th international conference on zeolites (naples). London: Heyden; 1980. p. 649–60.
[68] Derouane EG. In: Naccache C, Ben Taarit Y, Vedrine JC, Coudurier G, Praliaud H, Imelik B, editors. Studies in surface science and catalysis, vol. 5. Amsterdam: Elsevier; 1980. p. 5–18.
[69] Chang CD, Chu CTW. On the mechanism of hydrocarbon formation from methanol over zeolite catalysts: evidence for carbene intermediacy. J Catal 1982;74:203–6.
[70] Blaszkowski SR, van Santen RA. Theoretical study of C–C bond formation in the methanol-to-gasoline process. J Am Chem Soc 1997;119:5020–7.
[71] Olah GA. Higher coordinate (hypercarbon containing) carbocations and their role in electrophilic reactions of hydrocarbons. Pure Appl Chem 1981;53:201.
[72] Aramburo LR, De Smit E, Arstad B, Van Schooneveld MM, Sommer L, Juhin A, et al. X-ray imaging of zeolite particles at the nanoscale: influence of steaming on the state of aluminum and the methanol-to-olefin reaction. Angew Chem Int Ed 2012;51:3616–9.
[73] Kumar P, Thybaut JW, Svelle S, Olsbye U, Marin GB. Single-event microkinetics for methanol to olefins on H-ZSM-5. Ind Eng Chem Res 2013;52:1491–507.
[74] Clarke JKA, Darcy R, Hegarty BF, O'Donoghue E, Amir Ebrahimi V, Rooney JJ. Free-radicals in dimethyl ether on H-ZSM-5 zeolite—a Novel dimension of heterogeneous catalysis. J Chem Soc Chem Commun 1986;425.
[75] Chang CD, Hellring SD, Pearson JA. On the existence and role of free radicals in methanol conversion to hydrocarbons over H-ZSM-5: I. Inhibition by NO. J Catal 1989;115:282–5.
[76] Hunter R, Hutchings GJ, Pickl W. Methanol conversion to hydrocarbons over the zeolite catalyst H-ZSM-5 in the presence of oxygen and nitric-oxide—further evidence against a radical reaction-mechanism. J Chem Soc Chem Commun 1987;1369.
[77] Hunter R, Hutchings GJ, Pickl W. Mechanism studies on initial C–C bond formation in the zeolite ZSM-5 catalyzed methanol conversion reaction—evidence against a radical pathway. J Chem Soc Chem Commun 1987;843.
[78] Hutchings GJ, Holmen KJJA, Kolboe S, editors. Studies in surface science and catalysis, vol. 61. Amsterdam: Elsevier; 1991. p. 405–12.
[79] Cormerais FX, Perot G, Guisnet M. Selectivity of the dimethyl ether to hydrocarbons conversion on various zeolites. Zeolites 1981;1:141–4.
[80] Nagy JB, Gilson JP, Derouane EG. A ^{13}C-NMR investigation of the conversion of methanol on H-ZSM-5 in the presence of carbon monoxide. J Mol Catal 1979;5:393–7.
[81] Kagi D. In re: mechanism of conversion of methanol over ZSM-5 catalyst. J Catal 1981;69:242–3.
[82] Cormerais FX, Perot G, Chevalier F, Guisnet M. J Chem Res Synop 1980;362.
[83] Tajima N, Tsuneda T, Toyama F, Hirao K. A new mechanism for the first carbon-carbon bond formation in the MTG process: a theoretical study. J Am Chem Soc 1998;120:8222.
[84] Novakova J, Kubelkova L, Habersberger K, Dolejsek Z. Catalytic activity of dealuminated-Y and H-ZSM-5 zeolites measured by the temperature-programmed desorption of small amounts of preadsorbed methanol and by the low-pressure low reaction of methanol. J Chem Soc, Faraday Trans I 1984;80:1457.
[85] Dewaele O, Greers VL, Froment GA, Marin GB. The conversion of methanol to olefins: a transient kinetic study. Chem Eng Sci 1999;54:385.
[86] Blaszkowski SRN, Nascimento MAC, van Santen RA. Activation of C–H and C–C bonds by an acidic zeolite: a density functional study. J Phys Chem 1996;100:3463.
[87] Haw JF, Song W, Marcus DM, Nicholas JB. The mechanism of methanol to hydrocarbon catalysis. Acc Chem Res 2003;36:317–26.
[88] Schulz H. "Coking" of zeolites during methanol conversion: basic reactions of the MTO-, MTP- and MTG processes. Catal Today 2010;154:183–94.
[89] Hutchings GJ, Johnston P. Methanol conversion to hydrocarbons: investigation of the possible role of carbon monoxide in the formation of the initial carbon-carbon bond. Appl Catal 1990;67:L5–9.

[90] Dahl IM, Kolboe S. On the reaction mechanism for hydrocarbon formation from methanol over SAPO-34: I. Isotopic labeling studies of the co-reaction of ethene and methanol. J Catal 1994;149:458–64.
[91] Lesthaeghe D, Van der Mynsbrugge J, Vandichel M, Waroquier M, Van Speybroeck V. Full theoretical cycle for both ethene and propene formation during methanol-to-olefin conversion in H-ZSM-5. ChemCatChem 2011;3:208–12.
[92] Svelle S, Olsbye U, Joensen F, Bjørgen M. Conversion of methanol to alkenes over medium- and large-pore acidic zeolites: steric manipulation of the reaction intermediates governs the ethene/propene product selectivity. J Phys Chem C 2007;111:17981–4.
[93] Svelle S, Joensen F, Nerlov J, Olsbye U, Lillerud K-P, Kolboe S, et al. Conversion of methanol into hydrocarbons over zeolite H-ZSM-5: ethene formation is mechanistically separated from the formation of higher alkenes. J Am Chem Soc 2006;128:14770–1.
[94] Mole T, Bett G, Seddon D. Conversion of methanol to hydrocarbons over ZSM-5 zeolite: an examination of the role of aromatic hydrocarbons using 13carbon- and deuterium-labeled feeds. J Catal 1983;84:435–45.
[95] Mole T, Whiteside JA, Seddon D. Aromatic co-catalysis of methanol conversion over zeolite catalysts. J Catal 1983;82:261–6.
[96] Ceckiewicz S. Methanol conversion to hydrocarbons and dimethyl ether on decationized zeolite-T. Kinet Catal Lett React 1981;16:11.
[97] Olsbye U, Bjørgen M, Svelle S, Lillerud K-P, Kolboe S. Mechanistic insight into the methanol-to-hydrocarbons reaction. Catal Today 2005;106:108–11.
[98] Haw J, Marcus D. Well-defined (supra)molecular structures in zeolite methanol-to-olefin catalysis. Top Catal 2005;34:41–8.
[99] Tsai T-C, Liu S-B, Wang I. Disproportionation and transalkylation of alkylbenzenes over zeolite catalysts. Appl Catal A 1999;181:355–98.
[100] Sullivan RF, Egan CJ, Langlois GE, Sieg RP. A new reaction that occurs in the hydrocracking of certain aromatic hydrocarbons. J Am Chem Soc 1961;83:1156–60.
[101] Alwahabi SM, Froment GF. Conceptual reactor design for the methanol-to-olefins process on SAPO-34. Ind Eng Chem Res 2004;43:5112–22.
[102] Arstad BR, Nicholas JB, Haw JF. Theoretical study of the methylbenzene side-chain hydrocarbon pool mechanism in methanol to olefin catalysis. J Am Chem Soc 2004;126:2991–3001.
[103] Wang C-M, Wang Y-D, Liu H-X, Xie Z-K, Liu Z-P. Catalytic activity and selectivity of methylbenzenes in HSAPO-34 catalyst for the methanol-to-olefins conversion from first principles. J Catal 2010;271:386–91.
[104] Sassi A, Wildman MA, Ahn HJ, Prasad P, Nicholas JB, Haw JF. Methylbenzene chemistry on zeolite H-beta: multiple insights into methanol-to-olefin catalysis. J Phys Chem B 2002;106:2294–303.
[105] Song W, Fu H, Haw JF. Supramolecular origins of product selectivity for methanol-to-olefin catalysis on HSAPO-34. J Am Chem Soc 2001;123:4749–54.
[106] Espinoza RL. Oligomerization vs methylation of propene in the conversion of dimethyl ether (or methanol) to hydrocarbons. Ind Eng Chem Prod Res Dev 1984;23:449.
[107] Espinoza RL. Catalytic conversion of methanol to hydrocarbons: autocatalysis reconsidered. Appl Catal 1986;26:203–9.
[108] Espinoza RL, Stander CM, Mandersloot WGB. Catalytic conversion of methanol to hydrocarbons over amorphous or zeolitic silica-alumina. Appl Catal 1983;6:11–26.
[109] Dessau RM. On the H-ZSM-5 catalyzed formation of ethylene from methanol or higher olefins. J Catal 1986;99:111–6.
[110] Svelle S, Rønning PO, Olsbye U, Kolboe S. Kinetic studies of zeolite-catalyzed methylation reactions. Part 2. Co-reaction of ^{12}C propene or ^{12}C *n*-butene and ^{13}C methanol. J Catal 2005;234:385–400.
[111] Hill IM, Hashimi SA, Bhan A. Kinetics and mechanism of olefin methylation reactions on zeolites. J Catal 2012;285:115–23.
[112] Hill IM, Ng YS, Bhan A. Kinetics of butene isomer methylation with dimethyl ether over zeolite catalysts. ACS Catal 2012;2:1742–8.
[113] Dahl IM, Kolboe S. On the reaction mechanism for hydrocarbon formation from methanol over SAPO-34: 2. Isotopic labeling studies of the co-reaction of propene and methanol. J Catal 1996;161:304–9.

[114] Simonetti DA, Ahn JH, Iglesia E. Catalytic co-homologation of alkanes and dimethyl ether and promotion by adamantane as a hydride transfer co-catalyst. ChemCatChem 2011;3:704–18.
[115] Dessau RM, LaPierre RB. On the mechanism of methanol conversion to hydrocarbons over H-ZSM-5. J Catal 1982;78:136–41.
[116] Weitkamp J, Jacobs PA, Martens JA. Isomerization and hydrocracking of C_9 through C_{16} n-alkanes on Pt/H-ZSM-5 zeolite. Appl Catal 1983;8:123–41.
[117] Buchanan JS, Santiesteban JG, Haag WO. Mechanistic considerations in acid-catalyzed cracking of olefins. J Catal 1996;158:279–87.
[118] Simonetti DA, Ahn JH, Iglesia E. Mechanistic details of acid-catalyzed reactions and their role in the selective synthesis of triptane and isobutane from dimethyl ether. J Catal 2011;277:173–95.
[119] Froment GF. Kinetic modeling of hydrocarbon processing and the effect of catalyst deactivation by coke formation. Catal Rev 2008;50:1–18.
[120] Alwahabi SM, Froment GF. Single event kinetic modeling of the methanol-to-olefins process on SAPO-34. Ind Eng Chem Res 2004;43:5098–111.
[121] Kaeding WW, Butter SA. Production of chemicals from methanol: I. Low molecular weight olefins. J Catal 1980;61:155–64.
[122] Lesthaeghe D, Horré A, Waroquier M, Marin GB, Van Speybroeck V. Theoretical insights on methylbenzene side-chain growth in ZSM-5 zeolites for methanol-to-olefin conversion. Chem Eur J 2009;15:10803–8.
[123] Kaeding WW. Conversion of methanol to hydrocarbons: III. Methylation, ethylation, and propylation of benzene with methanol. J Catal 1988;114:271–6.
[124] Kaeding WW, Young LB, Chu C-C. Shape-selective reactions with zeolite catalysts: IV. Alkylation of toluene with ethylene to produce p-ethyltoluene. J Catal 1984;89:267–73.
[125] Kaeding WW. Shape-selective reactions with zeolite catalysts: V. Alkylation or disproportionation of ethylbenzene to produce p-diethylbenzene. J Catal 1985;95:512–9.
[126] Inui T, Phatanasri S, Matsuda H. Highly selective synthesis of ethane from methanol on a novel nickel silicoaluminophosphate catalyst. J Chem Soc Chem Commun 1990;205.
[127] Garwood WE, Kushnerick JD, Tabak SA. Catalytic process for oligomerizing ethene, US4717782; 1988.
[128] Khadzhiev SN, Kolesnichenko NV, Ezhova NN. Manufacturing of lower olefins from natural gas through methanol and its derivatives. Petrol Chem 2008;48:325–34.
[129] Best D, Wojciechowski BW. On identifying the primary and secondary products of the catalytic cracking of cumene. J Catal 1977;47:11–27.
[130] Wang C-M, Wang Y-D, Xie Z-K. Insights into the reaction mechanism of methanol-to-olefins conversion in HSAPO-34 from first principles: are olefins themselves the dominating hydrocarbon pool species? J Catal 2013;301:8–19.
[131] Ahn BJ, Armando J, Perot G, Guisnet M, editors. Comptes Rendus de l'Académie des Sciences Paris, vol. 288C. 1979. p. 245.
[132] Anderson JR, Foger K, Mole T, Rajadhyaksha RA, Sanders JV. Reactions on ZSM-5-type zeolite catalysts. J Catal 1979;58:114–30.
[133] Gayubo AG, Aguayo AT, Alonso A, Bilbao J. Kinetic modeling of the methanol-to-olefins process on a silicoaluminophosphate (SAPO-18) catalyst by considering deactivation and the formation of individual olefins. Ind Eng Chem Res 2007;46:1981–9.
[134] Lesthaeghe D, Van Speybroeck V, Marin GB, Waroquier M. The rise and fall of direct mechanisms in methanol-to-olefin catalysis: an overview of theoretical contributions. Ind Eng Chem Res 2007;46:8832–8.
[135] Kaiser SW. Small olefin interconversions, US4613721; 1985.
[136] Tau LM, Davis BH. Isotopic tracer studies of the conversion of methanol and ethene or propene to gasoline range hydrocarbons. Energy Fuel 1993;7:249–56.
[137] Svelle S, Rønning PO, Kolboe S. Kinetic studies of zeolite-catalyzed methylation reactions: 1. Coreaction of ^{12}C ethene and ^{13}C methanol. J Catal 2004;224:115–23.
[138] Rønning PO, Mikkelsen Ø, Kolboe S. Effect of co-feeding ethene or propene on the conversion of methanol over H-ZSM-5. Use of isotopic labeling for mechanistic studies. In: Proceedings of the 12th international zeolite conference. 2; 1998. p. 1057.

[139] Bjørgen M, Svelle S, Joensen F, Nerlov J, Kolboe S, Bonino F, et al. Conversion of methanol to hydrocarbons over zeolite H-ZSM-5: on the origin of the olefinic species. J Catal 2007;249:195–207.
[140] Bjørgen M, Olsbye U, Petersen D, Kolboe S. The methanol-to-hydrocarbons reaction: insight into the reaction mechanism from ^{12}C benzene and ^{13}C methanol coreactions over zeolite H-beta. J Catal 2004;221:1–10.
[141] Cui Z-M, Liu Q, Ma Z, Bian S-W, Song W-G. Direct observation of olefin homologations on zeolite ZSM-22 and its implications to methanol to olefin conversion. J Catal 2008;258:83–6.
[142] Li J, Qi Y, Xu L, Liu G, Meng S, Li B, et al. Co-reaction of ethene and methanol over modified H-ZSM-5. Catal Commun 2008;9:2515–9.
[143] Zhu Q, Kondo JN, Setoyama T, Yamaguchi M, Domen K, Tatsumi T. Activation of hydrocarbons on acidic zeolites: superior selectivity of methylation of ethene with methanol to propene on weakly acidic catalysts. Chem Commun 2008;5164–6.
[144] Struelens P. Dimethyl ether transformation in presence of ethylene with acid zeolites. Belgium: K.U. Leuven; 2012.
[145] Smits W. The catalytic conversion of ethylene to propylene. Unpublished results; 2010.
[146] Gorring RL. Diffusion of normal paraffins in zeolite T: occurrence of window effect. J Catal 1973;31:13–26.
[147] Froment GF, Dehertog WJH, Marchi AJ. Zeolite catalysis in the conversion of methanol into olefins. Catalysis 1992;9:1.
[148] Wilson ST, Lok BM, Messina CA, Cannan TR, Flanigen EM. Aluminophosphate molecular sieves: a new class of microporous crystalline inorganic solids. J Am Chem Soc 1982;104:1146–7.
[149] Flanigen EM, Lok BM, Patton RL, Wilson ST. In: Murakami Y, Iijima A, Ward JW, editors. New developments in zeolite sciences and technology. Amsterdam: Elsevier; 1986.
[150] Givens EN, Plank CJ, Rosinski EJ. Manufacture of light olefins, US4079096; 1978.
[151] Kaiser SW. Production of lights olefins, US4499327; 1985.
[152] Liang J, Li H, Zhao S, Guo W, Wang R, Ying M. Characteristics and performance of SAPO-34 catalyst for methanol-to-olefin conversion. Appl Catal 1990;64:31–40.
[153] Rubin MK. Synthetic crystal of MCM-35, US4981663; 1990.
[154] Zhu Q, Kondo JN, Inagaki S, Tatsumi T. Catalytic activities of alcohol transformations over 8-ring zeolites. Top Catal 2009;52:1272–80.
[155] Kokotailo GT, Lawton SL, Olson DH, Meier WM. Structure of synthetic zeolite ZSM-5. Nature 1978;272:437–8.
[156] Ahmed S. Methanol to olefins conversion over metal containing MFI-type zeolites. J Porous Mater 2012;19:111–7.
[157] Young LB. Conversion of toluene over zeolite ZSM-48, US4375573; 1983.
[158] Kokotailo GT, Schlenker JL, Dwyer FG, Valyocsik EW. The framework topology of ZSM-22: a high silica zeolite. Zeolites 1985;5:349–51.
[159] Martens JA, Tielen M, Jacobs PA, Weitkamp J. Estimation of the void structure and pore dimensions of molecular sieve zeolites using the hydroconversion of n-decane. Zeolites 1984;4:98–107.
[160] Schlenker JL, Rohrbaugh WJ, Chu P, Valyocsik EW, Kokotailo GT. The framework topology of ZSM-48: a high silica zeolite. Zeolites 1985;5:355–8.
[161] Haag WO, Lago RM, Weisz PB. The active site of acidic aluminosilicate catalysts. Nature 1984;309:589–91.
[162] Olson DH, Haag WO, Lago RM. Chemical and physical properties of the ZSM-5 substitutional series. J Catal 1980;61:390–6.
[163] Chu CTW, Chang CD. Isomorphous substitution in zeolite frameworks. 1. Acidity of surface hydroxyls in [B]-, [Fe]-, [Ga]-, and [Al]-ZSM-5. J Phys Chem 1985;89:1569–71.
[164] Jacobs PA, Martens JA, Weitkamp J, Beyer HK. Faraday Discuss Chem Soc 1981;72:353.
[165] Vedrine J, Auroux A, Coudurier G, Engelhard P, Gallez JP, Szabo G. In: Olson DH, Bisio A, editors. Proceedings of the 6th international zeolite conference. Surrey, UK: Butterworths; 1984. p. 479.
[166] Haag WO, Lago RM, Weisz PB. Faraday Discuss Chem Soc 1981;72:317–30.
[167] Rollmann LD. Systematics of shape selectivity in common zeolites. J Catal 1977;47:113–21.

[168] Anderson MW, Klinowski J. Direct observation of shape selectivity in zeolite ZSM-5 by magic-angle-spinning NMR. Nature 1989;339:200–3.
[169] Rollmann LD, Walsh DE. Shape selectivity and carbon formation in zeolites. J Catal 1979;56:139–40.
[170] Derouane EG, Vedrine JC. On the role of shape selectivity in the catalytic conversion of alcohols and simple hydrocarbons molecules on zeolite ZSM-5. J Mol Catal 1980;8:479–83.
[171] Katalytische Omzetting van Synthesegas in Lichte Olefins. Diensten voor Programmatie van het Wetenschapsbeleid; 1987.
[172] Dehertog WJH, Froment GF. Production of light alkenes from methanol on ZSM-5 catalysts. Appl Catal 1991;71:153–65.
[173] Prinz D, Riekert L. Formation of ethene and propene from methanol on zeolite ZSM-5: I. Investigation of rate and selectivity in a batch reactor. Appl Catal 1988;37:139–54.
[174] Li J, Wei Y, Liu G, Qi Y, Tian P, Li B, et al. Comparative study of MTO conversion over SAPO-34, H-ZSM-5 and H-ZSM-22: correlating catalytic performance and reaction mechanism to zeolite topology. Catal Today 2011;171:221–8.
[175] Rodewald PG. Manufacture of light olefins. 1978.
[176] Fleckenstein T, Litterer H, Fetting F. Selektive umwandlung von methanol in C_2–bis C_5-olefine an zeolith-katalysatoren. Chem Ing Tech 1980;52:816–8.
[177] Wu W, Guo W, Xiao W, Luo M. Methanol conversion to olefins (MTO) over H-ZSM-5: evidence of product distribution governed by methanol conversion. Fuel Process Technol 2013;108:19–24.
[178] Jacobs PA, Martens JA. Synthesis of high-silica aluminosilicates zeolites. Amsterdam: Elsevier; 1987.
[179] Venuto PB, Landis PS, Eley HPDD, Paul BW, editors. Advances in catalysis, vol. 18. New York: Academic Press; 1968. p. 259–371.
[180] Dejaifve P, Auroux A, Gravelle PC, Vedrine J, Gabelica Z, Derouane EG. Methanol conversion on acidic ZSM-5, offretite, and mordenite zeolites: a comparative study of the formation and stability of coke deposits. J Catal 1981;70:123–36.
[181] Sawa M, Niwa M, Murakami Y. Development of long-life dealuminated mordenite for methanol conversion to hydrocarbons. Chem Lett 1987;16:1637–40.
[182] Guisnet M, Magnoux P, Canaff C. In: Murakami Y, et al., editors. Proceeding of the 7th international zeolite conference, new developments in zeolite science technology. Tokyo: Elsevier; 1986. p. 701–7.
[183] Sawa M, Kato K, Hirota K, Niwa M, Murakami Y. Mordenite with long life and selectivity for methanol conversion to gasoline: mordenite modified by barium ion exchange, dealumination and chemical vapor deposition of silicon methoxide. Appl Catal 1990;64:297–308.
[184] Salvador P, Kladnig W. J Chem Soc, Faraday Trans I 1977;73:1153.
[185] Davidova N, Shopov D, Jaeger N, Schulz-Ekloff G. Conversion of methanol to hydrocarbons over metal-zeolite catalysts. React Kinet Catal Lett 1979;12:229–34.
[186] Hochevar S, Echevskii GV, Drzaj B, Ione KG. Methanol-based synthesis of hydrocarbons on bifunctional Y-type zeolites. React Kinet Catal Lett 1980;13:425–9.
[187] Sodesawa T. Methanol conversion to lower hydrocarbons over proton exchanged NaY zeolite catalysts. React Kinet Catal Lett 1986;32:251–5.
[188] Kikhtyanin OV, Mastikhin VM, Ione KG. Methanol conversion on aluminophosphates with zeolite structure. Appl Catal 1988;42:1–13.
[189] Barger PT. In: Guisnet M, Gilson J-P, editors. Zeolites for cleaner technologies, vol. 3. London: Imperial College Press; 2001. p. 239–60.
[190] Cui Z-M, Liu Q, Song W-G, Wan L-J. Insights into the mechanism of methanol-to-olefin conversion at zeolites with systematically selected framework structures. Angew Chem Int Ed 2006;45:6512–5.
[191] Teketel S, Olsbye U, Lillerud K-P, Beato P, Svelle S. Selectivity control through fundamental mechanistic insight in the conversion of methanol to hydrocarbons over zeolites. Microporous Mesoporous Mater 2010;136:33–41.
[192] Avidan AA. In: Studies in surface science and catalysis, vol. 36. 1988. p. 307–23.
[193] Wilson S, Barger P. The characteristics of SAPO-34 which influence the conversion of methanol to light olefins. Microporous Mesoporous Mater 1999;29:117–26.

[194] Chen NY, Degnan Jr TF, Smith CM. Molecular transport and reaction in zeolites, design and application of shape selective catalysts. New York: VCH Publishers; 1994.
[195] Schulz H, Siwei Z, Kusterer H, Tomoyuki lnui SN, Takashi T, editors. Studies in surface science and catalysis, vol. 60. Amsterdam: Elsevier; 1991. p. 281–90.
[196] Nawaz S, Kolboe S, Kvisle S, Lillerud KP, Stocker M, Holmen A, et al, editors. Studies in surface science and catalysis, vol. 61. Amsterdam: Elsevier; 1991. p. 421–7.
[197] Bleken F, Bjorgen M, Palumbo L, Bordiga S, Svelle S, Lillerud KP, et al. The effect of acid strength on the conversion of methanol to olefins over acidic microporous catalysts with the CHA topology. Top Catal 2009;52:218–28.
[198] Chang CD, Lang WH, Smith RL. The conversion of methanol and other O-compounds to hydrocarbons over zeolite catalysts: II. Pressure effects. J Catal 1979;56:169–73.
[199] Nawaz S, Kolboe S, Stöcker M, Curry-Hyde HE, Howe RF, editors. Studies in surface science and catalysis, vol. 81. Amsterdam: Elsevier; 1994. p. 393–8.
[200] Butter SA, Kaeding WW, Jurewicz AT. Conversion of alcohols, mercaptans, sulfides, halides and/or amines, US3894107; 1975.
[201] Chang CD, Lang WH, Silvestri AJ. Conversion of ethers, US3894106; 1975.
[202] Chang CD, Silvestri AJ, Smith RL. Production of gasoline hydrocarbons, US3928483; 1975.
[203] Chang CD, Kuo JCW, Lang WH, Jacob SM, Wise JJ, Silvestri AJ. Process studies on the conversion of methanol to gasoline. Ind Eng Chem Process Des Devel 1978;17:255–60.
[204] Suzuki K, Kiyozumi Y, Matsuzaki K, Shin S. Influence of the synthesis conditions of H-ZSM-5 on its physical properties and catalytic activity in methanol conversion: the water content of the reaction mixture. Appl Catal 1987;35:401–9.
[205] Wu X, Anthony RG. Effect of feed composition on methanol conversion to light olefins over SAPO-34. Appl Catal A 2001;218:241–50.
[206] Vedrine JC, Dejaifve P, Naccache C, Derouane EG, Seivama T, Tanabe K, editors. Studies in surface science and catalysis, vol. 7. Amsterdam: Elsevier; 1981. p. 724–38.
[207] Aguayo AT, Campo AESD, Gayubo AG, Tarrío A, Bilbao J. Deactivation by coke of a catalyst based on a SAPO-34 in the transformation of methanol into olefins. J Chem Technol Biotechnol 1999;74:315–21.
[208] Bos ANR, Tromp PJJ, Akse HN. Conversion of methanol to lower olefins. Kinetic modeling, reactor simulation, and selection. Ind Eng Chem Res 1995;34:3808–16.
[209] Guisnet M. Effect of coke molecules trapped in the micropores of zeolites as active species in hydrocarbon transformations. J Mol Catal A Chem 2002;182–183:367–82.
[210] Chen D, Rebo HP, Moljord K, Holmen A, Fuentes CHBaGA, editors. Studies in surface science and catalysis, vol. 111. Amsterdam: Elsevier; 1997. p. 159–66.
[211] Froment G, Dehertog W, Marchi A. Zeolite catalysis in the conversion of methanol into olefins. In: Catalysis, vol. 9, London: Royal Society of Chemistry; 1992. p. 1–64.
[212] Beeckman JW, Froment GF. Chem Eng Sci 1980;35:.
[213] Beeckman JW, Froment GF. Catalyst deactivation by active site coverage and pore blockage. Ind Eng Chem Fund 1979;18:245–56.
[214] Derouane EG. In: Imelik B, Naccache C, Coudurier G, Ben Taarit Y, Vedrine J, editors. Catalysis by acids and bases. Amsterdam: Elsevier; 1985.
[215] Chen D, Moljord K, Fuglerud T, Holmen A. The effect of crystal size of SAPO-34 on the selectivity and deactivation of the MTO reaction. Microporous Mesoporous Mater 1999;29:191–203.
[216] Chen D, Gronvold A, Moljord K, Holmen A. Methanol conversion to light olefins over SAPO-34: reaction network and deactivation kinetics. Ind Eng Chem Res 2006;46:4116–23.
[217] Mores D, Kornatowski J, Olsbye U, Weckhuysen BM. Coke formation during the methanol-to-olefin conversion: in situ microspectroscopy on individual H-ZSM-5 crystals with different brønsted acidity. Chem Eur J 2012;17:2874–84.
[218] Chen D, Rebo HP, Holmen A. Diffusion and deactivation during methanol conversion over SAPO-34: a percolation approach. Chem Eng Sci 1999;54:3465–73.

[219] Nishiyama N, Kawaguchi M, Hirota Y, Van Vu D, Egashira Y, Ueyama K. Size control of SAPO-34 crystals and their catalyst lifetime in the methanol-to-olefin reaction. Appl Catal A 2009;362:193–9.

[220] Lee Y-J, Baek S-C, Jun K-W. Methanol conversion on SAPO-34 catalysts prepared by mixed template method. Appl Catal A 2007;329:130–6.

[221] Anderson MW, Sulikowski B, Barrie PJ, Klinowski J. In situ solid-state NMR studies of the catalytic conversion of methanol on the molecular sieve SAPO-34. J Phys Chem 1990;94:2730–4.

[222] Qi G, Xie Z, Yang W, Zhong S, Liu H, Zhang C, et al. Behaviors of coke deposition on SAPO-34 catalyst during methanol conversion to light olefins. Fuel Process Technol 2007;88:437–41.

[223] Chen D, Rebo HP, Grønvold A, Moljord K, Holmen A. Methanol conversion to light olefins over SAPO-34: kinetic modeling of coke formation. Microporous Mesoporous Mater 2000;35–36:121–35.

[224] Chen D. Adsorption, diffusion and reactions in methanol to olefins on SAPO-34, Ph.D. thesis. Norwegian University of Science and Technology, Trondheim; 1998.

[225] Li J, Xiong G, Feng Z, Liu Z, Xin Q, Li C. Coke formation during the methanol conversion to olefins in zeolites studied by UV Raman spectroscopy. Microporous Mesoporous Mater 2000;39:275–80.

[226] Djieugoue M-A, Prakash AM, Kevan L. Catalytic study of methanol-to-olefins conversion in four small-pore silicoaluminophosphate molecular sieves: the influence of the structural type, nickel incorporation, nickel location, and nickel concentration. J Phys Chem B 2000;104:6452–61.

[227] Aguayo AT, Gayubo AG, Vivanco R, Olazar M, Bilbao J. Role of acidity and microporous structure in alternative catalysts for the transformation of methanol into olefins. Appl Catal A 2005;283:197–207.

[228] Yuen L-T, Zones SI, Harris TV, Gallegos EJ, Auroux A. Product selectivity in methanol to hydrocarbon conversion for isostructural compositions of AFI and CHA molecular sieves. Microporous Mater 1994;2:105–17.

[229] Bleken F, Bjørgen M, Palumbo L, SBordiga S, Svelle S, Lillerud KP, et al. Top Catal 2009;52:218.

[230] Bjørgen M, Olsbye U, Kolboe S. Coke precursor formation and zeolite deactivation: mechanistic insights from hexamethylbenzene conversion. J Catal 2003;215:30–44.

CHAPTER 6

Y Zeolites as a Major Component of FCC Catalysts: Main Challenges in the Modification Thereof

E.F. Sousa-Aguiar*,†
*Petrobras Research Centre (CENPES), Rio de Janeiro, Brazil †Federal University of Rio de Janeiro (UFRJ), Rio de Janeiro, Brazil

1 Introduction: The FCC Catalyst

1.1 History of the FCC Process

The process of fluid catalytic cracking (FCC) is undoubtedly at the heart of the refinery and may be considered one of the most important processes of the chemical industry. This process converts heavy atmospheric residues and vacuum distillates into value-added fractions such as gasoline, diesel, kerosene, and LPG.

At the beginning of the19th century, lamps for indoor illumination often used whale oil as a fuel. Petroleum-based illuminants such as lampions began to take the place of whale oil around 1850, giving birth to an emerging industry that produced kerosene via refining of crude oil. The growing automobile industry, by the end of this century, gave an alternative use to light fractions of oil, mainly gasoline. Soon, the supply of straight-run gasoline as fuel for automobile engines became insufficient. For that reason, the oil industry started looking for alternative processes aiming at increasing the production of lighter fractions. This fact was the driving force for the first cracking process, thermal cracking [1]. The thermal cracking method (also known as the "Shukhov cracking process") was invented by Russian engineer Vladimir Shukhov and patented in 1891 in the Russian empire, patent no. 12926, Nov. 27, 1891. Later, this process was modified by American engineer William Merriam Burton and patented as U.S. patent 1,049,667 on Jun. 8, 1908.

As newer and more powerful engines have been developed, higher octane gasoline has been requested. Moreover, aircraft engines needed fuel with octane ratings of nearly 100. Such features could not be achieved with the traditional thermal cracking process. Hence, the

Zeolites and Zeolite-like Materials. http://dx.doi.org/10.1016/B978-0-444-63506-8.00007-0

discovery of a more efficient cracking process by the French engineer Eugene Houdry [2] in the 1920s enabled the refineries to produce higher octane gasoline, also enabling an increase in the supply thereof. This process was called catalytic cracking ("the Houdry fixed-bed catalytic cracking") because cracking reactions were promoted by a solid catalyst [3]. The process comprised several separate reactors acting as a cyclic process, or rather, while in some reactors cracking was taking place, in others the catalyst underwent regeneration via coke burning with air. The concept of thermal cracking was further improved by Mobil's Thermofor Catalytic Cracking (TCC), in which a moving catalyst bed with a macro catalyst (spheres with the diameter of 1.5–2 mm) was used. The moving-bed system increased gasoline yields by 15%, over the fixed-bed system.

The knowledge of fluidized beds along with the preparation of fine particles of catalysts was the basis for the development of the FCC process [4,5]. The use of small particles offered advantages in heat transfer and mass diffusion over the large catalyst pellets used in both fixed bed and TCC. The first FCC unit started up in May 1942, at the Standard Oil New Jersey's refinery in Baton Rouge, Louisiana. The enormous demand for motor and aviation fuels due to World War II helped to establish this process. Indeed, between 1942 and 1945, 34 new units were put on stream.

The improvement of the FCC technology was carried out via several types of reactors. In fact, the first one was an up-flow Model I FCC, followed by the down-flow Model II FCC. Also, stacked FCC design, Esso model 4 and the straight-riser design were developed. Eventually, the side-by-side design with straight riser, better for large units, was commercialized in 1950.

1.2 The FCC Catalyst

The first commercial use of catalytic cracking occurred in 1915 when A. M. McAfee developed a batch process using aluminium chloride (which was a traditional Friedel-Crafts catalyst known since 1877) to promote the cracking of heavy petroleum oils. Nevertheless, the prohibitive cost of the catalyst prevented the widespread use of McAfee's process at that time.

Catalysts for cracking are historically subdivided into the following three main groups:

- *Natural catalysts*—Clays, such as bentonite or montmorillonites, underwent acid treatment to remove Na and Mg and increase their acidity;
- *Amorphous catalysts*—these were gels of silica and alumina, resulting from the reaction of Na_2SiO_3 and $Al_2(SO_4)_3$. There were two types of such catalysts, low alumina (10–15% of Al_2O_3) and high alumina (20–30% of Al_2O_3);
- *Crystalline catalysts*—Microspheres (40–150 μm), having a crystalline zeolite as the main component.

As a matter of fact, Houdry was the first one to use acid-treated bentonites as cracking catalysts back in 1936. In the 1940s, silica-alumina catalysts were synthesized, and they

Table 1: Main features of the first zeolite-containing catalysts

Type	Main Feature
Coke	Zeolite-containing catalysts reduced coke formation
Gasoline	Zeolite-containing catalysts increased gasoline yield
RON	Zeolite-containing catalysts reduced RON due to higher HTR

greatly improved the performance of clay-based catalysts. The early synthetic amorphous silica-alumina catalysts contained about 13% wt. of Al_2O_3 (low alumina); however, in 1955, the content of Al_2O_3 had increased to about 25%.

In 1962, the most important catalytic breakthrough of the FCC process took place when a component, known as Zeolite-Y, was added to the active alumina catalyst. It has been found that by adding small amounts of zeolite into the matrix of the older silica-alumina catalyst, a new catalyst was produced. Such a catalyst had an outstanding performance, much better than any catalyst before. The zeolite catalyst greatly improved the gasoline yield (Table 1). The first commercial zeolite catalysts were introduced in 1964, and zeolite catalysts continue being used today.

Remember, however, that gasoline produced via the pristine zeolite-containing catalysts presented a lower research octane number (RON). This was later related to the fact that zeolites, having a very high density of acid sites, were increasing the hydrogen transfer reaction (HTR), as described later [6]:

$$3\text{ Olefin} + \text{naphthene} \rightarrow \text{aromatic} + 3\text{ paraffin}$$

Such a reaction is detrimental to the gasoline quality when RON is concerned. For that reason, zeolites have undergone several types of modifications which aim at reducing hydrogen transfer (HT), thereby increasing the gasoline RON.

1.3 The Current FCC Catalysts

Although a very high number of formulations may be found when dealing with FCC catalysts now, regardless of their formulation, the catalyst components may be classified as the following [7]:

(a) *Active component*—responsible for most of the catalyst activity, are usually faujasites (or Y-zeolites). Currently, another zeolite has also been used (ZSM-5).
(b) *Matrices*—they promote bottoms conversion, increasing the overall activity of the catalyst. They must present a certain tolerance to Ni and V, to nitrogen-containing compounds and, whenever possible, promote DESOX effects. They must also present binding properties.

- Active matrix—normally, the active matrix is an alumina [8], which is acidic, although less acidic than the zeolite; on the other hand, aluminas display much larger pores than zeolites, and therefore they promote precracking of heavier molecules;
- Inert matrix—normally, the inert matrix is a clay, such as kaolin, used to confer physical properties;
- Synthetic matrix—the synthetic matrix is actually a binder that keeps all other components bound together in the final catalyst.

(c) *Functional ingredients*—ingredients added to the catalyst to perform a specific function (ie, traps to capture vanadium compounds).

FCC catalysts are in constant modification, new commercial catalysts being delivered to the market every 6 months or less. Indeed, new formulations are constantly being developed, aiming at:

- Disfavoring coke formation;
- Promoting bottoms conversion;
- Presenting high metal resistance (V, Ni, and Fe);
- Displaying an outstanding performance in terms of octanes, as compounds containing Pb may no longer be used.

1.4 The Preparation of FCC Catalyst

As previously mentioned, FCC catalysts are a multicomponent species, whose preparation is nontrivial. The type of the component used, the ingredients used in the preparation thereof, the order of addition of such ingredients, and how they respond to steam calcination will determine the final characteristics of the catalyst. Although different technologies may have different steps in the preparation of the catalyst, the conventional steps may be didactically divided into:

(a) Zeolite preparation;
(b) Zeolite modifications;
(c) Particle formation (spray drying);
(d) Ion exchange;
(e) Washing;
(f) Drying;
(g) Calcination.

Actually, the FCC catalyst particle is very similar to a concrete particle. Concrete is a composite construction material made primarily with gravel, cement, and sand, whose relative sizes are 15,250, and 5000. Regarding the FCC catalyst particle, one has a binder, alumina, and zeolite/clay, which have approximately the same relative sizes (Fig. 1), forming a particle with considerable hardness.

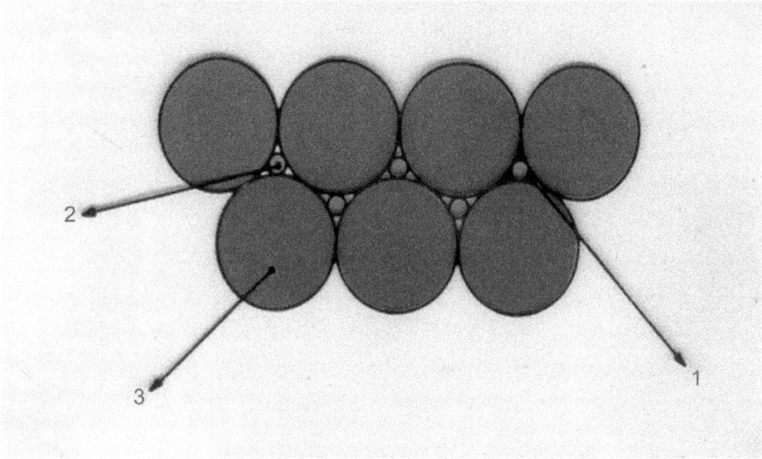

Fig. 1
FCC particle: the concrete model. 1. Binder; 2. Alumina; 3. Zeolite/clay.

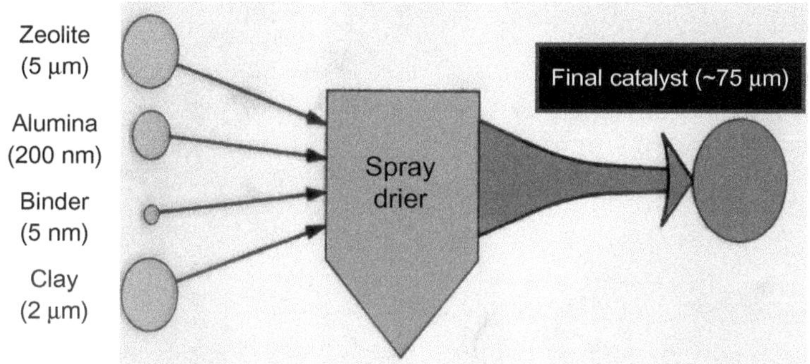

Fig. 2
Formation of the FCC catalyst particle: particles of different sizes generate the final catalyst via spray drying.

As demonstrated in Fig. 2, the different components, that is to say, the alumina, the zeolite, and the clay are basically powders which have different average particle sizes. They are often suspended in a vessel, along with the binder, generally produced "in situ." After suspending all the components, one must form the catalyst particle. The particle formation requires the use of any particle formation technologies. Normally, spray drying is the chosen one, as it may provide the producer with a relatively narrow particle size distribution and good resistance to attrition, which is one of the major concerns of the FCC catalyst performance (Fig. 2). Also, many technologies require a final calcination step in order to increase the catalyst hardness. However, calcination must be carried out carefully, as the zeolite component is very sensitive to steam calcination.

2 The Zeolite Component

There are several reasons why zeolites have become the preferred catalyst for many reactions of industrial importance. In the cracking process, the following characteristics deserve special attention:

(a) Thermal stability:
Surely, one of the most important reasons behind the selection of zeolites as catalysts for high temperature reactions is their good thermal stability. Most of the zeolites remain unaffected by temperatures as high as 650°C. For zeolites presenting high silica-to-alumina ratios (SARs), structural collapse becomes significant only at temperatures as high as 1000°C. Further, the introduction of rare earth (RE) ions in the zeolite also increases the thermal stability. Therefore, such zeolites are suitable for FCC processes, where high temperatures and other harsh operating conditions prevail.

(b) Acidity:
Zeolites exhibit much higher acidity than the earlier amorphous silica-alumina catalysts. Also, zeolites present both Bronsted and Lewis acid sites. The density of these acid sites is related to the SAR in the zeolite framework. Such density increases with the aluminum content. Nevertheless, as the SAR increases and the density of sites decreases, the strength of isolated sites increases, attaining a maximum. In general, the following parameters affect the acidity of a zeolite:
1. The nature or type of the acid sites.
2. The density or concentration of the acid sites.
3. The strength distribution of the acid sites.
4. The location of the acid sites within the zeolite framework.
5. The geometric distribution of the acid sites over the zeolite crystals.

An important parameter governing the acidity per unit cell in the zeolite framework is the unit cell size (UCS), represented by the symbol A_0. The value of A_0 is related to the SAR of the zeolite.

(c) Shape selectivity [9–11]:
The dimensions of the pores within the zeolite frameworks are often similar to the sizes of the molecules participating in the reactions which they catalyze, generating a strong influence on the selectivity of such reactions. The main types of selectivity are the following:
- *Reactant selectivity* occurs when only part of the reactant molecules is small enough to diffuse through the catalyst pores.
- *Product selectivity* occurs when some of the products formed within the pores are too bulky to diffuse out as observed products. They are either converted to less bulky molecules or eventually deactivate the catalyst by blocking the pores.
- *Restricted transition state selectivity* occurs when certain reactions are prevented because the corresponding transition state would require more space than is available

in the cavities. Neither reactant nor potential product molecules are prevented from diffusing through the pores. Reactions requiring smaller transition states proceed unhindered.

- *Molecular traffic control* may occur in zeolites with more than one type of the pore system. Reactant molecules here may preferentially enter the catalyst through one of the pore systems, while products diffuse out by the other.

(d) Concentration effects:

Concentration effects are related to the concepts of hydrophilicity or the hydrophobicity of the zeolites, affecting the reaction selectivity. Such properties may be fine-tuned when zeolites are being synthesized, producing specific zeolites for a given purpose.

As already mentioned, the zeolitic component is the most important component in the FCC catalyst, being able to provide the catalyst with both the convenient activity and selectivity. In FCC catalysts, the zeolite employed is faujasite, or rather, the faujasite zeolite (FAU) structure. The Si/Al ratio of the prepared Y-zeolite is an important feature of such material, in that the higher the Si/Al ratio, the more stable against acid and hydrothermal dealumination the zeolite is.

The zeolite is synthesized in its Na-form, which is a nonactive material, since acidic sites (Broensted acid sites) responsible for the formation of carbocations and hence cracking reactions are not present. Thus, the NaY zeolite must undergo ion exchange with ammonium (or rare earth ions) to prepare the ammonium/RE form of the zeolite which will generate the acidic form HY upon calcination.

2.1 REY Zeolites

The introduction of RE elements in the zeolitic component, via ion exchange, followed by a calcination step, is one of the most important modifications carried out in the process of FCC catalyst preparation, as it increases both the stability of the zeolite and the overall activity of the catalyst [12]. Furthermore, calcination promotes framework dealumination, changing the acidic and textural characteristics of the zeolite.

Zeolites with high La and Ce contents, which have been calcined in the presence of steam or not, have been studied using Rietveld structure refinement [13]. As depicted in Fig. 3, several crystallographic sites may be found in the Y-zeolite framework.

The results in Table 2 show that:

- After ion exchange, RE are located in sites 4 and 5 in the supercage;
- However, after calcination, migration to the sodalite cage and hexagonal prism takes place;
- Such migration occurs regardless of the type of the RE atom (La or Ce).

Furthermore, ^{27}Al-MAS/NMR confirms that dealumination is taking place upon calcination; however, the structure collapse is not observed. Interestingly, as previously reported [14],

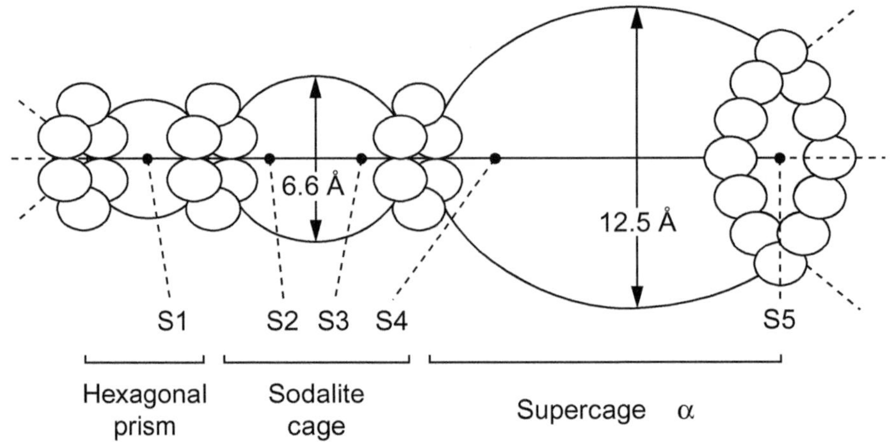

Fig. 3
The crystallographic sites for nonframework atoms in Y-zeolites.

Table 2: Location of different cations in crystallographic sites of the Y zeolite framework

S1	S2	S3	S4	S5	
	20.2 Na	32 H_2O	0.8 La	5.0 La	BC
9.8 Al	10.4 La	32 H_2O	20.2 Na		AC

BC, before calcination; AC, after calcination.

the framework collapse temperature increases with the introduction of RE, regardless of the type of the RE cation introduced (Fig. 4).

Another important issue regards the acidity of RE-containing Y zeolites. Studies [15] with zeolites with the same content of different cations (La, Nd, Sm, Gd, and Dy), verified that the ionic radius of the cation represents an important role in the Brönsted acidic sites formation. The IR spectrum (OH zone) clearly shows that RE introduction generates a band between 3490 and 3530 cm^{-1} (OH interacting with RE). Such a band undergoes a shift to higher wavenumbers as the ionic radius increases (Fig. 5).

A very intense band between 3496 and 3530 cm^{-1} is assigned to the existence of OH groups associated with RE cations. A shift in this band after pyridine adsorption (or lutidine) is observed, which has been attributed to an increase in the acid strength of the OH groups associated to rare-earth cations. Indeed, a linear correlation between the ionic radius of the rare-earth cation and the wavenumber of the band can be observed.

Some very interesting conclusions may be drawn from this series of experiments and characterization. In fact, RE do migrate to more internal sites upon calcination. Once located inside the sodalite cage and the hexagonal prism, groups RE-O-RE are formed. Such groups

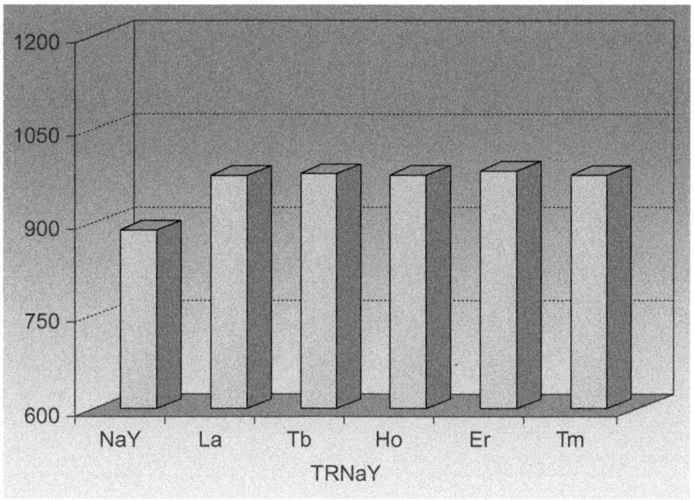

Fig. 4
Collapse temperature for Y-zeolites exchanged with different types of RE-cations.

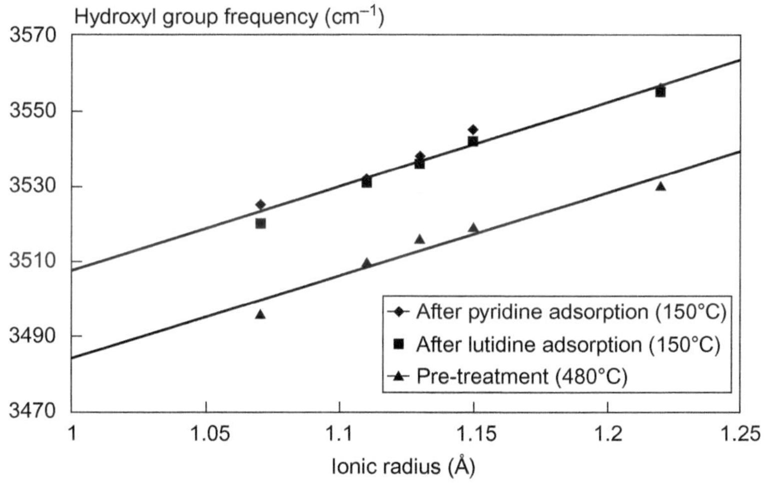

Fig. 5
Shift in the 3496/3530 cm^{-1} band after pyridine/lutidine adsorption for zeolites containing different RE-cations.

are capable of stabilizing the zeolite framework, by generating bridges with the oxygen atoms thereof. Moreover, the degree of stabilization depends on the degree of RE-exchange (Fig. 6).

Concerning acidity, $RE(H_2O)_n$ cations undergo hydrolysis upon calcination, generating Brönsted acidic sites. Such a hydrolysis reaction allows one to explain the higher acid site concentration in CREY (Calcined RE-Y) zeolites and their higher activity. The Brönsted sites seem to be a function of the type of RE cations introduced in the zeolite.

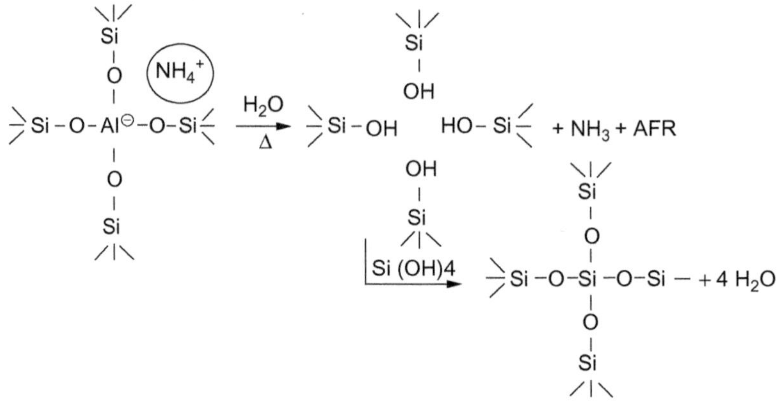

Fig. 6
NH_4Y zeolites. Reactions taking place upon calcination.

Another important issue concerns activity. Previously published papers [16–18] have also shown, using *n*-heptane cracking as a probe molecule:

- Activity depends on the degree of ion exchange;
- A maximum is attained for 40% of Na exchange;
- The following reactions are apparently taking place:

$$Ce^{3+} + 2H_2O \rightarrow \left[Ce(OH)_2\right]^+ + 2H^+$$

$$2Ce^{3+} + 2H_2O \rightarrow \left[Ce - OH - Ce\right]^{5+} + H^+$$

These reactions explain why a maximum activity is attained, depending on the degree of exchange. Nevertheless, such maximum activity depends on both the type of the RE cation and the type of the model reaction.

In terms of cracking, the presence of RE in the zeolite affects the capacity of transferring hydrogen. The reaction of HT, in the FCC process, is very important since:

- it competes with other important reactions such as cracking itself, isomerization and alkylation;
- it transforms olefins into paraffins, diminishing the gasoline RON;
- it is, therefore, determining for the process selectivity.

The presence of RE does increase Brönsted acidity, hence RE affects both activity and selectivity. The higher the ionic radius of the RE atom, the higher the acidity generated in the zeolite. Because HT increases with the increasing zeolite acidity, the higher the atomic radius of the RE atom, the higher the HT. Generally, RE favor HT, reducing RON by reducing olefin concentration in the gasoline. Nevertheless, because RE also affects zeolite stability, evaluation may be rather misleading sometimes. Also, it must be borne in mind that industrial results show that RE seem to have little influence on the gasoline motor octane number (MON). Finally, it is worth mentioning that REs also play an important role in preventing metals deactivation as they are a very effective vanadium trap.

REs are facing a terrible problem of increasing price and decreasing availability [19]. Thus, the removal of these ingredients from the FCC catalyst composition has been sought. However, the total removal of rare-earth elements from the zeolite component of an FCC catalyst would result in a considerable detrimental effect in most FCC operations due to the lower activity and worsened product yield slate, not to mention a problem of higher vanadium deactivation. For that reason, other elements such as Ca and Cr have been tried, but without encouraging results.

Recently, producers have announced the successful commercialization of Zero/Low Rare Earth Catalyst Technologies [20]. Bear in mind that some special features regarding this catalyst must be present, such as:

- very high amounts of the selective matrix;
- high accessibility;
- high zeolite levels for increased activity;
- high SAR, which results in superior zeolite stability;
- feedstocks with low vanadium contents.

2.2 NH_4Y Zeolites

As mentioned, NH_4Y zeolites must also undergo thermal treatments to generate Broensted sites upon calcination. Calcined ammonium-exchanged Y-zeolites are often known as USY (ultrastable Y-zeolites). Calcination is, however, a very complex step as several reactions may take place, as shown in Fig. 6. Furthermore, zeolites undergo structural modifications upon calcination.

Calcination generates acidic sites, but also promotes dealumination, generating aluminium species known as EFAL (Extra Framework Alumina). The removal of framework aluminium changes the UCS of the zeolite. The number of aluminium atoms in the unit cell may be calculated using the following correlation [21]:

$$N_{\mathrm{Al}} = 112.4(A_0 - 24.233)$$

where A_0 stands for the unit cell parameter, measured by X-ray diffraction (XRD).

Also, calcination generates mesoporosity, as shown in Fig. 7, probably because part of the framework is destroyed as dealumination proceeds.

Extra-framework Al species are a matter of controversy. In principle, they may:

- neutralize acidic sites according to the following reaction;

$$\mathrm{Al(OH)_3 + H^+[Z]^- \rightarrow Al(OH)_2{}^+[Z]^- + H_2O}$$

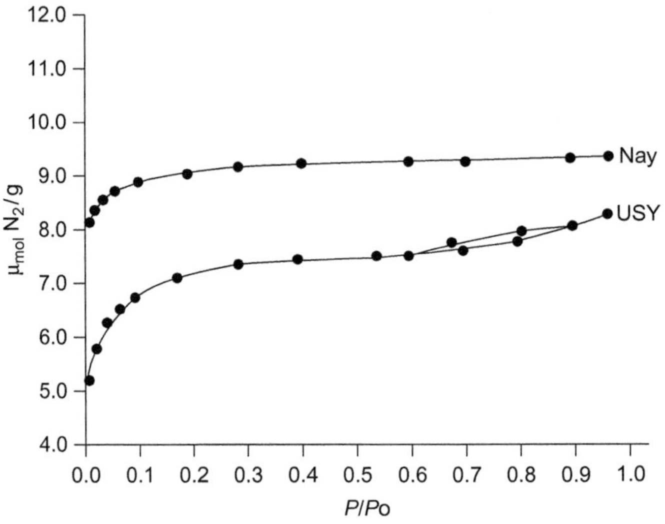

Fig. 7
Nitrogen adsorption isotherms. NaY—microporous, USY—mesoporous.

- undergo condensation (nAl(OH)$_3$ → Boehmite), blocking pores;
- migrate to the surface.

Results depicted in Table 3 are rather enlightening. Si/Al ratios have been measured using different techniques, namely fluorescence X-ray analysis ((FRX) which gives the overall composition of the zeolite), NMR (which gives the framework composition), and X-ray photoelectron spectroscopy ((XPS) which gives the surface composition).

Steam calcination increases the framework Si/Al ratio, since the aforementioned dealumination takes place. However, surface Si/Al drops considerably, indicating that the aluminium species have migrated to the surface. Acid treatments remove EFAL species up to a certain level, with such removal more pronounced on the surface.

Several factors affect the characteristics of the zeolite after calcination. Among them, the following deserve attention:

- Si/Al ratio of the parent zeolite;
- Na content of the exchanged zeolite;

Table 3: Si/Al ratios for zeolites after distinct types of treatment

Zeolite	Treatment	Si/Al Total	Si/Al NMR	Si/Al XPS
A	Steam calcination	2.3	8.2	0.7
B	A+acid treatment	6.0	8.6	8.1
C	B+acid treatment	7.5	8.6	13.2

- Steam partial pressure during calcination;
- Temperature of calcination;
- Calciner geometry (deep bed versus shallow bed).

2.3 The Concept of Accessibility

Recent literature has pointed out the importance of accessibility in cracking catalysts [22–28]. Accessibility is a parameter related to the mass-transfer characteristics of an FCC catalyst, which may be a selectivity determining step. Indeed, the geometric disposition of the components in the final catalyst does affect the selectivity in the cracking process. In fact, reducing diffusional effects improves the gasoline selectivity.

One of the first quantitative attempts to measure accessibility was the Albemarle Accessibility Index (AAI) [23–26], which measures the dynamic diffusion of high-molecular-weight molecules into an FCC catalyst particle. Among the characteristics that might have an effect on the accessibility of zeolites, the crystallite size, which affects the external surface area directly, can be distinguished. Thus, papers concerning the importance of small-crystallite zeolites on the cracking of bulky molecules are being revisited [27–29]. It has been suggested, for instance, that the primary cracking of gas oil molecules takes place on the external surface of zeolites [30]. Based on such assumption, it has been proposed that smaller-crystallite zeolites should be synthesized to increase the catalytic activity [31]. In fact, smaller-crystallite zeolites were more active but less stable. Following the same line, different crystallite zeolites have been incorporated in a matrix, observing that those containing the smaller-crystallite zeolites were indeed more active in the cracking of gas oil as well as more selective to gasoline and light-cycle oil [32]. Trying to overcome the problems of stability of the smaller crystallite, smaller-crystallite zeolites with a higher Si/A1 ratio have been synthesized, thereby increasing the advantages of such materials in the cracking of heavy gas oil [33]. With increased stability by ion exchanging the zeolites with high rare-earth contents a similar effect has been obtained.

The role of the zeolite crystallite size has been clearly demonstrated when different model molecules have been used (*n*-heptane and 1,3,5-tri-isopropylbenzene) (Fig. 8).

The cracking of 1,3,5-triisopropylbenzene [34] seems to be a convenient model reaction to evaluate the contribution of zeolitic external surface area in the cracking of large molecules such as asphaltenes. Other smaller molecules, albeit studied exhaustively as *n*-heptane, are not able to distinguish differences in activities presented by zeolites with approximately the same bulk and external compositions but differing in the crystallite size and external surface area. The cracking of 1,3,5-triisopropylbenzene carried out over two zeolites of 63- and 193-nm average crystallite sizes revealed that much higher rates of the reaction are obtained with the smaller-crystallite sample (Fig. 9).

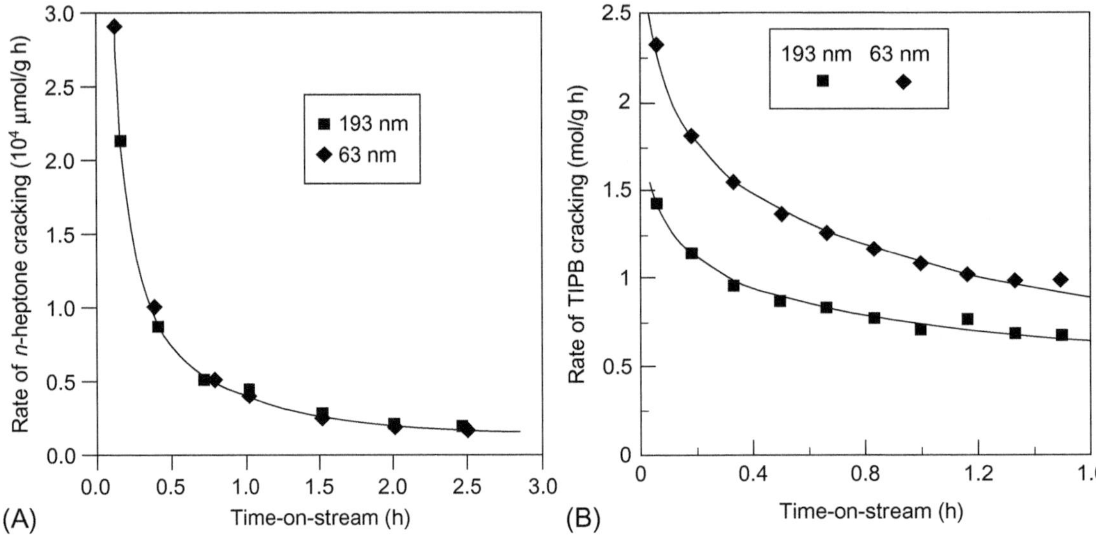

Fig.8

Influence of the crystallite size of RE-containing zeolites on the cracking of *n*-heptane (A) and 1,3,5-triisopropylbenzene (B).

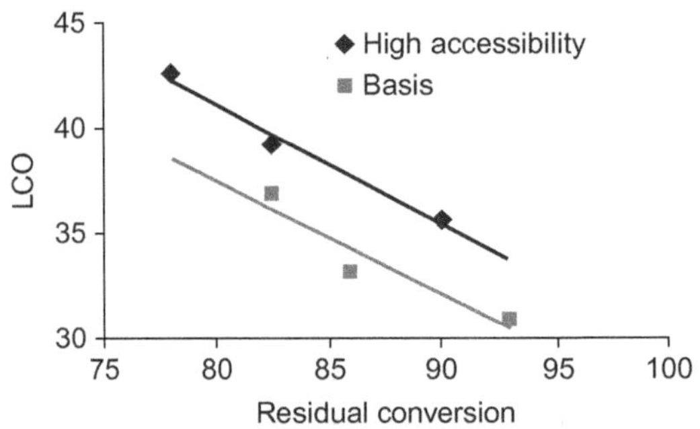

Fig. 9

The importance of high accessibility on the formation of LCO in the cracking of gas oil.

2.4 Zeolites as Functional Ingredients

The demand for light olefins such as ethylene and propylene has increased a lot in the last decade. Steam cracking of hydrocarbons has been the major source of light olefins for more than half a century. Nevertheless, recent studies have reported that ethylene and propylene can also be commercially produced through the cracking of hydrocarbons over modified ZSM-5 zeolites [35]. In that case, ZSM-5 would behave as a functional ingredient in the FCC process, or rather, as an ingredient added with a specific function. As a functional ingredient, ZSM-5 could be embedded in the final catalyst or used as a separate particle to be blended with the traditional catalyst.

ZSM-5 is a 10-membered sieve containing channel openings of 5.1–5.6 Å, which are smaller than the 7–8 Å cage openings for a Y zeolite. Hence, access of molecules larger than monomethylaliphatics into ZSM-5 is restricted, and reactions of molecules with critical diameters greater than 6 Å are severely diffusion limited. Furthermore, ZSM-5 contains strong Broensted acid sites necessary for cracking hydrocarbons, being practically unable to yield coke as easily as a Y-zeolite.

For that reason, ZSM-5 has been used as an FCC additive for boosting the gasoline octane number. Interestingly, the use of ZSM-5 also increases the C3 and C4 LPG selectivity with a concurrent decrease in the gasoline yield. The LPG that is yielded consists predominantly of olefins, whereas there is essentially no change in the dry gas (C2−) or coke selectivity. There is, however, a significant change in the gasoline composition, since:

- C7+ straight chain and branched olefins and paraffins decrease,
- C5 hydrocarbons increase and the average molecular weight of gasoline decreases,
- the concentration of aromatics and naphthenes increases.

Bear in mind that the rate of olefin cracking is more than two orders of magnitude higher than paraffin cracking, which considerably reduces paraffin cracking when such reactive olefins are present. Also, one may consider that paraffin may be yielded via olefin hydrogenation. In the presence of ZSM-5, primary olefins are cracked, hence the concentration of reactant olefins needed to produce paraffins is reduced. This mechanism implies a reduction of low-octane paraffins in gasoline as a direct result of olefin cracking [36].

Both the decrease in the gasoline molecular weight and the increase in C5 hydrocarbons are important reasons for the increase in the gasoline octane number. Also, as the total volume of gasoline is reduced owing to the decrease in C7+ aliphatics, the concentration of aromatics and naphthenes in the gasoline is higher, which helps to boost the gasoline octane number.

Some recent studies [37–41] report that uniformly nanosized ZSM-5 zeolites with sizes less than 100 nm have shown better catalytic and adsorptive properties than micron-sized ones because of the efficient mass transport of guest molecules.

In terms of chemical modifications, treatment of ZSM-5 with phosphorus compounds [42,43] seems to be an interesting route to enhance the selectivity to light olefins in cracking reactions. It has been demonstrated that after phosphorus treatment, the strong acid sites of the original zeolite are replaced by an increased number of weaker acid sites, whose concentration increases after steam treatment. Finally, the combined treatment phosphorus/REs [44] results in an improvement in both stability and activity. Apparently, REs reduce aromatics formation on the external surface area of the zeolite, whereas phosphorus reduces the loss in activity caused by dealumination.

3 Conclusions

FCC remains one of the most important chemical processes in the world, having existed for many years. Among the several challenges that the refiner must face are the following:

- Increasing stringent environmental regulation;
- Growing demand for cleaner fuels;
- Globalization;
- Increase in the production of derivatives from declining quality oil;
- Uncertainty about the consumer's choice;
- Growing pressure of several segments of the society aiming at the reduction of greenhouse gases;
- Maintenance of profitability.

The catalyst is certainly an outstanding tool to help fulfill all these requirements. The FCC catalyst is a multicomponent product, comprising a zeolite, matrices, a binder, and functional ingredients. The zeolite is, undoubtedly, the most important component and the modifications thereof will fine-tune the products slate of the FCC process. Controlled steam calcination, rational use of REs, increase in the Si/Al ratio, and increased accessibility are some of the factors that will have to be carefully examined to improve the catalyst performance.

Finally, zeolites may be used as functional ingredients. ZSM-5-based additives convert linear or near-linear gasoline range olefins into LPG olefins in combination with increasing the gasoline octane number. However, to maximize ZSM-5 properties, modifications with phosphorus compounds, along with convenient calcination, must be carried out.

References

[1] Sung HC, Brown GG, White RR. Thermal cracking of petroleum. Ind Eng Chem 1945;37(12):1153–61.

[2] American Chemical Society National Historic Chemical Landmarks. The houdry process for catalytic cracking. http://portal.acs.org/portal/PublicWebSite/education/whatischemistry/landmarks/houdry/index.htm [accessed 27.04.13].

[3] Jones D, Pujadó P. Handbook of petroleum processing. New York: Springer Science; 2008. 239.

[4] Avidan A, Edwards M, Owen H. Innovative improvements highlight FCC's pas and future. Oil Gas J 1990;88(2).

[5] Scherzer J. Octane-enhancing zeolitic FCC catalysts: scientific and technical aspects. Boca Raton, FL: CRC Press; 1990, ISBN: 0-8247-8399-9.

[6] Sedran U. Laboratory testing of FCC catalysts and hydrogen-transfer properties evaluation. Catal Rev Sci Eng 1994;36:405–31.

[7] Venuto PB, Habib ET. Fluid catalytic cracking with zeolite catalysts. New York: Marcel Dekker; 1979.

[8] Al-Khattaf S. The influence of alumina on the performance of FCC catalysts during hydrotreated VGO catalytic cracking. Energy Fuel 2003;17:62–8.

[9] Smit B, Maesen T. Towards a molecular understanding of shape selectivity. Nature 2008;451:671–8.

[10] Csicsery S. Catalysis by shape selective zeolites-science and technology. J Pure Appl Chem 1986;58(6):841–56.

[11] Jae J, Tompsett GA, Foster AJ, Hammond KD, Auerbach SM, Lobo RF, et al. Investigation into the shape selectivity of zeolite catalysts for biomass conversion. J Catal 2011;279(2):257–68.
[12] Sousa-Aguiar EF, Trigueiro FE, Zotin FMZ. The role of rare earth elements in zeolites and cracking catalysts. Catal Today 2013;218–219:115–22.
[13] Nery JG, Mascarenhas YP, Bonagamba TJ, Mello NC, Souza-Aguiar EF. Location of cerium and lanthanum cations in CeNaY and LaNaY after calcination. Zeolites 1997;18:44–9.
[14] Trigueiro FE, Monteiro DFJ, Zotin FMZ, Sousa-Aguiar EF. Thermal stability of Y zeolites containing different rare earth cations. J Alloys Compd 2002;344(1-2):337–41.
[15] Falabella Sousa-Aguiar E, Camorim VLD, Zotin FMZ, dos Santos RLC. A Fourier transform infrared spectroscopy study of La-, Nd-, Sm-, Gd- and Dy-containing Y zeolites. Microporous Mesoporous Mater 1998;25:25–34.
[16] Lemos F, Ribeiro FR, Kern M, Giannetto G, Guisnet M. Influence of lanthanum content of LaHY catalysts on their physico-chemical and catalytic properties: Comparison with CeHY catalysts. Appl Catal 1988;39:227–37.
[17] Lemos F, Ribeiro FR, Kern M, Giannetto G, Guisnet M. Influence of the cerium content of CeHY catalysts on their physicochemical and catalytic properties. Appl Catal 1987;29(1):43–54.
[18] Mendes A, Caeiro J, Ribeiro FR. Influence of lanthanum exchange on the catalytic properties of HY zeolite in cyclohexene transformation. React Kinet Catal Lett 1989;38(1):175–80.
[19] Bradsher K. New York Times, May 2nd, 2011.
[20] Abrams S. Oil Gas J 2011;2(14):1–2.
[21] Fichtner-Schmittler H, Lohse U, Engelhardt G, Patzelova V. Unit cell constants of zeolites stabilized by dealumination determination of Al content from lattice parameters. Cryst Res Technol 1984;19:K1.
[22] Jonker RJ, O'Connor P, Nicolaas H, Wijngaards J. US 6828153 B2; 2004.
[23] Olthof FP, Smeink R, Coopmans J, O'Connor P. Accessible catalysts for short contact time cracking. In: Occelli ML, O'Connor P, editors. Fluid cracking catalysts. New York: Marcel Dekker; 1998. p. 85.
[24] Fletcher R, Hakuli A, Imhof P. Oil Gas J 2002;2:54–9.
[25] Rainer D, Rautiainen E, Nelissen B, Imhof P, Vadovic C. Simulating iron-induced FCC accessibility losses in lab-scale deactivation. Stud Surf Sci Catal 2004;149:165–76.
[26] O'Connor P, Berends E. US 6930067; 2005.
[27] Falabella Sousa-Aguiar E, Murta-Valle ML, Sobrinho EV, Cardoso D. Cracking of 1,3,5-triisopropylbenzene over deeply dealuminated Y zeolite. Stud Surf Sci Catal 1995;97:417–22.
[28] Al-Khattafa S, Atiasb JA, Jaroschc K, de Lasa H. Diffusion and catalytic cracking of 1,3,5 tri-iso-propyl-benzene in FCC catalysts. Chem Eng Sci 2002;57:4909–20.
[29] Costa AF, Cerqueira HS, Sousa-Aguiar EF, Ludvig MM. Performance of FCC catalysts prepared with sub-micron Y zeolite. Stud Surf Sci Catal 2004;154(Part C):2296–301.
[30] Thomas CL, Barmby DS. The chemistry of catalytic cracking with molecular sieve catalysts. J Catal 1968;12:341–6.
[31] Maselli J, Peters A. Preparation and properties of fluid cracking catalysts for residual oil conversion. Cat Rev Sci Eng 1984;26:525.
[32] Rajagopalan K, Peters A, Edwards G. Influence of zeolite particle size on selectivity during fluid catalytic cracking. Appl Catal 1986;23:69.
[33] Camblor MA, Corma A, Martínez A, Mocholí FA, Pérez Pariente J. Catalytic cracking of gasoil: Benefits in activity and selectivity of small Y zeolite crystallites stabilized by a higher silicon-to-aluminium ratio by synthesis. Appl Catal 1989;55:65–74.
[34] Falabella Sousa-Aguiar E, Murta Valle ML, Silva MP, Silva DF. Influence of external surface area of rare-earth containing Y zeolites on the cracking of 1,3,5-triisopropylbenzene. Zeolites 1995;15:620–3.
[35] Rahim N, Karimzadeh R. Catalytic cracking of hydrocarbons over modified ZSM-5 zeolites to produce light olefins: A review. Appl Catal A Gen 2011;398(1–2):1–17.
[36] Zhao X, Roberie TG. ZSM-5 additive in fluid catalytic cracking. 1. Effect of additive level and temperature on light olefins and gasoline olefins. Ind Eng Chem Res 1999;38(10):3847–53.

[37] Tosheva L, Valtchev VP. Nanozeolites: Synthesis, crystallization mechanism, and applications. Chem Mater 2005;17:2494.
[38] Persson AE, Schoeman BJ, Sterte J, Otterstedt JE. The synthesis of discrete colloidal particles of TPA-silicalite-1. Zeolites 1994;14:557.
[39] Tago T, Nishi M, Kouno Y, Masuda T. New method for preparing mono-dispersed nano-crystalline silicalite via hydrothermal synthesis in water/surfactant/oil solution. Chem Lett 2004;33:1040.
[40] Madsen C, Jacobsen CJH. Nanosized zeolite crystals—convenient control of crystal size distribution by confined space synthesis. Chem Commun 1999;673.
[41] Watanabe R, Yokoi T, Tatsumi T. Extension of size of monodisperse silica nanospheres and their well-ordered assembly. J Colloid Interface Sci 2011;360(1):1–7.
[42] Blasco T, Corma A, Martínez-Triguero J. Hydrothermal stabilization of ZSM-5 catalytic-cracking additives by phosphorus addition. J Catal 2006;237(2):267–77.
[43] Lee J, Hong UG, Hwang S, Min Hye Y, In Kyu S. Catalytic cracking of C5 raffinate to light olefins over lanthanum-containing phosphorous-modified porous ZSM-5: Effect of lanthanum content. Fuel Process Technol 2013;109:189–95.
[44] Wang Q. EP0801027 A1, 1997.

CHAPTER 7

Supported Zeolite and MOF Molecular Sieve Membranes: Preparation, Characterization, Application

J. Caro
Leibniz University Hannover, Hannover, Germany

1 Fundamentals: Definitions and Phrases in Membrane Science and Technology

Table 1 summarizes terms and phrases used in membrane science and technology. To have sufficiently high fluxes, the molecular sieve separation layer (zeolite, metal-organic framework) has to be in the μm-scale. Because such thin layers are not self-supporting for mechanical reasons, inorganic membranes are usually prepared as supported membranes (Fig. 1).

2 Gas Separation by Membranes: Competition of Organic and Inorganic Membranes

Gas separation by membrane permeation can be very energy-efficient and it usually requires lower investment costs than the competing technologies' pressure swing adsorption and cryogenic distillation. Correspondingly, permeation technologies based on organic polymer membranes have been developed for nitrogen production, for example. And hydrogen recovery has been developed based on organic polymer membranes.

The separation factor of these organic polymer membranes is typically located in a moderate range, of around 5 and 10, but rarely higher than 20. As a rule of thumb and proven by recent publications, the membrane selectivity can be approximated as the product of the adsorption selectivity and diffusion selectivity [2]. This chapter provides a wealth of information on diffusion inside micro- and mesoporous structures using concepts and ideas that originate from Maxwell and Stefan. A molecular-level understanding of diffusion in a variety of materials such as zeolites, MOFs, covalent organic frameworks (COFs), carbon nanotubes, and cylindrical silica pores is provided with the aid of extensive data sets of molecular

Zeolites and Zeolite-like Materials. http://dx.doi.org/10.1016/B978-0-444-63506-8.00008-2

Table 1: Terminology for membranes and membrane processes following IUPAC definitions [1]

Term	Definition, Unit
Flux of component N_i, amount of permeate i (number of moles, volume, or mass) passing the membrane per time and surface area normal to the thickness direction	$mol\,m^{-2}\,h^{-1}$ or $m^3\,(STP)\,m^{-2}\,h^{-1}$
Permeance = pressure-normalized flux = permeate per time, area and pressure difference	$mol\,m^{-2}\,h^{-1}\,Pa^{-1}$ or $m^3\,(STP)\,m^{-2}\,h^{-1}\,bar^{-1}$
Permeability[a] = thickness-normalized permeance = permeability multiplied by membrane thickness, divided by area, time, pressure difference	$mol\,m\,m^{-2}\,h^{-1}\,Pa^{-1}$ or $m^3\,(STP)\,m^{-2}\,bar^{-1}$
Permeate, retentate	Permeate = stream that goes through the membrane and leaves the membrane module. Retentate = stream that has been depleted of permeate. Retentate is the part of the feed that leaves the membrane module without passing through the membrane
Ideal selectivity[b] $\alpha_{i,j}^{ideal}$ or permselectivity	Calculated as the ratio of the single component fluxes $\alpha_{i,j}^{ideal} = N_i / N_j$
Mixture separation factor[c] $\alpha_{i,j}^{real}$	Calculated from molar gas composition in permeate and retentate $\alpha_{i,j}^{real} = (y_i : y_j)/(x_i : x_j)$ with y and x as mole fractions i and j the permeate and retentate

[a]Assuming a linear Henry-like absorption isotherm, for polymer membranes the permeability P_i ($mol\,m\,m^{-2}\,h^{-1}\,Pa^{-1}$) can be expressed as product of solubility S ($mol\,Pa^{-1}\,m^{-1}$) and diffusivity D ($m^2\,h^{-1}$). For pore membranes, however, this simple relation is not valid because the limited pore volume the amount adsorbed does not increase linearly with the pressure (Henry-like behavior only for permanent gases at relative high temperatures) and the adsorption isotherm is usually curved (Langmuir-like).

[b]Instead of the ratio of fluxes N_i/N_j, the ideal selectivity $\alpha_{i,j}^{ideal}$ can be calculated as well as the permeance or permeability ratio of the components i and j, respectively.

[c]If only a negligible amount of the feed passes the membrane, the feed composition is almost the retentate composition. Therefore, also the ratio of the mole fractions permeate and feed can be used.

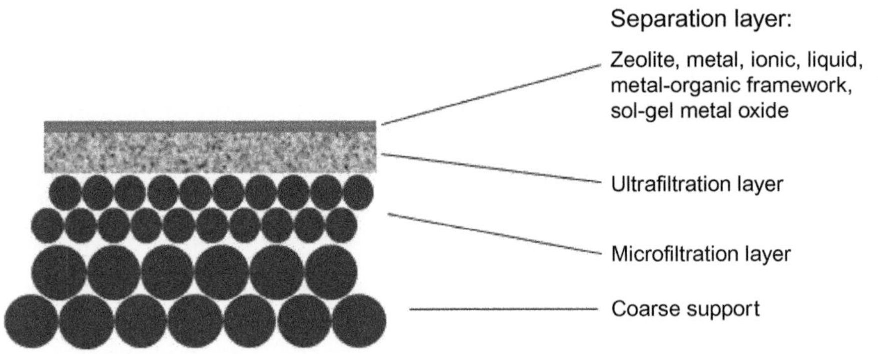

Fig. 1

Principle of a supported membrane: the μm-thick separation layer is deposited on a macroporous ceramic or metallic support. To reduce the pressure drop across the support, that is, to minimize the flow resistance, asymmetric supports are often used.

dynamic simulations. For a binary mixture, these selectivities are the ratio of the amount adsorbed (if normalized on the volume and pressure, the amount is called solubility) and the diffusion coefficients (diffusivity) of the two components in the polymer of which the membrane is made from. Unfortunately, often single components instead of mixed-gas adsorption and diffusion data are used.

The flexible organic polymers can easily be fabricated as hollow fibers or spiral wound membranes; this feature is an important advantage of organic over inorganic membranes. However, (i) their selectivities are rather moderate, (ii) they can be damaged or even destroyed by solvents, and (iii) they are not thermally stable at temperatures necessary for regeneration by calcinations (which simply results in combustion of the organic residues). Inorganic membranes like metals, carbons, or oxides do not show these disadvantages. However, they suffer from brittleness and thermal expansion problems. Molecular sieve membranes show pores in the region of a few Å and allow the separation of small molecules, such as H_2, He, H_2O, NH_3. However, several types of molecular sieve membranes have no sharp cut off because of a pore size distribution, such as the X-ray amorphous metal oxide (silica) or carbon membranes. Further, these noncrystalline membrane materials suffer from instability at high temperatures and in oxygen and/or steam atmosphere. As crystalline materials, zeolites and MOFs are more stable against degradation than the X-ray amorphous materials, and they show a uniform pore size without any distribution, which is determined by their crystallographic structure.

Despite their advantages, inorganic membranes have not so far found major industrial importance in gas separation. However, about 20 years ago, there was much enthusiasm about the development of supported thin zeolite layers as membranes for gas separation. We have to state that we underestimated the difficulties in the scale-up, and we have to realize that there is so far no major industrial use of inorganic membranes in gas separation. Fig. 2 shows a so-called Robson plot. This plot was proposed by Robson to demonstrate the performance of a membrane [6]. On the ordinate is plotted the selectivity of a gas pair, on the abscissa the permeability of the component that is separated through the membrane. Fig. 2 shows two highlights of recent zeolite membrane research: supported SAPO-34 and DD3R zeolite membranes. However, despite their performance in the CO_2/CH_4 separation (earth gas sweetening to fulfill pipeline-grade conditions) under lab conditions is far above the organic polymers membranes, no scale-up into the industrial scale is planned.

An exception is the recent successful application of supported zeolite LTA membranes in the dewatering of bioethanol by steam permeation (Fig. 3).

With the new material class of MOFs, a new nanoporous material is available and the question arises as to whether MOFs can solve the problems connected with the development of zeolite membranes. Although at first glance, zeolites and MOFs are very similar, there are significant differences between zeolites and MOFs which can either facilitate or

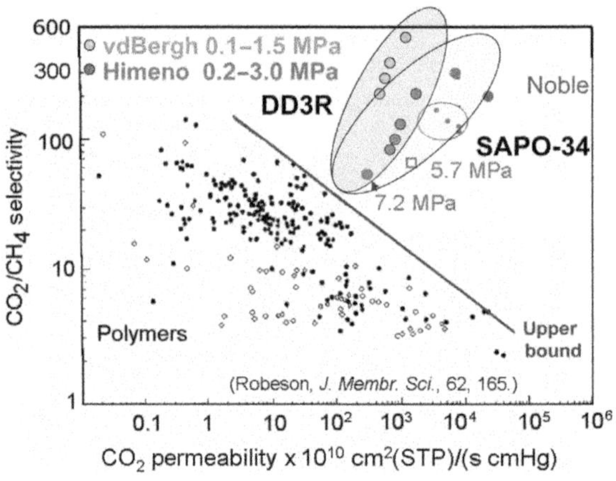

Fig. 2
Two very promising zeolite membranes for a technical application are SAPO-34 and DD3R. A supported SAPO-34 membrane has been developed using the seeding technology which can separate a CO_2/CH_4 mixture with a selectivity of 115 and a reasonable CO_2 permeance of $4\times10^{-7}\,mol\,m^{-2}s^{-1}Pa^{-1}$ at the feed pressure of 70 bar [3]. Also a supported DD3R zeolite membrane can separate CO_2 from CH_4 with a selectivity of over 4000 at 225 K and 1 bar of the equimolar feed [4,5]. The Robeson upper bound showing the limit of polymer membrane performance is from 1991. *From Van den Bergh J, Zhu W, Kapteijn F, Moulijn JA, Yajima K, Nakayama K, et al. Separation of CO_2 and CH_4 by a DDR membrane. Res Chem Intermediates 2008;34:467–74, with permission.*

Fig. 3
Four-channel alumina tubes, inside coated with an LTA zeolite layer in the Na state (4A) as used in Lithuania and four other plants recently built by GFT Membrane Systems GmbH Homburg/Saar and equipped with LTA membranes made by Fraunhofer IKTS Dresden/Hermsdorf, Germany (former HITK/inocermic). Ethanol in feed: 85 wt.%, ethanol in retentate: 99.8 wt.%. [7].

complicate the preparation of MOF membranes. In the following discussion, we outline the progress, challenges, and problems in the development of zeolite and MOF membranes will be outlined. A big step forward is the knowledge-based membrane development. Using the concept derived for polymer membranes—Membrane selectivity = adsorption selectivity × diffusion selectivity—the permeation behavior of zeolite and metal-organic framework membranes can be predicted (Fig. 4).

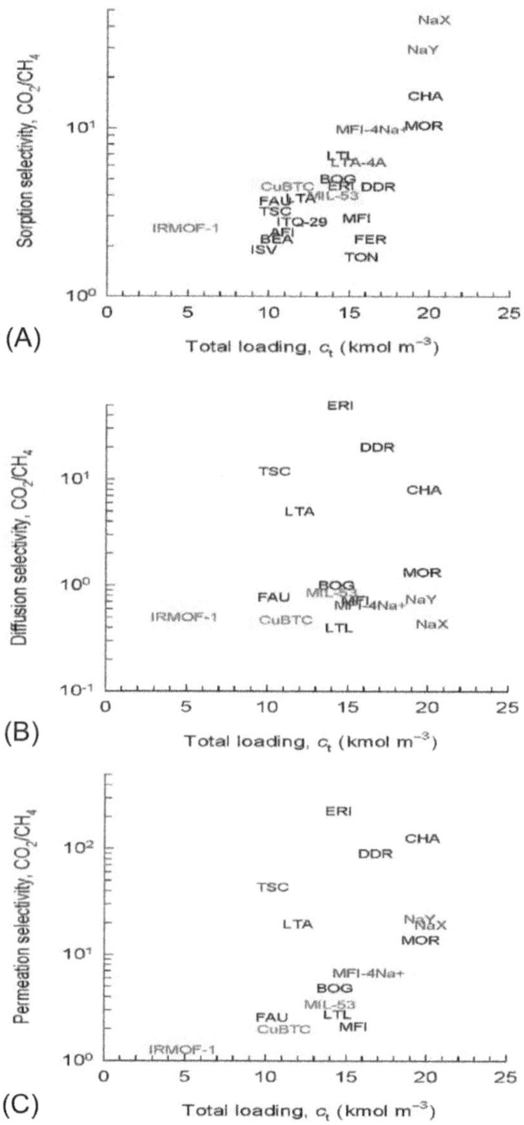

Fig. 4

Selectivities for separation of CO_2/CH_4 mixtures using a variety of microporous structures. (A) Adsorption selectivity (from GCMC simulations for binary mixtures taking $f_1 = f_2 = 0.5$ MPa at the upstream face), (B) diffusion selectivity ($=D_{1,self}/D_{2,self}$, evaluated at the total mixture loading at the upstream face), and (C) resulting permeations selectivity (=product of the adsorption and diffusion selectivities) [2]. The graphs were constructed based on the data from [8].

3 Zeolite and MOF Molecular Sieve Membranes: Similarities and Differences

As mentioned earlier, there is some progress in the development of zeolite membranes. However, especially the long-term stability, T-cycling, regeneration and the difficult housing have prevented a major industrial use of zeolite membranes. However, there is a healthy optimism that MOFs as a new star at the sky of nanoporous materials can solve these problems [9]. As hybrid organic-inorganic material, MOFs consist of metal cations or cationic oxide clusters that are linked by organic molecules thus forming a crystalline network. MOFs, often also termed as coordination polymers. As a coordination network, the MOFs are mechanically less stiff and brittle as zeolites [10]. As a consequence, MOFs frequently show a high degree of framework flexibility, which results in such effects as "breathing," "gate opening," and "linker dynamics" [11]. The critical review deals with a strange behavior of crystalline matter: its reversible swelling with large amplitude. Its consequences concern specific applications related to gas storage and separation by adsorption and permeation, sensors, catalysis. The examples given are classified according to the dimensionality of the network, and they present the general requirements and the structural rules which govern the existence of the phenomenon "breathing." The organic linker molecules have a high potential for specific functionalization allowing (i) the tuning of adsorptive interaction and (ii) the control of the pore size and accessible pore volume via the size of the functional groups. However, on the other hand, some MOFs show a low energy barrier for hydrolysis. As an example, MOF-5, which is one of the most known MOF structures, consists of Zn_4O^{6+} clusters interconnected by terephthalate anions (1,4-benzenedicarboxylate, BDC) which gives the formula $Zn_4O(BDC)_3$. MOF-5 is crystallized in organic solvents. However, if in contact with water or steam, MOF-5 decomposes into zinc hydroxide and terephthalic acid (Fig. 5).

The zeolitic imidazolate frameworks (ZIFs) form a subfamily of MOFs. In ZIFs, metal cations (eg, Zn^{2+}, Co^{2+}, and Cu^{2+}) are tetrahedrally coordinated by imidazolate linkers frequently resulting in a relative thermally and hydrothermally stable network showing zeolite topologies. A well-known ZIF structure is ZIF-8, in which 2-methylimidazolate anions are linked to Zn^{2+} cations resulting in an expanded SOD structure. Another well-known MOF stricture is ZIF-7 where the Zn^{2+} ions are interconnected by benzimidazolate anions (Fig. 6).

Olson and coworker showed for ZIF-8, that at 30°C propene adsorbs much faster than propene (Fig. 7). The pore size diameter of ZIF-8 adsorbents could be fine-tuned by Olson using chlorination, bromination, and methylation. Since propene is about 0.2–0.3 Å smaller in its critical diameter than propane, the ratio of the diffusivities of D(propene): D(propane) = 125 [13]. Small changes, big effects, Ref. [13] shows the surprising results that modifying the pore size of the MOF ZIF-8 by linker substitution, the sorption uptake rates of propane and propene can be controlled dramatically. Propene is only slightly (0.2 … 0.3 Å) smaller than propane, but its diffusivity can be 100 times higher.

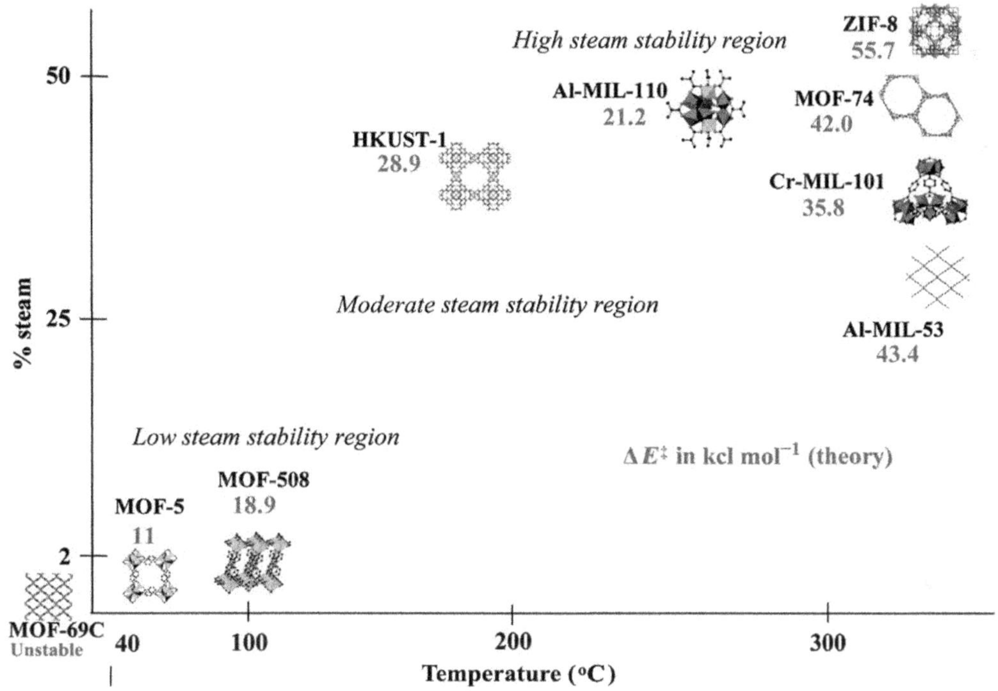

Fig. 5
Steam stability map for different MOFs. The position of the structure for a given MOF represents its maximum structural stability by XRD measurement, while the energy of activation for ligand displacement by a water molecule determined by molecular modeling is represented by the number below the structure code (in kcal mol^{-1}). *Taken from Low JJ, Benin AI, Jakubczak P, Abrahamian JF, Faheem SA, Willis RR. Virtual high throughput screening confirmed experimentally: porous coordination polymer hydration. J Am Chem Soc 2009;131:15834–42, with permission.*

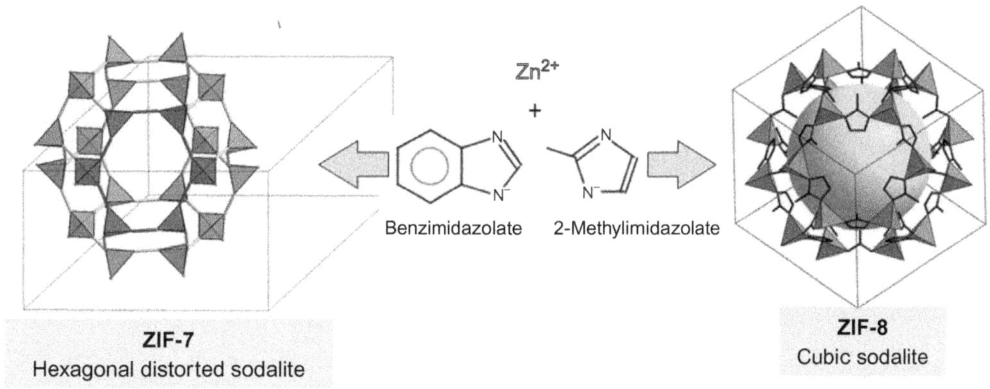

Fig. 6
SOD frameworks of ZIF-7 and ZIF-8.

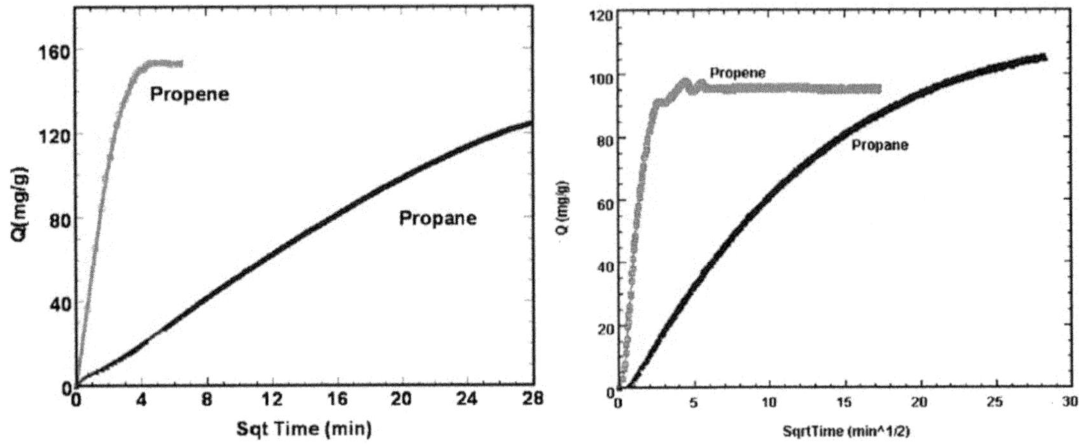

Fig. 7

Propene and propane sorption uptake by a powder sample of ZIF-8 (Zn-methylimidazolate). *Reproduction from Li K, Olson DH, Seidel J, Emge TJ, Gong H, Zeng H, et al. Zeolitic imidazolate frameworks for kinetic separation of propane and propene. J Am Chem Soc 2009;131:10368–9, with permission.*

All the tools from the preparation of zeolite membranes such as seeding, microwave heating, ceramic porous supports, or intergrowth-supporting additives can be applied for the synthesis of MOF membranes. Therefore, in a relative short time the first MOF membranes with molecular sieve properties have been developed. Highlights are the highly permeable ZIF-8 membrane for CO_2/CH_4 separation [14], the copper-net supported HKUST-1 membrane with H_2/N_2 selectivity of about 7 [15], the $Co_3(HCOO)_6$ membrane with a CO_2/CH_4 selectivity of up to 16 [16], a small pore MOF with 0.32 nm pore size for H_2/N_2 separation on alumina support [17], or the supported ZIF-69 membrane with a broad selectivity for different gas mixtures [18]. However, these MOF membranes show still a lower CO_2/CH_4 selectivity than the best zeolite membranes SAPO-34 [19] and DDR [20] membranes with separation factors above 115 and 4000, respectively. In our group we developed ZIF-7 [21], ZIF-8 [22], ZIF-22 [23], and ZIF-90 [24] membranes for the molecular sieving of hydrogen. These ZIF membranes have been operated up to 200°C and they were found to be stable at this temperature even in the presence of up to 3 vol.% steam.

Fig. 8A shows a typical cross section of a supported ZIF-8 membrane with "molecular sieving" properties; Fig. 8B shows the resulting permeation results for single-component gases. In the case of MOF membranes we must be careful with the expression "molecular sieving." Although the size of the pore window of ZIF-8 is estimated to be 0.34 nm from crystallographic data, several reports have shown that the framework structure of ZIF-8 is in fact more flexible rather than static and even large molecules like CH_4 (critical diameter 0.38 nm) can enter the pore network [13,25]. In another paper on chromatography, Lubbers et al. demonstrated that *n*-alkanes (critical diameter 0.43 nm) become easily adsorbed by ZIF-8 [26]. However,

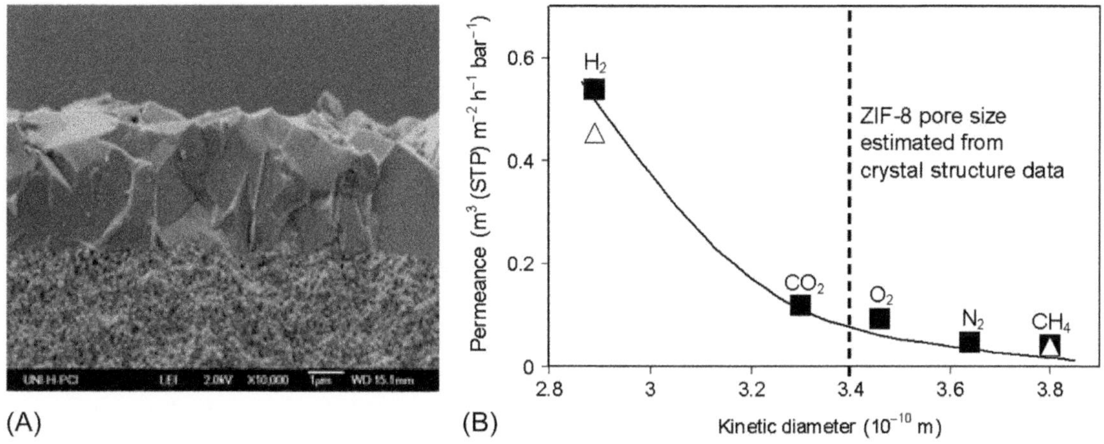

Fig. 8
(A, B) Typical supported ZIF membrane and a typical dependence of the permeance as a function of the kinetic diameter of the permeating molecules [22].

in another chromatographic study it was shown that more bulky branched alkanes (critical diameter > 0.54 nm) are not able to pass the narrow pore windows of ZIF-8, whereas the linear ones are adsorbed [27]. The consideration of framework flexibility is crucial to predicting correct separation figures for MOF membranes. By ignoring framework flexibility, Sholl et al. predicted a H_2/CH_4 separation factor of 10^7 on a ZIF-8 membrane [28], the experimental value measured is about 10–15 [22] showing that the permeation of methane is not negligible. Permeation studies on ZIF-22 [23] and ZIF-90 [24] membranes showed as well that no sharp cut-off exists for hydrocarbons with critical diameters larger than the crystallographic pore size. A more appropriate expression for the terminus "breathing" might be "gate opening." In a recent paper of the Kapteijn's group, it was shown that ethane is adsorbed by ZIF-7 at a lower pressure in comparison to the only slightly lighter and stiffer ethylene [29].

In the first molecular dynamic studies, the framework flexibility is considered in the calculation of diffusive transport [30]. Since the separation behavior of certain MOF membranes cannot be easily predicted by the molecular sieve effect assuming a rigid pore size, experimental and theoretical tools have been developed to predict the gas separation behavior of a pore membrane based on the interplay of mixed-gas diffusion and mixed-gas adsorption.

4 Membrane Preparation by In Situ Hydrothermal Synthesis Without Seeding

The challenge can be described as follows: To get a thin and perfect molecular sieve (zeolite, MOF) layer, nucleation and crystal growth has to start on as many positions on the support as possible. The more nucleation centers on the support surface, the thinner can be the dense zeolite or MOF layer. If there are only a few nucleation centers, only these islands grow to a zeolite/MOF particle. Surface OH groups are considered to facilitate nucleation on γ-alumina, titania, or silica supports. The density of surface OH can be increased by acid treatment. Another problem with the direct membrane crystallization is the mismatch of the electric surface charges (zeta potentials) between support and the precursor molecules of the zeolite or MOF from the solution.

Usually, the oxide-ceramic supports show at the synthesis conditions of zeolite membrane crystallization (pH > 7, T > 100°C) a negative surface charge (zeta potential). However, the aluminosilicate precursor macromolecules in the synthesis batch show a negative zeta potential as well. To facilitate the condensation of the zeolite precursors at the support, a reverse of the charge of the support can be obtained by a treatment of the ceramic support with aminopropyl-triethoxysilane (APTES) before synthesis [31] as shown in Fig. 9. The three ethoxy groups react with surface hydroxyl groups of the support, resulting in an anchor to the support amino group with a positive zeta potential. The reverse in charge from a negative to a positive surface loading can be also obtained by van der Waals adsorption of positively charged macromolecules [32].

As shown in Fig. 9, APTES treatment of the supporting α-alumina ceramics also improves the quality of a zeolite faujasite (FAU) membrane as shown in Fig. 10.

It can be helpful to repeat the APTES treatment several times: First the support is APTES-treated, then the growing zeolite layer has also become APTES-treated. The benefit of the technique is that after the APTES treatment, the surface of the already grown zeolite layer again becomes positively charged which facilitates the docking of the negatively charged precursor macromolecules of the zeolite synthesis solution. Fig. 11 shows as an example that the growth of an LTA-membrane layer has been interrupted, and

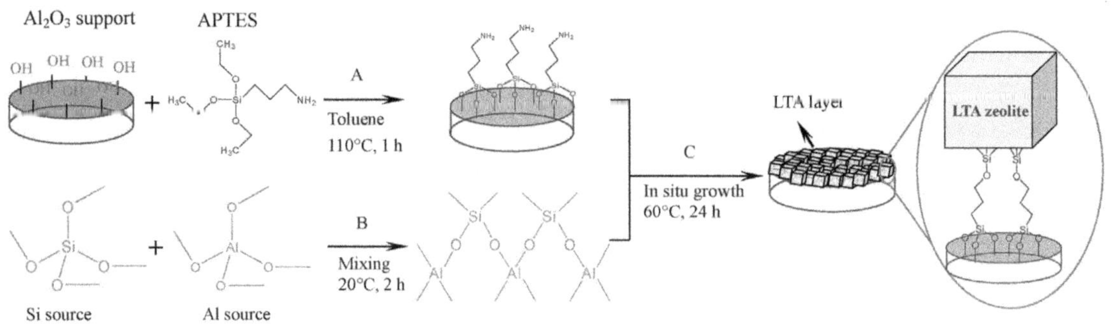

Fig. 9
Pretreatment of the support by 3-aminopropyltriethoxysilane (APTES) as a covalent linker between LTA zeolite membrane and the alumina support. *From Huang A, Liang FY, Steinbach F, Caro J. Preparation and separation properties of LTA membrane by using APTES as covalent linker. J Membr Sci 2010;350:5–9.*

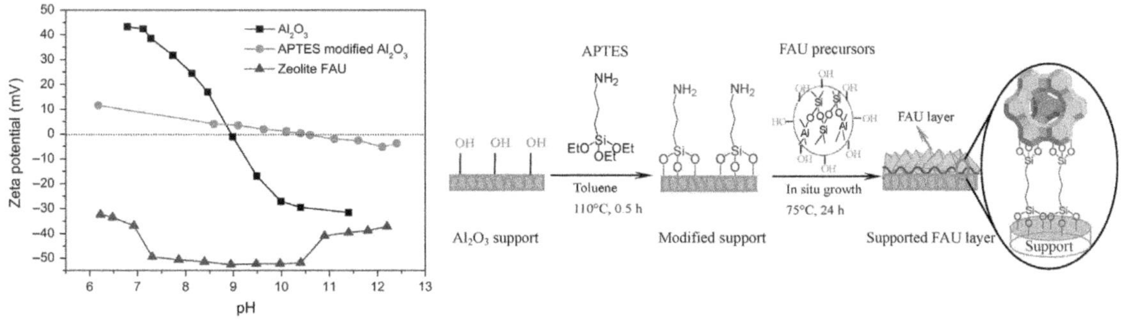

Fig. 10
Treatment of the α-alumina support by APTES as shown on the right, switches the zeta potential of the support from negative to neutral/slightly positive at the pH of zeolite FAU membrane formation. As a result, the negatively charged precursor macromolecules from the FAU zeolite synthesis solution can better adsorb and crystallize to a dense membrane layer [33].

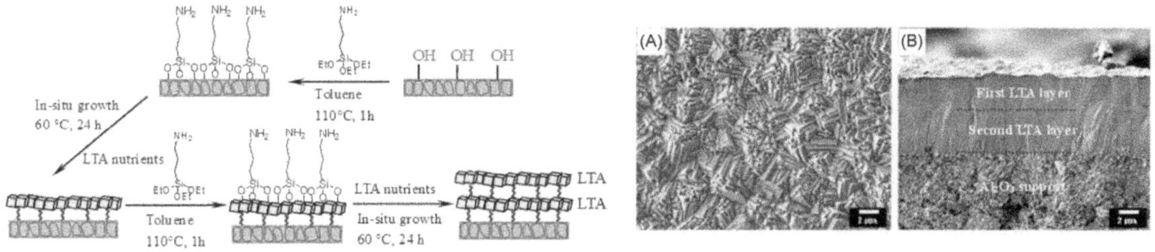

Fig. 11
After support treatment by APTES, the LTA membrane growth has been interrupted, the surface of the LTA layer was APTES-treated and membrane synthesis has been continued [34]. (A) shows the top view, (B) the cross section of a two-layer LTA membrane.

after APTES treatment layer growth has been continued. The multilayer membrane showed a better permeation quality. Nevertheless, if nucleation and crystal growth start on only a few positions of the support, thick layers are necessary to obtain a closed membrane (Fig. 12). The tuning of the surface charge is not only helpful in the direct in situ growth without seeding but also in the attachment of seeds to the support surface (see 5. Membrane Preparation).

5 Membrane Preparation by Secondary Growth Using Seeds

The concept of this widely used technique consists in the decoupling of (i) nucleation and seed formation at high supersaturation, and (ii) growth of the seed crystals to a continuous layer at low supersaturation. Usually, in the crystal growth step a new nucleation can be avoided and only the seeds grow to the molecular sieve layer. Therefore, the layers obtained by secondary growth are less polycrystalline. If the support surface was covered by a homogeneous and dense seed layer, relative thin zeolite and MOF layers can be obtained.

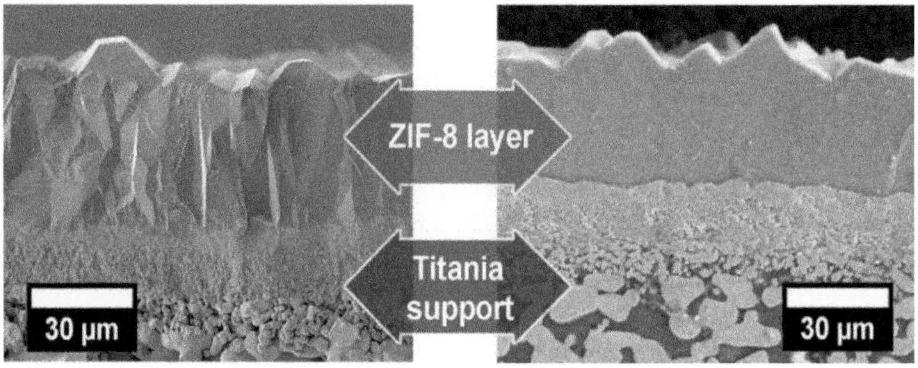

Fig. 12
Cross section of a ZIF-8 membrane on a macroporous titania support. *Left*: Scanning electron microscopy. *Right*: Energy dispersive X-ray spectroscopy (EDXS) in mapping mode, showing the sharp transition between the MOF layer and the support [22].

This seeding or secondary growth technique can be combined with a quick heating up of the autoclaves by microwaves. In the case of microwave heating, a new nucleation at heterogeneous surfaces (support, autoclave walls) is kinetically suppressed by the quick heating up, and only the seeds grow to the layer. As an example, Fig. 13 shows that at about 1 µm thin supported ZIF-7 molecular sieve layer could be obtained by combined seeding and microwave heating.

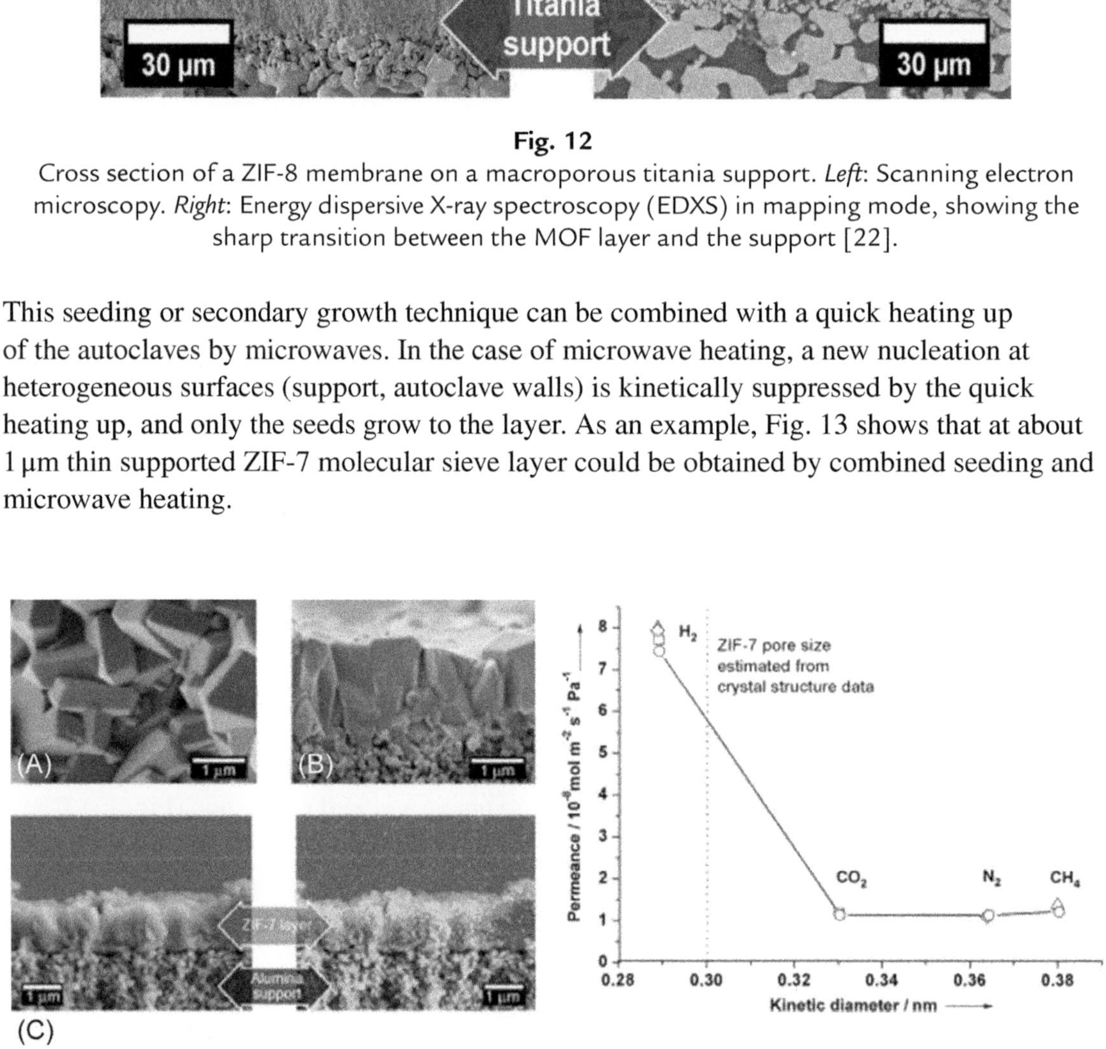

Fig. 13
Left: 1 µm thin MOF ZIF-7 layer on an alumina support prepared by seeding (secondary growth) in combination with microwave heating [35]. *Right*: The results of single gas permeation tests showing that the ZIF-7 membrane is hydrogen-selective. Note that the fluxes of CO_2, N_2, and CH_4 originate from framework flexibility rather than from membrane imperfections 8 (cracks, pin holes, etc.) [35].

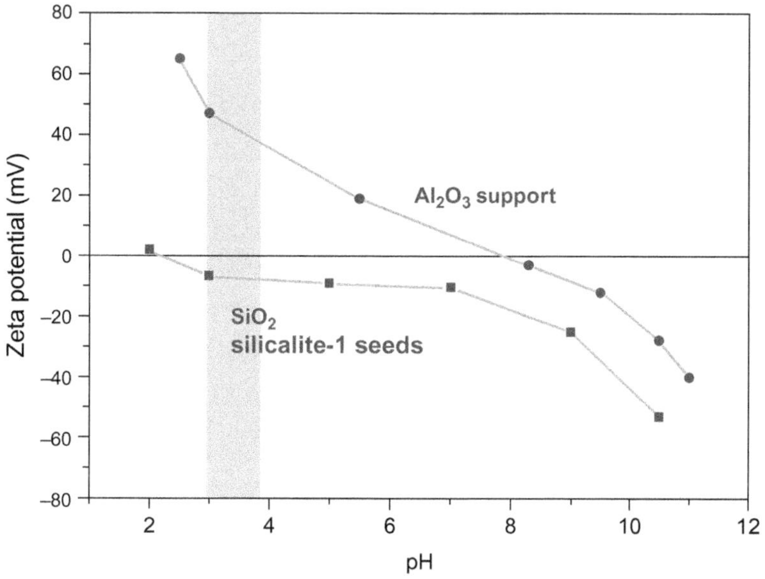

Fig. 14
Typical function of the zeta potential of a α-alumina support and a silicalite-I seed crystal. At a pH between 3 and 4, seeds and support attract electrostatically.

There are different concepts to bring seed crystals to the surface of the support. The usual way is to have different surface charges (zeta potentials) of the support and seed crystallite. By variation of the pH of the dipping solution, a pH is looked for, where support and seed crystallites have opposite charges so that they attract each other electrostatically (Fig. 14). The concept was developed by Mintova et al. [36,37]. If the surface chemistry of support and seed is too similar that no pH with opposite surface charges exist, the physical van-der-Waals adsorption of positively charged macromolecules can result in a charge reversal.

In another concept, a direct covalent bonding between the support and the zeolite film as well as between the crystallites in the growing zeolite film can be established by adding water-soluble bi-dentate additives such as di-isocyanates as proposed by Yoon [38]. On the one hand, the OH groups of the support and the zeolite precursor react and result in a strong attachment of the film to the support. On the other hand, neighboring crystallites in the growing film are linked to each other as well. Fig. 15 shows the attachment of seeds to an oxidic support surface [39].

By the consistent application of these concepts, any zeolite membrane can be prepared.[1] As an example, recently different molecular sieve membranes with the structure LTA could be

[1] As an indication of a "successful" preparation, it is commonly accepted that the membrane shows molecular sieving which is different from the Knudsen transport.

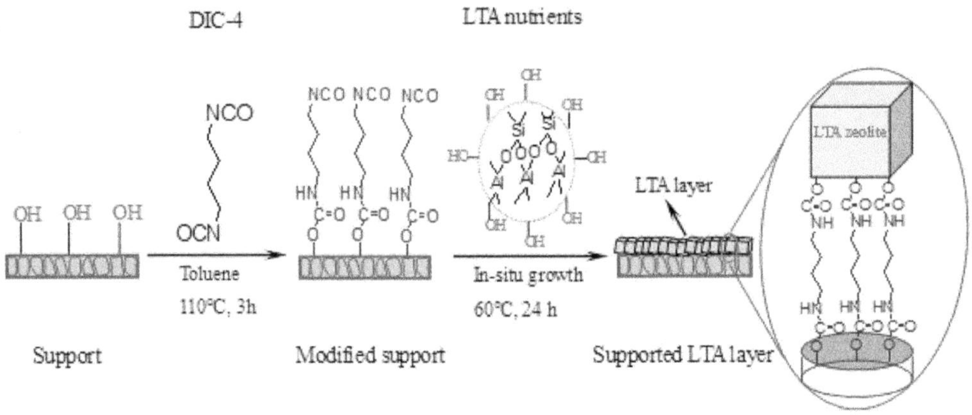

Fig. 15
Covalent attachment of seed crystals to a ceramic surface by di-isocyanate linker between seed crystallite and support [39].

successfully prepared as aluminosilicate [39], and cation-free as (germanium) silicate (ITQ-29) [40] and aluminophosphate [41,42].

6 Knowledge-Based Membrane Development: Prediction of Separation Behavior of Zeolite and MOF Molecular Sieve Membranes

Very recently, two independent approaches have been developed to predict the performance of molecular sieve membranes assuming that the permeation selectivity of a membrane for a gas mixture can be approximated by the product of mixed-gas adsorption selectivity and mixed-gas diffusion selectivity. However, the determination of mixed-gas equilibria and mixed-gas diffusion coefficients is challenging and time consuming.

- Using time- and space-resolved infrared (IR) microscopy (IRM), from the in situ study of sorption uptake of a gas mixture by a big MOF single crystal (size > 50 μm, because of the wave length of IR), the mixed-gas adsorption isotherms and diffusion coefficients can be determined [43] (Fig. 16).
- By straightforward molecular dynamics (MD) and grand canonical Monte Carlo (GCMC) simulations, the adsorption and diffusion selectivities can be calculated as well [2,44], and they were found to coincide very well with the experimental IR data. In view of the increasing vast number of MOFs that have been synthesized, there is a need for a systematic screening of potential candidates as adsorbent in pressure swing adsorption or as membrane material. By molecular simulations, the best MOFs are identified and compared with the classical porous materials. The separation performance is quantified by adsorption selectivity, working capacity, diffusion selectivity, and membrane permeability. The practical utility of the suggested screening is demonstrated by comparison with experimental data.

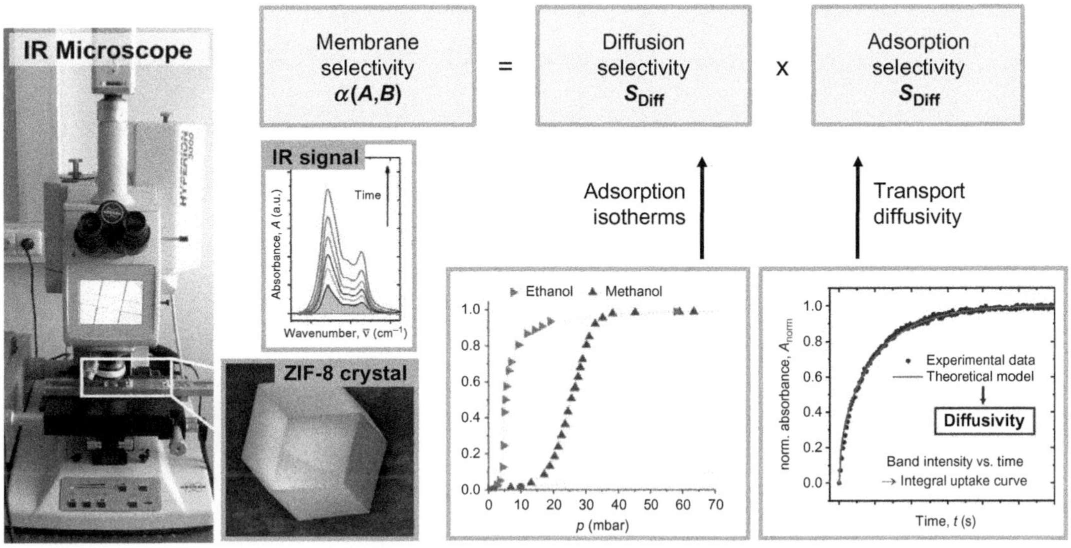

Fig. 16
Principle of IR microscopy for the determination of mixed-gas diffusion coefficients and adsorption equilibria. A big single crystal (here ZIF-8 of 200 μm size) is brought in the IR beam and gas adsorption starts. The IR absorbance is molecule-selective; this means that the simultaneous adsorption of an ethanol/methanol mixture can be followed as a function of time and position. In the example, ethanol adsorbs faster than methanol from a binary mixture. From the amount adsorbed by the crystal as a function of time—and additionally from concentration profiles—diffusion coefficients and mixed-gas adsorption equilibria can be determined. These data are the basis for the evaluation of the membrane performance following "Membrane selectivity = diffusion selectivity × adsorption selectivity."

In three case studies, this approach was successful in predicting the separation selectivity of a ZIF-8 membrane using the adsorption and diffusion selectivities as determined by IRM, MD, and GCMC. In the case of the CO_2/CH_4 permeation through a supported ZIF-8 membrane, a CO_2/CH_4-selectivity of 3.5 was measured at 23°C [45] (Fig. 17). This separation factor can be interpreted as the product of the CO_2/CH_4 adsorption selectivity of 3 multiplied by the CO_2/CH_4 diffusion selectivity of 1.5 giving a predicted membrane selectivity of 4.5, which is in good agreement with the measured value of 3.5. In another case study, the separation of equimolar ethylene/ethane mixtures through a supported ZIF-8 was studied [35]. At room temperature and low feed pressures, an ethylene/ethane separation factor of 3 was found, which can be explained by the product of an ethylene/ethane adsorption selectivity of 0.5[2] multiplied by an ethylene/ethane diffusion selectivity of 4.8

[2] Because ethane is stronger adsorbed than ethylene, the adsorption selectivity ethylene/ethane is <1. Whereas ethylene is stronger adsorbed than ethane on cation-containing polar adsorbents, nonpolar zeolites and MOFs prefer ethane instead of ethylene.

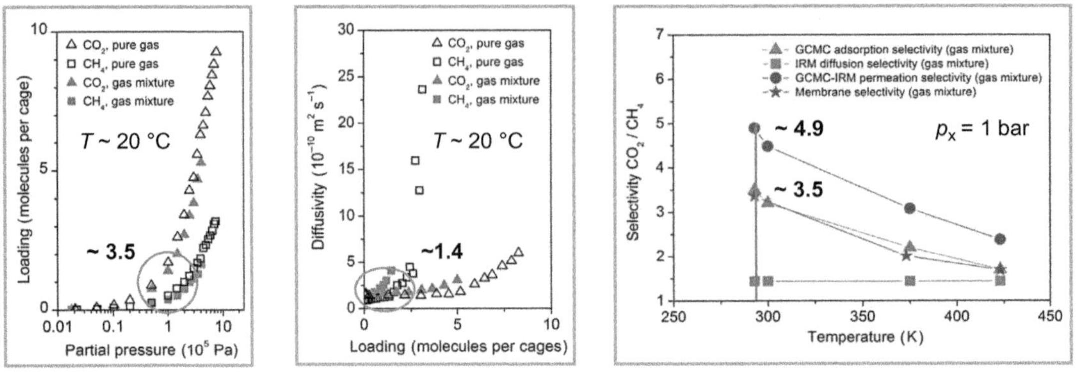

Fig. 17
Example for the prediction of a membrane performance on the basis of IR studies for the mixed-gas system CO_2/CH_4 on a ZIF-8 membrane. IR studies gave an adsorption selectivity of $CO_2/CH_4 = 3.5$ and a diffusion selectivity $CO_2/CH_4 = 1.4$. This results in a predicted membrane selectivity $CO_2/CH_4 = 4.9$. Also Grand Canonical Monte Carlo calculations predict this selectivity. Independent permeation studies showed for the mixture CO_2/CH_4 a selectivity of about 3.5 [45].

giving a prediction of 2.4 in favor of ethylene. One can learn from the latter example, that for improving the ethylene/ethane separation performance, a membrane with an improved ethylene adsorption should be developed. In the third case study, the simple model correctly predicted the H_2/CH_4-selectivity of around 15 for a ZIF-8 membrane by determining the diffusion selectivity (approximated for low loadings from the self-diffusivities) and adsorption data from the GCMC [46].

7 Scale Up: From the Lab to the Plant

As Fig. 2 shows, the first MOF membrane layers were still relative thick. By using seeding technique in combination with microwave heating, the thickness of the MOF ZIF-7 layer could be reduced to the μm-size [35] as it is state of the art for the best supported zeolite membranes like the MFI structure silicalite I [47]. The academia-industry team reports the results of a few-year cooperation. A breakthrough in the preparation of zeolite membranes is reported, due to the ultra-thin defect-free MFI layer the fluxes are one to two orders of magnitude higher than previous literature reports. The C_4, C_6, and xylene isomers have been tested under practice-relevant conditions. A high-permeance and high-selectivity MFI silicalite I membrane has been developed by seeded growth using an oriented seed layer avoiding all events that lead to a loss of the preferred orientation. The polycrystalline films are about 1 μm thick with single crystal grains of the same size.

Nevertheless, crystallization is a technique that results in relative thick films with grain boundaries that might be a source of stress to the polycrystalline layer upon the thermal expansion. Possible alternatives are PVD and CVD techniques, deposition of sol-gel prepared nanoparticles by dip-coating, or combined slip-casting and dip-coating techniques followed by pyrolysis steps giving C or SiC [48] layers of high chemical stability.

The layer-by-layer liquid phase deposition of the building blocks of a MOF on a support by repeated dip-coating potentially allows the automatized preparation of ultra-thin high-flux MOF membranes. Immersing a substrate alternatively in a metal salt solution and in a linker solution produces homogeneous and well-oriented films of about 40–100 nm thickness [49]. MOFs allow the step-by-step formation with alternating dip-coating of a substrate into a metal salt and linker solution. This process can be in situ monitored by using surface Plasmon resonance [50]. Nucleation, formation of secondary building units, and growth rate can be followed in situ. Very recently the first MOF molecular sieve membrane was successfully prepared by the layer-by-layer deposition [51]. After 90 cycles, the ether functionalized pillared-layer MOF of the type $[Cu_2L_2P]$ (L = ligand in layer, P = pillar ligand) showed a CO_2/CH_4 selectivity of about 4.5 (which is an anti-Knudsen separation) (Fig. 18).

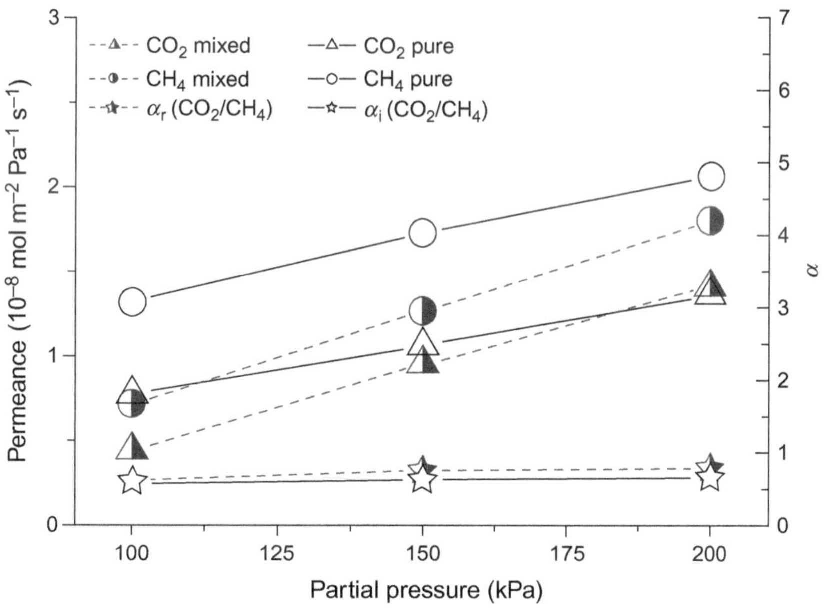

Fig. 18
Permeances of pure and equimolar mixed CO_2 and CH_4 measured for the $[Cu_2(ndc)_2(dabco)]_n$ membrane after 90 dipping cycles at room temperature (T = 298 K) as function of pressures at the feed side (total pressures for pure gases, partial pressures for the gas mixture). The ideal and mixed-gas separation factor α_i and α_r were calculated from the corresponding ratio of the CO_2/CH_4 permeances [51].

8 Prediction for the Next 5 Years: Mixed Matrix Membranes Containing MOFs

Mixed matrix membranes (MMM) consist of a nanoparticle filler like zeolite, metal-organic framework ionic liquid or carbon, in a continuous polymer phase thereby combining the molecular sieving or another property of the filler with the established processability of the polymers in one membrane. The concept of zeolite-based mixed matrix membranes is followed for more than 30 years. However, in most cases these zeolite-based MMMs did not show an improvement of the selectivity because of an imperfect embedding of the zeolite crystals into the polymer matrix as shown schematically in Fig. 19.

Three developments in particular triggered a renaissance of MMM research:

1. Progress in nanosciences helps to develop synthesis and handling of nanoparticles, like the room-temperature synthesis of ZIF-8 [52].
2. Developments in car tire fabrication allow the covalent chemical bonding of SiO_2 linkers to the rubber matrix, this concept can be taken over to MMM fabrication.
3. MOFs match much better to an organic polymer matrix than the inorganic zeolite.

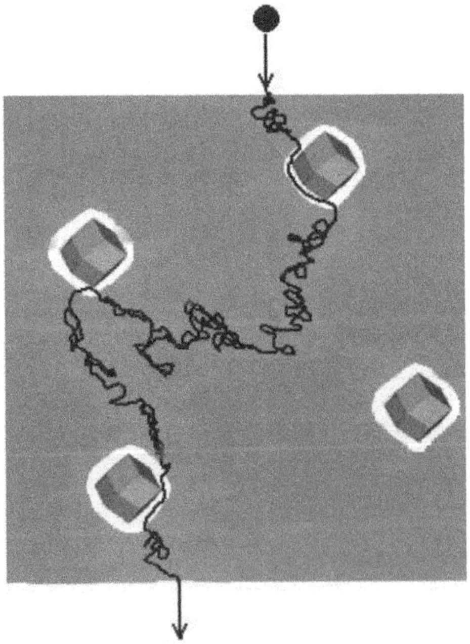

Fig. 19
Schematic drawing of a diffusing gas molecule through an imperfect MMM. The cavities around the ZIF-8 nanoparticles act as nonselective highways for gas transport. Because of this interfacial void, the diffusing molecule does not interact with the embedded filler particle. Thus, the effective thickness of the polymer membrane is reduced without change of selectivity.

For high permeation fluxes, a high ratio of membrane area per permeator volume is required, which can be obtained by hollow fiber or spiral wound membranes. Besides the spiral wound geometry in reverse osmosis, hollow fibers are most often the geometry of organic polymer membranes in gas separation. There were ambitious developments of inorganic hollow fibers [53–55]. However, these inorganic hollow fibers did not fulfill all expectations, because their brittleness causes problems at thermal expansion. It could be a promising concept, therefore, to combine the established spinning technology with a hybrid membrane material that contains molecular sieve nano particles in a basic polymer [56] especially MOFs [57]. However, for zeolite-based mixed matrix membranes, the separation results were not convincing, even if the inorganic zeolite has been surface-functionalized to match better with the organic polymer. We have a completely different situation with MOFs in a polymer matrix where the organic linkers can couple with the polymer matrix and relative high wt.% of MOF in the polymer up to 50% can be obtained [58]. These MOF-based mixed matrix membranes show remarkable higher fluxes and selectivities in comparison with the neat polymer [59–62].

One of the most stable MOFs is the structure ZIF-8: Zn(methylimidazolate)$_2$ [12]. The manufacture of ZIF-8 based MMM has already been tried by different groups using different polymers. Zhang et al. [63], for example, made 6FDA-DAM/ZIF-8 mixed matrix membranes, which showed high performance in the separation of C_3H_6 and C_3H_8 (Fig. 20). Ordoñez et al. [58] mixed ZIF-8 and

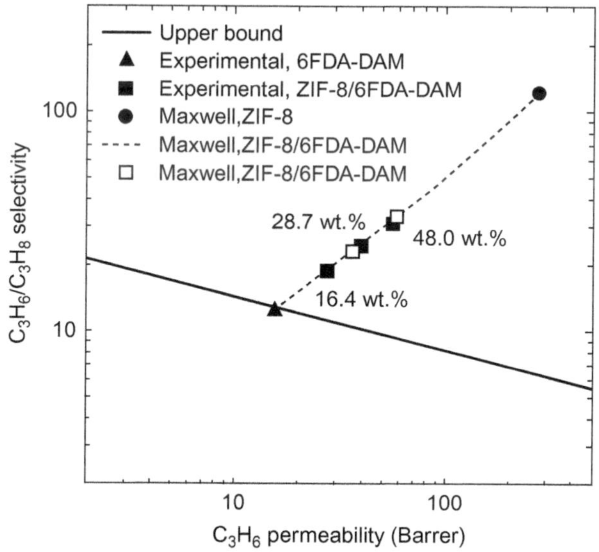

Fig. 20

Experimental and Maxwell model predicted permeation properties of ZIF-8/polyimide mixed matrix membrane. The straight line is the upper bound trade-off between selectivity and permeability in the Robson plat for polymer membranes. The filled dot (•) is an estimate for the pure ZIF-8 membrane. *From Zhang B, Dai Y, Johnson JR, Karvan O, Koros WJ. High performance ZIF-8/6FDA-DAM mixed matrix membrane for propylene/propane separations. J Membr Sci 2012;389:34–42, with permission.*

Matrimid® for several gas separations. They found increased permeability values for H_2, CO_2, O_2, N_2, CH_4, and C_3H_8 at loadings of 50wt.% 100nm ZIF-8 and improved ideal selectivities of gas pairs containing small molecules, such as H_2/O_2, H_2/CO_2, H_2/CH_4, CO_2/CH_4, CO_2/C_3H_8, and H_2/C_3H_8. Song et al. [64] made mixed matrix membranes with 20wt.% 60nm ZIF-8 particles and Matrimid®, which showed enhanced permeabilities of the pure gases H_2, CO_2, O_2, N_2, and CH_4 with negligible losses in selectivity. First pervaporation studies have been done by Liu et al. [65] who tested a 10wt.% ZIF-8/PMPS mixed matrix membrane for water/alcohol separation (Fig. 21). Fig. 22 shows that for the mixed matrix membrane NH_2-MIL-53 (Al) as MOF and polysulfone as polymer [62] the CO_2/CH_4 separation factor passes a maximum 25 wt.% polymer filling by the MOF. Fig. 23 shows for the gas mixture H_2/CO_2 on a mixed matrix membrane of ZIF-7/polybenzimidazole that the hydrogen permeability can be dramatically increased at constant H_2/CO_2 selectivity if the ZIF-7 content in the mixed matrix membrane is increased [66].

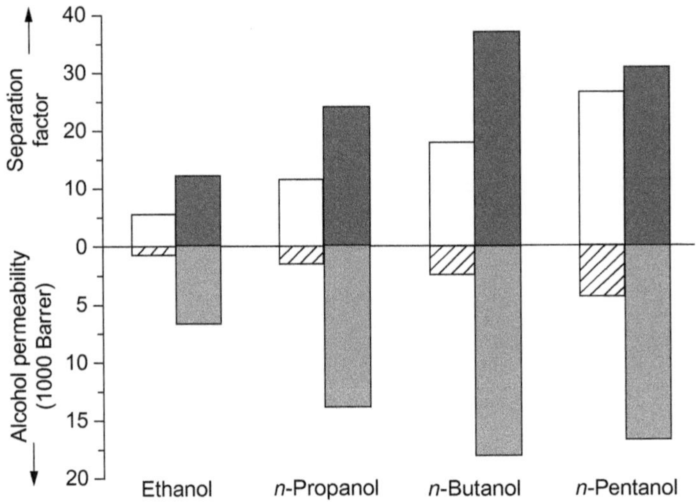

Fig. 21

Separation factor and alcohol permeability of polymethylphenylsiloxane (PMPS) (open and line filled columns) and ZIF-8-PMPS (*gray* and *light gray* columns) membranes for aqueous solutions of C_2–C_5 alcohols (1.0wt.% alcohols, 80°C). *From Liu X, Li YS, Zhu G, Ban Y, Xu L, Yang WS. An organophilic pervaporation membrane derived from metal–organic framework nanoparticles for efficient recovery of bio-alcohols. Angew Chem Int Ed 2011;50:10636–9, with permission.*

9 Conclusions

- Supported LTA membranes have been successfully installed during the last 10 years worldwide in about 25 plants for the dewatering of (bio)ethanol and i-propanol. Whereas there are outstanding separation performances of a new generation of zeolite membranes type SAPO-34 and DD3R on the lab scale, no short-term large-scale industrial application of zeolite membranes for gas separation can be predicted.

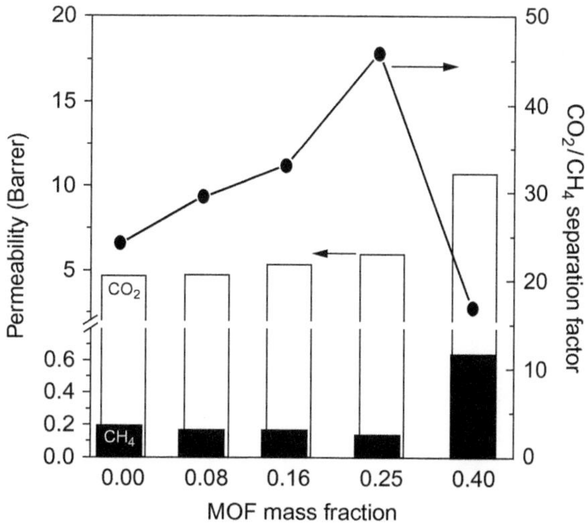

Fig. 22

Separation performance of mixed matrix membranes with different MOF loadings at 308 K for a mixture CO_2:CH_4 = 1:1 at retentate = feed pressure: 4 bar, permeate pressure 1 bar. CH_4 (*black*) and CO_2 (*white*) columns are the permeabilities (*left* ordinate). The CO_2/CH_4 separation factor is the right ordinate. *From Zornoza B, Martinez-Joaristi A, Serra-Crespo P, Tellez C, Coronas J, Gascon J, et al. Functionalized flexible MOF as filler in mixed matrix membranes for highly selective separation of* CO_2 *from* CH_4 *at elevated temperatures. Chem Commun 2011;47:9522, with permission.*

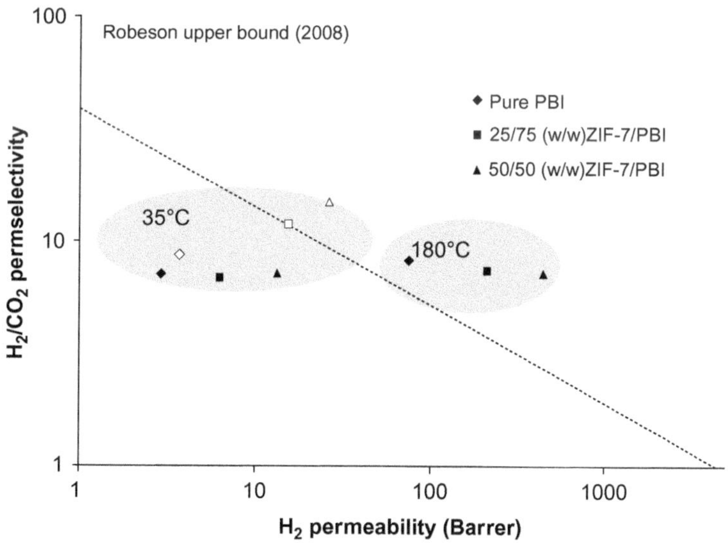

Fig. 23

H_2/CO_2 separation performance of pure PBI and ZIF-7/polybenzimidazole (PBI) nano-composite membranes compared to the Robeson upper bound (open symbols: pure gas separation performance; closed symbols: mixed-gas separation performance). *From Yang T, Xiao Y, Chung TS. Poly-/metal-benzimidazole nano-composite membranes for hydrogen purification, Energy Environ Sci 2011;4:4171–80, with permission.*

- Our knowledge to prepare zeolite membranes has decisively grown during the last 10 years. The toolbox for the preparation of improved zeolite membranes comprises the use of chemically modified ceramic supports, seeding by using different methods to attach the seeds to the support surface, intergrowth supporting additives which modify the zeta-potential of support, microwave heating, etc.
- The rigorous application of this toolbox which was developed for zeolite membrane synthesis, allowed the quick development of the first metal-organic framework (MOF) membranes in a short time.
- Several structure-related properties recommend MOFs as ideal membrane materials, especially linker substitution allows the fine-tuning of the pores size and the engineering of adsorption interactions. On the other hand, framework flexibility and linker flip-flop do not allow a sharp cut off in membrane separation.
- Despite much progress on the lab-scale, the scaling up of a discontinuous zeolite and MOF membranes synthesis in autoclaves is still a challenge; maybe layer-by-layer deposition of MOFs from the liquid or gas phase can solve this problem.
- As a rough estimate, the permeation through a molecular sieve membrane can be described by the permeation model for organic polymer membranes: the membrane selectivity is the product of adsorption selectivity and diffusion selectivity. This means that a knowledge-based molecular sieve membrane development based on independent adsorption and diffusion data is possible.
- A next-future application of MOFs is predicted for MOF-organic polymer-based mixed matrix membranes in hollow fiber or spiral wound geometry. Established membrane production technologies such as hollow fiber spinning or foil casting can be used for the preparation of mixed matrix membranes. Further, MOFs match much better with organic polymers than zeolites, for example, even if the latter have been hydrophobized, such as by silylation.

References

[1] Koros WJ, Ma YH, Shimidzu T. Terminology for membranes and membrane processes. Pure Appl Chem 1996;68:1479–96.

[2] Krishna R. Describing the diffusion of guest molecules inside porous structures. J Phys Chem C 2009;113:19756–81.

[3] Li SG, Falconer JL, Noble RD. Improved SAPO-34 membranes for CO_2/CH_4 separations. Adv Mater 2006;18:2601–3.

[4] Van den Bergh J, Zhu W, Kapteijn F, Moulijn JA, Yajima K, Nakayama K, et al. Separation of CO_2 and CH_4 by a DDR membrane. Res Chem Intermediat 2008;34:467–74.

[5] Himeno S, Tomita T, Suzuki K, Nakayama K, Yajima K, Yoshida S. Synthesis and permeation properties of a DDR-type zeolite membrane for separation of CO_2/CH_4 gaseous mixtures. Ind Eng Chem Res 2007;46:6989–97.

[6] Robson LM. The upper bound revisited. J Membr Sci 2008;320:390–400.

[7] Richter H, Voigt I, IKTS, personal information: several new plants have been erected by the German consortium with IKTS (former hitk/inocermic) NaA (4A) LTA membranes with the engineering of GFT Membrane Systems GmbH (Homburg/Saar, Germany). The capacities are 30,000 l/day, 60,000 l/day, 80,000 l/day, 100,000 l/day, and 200,000 l/day.

[8] Krishna R, van Baten JM. Using molecular simulations for screening of zeolites for separation of CO_2/CH_4 mixtures. Chem Eng J 2007;133:121–31.
[9] Gascon J, Kapteijn F. Metal-organic framework membranes–high potential, bright future? Angew Chem Int Ed 2010;49:1530–2.
[10] Tan JC, Cheetham AK. Mechanical properties of hybrid inorganic-organic framework materials: establishing fundamental structure-property relationships. Chem Soc Rev 2011;40:1059–80.
[11] Ferrey G, Serre C. Large breathing effects in three-dimensional porous hybrid matter: facts, analyses, rules and consequences. Chem Soc Rev 2009;38:1380–99.
[12] Low JJ, Benin AI, Jakubczak P, Abrahamian JF, Faheem SA, Willis RR. Virtual high throughput screening confirmed experimentally: porous coordination polymer hydration. J Am Chem Soc 2009;131:15834–42.
[13] Li K, Olson DH, Seidel J, Emge TJ, Gong H, Zeng H, et al. Zeolitic imidazolate frameworks for kinetic separation of propane and propene. J Am Chem Soc 2009;131:10368–9.
[14] Venna SR, Carreon MA. Highly permeable zeolite imidazolate framework-8 membrane for CO_2/CH_4 separation. J Am Chem Soc 2010;132:76–8.
[15] Guo H, Zhu G, Hewitt IJ, Qiu S. Twin copper source growth of metal-organic framework membrane: $Cu_3(BTC)_2$ with high permeability and selectivity for recycling H_2. J Am Chem Soc 2009;131:1646–7.
[16] Zou X, Zhang F, Thomas S, Zhu G, Valtchev V, Mintova S. $Co(HCOO)_6$ microporous organic framework membrane for separation of CO_2/CH_4 mixture. Chemistry 2011;17:12076–83.
[17] Ranjan R, Tsapatsis M. Microporous metal organic framework membrane on porous support using the seeded growth method. Chem Mater 2009;21:4920–4.
[18] Liu Y, Zeng G, Pan Y, Lai Z. Synthesis of highly c-oriented ZIF-69 membranes by secondary growth and their gas permeation properties. J Membr Sci 2011;379:46–51.
[19] Li SG, Falconer JL, Noble RD. Improved SAPO-34 membranes for CO_2/CH_4 separations. Adv Mater 2006;18:2601–3.
[20] Van den Bergh J, Zhu W, Kapteijn F, Moulijn JA, Yajima K, Nakayama K, et al. Separation of CO_2 and CH_4 by a DDR membrane. Res Chem Intermediat 2008;34:467–74.
[21] Li YS, Liang FY, Bux H, Yang WS, Caro J. Zeolitic imidazolate framework ZIF-7 based molecular sieve membrane for hydrogen separation. J Membr Sci 2010;354:48–54.
[22] Bux H, Liang FY, Cravillon J, Wiebcke M, Caro J. Zeolitic imidazolate framework membrane with molecular sieving properties by microwave-assisted solvothermal synthesis. J Am Chem Soc 2009;131:16000–1.
[23] Huang A, Bux H, Steinbach F, Caro J. Molecular sieve membrane with hydrogen selectivity: ZIF-22 in LTA topology prepared with 3-aminotriethoxysilane as covalent linker. Angew Chem Int Ed 2010;49:4958–61.
[24] Huang A, Dou W, Caro J. Steam-stable zeolitic imidazolate framework ZIF-90 membrane with hydrogen selectivity through covalent functionalization. J Am Chem Soc 2010;132:15562–4.
[25] Zhou W, Wu H, Udovic TJ, Rush JJ, Yildirim T. Quasi-free methyl rotation in zeolitic imidazolate framework-8. J Phys Chem 2008;112:12602–6.
[26] Luebbers MT, Wu T, Shen L, Masel RI. Effects of molecular sieving and electrostatic enhancement in the adsorption of organic compounds on the zeolitic imidazolate framework ZIF-8. Langmuir 2010;26:15625–33.
[27] Chang N, Gu ZY, Yan XP. Zeolitic imidazolate framework-8 nanocrystal coated capillary for molecular sieving of branched alkanes from linear alkanes along with high-resolution chromatographic separation of linear alkanes. J Am Chem Soc 2010;132:13645–7.
[28] Haldoupis E, Nair S, Sholl DS. Efficient calculation of diffusion limitations in metal organic framework materials: a tool for identifying materials for kinetic separations. J Am Chem Soc 2010;132:7528–39.
[29] Gücüyener C, van den Bergh J, Gascon J, Kapteijn F. Selective ethane adsorption on the metal organic framework ZIF-7 through a gate-opening mechanism. J Am Chem Soc 2010;132:17704–6.
[30] Seehamart K, Nanok T, Krishna R, van Baten J, Remsungnen T, Fritzsche S. A molecular dynamics investigation of the influence of framework flexibility on self-diffusivity of ethane in Zn(tbip) frameworks. Microporous Mesoporous Mater 2009;125:97–100.
[31] Huang A, Liang FY, Steinbach F, Caro J. Preparation and separation properties of LTA membrane by using APTES as covalent linker. J Membr Sci 2010;350:5–9.

[32] Huang A, Caro J. Cationic polymer used to capture zeolite precursor particles for the facile synthesis of oriented zeolite LTA molecular sieve membrane. Chem Mater 2010;22:4353–5.
[33] Huang A, Wang N, Caro J. Seeding-free synthesis of dense zeolite FAU membranes on 3-aminopropyltriethoxysiolane-functionalized alumina supports. J Membr Sci 2012;389:272.
[34] Huang A, Wang N, Caro J. Synthesis of multi-layer zeolite LTA membranes with enhanced gas separation performance by using 3-aminopropyltriethoxysilane. Microporous Mesoporous Mater 2012;164:294.
[35] Li YS, Liang FY, Bux H, Feldhoff A, Yang WS, Caro J. Molecular sieve membrane: supported metal-organic framework with high hydrogen selectivity. Angew Chem Int Ed 2010;49:548–51.
[36] Mintova S, Hedlund J, Valtchev V, Schoeman B, Sterte J. ZSM-5 films prepared from template-free precursors. J Mater Chem 1998;10:2217.
[37] Hedlund J, Noack M, Kölsch P, Creaser D, Caro J, Sterte J. ZSM-5 membranes synthesized without organic templates using a seeding technique. J Membr Sci 1999;159:263–73.
[38] Chun YS, Ha K, Lee YJ, Lee JS, Kim HS, Park YS, et al. Diisocyanates as novel molecular binder for monolayer assembly of zeolite crystals on glass. Chem Commun 2002;(17)1846–7.
[39] Huang A, Caro J. Facile synthesis of LTA molecular sieve membranes on covalently functionalized supports by using diisocyanates as molecular linker. J Mater Chem 2011;21:11424–9.
[40] Huang A, Caro J. Steam-stable hydrophobic ITQ-29 molecular sieve membrane with H_2 selectivity prepared by secondary growth using Kryptofix 222 as SDA. Chem Commun 2010;46:748–50.
[41] Huang A, Liang FY, Steinbach F, Gesing TM, Caro J. Neutral and cation-free LTA-type aluminophosphate ($AlPO_4$) molecular sieve membrane with high hydrogen permselectivity. J Am Chem Soc 2010;132:2140–1.
[42] Huang A, Caro J. Highly oriented, neutral and cation freeAlPO LTA: from seed crystal monolayer to a molecular sieve membrane. Chem Commun 2011;47:4201–3.
[43] Chmelik C, Bux H, Caro J, Heinke L, Hibbe F, Titze T, et al. Mass transfer in a nanoscale material enhanced by an opposing flux. Phys Rev Lett 2010;104:085902–5.
[44] Krishna R, van Baten JM. In silico screening of metal-organic frameworks in separation applications. Phys Chem Chem Phys 2011;13:10593–616.
[45] Bux H, Chmelik C, van Baten JM, Krishna R, Caro J. Novel MOF-membrane for molecular sieving predicted by IR-diffusion studies and molecular modeling. Adv Mater 2010;22:4741–3.
[46] Hertäg L, Bux H, Caro J, Chmelik C, Remsugnen T, Knauth M, et al. Diffusion of CH_4 and H_2 in ZIF-8. J Membr Sci 2011;377:36–41.
[47] Hedlund J, Sterte J, Anthonis M, Bons A-J, Carstensen B, Corcoran N, et al. High-flux MFI membranes. Microporous Mesoporous Mater 2002;52:179–89.
[48] Elyassi B, Sahimi M, Tsotsis TT. Silicon carbide membranes for gas separation applications. J Membr Sci 2007;288:290–7.
[49] Zacher D, Yusenko K, Bétard A, Henke S, Molon M, Ladnorg T, et al. Liquid-phase epitaxy of multicomponent layer-based porous coordination polymer thin films of [M(L)(P)0.5] type: importance of deposition sequence on the oriented growth. Chem Eur J 2011;17:1448–55.
[50] Shekkah O, Wang H, Zacher Z, Fischer RA, Wöll C. Growth mechanism of metal-organic frameworks: insights into the nucleation by employing a step-by-step route. Angew Chem Int Ed 2009;48:5038–41.
[51] Bétard A, Bux H, Henke S, Zacher D, Caro J, Fischer RA. Fabrication of a CO_2-selective membrane by stepwise liquid-phase deposition of an alkylether functionalized pillared-layer metal-organic framework $[Cu_2L_2P]$ on a macroporous support. Microporous Mesoporous Mater 2012;150:76–82.
[52] Cravillon J, Münzer S, Lohmeier SJ, Feldhoff A, Huber K, Wiebcke M. Rapid room-temperature synthesis and characterization of nanocrystals of a prototypical zeolitic imidazolate framework. Chem Mater 2009;21:1410–2.
[53] Wu Z, Wang B, Li K. Functional LSM–ScSZ/NiO–ScSZ dual-layer hollow fibres for partial oxidation of methane. Int J Hydrogen Energy 2011;36:5334–41.
[54] Liu S, Gavalas G. Preparation of oxygen ion conducting ceramic hollow-fiber membranes. Ind Eng Chem Res 2005;44:7633–7.
[55] Wang HH, Tablet C, Schiestel T, Caro J. Hollow fiber membrane reactors for the oxidative activation of ethane. Catal Today 2006;118:98–103.

[56] Chung TS, Jiang LY, Li Y, Kulprathipanja S. Mixed matrix membranes (MMM) comprising organic polymers with dispersed inorganic fillers for gas separation. Prog Polym Sci 2007;32:483–507.

[57] Hu J, Cai HP, Ren HQ, Wei YM, Xu ZL, Liu HL, et al. Mixed-matrix membrane hollow fibers of $Cu_3(BTC)_2$ MOF and polyimide for gas separation and adsorption. Ind Eng Chem Res 2010;49:12605–12.

[58] Ordonez MJC, Balkus KJ, Ferraris JP, Musselman ICH. Molecular sieving realized with ZIF-8/Matrimid mixed matrix membranes. J Membr Sci 2010;361:28–37.

[59] Perez EV, Balkus KJ, Ferraris JP, Musselman IH. Mixed-matrix membranes containing MOF-5 for gas separation. J Membr Sci 2009;328:165–73.

[60] Basu S, Maes M, Cano-Odena A, Alaerts L, de Vos DE, Vancelecom IFJ. Solvent resistant nanofiltration (SRNF) membranes based on metal-organic frameworks. J Membr Sci 2009;344:190–8.

[61] Bae TH, Lee JS, Qiu WL, Koros WJ, Jones CW, Nair S. A high-performance gas-separation membrane containing submicrometer-sized metal-organic framework crystals. Angew Chem Int Ed 2010;49:9863–6.

[62] Zornoza B, Martinez-Joaristi A, Serra-Crespo P, Tellez C, Coronas J, Gascon J, et al. Functionalized flexible MOF as filler in mixed matrix membranes for highly selective separation of CO_2 from CH_4 at elevated temperatures. Chem Commun 2011;47:9522.

[63] Zhang B, Dai Y, Johnson JR, Karvan O, Koros WJ. High performance ZIF-8/6FDA-DAM mixed matrix membrane for propylene/propane separations. J Membr Sci 2012;389:34–42.

[64] Song Q, Nataraj S, Roussenova M, Tan J, Hughes D, Li W, et al. Zeolitic imidazolate framework (ZIF-8) based polymer nanocomposite membranes for gas separation. Energy Environ Sci 2012;5:8359–69.

[65] Liu X, Li YS, Zhu G, Ban Y, Xu L, Yang WS. An organophilic pervaporation membrane derived from metal–organic framework nanoparticles for efficient recovery of bio-alcohols. Angew Chem Int Ed 2011;50:10636–9.

[66] Yang T, Xiao Y, Chung TS. Poly-/metal-benzimidazole nano-composite membranes for hydrogen purification. Energy Environ Sci 2011;4:4171–80.

CHAPTER 8

Further Steps of Zeolites Toward Industrial Applications: A Short-Range Outlook

L.M. Kustov*,†, A. Kuperman‡, A. Kustov*,†

**N.D. Zelinsky Institute of Organic Chemistry, Russian Academy of Sciences, Moscow, Russia †Moscow State University, Moscow, Russia ‡Chevron Corporation, Richmond, United States*

1 Introduction

This short-range review of zeolite applications in various processes of commercial interest can start where Edith Mary Flanigen ended in her excellent review of 1980 [1]: "The future of zeolites looks bright". Indeed, if one takes a look at the ascending curve of new zeolite structures that continue to appear every year and takes into account a great pool of new zeolite-like and similar new materials, including organic-inorganic hybrid systems and hierarchical materials, though they are not zeolites by definition, then one will come to the same conclusion: "The future of zeolites and zeolite-like materials is shining and is going to be prosperous in the first centenary of the new millennium." We will focus in this chapter mainly on the achievements made after 2000, while just shortly summarizing the progress made by the end of the last century.

2 General Trends

The trends in zeolite chemistry and catalysis were governed by a few general trends in chemistry and applied catalysis over the last 10–20 years, including the shifts in the paradigms related to raw materials, energy, environmental concerns, product demand, and the like.

- The *shift in raw materials* as a result of diversification of carbon sources from the triade known for about 100 years (oil, coal, and gas) to include renewable raw materials: vegetable oil, cellulose, and other carbohydrates, lignin, biomass, carbon dioxide as the major greenhouse gas, and other available feedstock. Also, even if we consider the traditional raw materials, the focus is shifting from oil to gas processing (gas to liquid (GTL) and other processes) and revisiting new technologies for coal conversion (coal to liquid). Oil processing had to change the traditional structure as well, because more and

Zeolites and Zeolite-like Materials. http://dx.doi.org/10.1016/B978-0-444-63506-8.00009-4

more heavy oil is produced and because it contains more and more sulfur, nitrogen, and condensed compounds. Therefore, the traditional zeolites cannot serve with the same efficiency to convert heavy raw materials and renewable feedstock also represented by compounds with a high-molecular weight and a large size, as they are used to serve to transform light hydrocarbons produced by cutting a part of oil at refineries. The new zeolite-like materials should have a different porous structure to incorporate mesopores to accommodate such bulky molecules. They should have a better organization at the nanoscale level, for instance; the acidic (or basic) functions of a certain strength should be placed, if necessary, at a certain distance from the ideally designed metal nanoparticles of a certain size (or nanoparticles of active metal oxide, metal sulfide, etc.) to act in perfect harmony together with the tailor-wise designed zeolite-like structure playing the role of a nanoreactor.

- The *shift in the energy scene* is also very significant and dictates the new rules for the designers of zeolite materials. Indeed, though even a smart zeolite chemist can hardly imagine their use in such renewable energy-producing sources as wind, geothermal, tides, geysers, he or she can be sure that zeolite-like materials will one day find applications in solar energy–harvesting devices, light-emitting diodes (LEDs), fuel cells, batteries, and high-efficiency optoelectronic devices (optical switches, etc.). No doubt, smart zeolite-based catalysts will be in permanent demand for the energy sector, no matter if they are used for biogas or biomass processing or as materials for the components of the integrated devices producing energy. The use of hydrogen as the energy carrier, though not to come tomorrow, is another future area of potential zeolite application. Production (from biogas, biomass, and water) and storage of hydrogen as well as the transformation of H_2 into energy (as a heat or electricity) are the pathways for the next generation of zeolite materials.
- The *shift in environmental issues* and the turn to and the acceptance of "green" chemistry and its 12 principles as a *must* for future development, paving the way from the ideas of the chemical process and laboratory studies all the way toward the industrial scale, also affects the zeolite scene. Euro-5 and Euro-6 specs and similar grades to come into practice in other developed and developing countries are the legislative press for gasoline and diesel producers, because they limit the contents of sulfur, benzene, and aromatics, oxygenates to the level that is hardly achievable by using modern processing schemes. On the other hand, the problem of the carbon footprint and CO_2 conversion into valuable and desirable large-scale products may be solved, inter alia, by using some of the specially designed zeolite catalysts and zeolite-based CO_2 capture materials. Finally, the field of remediation of water resources (ground water, sewage water, etc.) and soil remediation from pollutants (perchlorobiphenyls, pesticides, Cr, Hg, Cu, As, and other ecotoxicants) is calling for the more efficient materials and methods. The recently discovered nano-zero-valent-iron particles (freestanding or supported by carbon or silica carriers) provide some solution to these remediation problems. However, to the best of

our knowledge, no such systems have been designed on the basis of zeolite-like supports. Not only clean water, but clean air as well should be the focus of zeolite science. Quite a number of zeolite catalysts are used for $DeNO_x$ and $DeSO_x$ in car exhausts as well as exhausts of heat plants. Zeolites, due to their unique adsorption properties, can be used as concentrators of toxic components of exhausts and, after supporting an active phase like metal nanoparticles, as efficient catalysts for removal of diverse volatile compounds from the air.

- We should also consider *product-driven changes* in zeolite area, both from the aforementioned point of view of the design of new zeolite-like materials with a proper balance between micropores and mesopores, like hierarchical systems and large-pore zeolite-like materials, and from the point of view of the shift in the demand for chemical products. For example, the focus in oil processing is gradually shifting from light gasoline-range products toward diversified products (jet fuels, arctic fuels, and special types of diesel) and hydrocracking and the ring-opening processes are becoming more important. Also, the selectivity already became the most important characteristic of any process and any catalyst, and the needs of the modern petrochemistry and fine chemistry form some new reality as to which products should be produced (monomers, components of new specs of fuels, etc.).
- One should also keep in mind the *process-driven changes* in the zeolite science. Most industrial processes reached a saturation level in their development. A significant, not incremental, improvement can be anticipated in the case when the chemical process is organized in synergy with the action of the catalyst. Thus, the reactor design becomes no less important than the catalyst design. In some cases, the reaction occurring inside the pore on the active site may propagate into the gas phase, for instance, living polymerization or some examples of oxidation reactions when the product formed on the heterogeneous catalyst undergoes further transformations in the homogeneous mode. Sometimes, introduction of certain agents may shift the equilibrium in the system and thereby increase dramatically the yield of the product that is not expected to be formed in large amounts because of the thermodynamic limitations. Examples of this kind include the use of zeolites as water adsorbents in processes occurring with the formation of water molecules (synthesis of ethers, dimethylcarbonate, etc.). Nonequilibrium conditions can be also achieved by creating gradients of concentrations or temperature inside the catalyst or inside the reactor.

The grounds of catalysis by zeolites were clearly and conclusively presented in four chapters included in the book (actually, a handbook for zeolite scientists) entitled *Introduction to zeolite science and practice* [2]. These chapters were focused on (1) acid-catalyzed hydrocarbon conversion (by Martens and Jacobs), (2) hydrocarbon processing (industrial processes) (by Maxwell and Stork), (3) organic syntheses by zeolite catalysts (by Hoelderich and van Bekkum), and (4) selective oxidation reactions by transition metals introduced into zeolites (by Bellussi and Rigutto). The chapter by Martens and Jacobs also gives a comprehensive analysis of carbocation and other mechanisms of hydrocarbon transformations on acidic forms of zeolites; therefore, we are not going to dwell on these questions. Keep in mind the major conclusions related to the shape-selectivity effects in zeolite-like materials related to the confinement of the reactant (reactant selectivity), product (product selectivity), and transition state (transition state selectivity) in the channels and cavities of the particular zeolites used in a specified catalytic process. Examples of the shape selectivity include such reactions as iso-dewaxing, light paraffin conversion into aromatics, selective oxidation processes, methanol-to-olefins, deep catalytic cracking, and selective catalytic NO_x reduction. The main feature of zeolites is the size of the rings forming the channels/cavities. This size includes 8-, 10-, 12-membered rings (most of the known and widely used zeolites), as well as 14- and 18-membered rings (UTD-1, SSZ-44, AlPO-8, VPI-5) and 20-membered rings (cloverite). The mesoporous materials (MCM-41, MCM-48, SBA-15 and others) typically have a large pore opening of 2–3 nm, but the effects of shape selectivity are described for these structures as well.

Since 2001, the zeolite catalysis made a leap-frog jump after the introduction of mesoporous materials, metal-organic frameworks, and hierarchical systems, as well as after the turnaround in the paradigm in technologies and feedstock availability and future trends.

It is worth noting that a number of excellent reviews have been published in the recent literature describing the state of the zeolite science and summarizing the achievements and progress made since the discovery of zeolites and especially focusing on the new applications of zeolite-like materials. The reader may wish to learn more about the industrial applications of zeolites in [3–5].

The paper written by Martens and his colleagues [6] provides a new insight into zeolite materials. The authors propose the new terms "Zeogrids" and "Zeotiles" for the hierarchical materials built from assembled MFI zeolite (ZSM-5 type high-silica zeolite family) precursor units. The secondary porosity in such materials is created by the way of the self-assembly of nanoparticles formed during the MFI zeolite synthesis in the presence of supramolecular templates. The aggregated species of this kind are termed "nanoslabs." Zeogrids are considered to be layered materials with lateral spacings between nanoslabs creating galleries to be regarded as supermicropores. Zeotiles represent a diversity of tridimensional nanoslab assemblies with mesopores. Zeotile-1, -4, and -6 contain hexagonal mesostructures. Zeotile-1 is characterized by the presence of triangular and hexagonal channels; Zeotile-4 exhibits hexagonal channels interconnected via slits. Zeotile-2 is an example of the cubic structure with gyroid-type mesopores. The Zeogrids and Zeotiles behave in adsorption, separation, and catalysis quite differently from zeolites and mesoporous materials produced from unstructured silica. The efficient instrument to estimate the shape-selectivity effects in such systems is the adsorption of n- and *iso*-alkanes. The mesoporosity of Zeotiles is proposed for application in chromatographic separation of biomolecules. Zeotiles demonstrate promising separation efficiency in CO_2 sequestration. Also, due to the simplicity of the synthesis that does not include the hydrothermal stages, Zeogrids are robust materials for membrane synthesis. The outstanding performance of Zeogrid membranes in gas separation, nanofiltration, and pervaporation is observed. The catalytic properties of these unique materials may also be superior to conventional zeolites due to the presence of mesopore arrays. For instance, the Beckmann rearrangement of cyclohexanone oxime on Zeogrids and Zeotiles proceeds similar to the reaction catalyzed by silicalite-1 zeolite. The acid function or redox catalytic activity can be introduced in Zeogrids and Zeotiles via incorporation of Al and Ti atoms in the nanoslabs during the synthesis. Zeogrids demonstrate quite high activity in hydrocracking, catalytic cracking, alkylation, and olefin epoxidation reactions. Zeogrids and Zeotiles may be combined into a new family of hierarchical silicate materials.

Hierarchical (or mesoporous) zeolites became the focus of the review by Christensen et al. [7]. The main reason behind the development of hierarchical zeolites is to achieve heterogeneous catalysts with an improved porous structure and thereby enhanced performance in alkylation of benzene with alkenes, alkylation, and acylation of other compounds, methanol conversion into hydrocarbons, aromatization processes, isomerization of paraffins, cracking of diverse substrates and raw materials (naphtha, aromatic compounds, hexadecane, vacuum gas oil, and some polymers), and hydrotreating. The reactions that are of interest from the point of view of fine chemicals synthesis occurring on hierarchical zeolites include aldol condensation, esterification, acetalization, olefin epoxidation, and Beckmann rearrangement.

In the sections to follow we will give an outlook of the recent, most important results related to different aspects of zeolite catalysis; therefore, the chapter was structured by the principle of processes and chemistry combining different catalysts and approaches.

3 Hydrocarbon Processing

3.1 Fluid Catalytic Cracking

Catalytic cracking is now considered a mature process, though evolutionary improvements of the conventional Y-type zeolites used in the most efficient version of the catalytic cracking, that is, fluidized-bed catalytic cracking (FCC), are reported every year. Such improvement, for instance, concerns diverse additives to the basic cracking catalyst composition (Y zeolite) and include HZSM-5 zeolites that increase the activity, small amounts of noble metals that accelerate the regeneration stage as well as modifying additives to the silica-alumina and clay as binders. This process is discussed in this book in the chapter by Falabella.

ZSM-5 zeolites were shown to perform shape-selective cracking (called Selectoforming with a later improved M-forming process by Mobil) [8].

The studies of cracking of individual hydrocarbons help to achieve the desirable effects. One of the recent publications [9] is focused on the use of ferrierite (H-FER) or H-FER/HZSM-5 blends in catalytic cracking to enhance the production of light olefins using catalytic cracking of *n*-hexane at 500°C as a model. The final evaluation was made using the high-olefin catalytic cracking (HOCC) of a Brazilian gasoil feedstock at 600°C. For the HOCC tests, both H-FER and the H-FER/HZSM-5 blends were mixed with a reference commercial FCC equilibrium catalyst so that the total zeolite content ranged from 0.5 to 8 wt.%. For *n*-hexane and HOCC catalytic tests, the use of ferrierite led to the improvement of olefin production. The H-FER zeolite demonstrated higher selectivities toward propylene and butene than HZSM-5. On the other hand, HZSM-5 showed a higher ethylene selectivity. The HZSM-5 zeolite demonstrated higher ethylene and propylene yields than H-FER with HOCC, but the application of the HZSM-5/H-FER composites resulted in enhanced yields of the overall amount of light olefins (ethylene, propylene, and C_4 olefins) compared to pure HZSM-5 zeolite. This synergy observed for the combination of H-FER and HZSM-5 was explained by the fact that the H-FER zeolite has a lower hydrogen transfer ability.

Some of the hierarchical materials were tested in the cracking of gasoil [10]. MCM-36 was found to be a highly active material prepared by the pillaring of a layered zeolite precursor, MCM-22(P). During the preparation, significant destruction of the layered structure occurred, and the framework silicon/aluminum (Si/Al) ratio, which affects the acidity and catalytic activity, somewhat changed. Performing the pillaring process under mild conditions allowed the authors to prevent the destruction. The new materials demonstrated a higher catalytic activity (by some 20–30%) in gasoil cracking compared to the conventional MCM-36 materials.

3.2 Hydrocracking

Hydrocracking is becoming even more important than FCC, especially taking into account the tendency to use heavier oil components, asphaltenes, and similar high-molecular-weight raw materials. Martens contributed with a great number of excellent publications in this area, and the recent studies are focused on both mechanistic aspects of hydrocracking and the application of new zeolite and zeolite-like structures. The Ni-Mo and Ni-W compositions supported on zeolites demonstrate good results in the hydrocracking of heavy oil components.

Chevron introduced a novel concept via the Paragon process that combines high-pressure hydrocracking with the low-pressure step using ZSM-5 zeolite as a shape-selective catalyst. The second stage provides conditions for the selective cracking of normal and slightly branched paraffins to produce C_3–C_8 olefins to further boost the octane number [11].

A large tetramethylbranched alkane (2,6,10,14-tetramethylpentadecane) was used as a model for investigating the hydrocracking reaction pathways [12]. Hydroconversion of this substrate over Pt/SiO_2–Al_2O_3 and Pt/USY zeolite catalysts resulted in two parallel reactions: methyl shift and hydrocracking. The zeolite altered the four methyl-branching positions, while the silica-alumina favored only the methyl shift of the inner methyl groups. It was concluded that rival β-scission reactions occurred at significantly different reaction rates. Carbon-carbon bonds next to inner branchings were preferentially cracked. The authors draw the conclusion that the kinetic models established for short alkanes do not apply for heavy multibranched alkanes.

3.3 Dewaxing

Dewaxing is the process aimed at the conversion of linear long-chain hydrocarbons while leaving intact the branched isomers. The different options of the technology were developed by Mobil, Chevron, and Akzo and are based on the shape-selectivity effects of ZSM-5 zeolite. Other types of medium-pore size zeolites (ZSM-23, B-ZSM-5) have been also used.

The group that Martens studied was IZM-2 (IFPEn Zeolitic Material number 2), which is a silica-alumina zeolite composition with an as-yet unknown structure [13]. The synthesis of this material was performed using 1,6-bis(methylpiperidinium)hexyl dibromide as a template. The resulting IZM-2 particles represent intergrowths of 20–60 nm nanocrystals of various orientations and are definitely microporous. Evaluation of the pore architecture used two test reactions, namely (1) m-xylene isomerization and disproportionation and (2) *n*-decane isomerization-hydrocracking, which revealed the presence of two types of pores of different diameters. The catalytic experiments unambiguously showed the presence of 10-membered ring pores that are present next to wider pores, presumably 12-membered ring pores. Pt/IZM-2 demonstrated superior activity in the hydroisomerization of *n*-hexadecane, especially selective in the formation of multibranched isomers (Fig. 1).

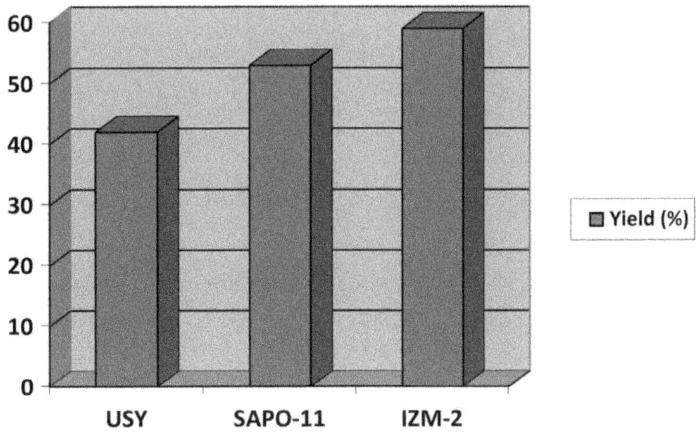

Fig. 1
Yield of multibranched iso-C_{16} parrafins (in %) in hydroisomerization of cetane on different zeolites.

4 Production of Octane Boosters

4.1 Alkylation of C_4

Isobutane alkylation with butenes is the process of producing high-octane products, such as trimethylpentanes (TMP) with 2,2,4-trimethylpentane being most desirable product with its 100% octane number. Up to now homogeneous catalysts were used on the industrial scale for this reaction with concentrated sulfuric acid or HF being the unprecedented catalysts. The disadvantages of these catalytic systems are well known and include, inter alia, corrosion problems, which are of environmental concern because of the necessity of utilization of sulfuric and fluoride wastes.

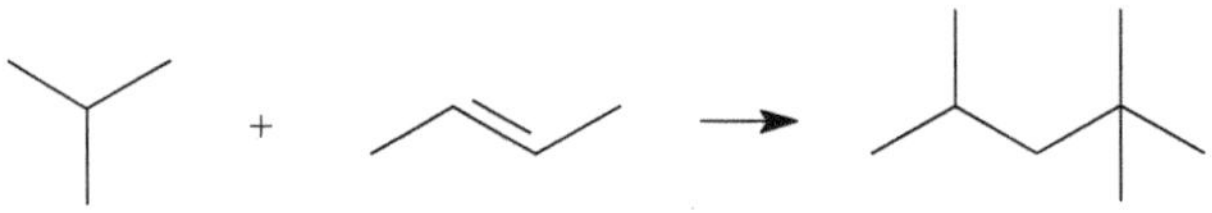

Multiple attempts to use zeolites as robust heterogeneous catalysts do not move further from the lab scale and demonstration units. Almost all the large industrial companies (Shell, ABB Lummus, Chevron, Haldor Topsoe, just to mention a few) tried to solve the problem of poor catalyst stability and catalyst regeneration, but all with limited success. Usually mixed rare-earth-hydrogen forms of zeolite Y demonstrate quite good initial activity, but after 4–8 h of time-on-stream in a flow setup, the catalyst tends to deactivate and requires regeneration at temperatures higher than 400°C, which is not quite a fit to the alkylation process temperature of about 150–200°C. Attempts to use solvents for regeneration seem to be quite an expensive solution. Research at Shell [14] resulted in the achievements: first, Beta type zeolite structure (BEA) zeolites turned out to be somewhat better than faujasites, and second, regeneration can be performed by purging the deactivated catalyst in hydrogen, though noble metals should be added as a cocatalyst for the regeneration stage, but still the catalyst lifetime is very short.

Catalysts based on BEA zeolite exchanged with lanthanum were proposed in [15]. It was found that these catalysts have a lower activity than the Y-type zeolites both in the protonic form and in lanthanum forms. Note that the total acidity measured by pyridine temperature-programmed desorption is similar in the BEA and Y zeolites. However, the BEA zeolite contains a significant amount of stronger acid sites, and as a result, the deactivation process occurs faster. The addition of platinum to the BEA zeolite is useful to regenerate the catalyst using hydrogen at 80°C, although long times are needed to accomplish this regeneration.

4.2 Paraffin Isomerization

Paraffin skeletal isomerization is the second of the reaction-producing high-octane hydrocarbons, though TMPs cannot be obtained by using diverse isomerization catalysts, even the Pt/Cl–Al_2O_3, has been used on the industrial scale in *n*-butane isomerization (hydrogen once through Butamer process) for more than 40 years. Diverse zeolite catalysts have been tested in the isomerization of both light alkanes (C_4–C_6) and longer-chain hydrocarbons (C_7–C_{16}). Pt/H-MOR applied in the total isomerization of paraffins (TIP) process is one of the most active commercial catalysts, although it is known that the initial activity of the H-MOR containing no platinum is about the same as that of Pt/H-MOR. The role of platinum is mostly related to the stabilization of the catalyst activity by preventing the formation of heavier products on the catalyst. Other zeolites and zeolite-like materials such as mazzite, omega, and BEA zeolite showed the activity comparable to that of H-MOR-based catalysts [16]. The only advantage of H-MOR seems to be the large-scale production of this zeolite. Also, the H-BEA zeolite is somewhat less thermally stable compared to H-MOR zeolite.

Quite a few attempts have been made to use HZSM-5 and HZSM-11 zeolites in paraffin isomerization. For instance, the effect of the Si/Al ratio on the properties of Ir/Pt-HZSM-5 zeolite in *n*-pentane isomerization was studied [17]. The increase in the Si/Al ratio decreased the number of strong Brönsted and Lewis acid sites and the catalytic activity in *n*-pentane isomerization. Ir/Pt-HZSM-5 with a Si/Al ratio of 23 showed the highest activity toward *n*-pentane isomerization. The optimum conditions of *n*-pentane isomerization over Ir/Pt-HZSM-5 gave the following values of the *n*-pentane conversion, isopentane selectivity, and isopentane yield: 63.0%, 98.2%, and 61.9%, respectively.

Among the new types of microporous and mesoporous materials close to zeolites, one should mention SBA-15 and MCM-41, but these materials are less acidic compared to mordenite or BEA zeolite and require modification with some acidic additives. SBA-15 ordered mesoporous silica materials containing up to 40 wt.% of HPWO (phosphotungstic heteropolyacid) were synthesized via an original method involving the introduction of HPWO in an acidified solution of P123 tri-block copolymer used as the SBA-15 mesostructuring agent [18]. The samples with similar HPWO loadings were also prepared for comparison by impregnation of SBA-15. Physicochemical characterization showed that even after calcination, HPWO in the direct synthesis samples was better dispersed than that in the impregnated samples. SBA-15-supported HPWO materials were converted into metal-containing bifunctional catalysts. The direct synthesis samples were superior to impregnated samples in the conversion of *n*-decane achieving the high yields of skeletal isomers (>70%), of multibranched skeletal isomers, and of branched cracked products, and hydrocracking reactions limited to primary cracking. The HPWO/SBA-15 samples showed a broader optimum reaction temperature window for isomerization and a higher *n*-decane dibranching selectivity as compared to ultrastable Y zeolites.

A tungstophosphoric acid (HPWO) grafted Pt/MCM-41 catalyst was also synthesized and tested in paraffin isomerization [19]. The Keggin structure of the supported heteropoly acid $H_3PW_{12}O_{40}$ was rather stable after calcination. The conversion of *n*-heptane in the isomerization reaction over Pt/$H_3PW_{12}O_{40}$/MCM-41 catalyst linearly increased from 5.4%

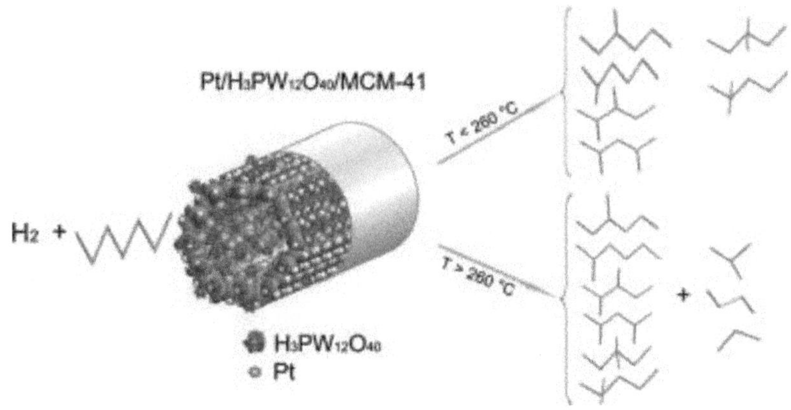

Fig. 2
The application of supported heteropolyacid on MCM-41 in paraffin isomerization. *Figure from reference Yang X, Wang JA, Chen L et al. Catal Commun 2012;28:202 with permission.*

to 79.2% with the reaction temperature rising from 220°C to 320°C. The isomerization selectivity was 100% at temperatures not exceeding 260°C. At higher temperatures, cracking products were observed: *n*-butane, isobutane, and propane. The ratio of multibranched to monobranched isomers was between 0.60 and 0.85, indicating that the mesostructured catalyst exhibits unique shape selectivity in the skeletal isomerization of *n*-heptane (Fig. 2).

The modification of this catalyst with zirconium introduced in the MCM-41 structure [20] resulted in a further improvement of the isomerization activity. Thus, the catalytic properties of the 1%Pt/25%$H_3PW_{12}O_{40}$/Zr-MCM-41 in the hydroisomerization of *n*-heptane at atmospheric pressure are characterized by the enhanced selectivity to multibranched isoheptanes. 2-Methylhexane (2-MH) was predominant in the monobranched isoheptanes, and 2,3-dimethylpentane (2,3-DMP) was the prevailing compound in the multibranched hydrocarbons. Among the cracking products, butanes and propane were formed. The formation of the multibranched isoheptanes showed a correlation with the mesopore diameters of the catalysts.

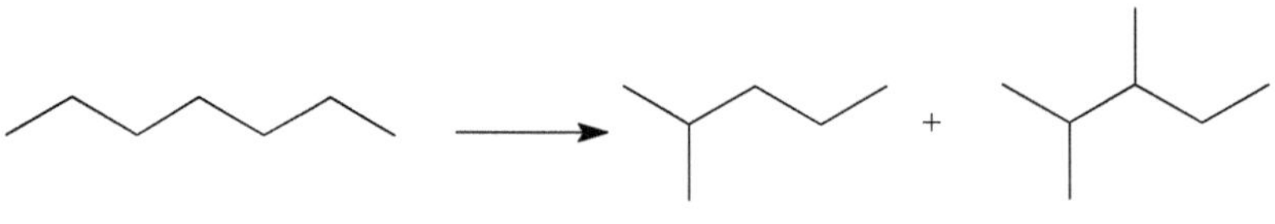

It is known that sulfated zirconia, especially promoted with Fe, Mn, and other cations exhibits a high activity in *n*-butane isomerization [21–23]. The attempt to prepare a composite material such as copper oxide-promoted sulfated zirconia (CuSZ) supported on MCM-41 by impregnation [24] seems to be successful. *n*-Hexane isomerization on the CuSZ/MCM-41 catalyst resulted in 2-MP, 3-MP and 2,3-DMB are the major isomerization products.

The product distribution resembles that reported for Pt based catalysts. The optimal CuO loading in these catalysts calcined at 700°C is about 3.2 wt.%.

4.3 Isomerization of Olefins

The grounds of this process based on the use of medium-pore zeolites, such as FER, were laid down by the research works at Shell [25] and Texaco [26]. Ferrierite exhibits a better isobutene selectivity in 1-butene skeletal isomerization, and produces less C_{5+} products and coke. The zeolites for this process have been commercialized by Mobil and British Petroleum (ISOFIN process) and Lyondell.

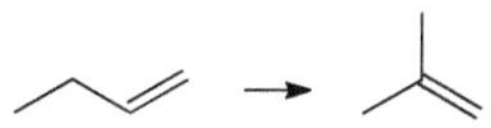

5 Upgrading Hydrocarbons

5.1 Aromatization of Alkanes C_6–C_8

The dehydrocyclization of *n*-hexane, *n*-heptane, or *n*-octane over Pt/KL (BaKL) zeolites has been known since the 1980s and is reviewed by Derouane [27]:

The zeolites of Linde type L (LTL) were shown to exhibit the shape-selectivity effect due to a specific confinement of *n*-hexane as a 6-ring pseudocycle inside the zeolite structure (Maxwell in Ref. [2]). Chevron researchers published the results on hexane aromatization on Pt/Cs-B-BEA zeolite and developed the AROMAX process [28]. The reaction proceeds at 460–480°C with a rather high selectivity close to 80%, but the catalyst is sulfur-sensitive and the commercial application can hardly be envisioned.

5.2 Ring Opening

The new, more stringent legislation introduced in the United States and Western Europe presumes more severe requirements on motor fuels, including gasoline and diesel. One of the goals is the reduction of the soot formation from diesel engines and enhancement of the cetane number of diesel (the content of normal long-chain paraffins). One of the approaches to the improvement of the cetane number of diesel fuel is the ring opening of polycyclic aromatic and naphthenic compounds.

Analysis of the scarce literature related to decalin conversion shows that its selective transformation into *n*-decane is an extremely difficult task. Let us try to theoretically estimate the probability of decalin conversion into *n*-decane. In this case, we assume that only ring opening occurs and no side reactions proceed. Also, we do not take into account any thermodynamic considerations and the steric effects. Thus, the estimate obtained is only a theoretical estimate. The approach is illustrated in Fig. 3. The theoretical probability to open the ring of cyclohexane to yield *n*-hexane is 100%, whereas for methylcyclopentane (MCP) this probability diminishes to 40%. In the case of decalin, we have to consider a two-step ring opening and multiply the corresponding probabilities. Then after summing up the probabilities of each variant, we arrive at the value of 24%. We remind readers that this probability does not take into account the thermodynamics and sterical hindrances for ring opening of different C–C bonds in the decalin molecule (especially the internal C–C bond in the ring). Also, this value will further decrease if we take into account the kinetic factors and parallel side reactions occurring in the system (cracking, isomerization, and dehydrogenation). Thus, even in the case of cyclohexane, when the theoretical probability is 100%, the yield of only 60% was attained. What, then, can be the real yield of *n*-decane when the overall theoretical probability of decalin conversion into *n*-decane is only 24%?

Of course, this analysis does not take into account any thermodynamic or kinetic limitations of the processes and neglects all the possible side reactions, while considering only the ring-opening (hydrogenolysis) reaction. On the other hand, this pattern does not take into account the steric hindrances for hydrogenolysis of external

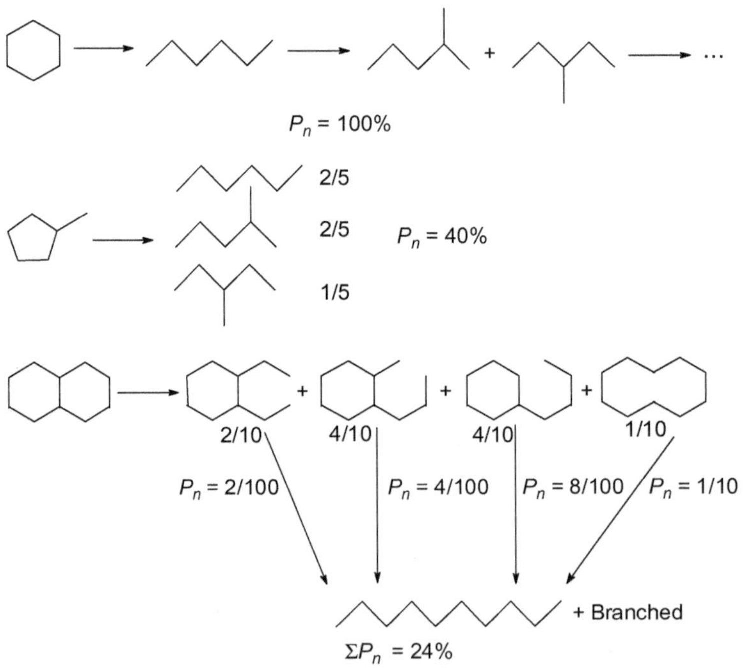

Fig. 3
Theoretical probability of ring opening to normal paraffins.

or internal C–C bonds. Thus, even such an idealized pattern shows that the highest probability to obtain *n*-decane from decaline is only 24%.

The thermodynamical data for the simplest reaction of cyclohexane hydrogenolysis into *n*-hexane are presented in Table 1. These data show that the high yields can be theoretically expected at quite low temperatures.

The information on ring opening of naphthenes, such as cyclohexane and higher cyclic compounds is scarce. The ring-opening conversions of naphthenes are reported by Exxon researchers (Daage and coauthors), Universal Oil Products (UOP) (L. Galperin), the team from Abo Akademi, Finland (Kubicka) and some others [29–36].

On the contrary, the transformations of MCP are studied in much detail. In most studies, this reaction was used; however, only as a model reaction and the goals of these studies were limited by solution of purely fundamental problems. For instance, in papers by Sachtler and coauthors [37], the MCP conversion via ring opening or ring enlargement (to cyclohexane) was considered to be a tool for studying the state of supported platinum and the nature and properties of electron-deficient Pt particles. The conversions were very low in order to follow merely the primary steps of this reaction and to measure the kinetic parameters. Dossi et al. [38] and Ostgard et al. [39] also studied MCP conversion on Pt/KL zeolites.

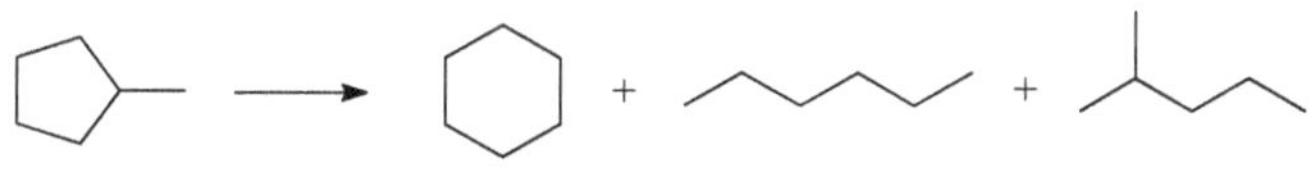

Kubicka and Okal [40] showed that ring opening is a bifunctional reaction that requires both Brønsted acid sites and metal (presumably electron-deficient) nanoparticles with well-defined properties.

Table 1: Equilibrium composition for ring opening of cyclohexane (CH)

	Product Composition (wt%)		
Temperature (°C)	CH	H_2	$n\text{-}C_6H_{14}$
100	1.0	0.02	98.98
200	3.9	0.1	96.0
300	10.7	0.3	89.0
400	20.6	0.5	78.9
500	33.3	0.8	65.9
	CH	H_2	2-Methylpentane
100	0.5	0.01	99.49
200	2.9	0.1	97
300	7.8	0.2	92
400	16.6	0.4	83
500	29.3	0.7	70

The MCP conversion was also studied in detail in [41] on Pt/KL zeolite as a function of the hydrogen to substrate ratio and reduction temperature. The ring-opening selectivity did not change in the range of the hydrogen: substrate ratios of 8–200. The turnover frequency (TOF) value went through the maximum with increasing dilution of the substrate. Ring-opening activity was also reported for the conversion of toluene on NiHY zeolites [42]. It was shown that the activity enhances with increasing Ni content.

Ring opening of cyclic compounds was investigated in [43,44]. The contribution of ring opening was found to increase with increasing reaction temperature, while that of skeletal isomerization diminished.

The most interesting and commercially viable catalytic system based on 1%Pt/H-Beta zeolite was developed by Mobil Company [45]. This catalyst exhibits rather high activity in ring opening of C_6 cyclic compounds and is used in a combination with the isomerization catalyst to improve the C_6 downstream feed. In this case, however, a 1:3 mixture of normal and isoparaffins was formed at the conversion of ~80% at 230–270°C under high-pressure conditions.

The team led by Dmitry Murzin studied the kinetics of the decaline ring opening on conventional catalysts such as Pt/zeolites:

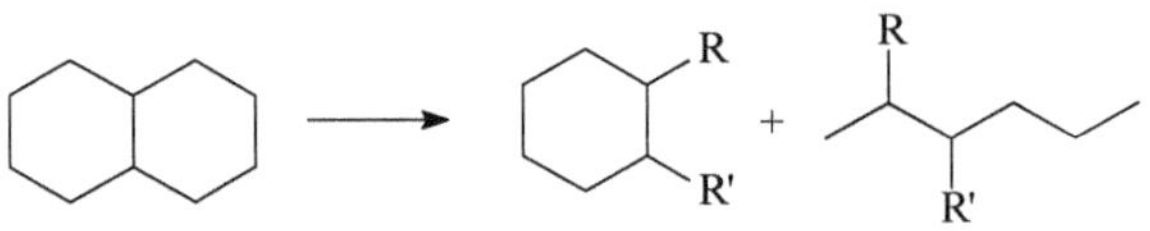

Also, they revealed the effect of the acidity of the carrier on the selectivities of different stages of the process [29]. The metal-support interactions on a series of catalysts of different acidities, including platinum-modified zeolites and H-MCM-41, are investigated by means of X-ray photoelectron spectroscopy, CO, and pyridine adsorption, and a model reaction (ring opening of decalin). The electronic properties of Pt are influenced by the acidity of the support, and the alteration of Pt properties increases with increasing acidity of the support, as can be seen from the changes in the Pt binding energy and stretching frequency of adsorbed CO. At the same time, the presence of platinum affects the acidic properties of the supports by reducing the strength of the acid sites. This is observed directly as the changes in desorption of pyridine from the acid sites and indirectly as the suppression of cracking reactions during the ring opening of decalin on the Pt-modified catalysts.

In another publication by the same authors [30], ring opening of decalin over Pt- and Ir-modified SAPO-5 and VPI-5 catalysts was carried out in a batch reactor at 573–623 K in the presence of hydrogen. Brønsted acidity was found to be essential for both isomerization and ring-opening reactions. The presence of a noble metal was found to be beneficial for isomerization and ring opening as well.

The activity and selectivity of platinum-modified zeolites (Beta, Y, Mordenite) were investigated in the ring opening of decalin at 473–543 K in the presence of hydrogen [31]. In the course of decalin transformation, skeletal isomerization, stereoisomerization, ring opening, and cracking took place. The presence of platinum resulted in an enhancement of isomerization and ring-opening rates, as compared to the proton forms of zeolites. The isomerization and ring-opening rates increased three and five times, respectively. The hydrogen pressure was found to suppress the secondary reactions and to prevent the catalyst deactivation. The isomerization and ring-opening reactions were not affected by the presence of hydrogen. Interactions between platinum and Brønsted acid sites suggested that the platinum nanoparticles were located in the channels of the studied zeolites. As a result of these interactions, the strength of Brønsted acid sites was reduced, and consequently, fewer cracking products were formed, at comparable conversions, over platinum-modified zeolites than over the corresponding parent zeolites. Similar results were obtained in [32]. The activity of H-Beta (Si/Al=25, H-Beta (Si/Al=75), HY (Si/Al=12), H-Mordenite (Si/Al=20), and H-[Al]MCM-41 and their deactivation were correlated to their acidity determined by pyridine adsorption, pore structure, and surface area changes. Skeletal isomerization, stereoisomerization, ring opening, and cracking were the prevailing reactions. Skeletal isomerization included several parallel and consecutive reactions resulting in a wide variety of products, which underwent ring opening, and subsequently cracking. Virtually no direct ring opening of decalin was observed; that is, decalin must first be isomerized to alkyl-substituted bicyclononanes and bicyclooctanes to make possible decalin ring opening. The main products were grouped according to their structural features to evaluate the product distribution, which was influenced by the catalyst acidity and the pore structure. The initial activity of the catalysts increased with rising acidity and reaction temperature, respectively. The deactivation of the investigated catalysts depended on their pore structure; HY was deactivated faster than Beta zeolites as it allowed the formation of large hydrocarbon intermediates inside the cavities.

The papers from Exxon researchers (Daage and others) [34,35] on ring opening of naphthenic molecules provide evidence for the multiple cleavages of both endo- and exocyclic C–C bonds. Selective ring opening requires that only one endocyclic C–C bond per naphthene ring be severed, preserving thereby reactant molecular weight. The products of selective ring opening are alkanes and alkylnaphthenes. Over hydrocracking catalysts the yield of alkanes with the same number of carbon atoms as the reactant naphthenes is unacceptably low as a result of extensive dealkylation of alkylnaphthenes and secondary cracking of alkanes. Alkylcyclopentanes, in contrast, selectively undergo ring opening by hydrogenolysis over a number of noble metal catalysts.

Under similar reaction conditions, the selective ring-opening rates of alkylcyclohexanes are, however, one to two orders slower than those of alkylcyclopentanes. The addition of a ring-contraction acidity function converts alkylcyclohexanes into more easily ring-opened alkylcyclopentanes, greatly facilitating selective ring opening.

Selective ring-opening rates and selectivities are optimal when ring isomerization occurs by a nonbranching ring contraction. Branching ring contraction, creating increased numbers of ring substituents, is detrimental to both ring-opening rates and selectivities. When an effective acidity function is coupled with a high-activity hydrogenolysis metal, such as iridium, the resulting bifunctional catalyst system greatly outperforms conventional hydrocracking catalysts for the selective conversion of naphthenes to alkanes.

MCP ring opening was studied over platinum-containing catalysts prepared on acid and basic (or neutral) zeolite supports [36]. The acid supports included mordenite, Y, MFI, BEA, and Omega zeolites. The basic (neutral) supports included zeolite LTL. It was demonstrated that the selectivity of MCP ring opening depends on the acid-base properties of the catalysts. In the presence of acid catalysts, the selectivity to ring opening is low and the main reaction is MCP isomerization to cyclohexane followed by dehydrogenation and formation of aromatics. On the basic catalysts, the selectivity to ring opening is very high (up to 98%), as far as the acid-catalyzed isomerization and cracking reactions are almost completely suppressed. The addition of potassium neutralizes the acid active sites of the bifunctional acid catalysts and converts them into monometallic basic catalysts, which are favorable for nonselective MCP ring opening with C_6 paraffin distribution close to the statistical distribution.

The catalytic conversion of cyclohexane over HZSM-5 in hydrogen or nitrogen over a series of H-zeolites loaded with platinum was studied in [46]. The distribution of the products formed via hydrodecyclization was found to depend definitely on whether or not dimerization of primarily formed hexyl-carbenium ions is sterically hindered in the zeolitic pore system. Reaction routes involve, as an essential step, the hydrogenation of intermediate carbenium ions, formed either primarily by reactant adsorption or cracking, by spillover hydrogen activated on the metallic catalyst component. The basically different product distribution obtained by decyclization of cyclohexane in a nitrogen atmosphere or over monofunctional HZSM-5 is attributed to disproportionation of primarily formed hexyl-carbenium ions and subsequent hydrogen transfer reactions, which results in a broad spectrum of paraffins and aromatics in the effluent and carbon-rich deposits on the catalyst surface causing fast catalyst deactivation.

A kinetic analysis for the ring-opening (using a high hydrogen concentration in the reaction medium) of methylcyclohexane over a catalyst based on HZSM-5 zeolite [47] was performed.

The combination of ring-contraction function and ring-opening function in the physical mixtures and segregated beds of Pt/HY and Ir/SiO_2 catalysts in order to optimize the production of the iso-alkanes with the highest octane number allowed the authors of [48] to find the optimal composition. The segregated catalyst beds provide a better control of the selective breaking of C–C bonds of isomers produced by ring contraction, which optimizes the octane number of the products.

Ring opening of cis-decalin on Ir/ and Pt/La-X zeolite catalysts was studied by the group of Weitkamp [49]. The bifunctional catalysts were tested at temperatures between 200 and 255°C and a hydrogen pressure of 52 atm. Both Ir/La-X and Pt/La-X showed a pronounced tendency for skeletal isomerization of decalin. An important isomer was formed by rearrangement, namely spiro[4.5]decane. The studies showed that mainly C–C bond cleavage of the ring-opening products occurs with one remaining naphthenic ring. In another study [50], two series of catalysts were tested based on zeolite Y in which the alkali cations from lithium to cesium were introduced by ion exchange. One series contained 3 wt.% of iridium, the other contained 3 wt.% of platinum. The fastest reaction was the stereoisomerization of cis- to trans-decalin. Other reactions of decalin were skeletal isomerization, ring opening to C_{10} hydrocarbons with one remaining naphthenic ring, double-ring opening to open-chain decanes and formation of hydrocarbons with the carbon number less than 9. For Ir/Na,HY and Pt/Na,HY, the yields of open-chain decanes close to 40% could be achieved. In the case of iridium-containing catalysts, the hydrocarbons with the reduced carbon number were produced with relatively large amounts of C_1+C_9 and C_2+C_8, for which the term "hammock-type curve" was coined. Interestingly, open-chain decanes formed on platinum tend to be less strongly branched compared to iridium catalysts.

Decalin ring opening was also studied on Ir/H,A-Beta zeolite catalysts (A stands for an alkali metal cation) [51]. The Ir content was 1–5 wt.%, and the alkali cation was varied from lithium to cesium. The yields of open-chain decanes were up to 44%, with selectivities close to 47%. The term "high-performance ring-opening catalysts" (HIPEROCs) was proposed (Fig. 4). Hydrogenolysis on the metal is considered as the main ring-opening mechanism on HIPEROCs. The role of Brønsted acid sites is a mild isomerization of six-membered into five-membered naphthenic rings which are much easier to open by hydrogenolysis.

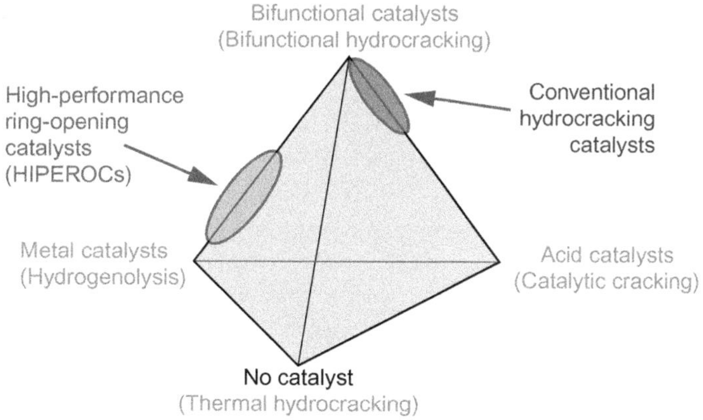

Fig. 4
Illustration of "high-performance ring-opening catalysts." *Figure from reference Santi D, Holl T, Calemma V, Weitkamp J. Appl Catal A Gen 2013;455:46 with permission.*

Perhydroindane was chosen as a substrate for ring opening in [52]. Bifunctional Pd/Na, H-Beta, Ir/Na,HY, and Pt/Na,HY zeolites with a low concentration of Brønsted acid sites were compared to the catalysts containing the same noble metals on the nonacidic support (silica). In the case of the bifunctional zeolite Pd/Na,H-Beta, skeletal isomerization of perhydroindan was the primary reaction followed by opening of one naphthenic ring, the formation of C_9 paraffins in low yields (6%) and hydrocracked products C_8-. The carbon number distribution of the latter was volcano-shaped with no C_1, C_2, C_7, and C_8 indicating a carbocationic hydrocracking of C_9 precursors with one naphthenic ring. For Ir/Na,HY and Pt/Na,HY ("high-performance ring-opening catalysts"), ring opening and hydrocracking to C_8- occurred by hydrogenolysis on the metal nanoparticles. Opening of the five-membered ring was found to be much faster than opening of the six-membered ring. The maximal selectivities to nonanes achieved on Ir/Na,HY and Pt/Na,HY were rather high, 49% and 54%, respectively, and significantly better than the selectivities to decanes observed previously with decalin as a model. The paraffins formed on Pt/Na,HY were much less branched than those formed on Ir/Na,HY, which was accounted for by the different hydrogenolysis mechanisms on these metals:

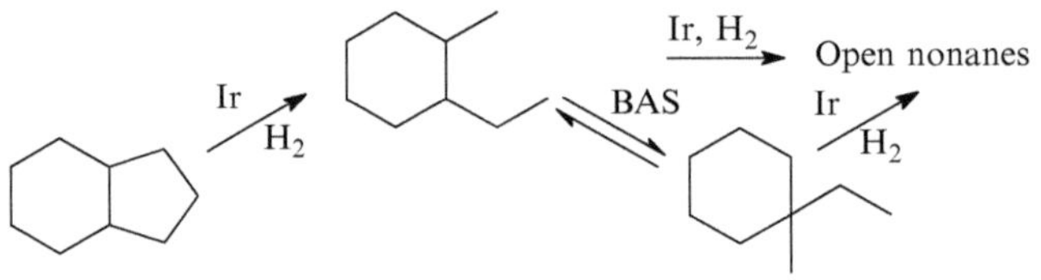

Bimetallic compositions (Pt-Ir/HY) showed a better performance in the ring opening of decalin at 300–350°C compared to the monometallic compositions [53,54]. Increasing the Ir content in Pt-Ir/HY was found to produce an increase of the hydrogenolytic activity and a decrease of the dehydrogenating activity. The acidic character of Ir oxide species increased the amount of strong acid sites of the support. On the other hand, the Pt content increase favored the formation of medium-strength acid sites. The catalysts with a higher content of Ir had the higher hydrogenolytic activity, while an increase in the Pt content decreased the hydrogenolytic activity. The presence of Pt promoted the dehydrogenation of cyclohexane. Higher yields of dehydrogenated products and increased conversion of cis-decalin were observed in the high-temperature region. The Ir(1.0%)Pt(1.0%) catalysts at 350°C and Ir(1.5%)Pt(1.5%) catalysts at 300°C seem to have the proper balance between the metal and acid functions that favors the ring opening of decalin (Fig. 5).

The Pt-Ir zeolite catalysts were compared to the Ni-Mo carbide supported onto HY and H-BEA zeolites [55]. The HY-supported Pt-Ir and Ni-Mo carbide catalysts showed higher ring-opening activity and selectivity than the corresponding H-BEA catalysts. The maximum yield of ring-opening products, 33.5%, was observed at 240°C for the HY-supported Ni-Mo carbide catalyst, though the selectivity was only 40%. For the Pt-Ir/HY catalyst, the selectivity was as high as 65% at 220°C, with the yield being only slightly lower than that for the Ni-Mo carbide system (Fig. 6). The yield and selectivity for the carbide catalysts

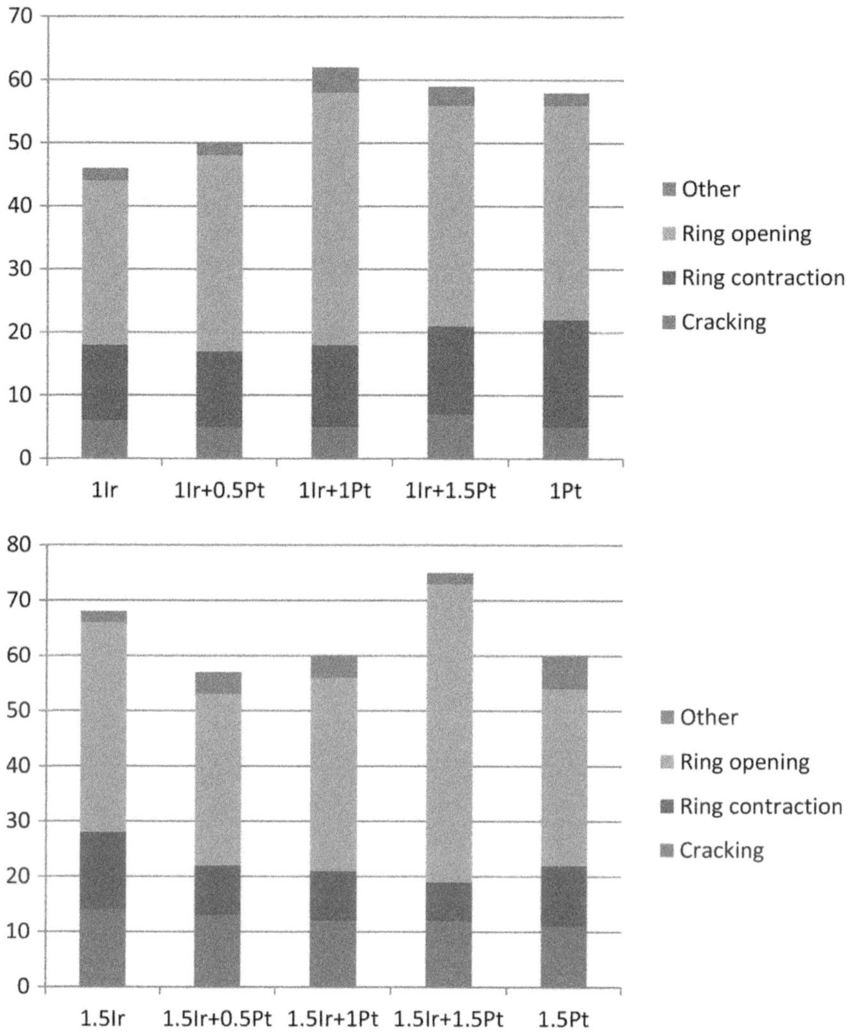

Fig. 5
Decaline conversion of Ir, Ir-Pt, and Pt catalysts based on HY zeolite (figures before the metal symbol denote the metal loading in wt.%).

supported on HY are comparable with the corresponding values for HY-supported noble metal catalysts. Thus the non-noble catalysts were proposed for the ring-opening reaction.

A heavier substrate (perhydrophenanthrene) was subjected to ring opening [56]. The effects of the acidity of Y and ZSM-5 zeolite catalysts, temperature, and catalyst-oil ratio were examined. The products from naphthenic ring-opening reactions of perhydrophenanthrene over zeolite catalysts turned out to be monocyclic or dicyclic cycloparaffins (cyclohexane, decalin), which are further converted into isoparaffins (2-methylpentane, methylhexanes) by side chain breaking, or into monocyclic aromatics (benzene, C_{1-4}-alkylbenzenes) by hydrogen transfer, or into dicyclic aromatics (tetrahydronaphthalene, naphthalene, and alkyl

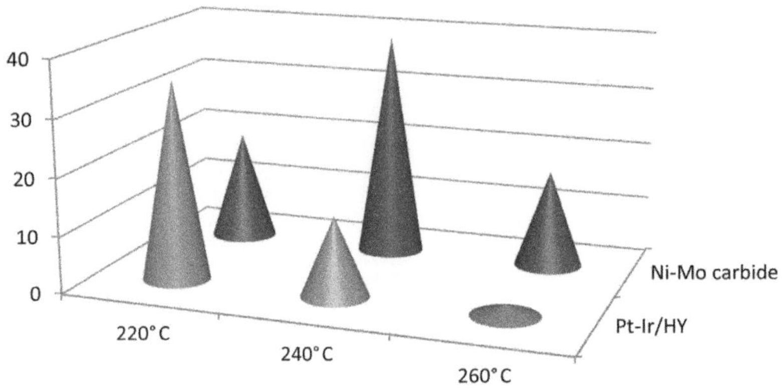

Fig. 6
Yield of ring-opening products: comparison of Pt-Ir/HY zeolite and Ni-Mo carbide catalysts.

naphthalenes) by deep hydrogen transfer. Products from dehydrogenation condensation reactions of perhydrophenanthrene are polycyclic aromatics with more than three rings (phenanthrenes, pyrenes) and coke. Due to the influence of naphthenic diffusion and adsorption on catalysts, the selectivity of the naphthenic ring-opening reaction is higher over Y catalyst than that over the ZSM-5 catalyst. The relative ratio of naphthenic ring-opening reactions to dehydrogenation condensation reactions is higher over the Y zeolite catalyst than that over the ZSM-5 catalyst. At the quite high-weight hourly space velocity of $10\,h^{-1}$, temperature of 475–550°C, catalyst-oil mass ratio of 3.0–9.0, reactions of bimolecular hydrogen transfer and dehydrogenation condensation contribute significantly, and the selectivity of the products from naphthenic ring-opening reactions decreases with the rising temperature or the catalyst-oil ratio over the Y zeolite catalyst.

The efforts to replace noble metals in the composition of ring-opening catalysts were partially rewarded in the case of carbided molybdenum-containing zeolite systems [57]. The studies of ring opening of naphthalene over Mo_2C/HY catalysts containing 7–27 wt.% Mo_2C showed a synergistic effect between the Mo_2C and the acidic support that cannot be obtained with a mechanical mixture of the two catalyst components. Bulk Mo_2C showed no selectivity for ring-opening products [58]. Supporting Mo_2C on HY zeolites significantly increased the hydrogenation of naphthalene and the subsequent formation of ring opening (RO) products, as compared to the HY zeolites alone or the bulk material [59]. The yield of ring-opening products depends on the SiO_2/Al_2O_3 ratio and the Mo_2C loading and reaches 33 wt.% for 7.4 wt.% Mo_2C/HY zeolite of moderate acidity ($SiO_2/Al_2O_3=12$). This value of the yield is comparable to the yields reported for noble metal catalysts on acidic supports.

The application of ring-opening catalysts in semi-industrial tests is described using upgrading of light cyclic oil as a feed and sulfided Ni-W/USY zeolite catalysts [60]. The process involved both hydrogenation and ring-opening reactions. Preliminary tests with tetralin as the model compound revealed high tetralin conversions at T=613 K, P=4.0 MPa, $LHSV$=2.0 h^{-1}, H_2/oil (v/v)=500. The conclusion has been drawn that the medium to strong zeolite acidity is the key factor in the ring-opening selectivity. The maximum high cetane ring-opening product yield of 39.1 wt.% is obtained at a C_{10} yield of 98.2 wt.% for the ultra-stable Y (USY) zeolite with a medium acidity. With the light cycle oil as a feed, the activity of the catalyst was shown to be higher than that of commercial catalysts, giving the cetane number increase by some 14 points at the diesel yield of 97.6 wt.%.

Upgrading of the aromatic feedstock (pyrolysis gasoline) over bifunctional Pt catalysts on MFI, BEA, and faujasite (FAU) zeolites, and an amorphous silica-alumina [61] at 5 MPa and 350–450°C was performed in different directions: the Pt/MFI catalyst was optimal for a steam cracker feedstock production, the Pt/BEA catalyst—for isoalkane-rich gasoline pool manufacture and the Pt/silica-alumina catalyst for severe aromatic reduction while controlling the extent of ring scission.

5.3 Hydroprocessing (Aromatics Saturation, Hydrodesulfurization (HDS), Hydrodenitrogenation (HDN))

Typically zeolites are used downstream of the hydrodesulfurization units to crack the low octane components (Octgain process by Mobil and ISAL by Intervep/UOP) (Maxwell, Stork, in Ref. [2]).

There are some examples when the zeolite carriers were used for supporting the well-known Mo-Ni active catalysts for HDS/HDN (hydrodesulfurization/hydrodenitrogenation) processes. Zeolite catalysts containing Mo-Ni active phases or noble metals turned out to be quite efficient for hydroprocessing of diverse feedstocks. Pt-zeolite catalysts exhibit a high catalytic activity in thiophene hydrodesulphurization and benzene hydrogenation [62]. Finely dispersed Pt nanoparticles (2–5 nm) located in zeolites demonstrated high efficiency in ultra-deep desulphurization of light cyclic oil. The good catalyst performance in the hydrogenation activity and sulfur resistance can be explained by the favorable pore space architecture and the location and the electronic state of the Pt clusters.

6 Gas to Liquid: The Place of Zeolites

6.1 Methane Aromatization

Starting with the early publications in 1980s, quite many attempts have been made to develop Mo/zeolite catalysts for methane aromatization:

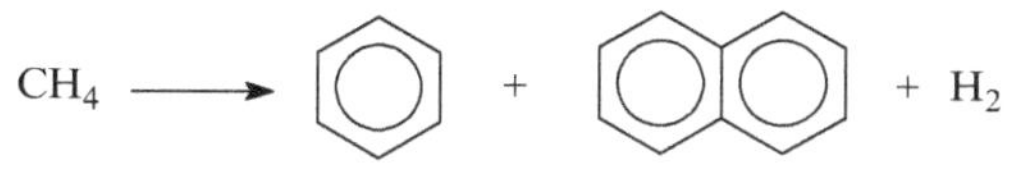

Although the Mo/HZSM-5 zeolite demonstrates the conversion close to 10–12% and the selectivity to benzene about 85–90%, its stability is poor and the quest for new Mo-containing zeolites as well as for additives to Mo/ZSM-5 materials is still under way. The studies of Mo/ZSM-5 and Mo/MCM-22 (4.67 wt.% Mo) [63] showed that the Mo catalyst based on microsized ZSM-5 is significantly more active, selective and stable in methane conversion compared with the catalyst based on the ZSM-5 sample with the particle size in the nanosized region. Physicochemical analysis indicated that Mo is located mostly on the external surface in the nanosized catalyst, while Mo species in microsized HZSM-5 interact with the internal Brönsted acid sites. These species are assumed to be the precursors for the formation of the most active carbided Mo sites.

Comparison of the Mo-IM-5 and Mo-ZSM-5 catalysts in nonoxidative aromatization of methane [64] showed that the former exhibits a higher methane conversion and higher benzene selectivity. The Mo-IM-5 catalyst was also more stable than Mo-ZSM-5. The catalytic behavior of Mo-IM-5 may be attributed to its unusual two-dimensional 10-member-ring channel system with the character of three-dimensional cavity.

Mo-modified catalysts Mo/TNU-9 [65] perform in the best way in methane aromatization at Si/Al = 50, 6 wt.% MoO_3 loading, and reaction temperature of 973 K. The Mo/TNU-9 catalyst showed a higher CH_4 conversion and selectivity to benzene compared with Mo/ZSM-5. The performance of Mo/TNU-9 can also be explained by the system of three-dimensional 10-member-ring channels with the presence of large 12-member-ring cavities.

The Mo-containing MFI-type core-shell HZSM-5-Silicalite-1 s (HZ5@S1) materials with various core-shell ratios prepared by the epitaxial growth of Silicalite-1 on HZSM-5 demonstrated the high shape selectivity to aromatics and stability in methane dehydroaromatization [66]. The silicalite-1 layer covering the HZSM-5 core can eliminate the external acid sites, and thus prevent the formation of active Mo species associated with Brønsted acid sites on the external surface of catalysts. However, the overgrowth of the Silicalite-1 shell may lead to a severe inhibition of the Mo species migration into zeolite pores and consequent anchoring on the Brønsted acid sites of the HZSM-5 core. Therefore, the catalytic performance is dependent on the core-shell ratio.

Mesoporous ZSM-5-S and ZSM-5-M samples prepared by using ordered mesoporous carbon (CMK-3) and disordered carbon rods (C-MCM-41) as the hard template, respectively, were compared with the conventional ZSM-5-C in methane aromatization [67] The Mo-ZSM-5-S and Mo-ZSM-5-M catalysts showed similar conversions of methane, but higher yields of aromatics compared with Mo-ZSM-5-C. The mesoporous catalysts were also more stable than Mo-ZSM-5-C. The formation of the secondary mesoporous system within the zeolite crystal, which may lead to easier access to the active sites for reactants and be favorable for the diffusion of larger molecules formed in the microporous channels during the methane aromatization reaction, account for the better stability of the mesoporous systems.

Boron (0.05–0.2 wt.%) and silver (0.5 wt.%) additives exhibit a positive effect on the performance of Mo/HZSM-5 catalysts [68]. Boron did not influence the conversion but improved the selectivity for C_2 hydrocarbons with increasing boron content, while the selectivity for aromatics was decreased when the boron content was higher than 0.05 wt.%. The presence of silver in Mo/HZSM-5 catalysts increased the conversion and selectivity for C_2 hydrocarbons and aromatics (Fig. 7).

Zn/HZSM-5 catalysts known as efficient systems for ethane and propane aromatization (the UOP Cyclar Process) also showed some activity in methane aromatization [69]. At 723 K, methane can easily be activated in the presence of ethylene in the feed and converted to higher hydrocarbons (C_2–C_4) and aromatics (C_6–C_{10}), through its reaction over rare metals modified Zn/HZSM-5 zeolite catalysts. The CH_4 conversion of 37.3% was obtained and the catalysts showed a longer lifetime than usual metal supported HZSM-5 zeolite catalysts without adding any rare-earth metals.

$$CH_4 + C_2H_4 \rightarrow C_3H_8$$

Attempts have also been made to improve the performance of the conventional Mo/HZSM-5 catalyst [70]. The particle size of HZSM-5 affects the activity of Mo/HZSM-5 in the nonoxidative CH_4 dehydroaromatization. Thus, ball-milling of the 5 wt.% Mo/HZSM-5 catalyst of an average particle size of 4 μm to produce the samples with the average particle sizes of 1.9, 1.0, 0.5, and 0.2 μm resulted in the lower Mo dispersion and lower acidity compared to the parent zeolite catalyst. The maximum benzene formation activities of the catalysts with the particle sizes of 1.9–0.5 μm were higher than that over the parent catalyst. All these data are indicative that the use of the zeolites with small particle sizes with less resistance toward the diffusion of formed benzene out of their short channels is effective in designing the Mo/HZSM-5 catalyst to ensure the high benzene formation activity.

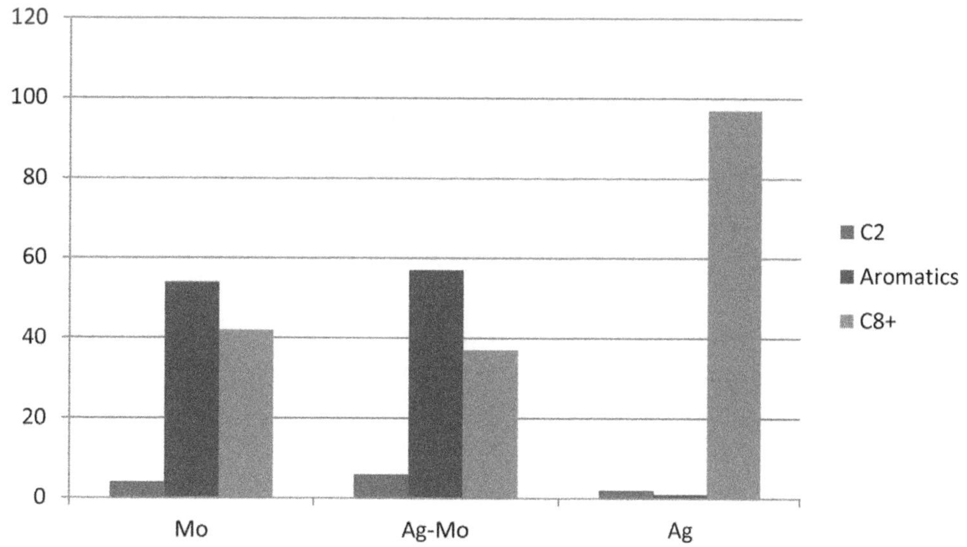

Fig. 7
The effect of silver on the selectivities of Mo/ZSM-5 catalysts in methane aromatization.

Ways to improve the catalytic activity and stability of Mo/HZSM-5 catalysts are described based on the catalyst and reaction as well as reactor design [71]. These ways include:

- Addition of coreactants,
- Coupling the methane aromatization with other reactions, such as Boudouard reaction and methane reforming or oxidative coupling of methane,
- The use of membrane reactors or fluidized bed reactors.

6.2 C_2–C_4 Aromatization and Dehydrogenation

The advances in recent research on nonoxidative aromatization of methane in the presence of propane over different modified HZSM-5 catalysts are reviewed in [72]. The thermodynamic consideration, the isotope verification, and the mechanism of the activation of methane in the presence of propane are discussed in the paper in detail.

Involvement of C_2–C_4 alkanes in the aromatic products can be achieved either by direct aromatization of the paraffinic molecules (similar to the Cyclar process) or via interaction of these alkanes with aromatic molecules, like benzene (dehydroalkylation).

Benzene alkylation with ethane into ethylbenzene is possible over a Pt/H-MFI bifunctional catalyst at temperatures 290–490°C [73].

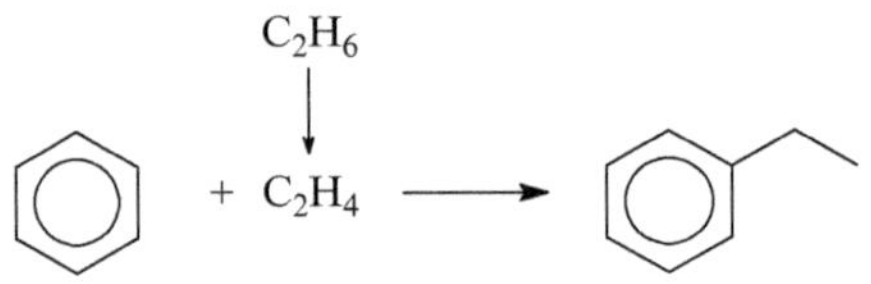

The temperature affects in a different way the thermodynamic equilibrium of two major reaction steps: (i) ethane dehydrogenation into ethene on Pt particles and (ii) benzene alkylation with ethene on the Brønsted acid sites. Thus, significant differences in the product distribution and catalyst performance are observed at different temperatures. With increasing temperature, ethane dehydrogenation is accelerated, whereas the ethylbenzene formation rate increases in much lesser extent. Therefore, a high concentration of ethene is observed at high temperatures (450–490°C), while the maximum concentrations of ethylbenzene are almost constant at all temperatures. Ethene, which is formed in excess at high temperatures, is converted via oligomerization and cracking steps into higher alkenes that alkylate benzene (or aromatic products). The alkylation reactions are accompanied by cyclization and dehydrogenation resulting in the formation of polyaromatics (coke precursors) and subsequent catalyst deactivation. On the contrary, at lower temperatures (290–410°C), ethene reacts preferably with benzene to produce ethylbenzene. Side reactions are suppressed in this case and high selectivity and catalyst stability are observed. The conclusion has been drawn on the basis of the analysis of the activity, selectivity, and stability patterns: the optimum temperatures for the direct benzene alkylation with ethane into ethylbenzene are 370–410°C.

Further improvement of the process may be achieved by selective removal of hydrogen from the reactor.

The deactivation of the Pt/H-MFI catalyst in the reaction of benzene alkylation with ethane proceeds in two modes [74]. The first mode corresponds to the selective deactivation of the Pt sites, which are highly active in the hydrogenolysis reactions. This deactivation occurs during the first ~4 h of the catalyst operation and results in the significant enhancement of the selectivity to ethylbenzene. The second, much slower, deactivation process is observed during tens of hours and is responsible for the deactivation of Pt and acid sites with similar rates.

Dehydrogenation of propane (or butane) that may be considered as the first stage of the aromatization process proceeds on microporous Pt/ZSM-5 and Pt/BEA and mesoporous Pt-SBA-15 materials [75]. Pt-ZSM-5 exhibits the highest propane conversion followed by Pt-BEA and Pt-SBA-15 indicating that the activity decreases with increasing pore size (ZSM-5 < Beta < SBA-15) of the support. The amount of coke formed is the highest on Pt-ZSM-5 and is the lowest on Pt-BEA. Pt-ZSM-5 shows a better selectivity and stability than Pt-SBA-15, despite a higher coke content. These findings demonstrate that the three-dimensional microporous materials are better catalytic supports for propane dehydrogenation than the mesoporous SBA-15 material.

The performance of a new Pt-Sn/slit-SAPO-34 catalyst in selective C_3–C_4 dehydrogenation to light olefins [76] was shown to be optimal at an atmospheric pressure, hydrogen to alkane molar ratio of 0.2, weight hourly space velocity of 5 h^{-1} and temperature about 585°C. The light paraffin conversion as high as 40% and the total olefin selectivity above 95% were achieved over Pt-Sn/slit-SAPO-34.

6.3 Fischer-Tropsch Process

Presently the major step in the GTL conversion is considered to be the Fischer-Tropsch synthesis of hydrocarbons from synthesis gas ($CO + H_2$). The classical processes produce waxes (solid hydrocarbons) that are further upgraded into the components of liquid fuels (gasoline, diesel). Diesel production by this technology seems to be the best solution. The synthetic diesel fuel has better characteristics compared to the fuel grades produced from oil (standard EN-590): the cetane number is 75 (versus 55 for the oil-derived diesel); the content of polynuclear aromatic compounds is 0.1 (versus 6%); the sulfur content is 0 ppm (versus 50 ppm). Such synthetic fuels can be used as additives to the oil-based diesel. GTL-diesel is used in Germany, Austria, and Sweden and the blends of the synthetic fuel and conventional diesel are used in France, Italy, and other countries.

The Shell technology—Shell Middle Distillate Synthesis—produces the diesel fraction by hydrocracking of solid waxes, which were the products of Fischer-Tropsch synthesis. An alternative may be the direct production of liquid products by Fischer-Tropsch (FT) synthesis in a single step by using carriers exhibiting strong acidity, like zeolites. The acidic carrier can perform several functions:

1. Ensuring the higher dispersion and therefore higher specific surface area of the supported active metal.
2. Forming new phases with a supported component.
3. Revealing the own catalytic activity in the conversion of intermediates and products formed in FT synthesis.

The position of the maximum on the curve corresponding to the Schulz-Flory distribution depends on the surface area of the carrier [77]: with increasing S, the maximum is shifted to the products with a lower molecular weight. The reasons behind this pattern are the high dispersion of the active catalyst [78] and internal diffusion control of the chain growth by the carrier (the so-called cage-effect typical for zeolites) [79]. The shape-selectivity effects are also observed in the FT synthesis on zeolite-based catalysts, with certain advantages characteristic for zeolites [80]:

- The high surface area;
- The uniform distribution of the acid sites;
- The narrow pore-size distribution affording the formation of active metal nanoparticles of the uniform size;
- The enhancement of the dispersion of the active metal;
- The shape-selectivity effect with respect to the hydrocarbons formed [81,82], decreasing the selectivity toward high-molecular-weight products.

The necessary prerequisite of the zeolite support in the case of Co-containing catalysts for modified FT synthesis is its activity in the secondary reactions of hydrocarbons at the temperatures typical for the FT synthesis (180–250°C). It should be noted that the active sites of the chain growth in the FT synthesis and the active sites of zeolites responsible for cracking and isomerization demonstrate different deactivation behavior. The acid sites deactivate faster due to coke formation and the separate regeneration of the acid sites is hardly possible. One of the ways to suppress the deactivation is supporting small quantities of metals active in hydrogenation (Pd) [83]. Palladium additives ensure the decrease of the olefin content in the products [84] and synergy between the acid sites and Pd nanoparticles due to hydrogen spillover from palladium. Three possible options for the arrangement of the acid and metal functions responsible for the secondary reactions and FT synthesis, respectively, are described in the literature:

A. Supporting the metal (cobalt) on the zeolite as a carrier by impregnation with solutions of the precursors. The drawback may be the destruction of the zeolite framework and a decrease of the number of acid sites as a result of the localization of a cobalt oxide precursor in the vicinity of the framework aluminum ions [85]. It is known that the highest yield of liquid hydrocarbons on Co-zeolite catalysts of FT synthesis is observed for zeolites with higher Si/Al ratios (up to 100) and at Co loading of 20–35 wt.% [86].

B. A mechanical (physical) mixture of the FT synthesis catalyst and zeolite [87]. The size and shape of the mixed particles affects the performance of the resulting mixture [88]. α-Olefins formed as primary FT synthesis products are isomerized or cracked on acid sites of the zeolite, which prevents the participation of these α-olefins in the further chain growth via readsorption. When the powdered components are mixed, the activity enhancement is most pronounced [89].
C. Preparation of so-called capsular FT synthesis catalysts [90], with the nucleus being the catalyst of hydrocarbon synthesis from CO and H_2 and the outer shell is the zeolite membrane (Fig. 8). The depth of the zeolite shell controls the transport of CO and H_2 to the active sites of the FT catalyst and of hydrocarbons from this catalyst. The disadvantage of this type of capsular catalysts is the complexity of the preparation. Also, the CO conversion on capsular catalysts is lower and the H_2/CO ratio is higher than on the mechanical mixture of the components, since the membrane limits the transport of the feed to the nucleus, while CO is passing through the membrane faster than hydrogen. Typically, the capsular catalysts demonstrate a very narrow product distribution (hydrocarbons C_9–C_{12}), unlike the mechanical mixture (usually C_8–C_{20}).

The composition of hydrocarbons formed on a hybrid catalyst is determined both by the properties of the FT catalyst and the zeolite. The accessibility of acid sites to *n*-paraffins in the case of 10%Co/zeolite catalysts [91] is governed by the zeolite architecture, which has even a more significant effect on the group and fraction composition of the products than the strength of acid sites. With increasing cobalt content:

- The activity of the catalyst increases;
- The temperature of the FT synthesis decreases;
- The extent of polymerization of the products grows [92];
- The hydrogenating function of the catalyst is enhanced (ie, the content of olefins in the products decreases [93];
- The reduction of the active component is easier.

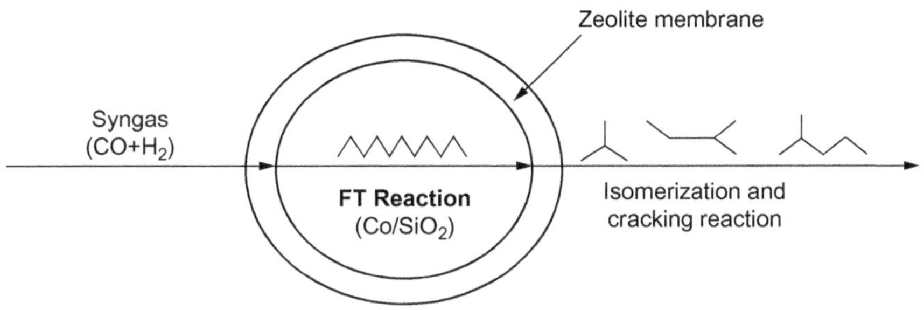

Fig. 8
A schematic presentation of the operation of the capsular catalyst in FT synthesis. *Figure from reference He J, Yoneyama Y, Xu B, Nishiyama N, Tsubaki N. Langmuir 2005;21:1699 with permission.*

The crucial factor affecting the molecular weight distribution is the dispersion of cobalt particles [94]: the higher the dispersion, the higher the content of liquid hydrocarbons in the products. The following parameters, in turn, determine the dispersion of cobalt on a zeolite catalyst [95]:

- Pore diameter of the zeolite;
- Cobalt loading in the catalyst;
- Interaction between the active component and the zeolite;
- Conditions of the thermal treatment and preliminary activation of the catalyst.

Besides the zeolite-based catalyst characteristics, the process conditions exert an effect on the product distribution. Usually the higher process temperatures are beneficial for the lower molecular weight of the synthetic hydrocarbons, whereas the pressure increase results in a growth of the extent of polymerization.

Cobalt on Beta zeolite catalysts [96] demonstrated a bifunctional character with the maximum activity at a specific concentration of the metal and particle size. The selectivity was strongly influenced by the zeolite Beta acidity, which promotes the formation of isoparaffins at 220°C, unlike conventional zeolites revealing a poor isomerization activity at this temperature. This difference was explained by the small size of Beta zeolite crystallites.

The optimum operating conditions for Co/HZSM-5 were determined as $T = 513$ K, space velocity $= 1.1\ h^{-1}$, Co on HZSM-5 = 11.6%, $H_2/CO = 1.7$ [97].

Except for cobalt systems, other metals also demonstrate activity in the Fischer-Tropsch process. Mo/HZSM-5 turned out to be active in FT synthesis [98]. The catalysts were tested at a low H_2/CO molar ratio ~ 1.0, this composition being typical for biomass gasification. Liquid hydrocarbons obtained on Mo/HZSM-5 at 573 K were presented by alkylaromatics and lower branched and cyclic alkanes. The formation of aliphatic hydrocarbons was close to zero. The gas products included C_1–C_4 alkanes. Higher alcohols and carboxylic acids (C_1–C_6) were observed in the aqueous phase. The formation of hydrocarbons on Mo/zeolite is accounted for by the bifunctional zeolite acidity and molybdenum metal activities via alcohols as intermediates. The zeolite Y was also found to be a good support for Mo in the FT reaction.

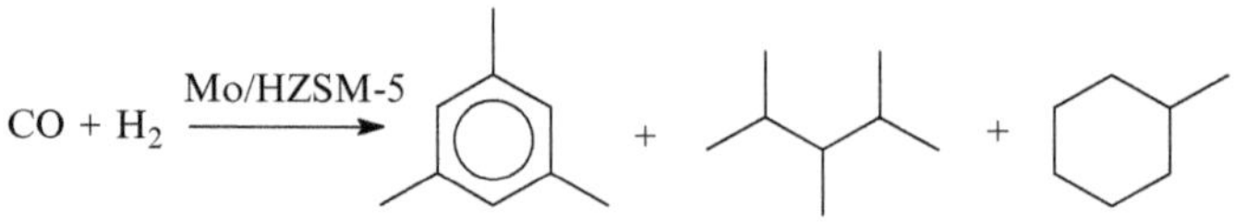

Syngas conversion offers a potential alternative to flaring or reinjection for monetization of associated gas. The Chevron's Gas Conversion Catalysis (GCC™) technology, a commercially viable hybrid syngas conversion catalyst comprising both a Fischer-Tropsch active metal and a zeolite component that provides the proper product distribution in favor of liquid hydrocarbons without the need for downstream hydroprocessing at high pressures is described in [99].

Ternary composites have also been used comprising a Fischer-Tropsch catalyst, a methanol synthesis catalyst, and a zeolite [100]. Two Fe-based catalysts (ie, one promoted by K and the other by Ru), two HY zeolites with different acidities, a commercial HZSM-5, and $Cu/ZnO/Al_2O_3$ (methanol synthesis catalyst) were tested in these composites. Dimethyl ether (DME), methanol, and hydrocarbons were formed. Addition of the $Cu/ZnO/Al_2O_3$ catalyst to a binary mixture of a Fischer-Tropsch catalyst and HZSM-5 results in the increase of the CO conversion by more than 20 times. The DME selectivity decreases as the conversion increases. Y zeolites and the Fischer-Tropsch synthesis catalyst promoted by Ru generated the most active composites. The role of zeolites in the ternary composite is assumed with the DME synthesis. First, methanol is synthesized from syngas on $Cu/ZnO/Al_2O_3$; then it is dehydrated by an acid catalyst to produce DME; and finally, DME initiates FT synthesis, which is then propagated by CO.

Ru,Ni-promoted Co/HZSM-5 catalysts were efficient in the FT synthesis of gasoline-range hydrocarbons [101]. The activity increased significantly upon the addition of a Ru or Ni promoter. The CO conversion increased in the order Co/HZSM-5 < Ru–Co/HZSM-5 < Ni–Co/HZSM-5. The Ru–Co/HZSM-5 catalyst is still active after 150 h on stream at 540 K. The Ni–Co/HZSM-5 catalyst also showed good catalytic performance even at a lower temperature of 524 K. The temperature rise leads to the increase of the content of isoparaffins.

Fischer-Tropsch synthesis can also be used for olefin production from syngas, in particular using Fe–Cu–K catalysts on ZSM-5 [102]. The Fe–Cu–K/ZSM-5 catalyst with a low Si/Al ratio (25) is more selective to olefins C_2–C_4 than other catalysts and ensures a higher olefin/(olefin + paraffin) ratio, presumably due to the facile formation of iron carbide in the course of the FT reaction.

10%Co/ITQ-2 catalysts prepared by reverse micellar synthesis using a surface silylated ITQ-2 delaminated zeolite [103] is characterized by the uniform Co^0 particle size distribution (5–11 nm). The authors revealed the effect of the Co particle size on the catalyst performance. At the Fischer-Tropsch synthesis conditions (493 K, 2.0 MPa), TOF increases from 1.2×10^{-3} to $8.6 \times 10^{-3}\,s^{-1}$ when $d(Co^0)$ increased from 5.6 to 10.4 nm, and then it remains constant up to a particle size of 141 nm. The Co surface reconstruction occurs during the reaction, no matter what the metal particle size is. Interfacial $Co^{\delta+}$ sites are likely the reason behind the decreased TOF observed for $d(Co^0) < 10$ nm (Fig. 9).

30%Co/SBA-15 catalysts with different ruthenium contents (0.05–0.5%) [104] demonstrate some advantages over other types of zeolite catalysts. The addition of Ru shifted the reduction temperature of cobalt ($Co_3O_4 \rightarrow CoO$ and $CoO \rightarrow Co^0$) to lower temperatures and suppressed the formation of Co^{2+} species. Hydrogen spillover from ruthenium to cobalt oxide clusters was postulated. With the increase of the ruthenium content, enhancement of the catalytic activity in FT synthesis was unraveled.

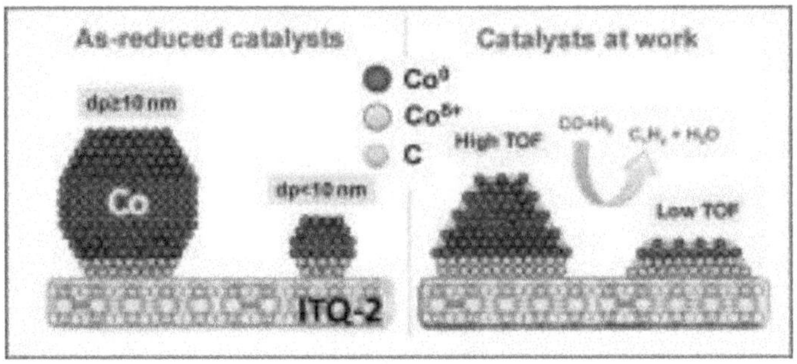

Fig. 9
Reconstruction of the Co catalyst during the FT synthesis reaction. *Figure from reference Prieto G, Martínez A, Concepción P, Moreno-Tost R. J Catal 2009;266(1):129 with permission.*

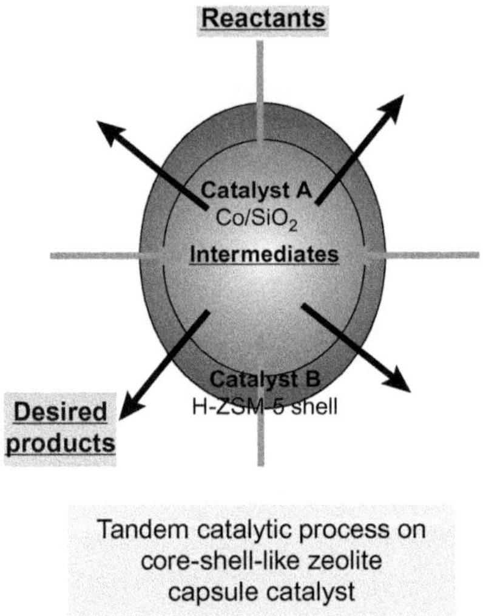

Fig. 10
The operation of the capsule-type catalyst. *Figure from reference Yang G, Xing C, Hirohama W et al. Catal Today 2013;in press with permission.*

Light isoparaffins can be produced from syngas on a hybrid (capsular) catalyst representing a mixture of the FT catalyst and zeolite [105]. The zeolite capsular catalyst with a core-shell structure was prepared using 10% Co/SiO_2 as core catalysts and HZSM-5 zeolite. The performance of the capsule-type catalysts (Fig. 10) in the one-step synthesis of light isoparaffins was better (a higher selectivity) than that of the conventional hybrid catalyst.

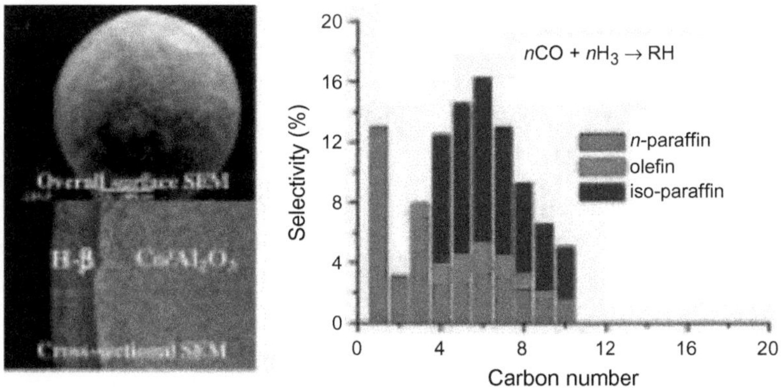

Fig. 11
The structure and performance of the core-shell FT catalyst. *Figure from reference Li X, He J, Meng M, Yoneyama Y, Tsubaki N, J Catal 2009;65(1):26 with permission.*

Another type of the encapsulated catalyst with the H–β zeolite shell over Co/Al_2O_3 pellets was described in [106]. The core-shell structure without defects was produced by the hydrothermal synthesis. The molar ratio of $C_{isoparaffin}/C_{n\text{-}paraffin}$ in the products increased by 64% for the encapsulated catalyst compared with the physically mixed components (Fig. 11).

The capsule catalyst Co/SiO_2-Z with zeolite HZSM-5 as the shell and the Co/SiO_2-Z-Pd capsule catalysts were used for isoparaffin synthesis from syngas at $H_2/CO=2/1$, 1.0 MPa and 533 K [107]. The isoparaffin and olefin selectivity increased for the HZSM-5-based capsule catalysts compared to the conventional Co/SiO_2 catalyst. Supporting Pd by sputtering caused the selectivity to isoparaffins to increase, and the selectivity to olefins to decrease compared to the samples prepared by incipient wetness, because metallic Pd supported by sputtering on the zeolite shell could hydrogenate olefins efficiently. The CO conversion, CH_4 and CO_2 selectivity on the catalyst Co/SiO_2-Z-Pd prepared by sputtering increased with a temperature rise from 513 to 553 K. At elevated temperatures, the isoparaffin selectivity also increased, whereas the olefin selectivity diminished.

7 Petrochemicals

7.1 Aromatics Alkylation With Alkanes, Alkenes, and Alcohols (Cumene, Dimethylnaphthalene, Ethylbenzene, and Diethylbenzene)

Diverse alkylated products are in strong demand as octane boosters, precursors for other petrochemical and bulk chemical products. Besides conventional zeolites traditionally used in alkylation processes (Y, Beta), quite a few novel structures have been proposed.

Preliminary coking was used to modify the alkylation activity of the MCM-49 zeolite. Lewis acid sites (73%) are mainly located in the supercages, while 38% of Brönsted acid sites are located in the surface pockets, 28% in the sinusoidal pores, and only 34% in the supercages. Alkylation of benzene with ethylene mainly occurs in the surface pockets, and the contribution from the sinusoidal pores is limited [108]. The generation of heavy aromatics in supercages results in the deactivation of MCM-49. The mild alkali treatment enlarged the opening of supercages without affecting the sinusoidal pores, thereby resulting in the improved diffusion of large molecules and higher catalytic activity in the alkylation of benzene with ethylene (Fig. 12).

The aromatics alkylation with olefins is a well-known technology, especially in the case of ethylbenzene (a Mobil-Badger process [109]) and cumene production [110]. Ethylbenzene synthesis can be catalyzed by diverse modified HZSM-5, BEA, rare-earth Y, and MCM-22 zeolites. In most cases, the selectivity is pretty high (99%), but the process must be carried out at a large excess of benzene and the conversion of the latter typically does not exceed 20–25% at 400°C and $WHSV = 3\,h^{-1}$. For cumene production, a few commercial processes have been developed by CD-Tech (Y zeolite), Lummus-Unocal (Y zeolite), Enichem (H-BEA), Mobil-Raytheon (MCM-22), Dow Chemical (dealuminated mordenite (MOR)), and UOP (a Q-Max process with MgSAPO-31).

2,6-Di-isopropylnaphthalene is an important precursor for polyesters and can be produced using dealuminated mordenites with unidimensional channels (Dow Chemical, Catalytica, Monsanto) with the preferential formation of 2,6-isomer with an

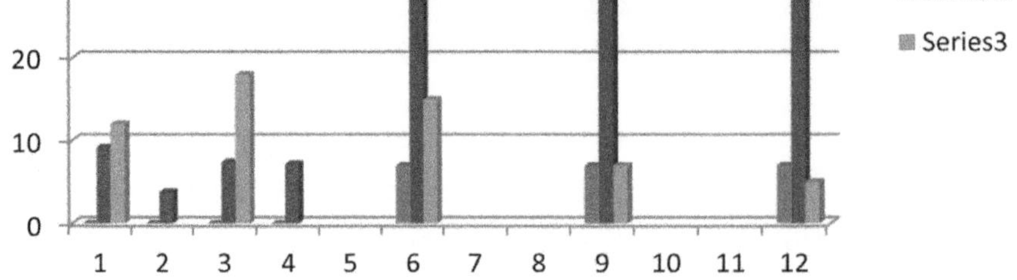

Fig. 12
Dependence of the ethylene conversion (in %) in sinusoidal pores (series 1), pockets (series 2), and supercages (series 3) versus the duration of the steam treatment of zeolite MCM-49 (in hours).

excess over the 2,7-isomer. Similar modified zeolites are also selective in alkylation (p,p′-di-isopropylation) of biphenyl [111].

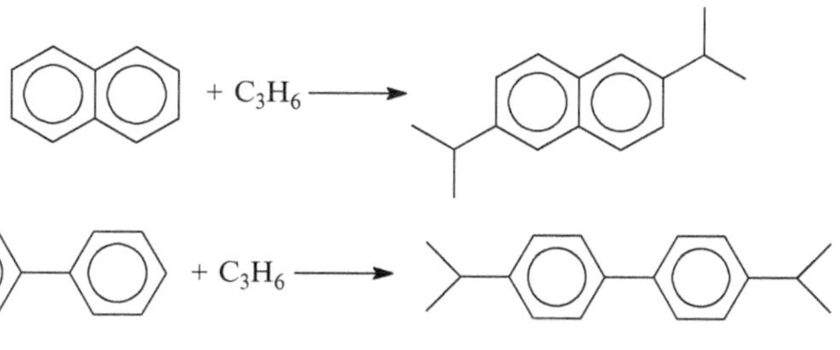

7.2 Methanol-to-Olefins and Methanol-to-Gasoline Processes (MTO/MTG)

In the Mobil methanol-to-gasoline (MTG) process, methanol first is converted into olefins that are then transformed into gasoline-range hydrocarbons on a HZSM-5 zeolite. Only medium-size zeolites are suitable for these reactions, because the smaller-size zeolites cannot accommodate aromatic hydrocarbons, whereas the large-pore zeolites produce heavier products [112]. An alternative to ZSM-5 zeolite is ZSM-11 with close geometries of the channels.

The Mobil methanol-to-olefins (MTO) process may seem to be an abridged version of the MTG process, but the distribution of olefins is governed not only by the process conditions, but largely by the architecture of the zeolite, with ZSM-5 being the best option for the modification. In this case, smaller-pore zeolites (ZSM-34, SAPO-34) show excellent results so far as the olefins formed should not be transformed into gasoline. The latter zeolite-like material developed and commercialized by UOP jointly with Norsk Hydro provides the selectivities to ethylene and propylene above 85% at the 100% methanol conversion [113].

The conversion of *n*-butane and methanol (methanol/*n*-butane molar ratio of 3) on HZSM-5 zeolites with different SiO_2/Al_2O_3 ratios and doped with Ni and SAPO-18 affords the yield of C_2–C_4 olefins 24.4% (11.5% of propene), with a selectivity of 43% at 575°C (Fig. 13) [114]. The catalyst survives at least 10 reaction-regeneration cycles.

DME homologation allows one to produce isobutane and triptane (2,2,3-trimethylbutane) on the acid forms of zeolites (BEA, FAU, MFI) via methylation and hydride transfer steps that favor isobutane and triptane formation because of the relative stabilities of transition states for chains along the preferred growth [115]. Large-pore zeolites (BEA, FAU) form triptane with a higher selectivity than mesoporous SiO_2–Al_2O_3, because confinement within large pores preferentially solvates the larger transition states for hydride transfer and methylation. This preserves the four-carbon backbone in triptane, over smaller transition states for alkoxide isomerization steps, which disrupts this backbone and causes growth beyond C_7 chains and subsequent β-scission to form isobutane. MFI forms isobutane and triptane with a much lower selectivity than mesoporous acids or large-pore zeolites. This is because smaller pores restrict the

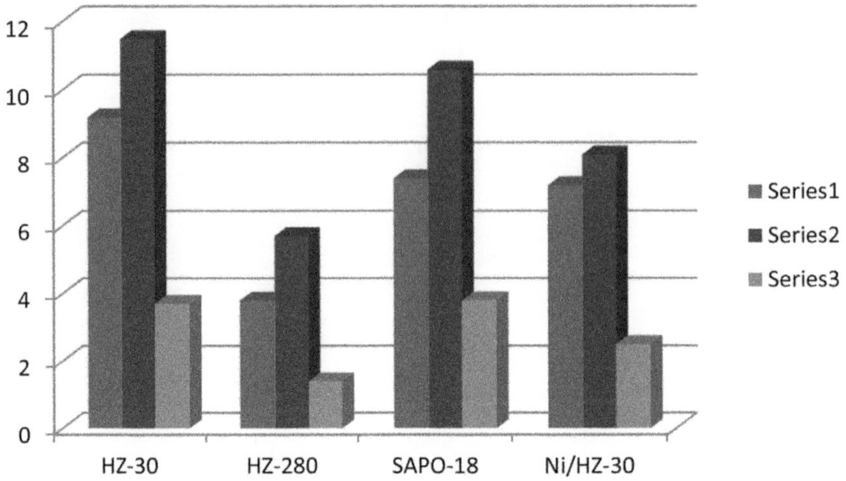

Fig. 13
Yields of olefins (1-ethene, 2-propene, 3-butenes) in the conversion of the butane + methanol mixture on different zeolites.

formation of bimolecular methylation and hydride transfer transition states required for chain growth and termination steps compared to those for monomolecular alkoxide isomerization. Thus, the selective formation of isobutane and triptane from C_1 precursors like DME is favored on acid sites as a result of the relative stability of methylation, hydride transfer, and isomerization transition states, but to a lesser extent when small confining voids and stronger acid sites preferentially stabilize monomolecular isomerization transition states. The observed effects of acid strength and confinement on rates and selectivities reflect the more effective stabilization of ion pairs on stronger acid sites and within solvating environments.

The oligomerization of light olefins into gasoline and middle distillates was the focus of intensive research in the 1980s; however, in the last years the interest toward this technology decreased and just a few applications have been implemented. The present market trends related to the demand of diesel to gasoline ratio in Europe are opening new opportunities for the development of an oligomerization process to effectively convert light olefins as products of FCC to middle distillate fractions, increasing the global availability of diesel. Belussi et al. [116] used a ZSM-5-based catalyst to study the deactivation rate as a function of space velocity using the real feedstock of the FCC unit.

The study of the catalytic performance of HZSM-5 and Zn/HZSM-5 in aromatization of DME [117] showed that the increase in the number and strength of acid sites in the HZSM-5 zeolite was beneficial for DME aromatization. The aromatization activity was enhanced by loading Zn into HZSM-5. The yields of the total and C_8 aromatics on HZSM-5 at 360°C were 50.0% and 28.6%, respectively, and over 2%Zn/HZSM-5 they were 66.2% and 39.0%, respectively (Fig. 14). The total aromatics yield increased to 78.0% in the case of Zn/HZSM-5 at 480°C.

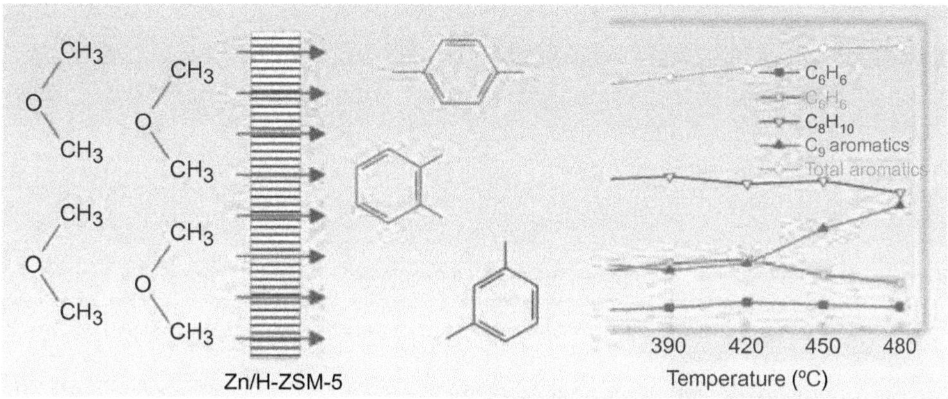

Fig. 14
Reaction pattern and product distribution in the conversion of DME into aromatics. *Figure from reference Fang Y, Tang J, Huang X et al. Chin J Catal 2010;31(3):264 with permission.*

8 Other Processes (Beckman Reaction, Partial Oxidation, etc.)

The previously cited chapter by Hoelderich and van Bekkum in the book *Introduction in zeolite science and practice* presented a good number of reactions of fine chemicals synthesis using diverse zeolites and zeolite-like materials. The analysis from 2001, however, was mostly limited by classical zeolites with the aluminosilicate framework with just a few examples of aluminophosphates, titanosilicates, and mesoporous materials. Nevertheless, such reactions catalyzed by zeolites have been discussed in the review of 2001: (1) isomerization (double-bond shift, isomerization of tricyclic molecules, like synthesis of adamantane, isomerization of terpenes, diverse rearrangements, conversion of aldehydes into ketones), (2) electrophilic substitution in arenes (alkylation of aromatics, including the synthesis of linear alkylbenzenes, alkylation and acylation of phenols, heteroarenes and amines, aromatics nitration and halogenation), (3) cyclization, including the formation of heterocycles, Diels-Alder reaction, (4) nucleophilic substitution and addition, (5) Meerwein-Ponndorf-Verley and Oppenauer reactions, and (6) several examples of two-step and multistep reactions. The major concerns related to the use of zeolite-like materials in organic chemicals syntheses were related to:

- The bulky size of many organic molecules that hinders the accessibility of the zeolite pores and active sites inside the channels and cavities and their diffusion in and out of the porous system of the mostly microporous materials;
- The presence of functional groups that first are involved in interactions (sometimes undesirable) with available acid sites of zeolites and second create further constraints for the diffusion and adsorption;
- Possibilities of occurrence of diverse side reactions, including coke formation.

It is difficult to choose just a few of the most interesting examples of the application of zeolite-like materials in bulk and fine chemicals synthesis, because the recent decade manifests an overwhelming number of such examples demonstrating the outstanding performance of diverse molecular sieves, especially mesoporous materials, in these areas. Here we selected only the publications that reflected the personal interests of the authors rather than the merits of the contributions in this ever-expanding field of research.

Partial oxidation reactions combine a number of processes catalyzed by zeolites modified by transition or noble metals. Some of the examples are included in Table 2.

The performances of various Ti-zeolite systems in partial oxidation reactions catalyzed by H_2O_2 or organic hydroperoxides are presented in [126]. Ti-ZSM-5 is also efficient in the direct conversion of isobutene and methanol into methyl *tert*-butyl ether. This quest is under way with the application of new classes of mesoporous and hierarchical materials.

Vanadium-containing zeolites also showed some promise in the selective oxidation of alkanes and phenol oxidation into diphenols. The V-ZSM-5 zeolite exhibited an excellent selectivity in oxidative dehydrogenation of propane into propene (95%), though the conversion was rather low (only 12.5%) and N_2O was used as an oxidant at 350°C [127]. With O_2 the selectivity dropped to 73% and the same activity can be achieved at a higher temperature (450°C).

The liquid-phase oxidation of hexane on Ti- and V-silicalites resulted in a mixture of alcohols, aldehydes, and ketones C_6 [128]. The same V-MFI compositions demonstrate the activity in phenol hydroxylation comparable to that of TS-1 and TS-2.

Table 2: Partial oxidation reactions catalyzed by zeolites

Substrate	Catalyst	Oxidant	Product	Ref.
C_2H_6	Co-BEA	O_2 (+NH_3)	$CH_3C{\equiv}N$	[118]
C_2H_4	Cu-Pd-Y	O_2	CH_3CHO	[119]
C_3H_6	Cu-Pd-Y	O_2 (+AcOH)	CH_3COCH_3 ($CH_2{=}CHCH_2OAc$)	[119,120]
C_3H_6	Cu-Y	O_2 (+NH_3)	$CH_2{=}CHCHO$	[121]
C_4H_6	V/zeolites	O_2	Maleic anhydride, furan	[122]
Aromatics	Fe-HZSM-5	N_2O	Phenols	[123]
Phenol	TS-1, Ti-BEA, Ti-ZSM-5, Ti-ZSM-12, Ti-MCM-41	H_2O_2	Diphenols	[124]
Alkenes	TS-1, Ti-BEA, Ti-ZSM-5, Ti-ZSM-12, Ti-MCM-41	H_2O_2	Epoxides	[124]
Cyclohexanone	TS-1, Ti-BEA, Ti-ZSM-5, Ti-ZSM-12, Ti-MCM-41	H_2O_2	Oxime	[125]

Gold deserves a special attention. It should be noted that the impetus given by the publications of Haruta in 1980s on low-temperature CO oxidation on Au nanoparticles [129] resulted in a flourishing field of research of the properties of Au NPs (and later Ag NPs) in diverse reactions of organic synthesis. Therefore, we will focus in the short review below on the most outstanding results (in our opinion) obtained in studying Au and Ag NPs on zeolite-like materials.

Epoxidation of alkenes with oxygen until recently was possible only on silver catalysts and only in the case of ethylene, while epoxidation of other alkenes (propene, styrene) requires organic peroxides or (in rare cases) hydrogen peroxide. Haruta et al. [130] found that epoxidation of propene using an Au/TS-1 (TS-1 is a titanosilicate with the MFI topology) is catalyzed efficiently only by gold particles with the size not exceeding 1–2 nm, whereas the smaller particles exhibit a poor activity because of their nonmetallic (molecular) nature.

The Au/TS-1 catalyst was used for propene and styrene epoxidation using an H_2+O_2 mixture, which produced H_2O_2 on Au NPs to oxidize the alkene [131]. Again the optimal size of Au NPs was 2 nm. There is an opinion [132] that the gold particles of a particular size consume a part of hydrogen from the mixture for production of propane; therefore, the optimal size of Au particles may be a little broader (2–10 nm) and they were shown to be semispherical (Fig. 15, Au_{10} NPs):

Chretien et al. supposed that a partial electron deficit on gold nanoparticles ($Au^{+\delta}$) favors the strong adsorption of propene and, as a result, enhances the catalytic activity [133].

The gold particles approximately of the same size (2–4 nm) demonstrated a high activity in oxidation of alcohols, including benzylic alcohols, into aldehydes, with the particles being

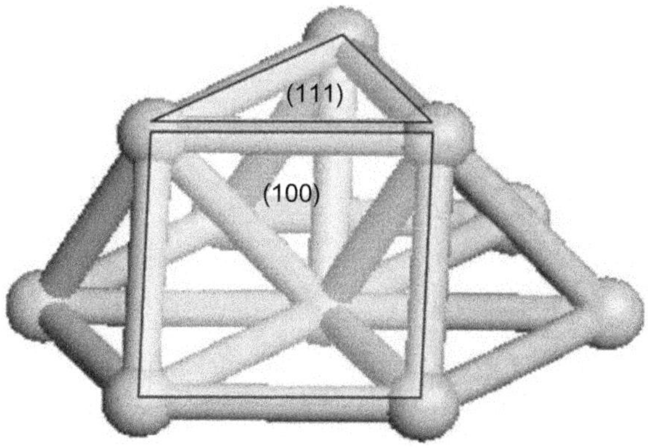

Fig. 15
Au_{10} nanoparticle.

stabilized on mesoporous carriers when the size of the NPs is controlled by the topology of the mesoporous material [134].

$$RCH_2OH \rightarrow RCHO$$

Au NPs on zeolite Y were also found to be highly active, selective, and reusable as a catalyst in the aerobic oxidation of benzyl alcohol under mild conditions [135].

A similar size effect was observed in cyclohexane oxidation into cyclohexanol and cyclohexanone, a commercially important process [136]:

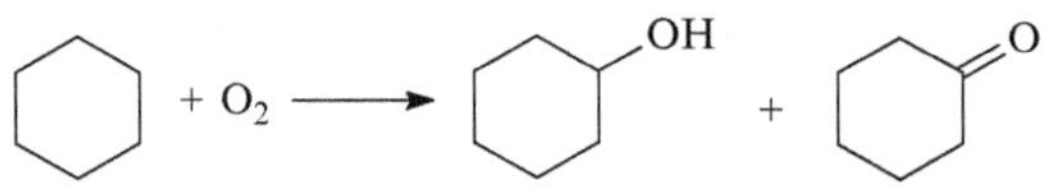

Mesoporous silicates serve as the most suitable host for the gold NPs with the size of 2–4 nm. The cyclohexane conversion may reach 20–30% at the selectivity to cyclohexanol up to 95%, unlike the commercial nickel naphthenate catalysts (conversion 4%, selectivity 70–85%).

A synergy of gold and silver in hydrogenation of esters, for instance, dimethyl oxalate with the formation of methyl glycolate, as well as in hydrogenation of acetic acid was revealed in [137]. Note that monometallic catalysts containing Au or Ag only are practically inactive in these reactions. A mesoporous carrier SBA-15 ensuring the formation of very small particles with a uniform distribution was found to be most efficient. The activation energy of the processes decreases from 140–160 kJ/mol for Ag/SBA-15 and Au/SBA-15 to 41 kJ/mol for Au-Ag/SBA-15. However, the size of Au NPs for the bimetallic system (2–2.5 nm) is 2–10 times smaller compared to the monometallic compositions (4.7 nm for Au/SBA-15 and 20-28 nm for Ag/SBA-15).

A silver-ion-exchanged HZSM-5 zeolite sample exhibited prominent catalytic performance in the partial oxidation of CH_4 at temperatures above 573 K, exceeding by far the activity of nonzeolitic Ag/SiO_2–Al_2O_3 and Ag/SiO_2 catalysts. The spectroscopic studies showed that the simultaneous existence of Ag^+ ions and small clusters of Ag particles leads to the partial oxidation of methane. The authors believe that dioxygen activated on small Ag metal clusters elaborates a surface oxide layer and the thus-formed species is decomposed and the oxygen activated in this way on the Ag metal spills over and reacts with methane activated by the Ag^+ ions. This catalyst may be promising in the conversion of natural gas into higher value-added chemicals and fuels.

The direct synthesis of formic acid by partial oxidation of methane using H_2O_2 at 373 K was accomplished on zeolite catalysts [138]. The yield (13.0%) and selectivity (66.8%) to formic acid were achieved using HZSM-5 zeolite. Tryphenylphosphine (Ph_3P) was used as a promoter.

Among other interesting metal-containing catalysts, one could mention Ni-, Cu-, and Fe-Y zeolites and their catalytic properties in oxidative dehydrogenation of ethane to ethylene, another reaction of commercial importance nowadays [139]:

$$C_2H_6 + \frac{1}{2}O_2 = C_2H_4 + H_2O$$

Acidic Ni-Y zeolite exhibits an ethylene productivity up to 1.08 g(C_2H_4) g(Cat)$^{-1}$ h^{-1} with a selectivity of ~75%. Acidic Cu- and Fe-Y zeolites are somewhat inferior with the ethylene productivity of 0.37 g(C_2H_4) g(Cat)$^{-1}$ h^{-1} and a selectivity of ~50%. Surprisingly, the authors concluded that the acidity of the zeolite favors both oxidative dehydrogenation of ethane (ODHE) productivity and ethylene selectivity.

CO_2-assisted dehydrogenation of ethane to ethylene was studied on gallium-based catalysts [140]. Ga_2O_3/HZSM-5 showed quite good stability, with the higher Si/Al ratio favorable for the resistance to deactivation. The Ga_2O_3/HZSM-5(Si/Al = 97) catalyst exhibits both high activity and stability for dehydrogenation of ethane, the ethane conversion reaches 15% and the ethylene selectivity is close to 94% with a stable performance for 70 h. The enhancement of the catalyst stability with increasing Si/Al ratio of the zeolite is caused by the decrease of the acidity of the catalysts, resulting in the suppression of side reactions, such as cracking and oligomerization. The effect of CO_2 on the dehydrogenation reaction is attributed to the reverse water gas shift reaction. The catalyst stability is also enhanced by the introduction of CO_2.

Chromium oxide catalysts based on pore-expanded mesoporous MCM-41 silica were used for nonoxidative ethane dehydrogenation [141]. The ethane conversion and coke formation depend on the chromia loading. The 5%Cr/PE-MCM-41 catalyst exhibited the best performance with the ethane conversion of 23% and ~99% ethylene selectivity at 650°C. Three different chromium species, Cr(II), Cr(III) and Cr(VI), with their contents varying with the reaction time-on-stream, were revealed. A correlation of the ethane dehydrogenation activity with the concentration of Cr(III) species as the main active sites was established.

The one-step synthesis of methyl-isobutyl ketone (MIBK) from acetone on palladium catalysts on NaX zeolite (1 and 3 wt.% of Pd) [142] revealed the high selectivity of 80% to MIBK for a H_2/acetone ratio of 6.2:

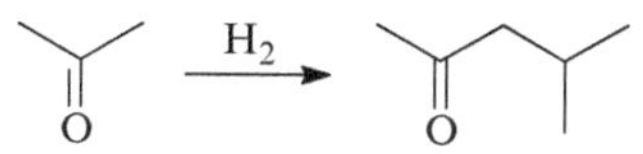

The optimal temperature providing for the maximum selectivity to MIBK was 523 K. Decreasing the reduction temperature from 633 to 573 K for PdNaX enhanced the hydrogenation activity and decreased the MIBK selectivity.

Selective oxidation of hydrocarbons/terpenes in the liquid phase was studied using three types of zeolite- and mesoporous silica-based catalysts: (1) transition metal complexes (metal phthalocyanines, copper acetate dimers, and Co/Mn acetate trimer) encapsulated in zeolite Y, (2) transition metal complexes (Mn-Salen) grafted on SBA-15, and (3) transition metal ions in framework positions of zeolites and mesoporous molecular sieves like Ti-silicates [143]. The metal complexes exhibited enhanced catalytic activity/selectivity after being grafted on the support. The formation of a dimer (for copper acetate) or geometric distortion in the zeolite cavities (for metal phthalocyanines) and the changes in the energy levels and redox potentials modify the catalytic activity in the selective oxidation of hydrocarbons of the encapsulated metal complexes. In the case of Mn-Salen grafted on SBA-15, the introduction of the acidic function onto the silica surface (by introducing SO_3H groups) leads to a more efficient reduction of Mn ions and enhanced catalytic activity in the selective epoxidation of limonene. When Ti ions are introduced in the framework positions, reactive metal-oxo species are formed upon the contact with H_2O_2 or O_2, which influence the mechanism of the cleavage of the O-O bond (heterolytic or homolytic) and the product selectivity.

Iron-modified zeolite L nanocrystals were applied for the oxidation of phenol in the presence of acetic acid with the 93.4% conversion in 30 min with selective formation of 77.5% of catechol and 22.5% of hydroquinone [144]. Acetic acid was oxidized by hydrogen peroxide to form peracetic acid, which served as a better oxidizing agent (Fig. 16).

The research on the synthesis and catalytic properties of the so-called ship-in-the-bottle materials that have started in late 1960s is under way. New Host-Guest Nanocomposite Materials were prepared by supporting transition metal (M = Mn(II), Co(II), Ni(II), and Cu(II)) complexes with an octahydro-Schiff bases: (H_4-N_4O_4) = 2,7,13,18-tetramethyl-3,6, 14,17-tetraazatricyclo-[17.3.1.1]-tetracosa-1(23),2,6,8(24),9,11,13,17,19,21-decaene-9,11,20,22-tetraol; H_4([H]$_8$-N_4O_4) = 2,7,13,18-tetramethyl-3,6,14,17-tetraazatricyclo-[17.13.1.1.]-tetracosa-1(23),8(24),9,11,19,21-hexane-9,11,20,22-tetraol) encapsulated in

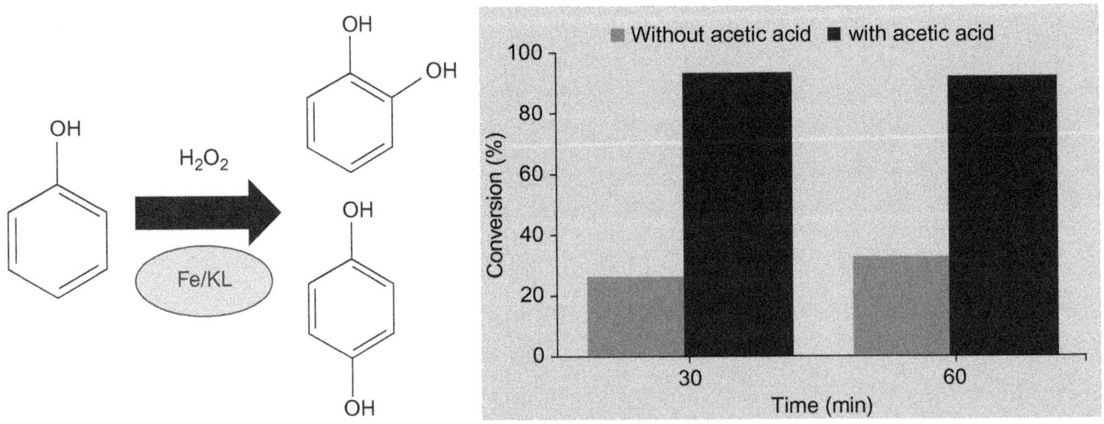

Fig. 16
Reaction pattern and conversion of phenol on Fe/KL zeolite. *Figure from reference Adam F, Wong J-T, Ng E-P. Chem Eng J 2013;214:63 with permission.*

nanopores of zeolite Y. Their high catalytic activity in oxidation of cyclohexane was observed [145]. Zeolite-encapsulated copper(II) complex was more active than the corresponding cobalt(II), manganese(II), and nickel(II) complexes (Fig. 17). The catalytic properties of the complexes are affected by their geometry and by the steric environment of the active sites. The materials are reusable, at least in a dozen of the tests.

Metathesis of olefins is a new area for zeolites, and one of the newcomers is a series of Mo/mordenite-alumina catalysts tested in 1-butene metathesis [146].

The Na exchange degree of mordenite significantly affected the product distribution and catalytic activity. The presence of acidic sites caused different anchoring modes and dispersion of Mo species on the zeolite and, as a result, the different reduction behavior of Mo species in the olefin atmosphere.

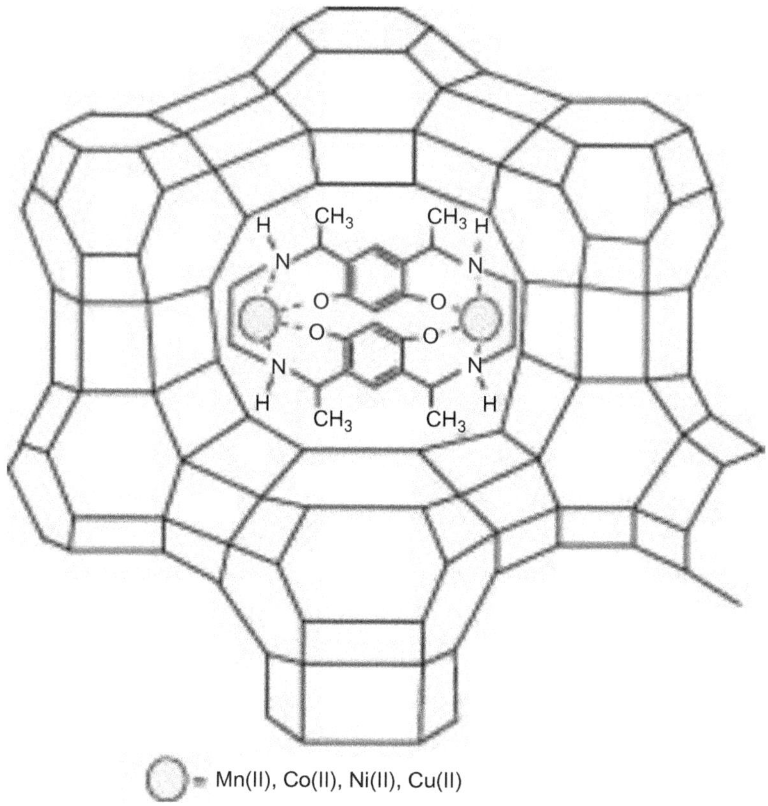

Fig. 17
Ship-in-the-bottle complexes in Y zeolite. *Figure from reference Salavati-Niasari M, Salimi Z, Bazarganipour M, Davar F. Inorg Chim Acta 2009;362(10):3715 with permission.*

Mixed WO_3/Al_2O_3/HY catalysts prepared by calcination of physically mixed WO_3, Al_2O_3 and HY zeolite showed unique behavior in the metathesis between ethene and 2-butene to produce propene [147]. Monomeric tetrahedrally coordinated surface tungstate species responsible for the metathesis activity were formed via the interaction with Brönsted acid sites of HY zeolite. Polytungstate clusters are supposed to be less active in the metathesis reaction. The best catalyst demonstrates the 2-butene conversion close to the thermodynamic equilibrium value (~64%) at 453 K. The catalysts are bifunctional [148]: they catalyze first isomerization of 1-butene to 2-butene and then cross-metathesis between 1-butene and 2-butene to produce propene and 2-pentene. 10%WO_3/Al_2O_3-70%HY exhibits the highest propene yield.

For the Mo/H-Beta zeolites, the formation of the $Al_2(MoO_4)_3$ phase and the decrease in the concentration of Brønsted acid sites explains the low catalytic activity of Mo/H-Beta in metathesis of ethylene and 2-butylene to propylene [149]. A promoting effect of Mg was revealed in the Mo/H-Beta–Al_2O_3 catalyst for cross-metathesis of ethene and butene-2 to propene [150]. The stability is improved at the Mg content of 1–2 wt.% due to the elimination of weak acid sites and suppression of the side olefin oligomerization reaction.

3%Mo/MCM-22-Al_2O_3 catalysts were evaluated in the metathesis reaction of ethene and butene-2 [151]. The γ-Al_2O_3 phase favored the formation of polymolybdate or multilayered Mo oxide, while $Al_2(MoO_4)_3$ species were formed over MCM-22.

A new environment-friendly synthesis of hydroxyphenylalkylketones was developed based on the use of a commercial pentasil-type zeolite ensuring both the high conversion of phenylacetate and the high selectivity toward *ortho*-hydroxyacetophenone [152].

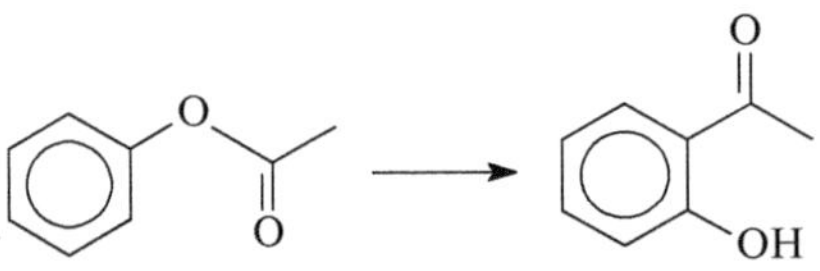

Excellent catalytic activity was found with longer-chain esters.

Pd(II) anchored into the organically modified MCM-41 exhibits a very high catalytic efficiency in Suzuki cross-coupling under mild reaction conditions [153]:

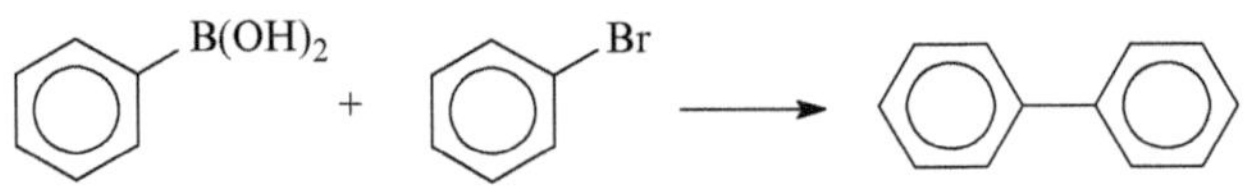

The selectivity to bi-aryl products modified by either electron-donor or electron-withdrawing substituents and recyclability of the catalyst make this catalyst a promising alternative to the homogeneous complexes.

The performance of CuBTC and FeBTC (MOF-type materials, BTC stands for benzene tricarboxylate) in Beckmann rearrangement of diverse aromatic and nonaromatic oximes was compared with that of now conventional zeolites Beta and USY [154]. The reactivity of the substrates of the Beckmann rearrangement increases in the order: camphor oxime < cyclohexanone oxime < indanone oxime < acetophenone oxime. The Beta and Y zeolites show a higher activity in the transformation of small aromatic (acetophenone oxime, indanone oxime) and nonaromatic oximes (cyclohexanone oxime) providing a 100% selectivity to the lactams, whereas CuBTC was the most active in the reaction of bulky oximes due to the absence of spatial constraints.

The production of ω-laurolactam by Beckmann rearrangement of cyclododecanone oxime on a solid acid catalyst in the vapor phase was reported in [155]. The acid-treated [Al,B]-BEA zeolite at 320°C and reduced pressures demonstrates the complete conversion combined with the excellent selectivity of 98%.

Crosslinked TS-1 zeolite materials are among the best catalysts for the gas-phase Beckmann rearrangement of cyclohexanone oxime to ε-caprolactam [156]. Crosslinking of the zeolite particles resulted in the additional mesoporosity and as a result, in an increased activity in the title reaction. Note that the deactivation of the catalysts is significantly decreased. The increased activity correlates with the newly created mesopore surface area.

An Al-form of MCM-41 exhibits the good catalytic performance in the Beckmann rearrangement of cyclohexanone oxime, with the active sites supposedly being the nest silanols (at structural defects) of very weak Brønsted acid character. Under mild reaction conditions, the selectivity to caprolactam of 100% was obtained [157].

Paracetamol (a widely used over-the-counter analgesic and antipyretic) was synthesized by the environmentally benign liquid-phase Beckmann rearrangement of 4-hydroxyacetophenone oxime over Al-MCM-41 modified with H_3PO_4 [158]. The conversion of 4-hydroxyacetophenone oxime to paracetamol increased with the H_3PO_4 loading in the catalysts with the maximum at 30 wt.%.

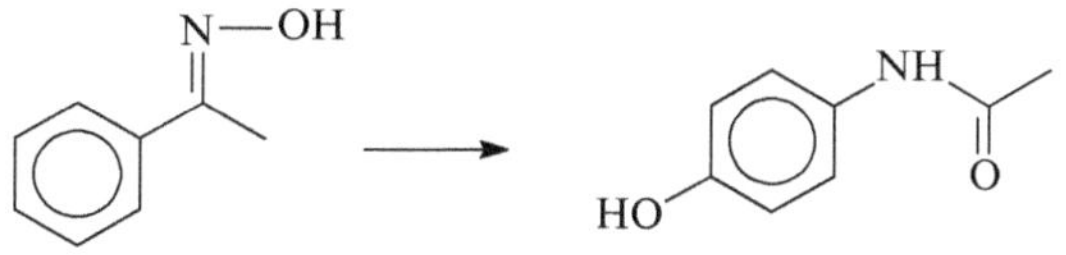

The rearrangement of epoxides into aldehydes was reported to occur with the conversion close to 100% at the selectivity of about 90% on TS-1 and Ti-BEA zeolites [159]:

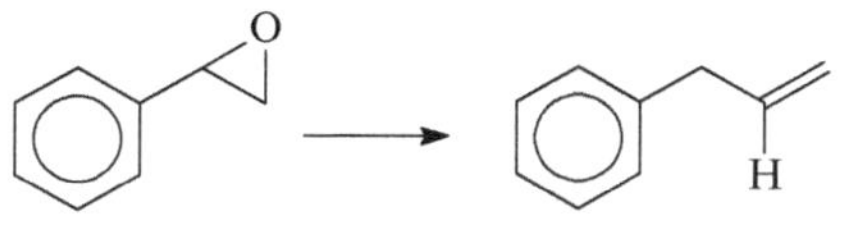

The conversion of methyl octanoate into hydrocarbons on the HZSM5 zeolite catalyst was used as a model for the production of hydrocarbon fuels from biodiesel [160].

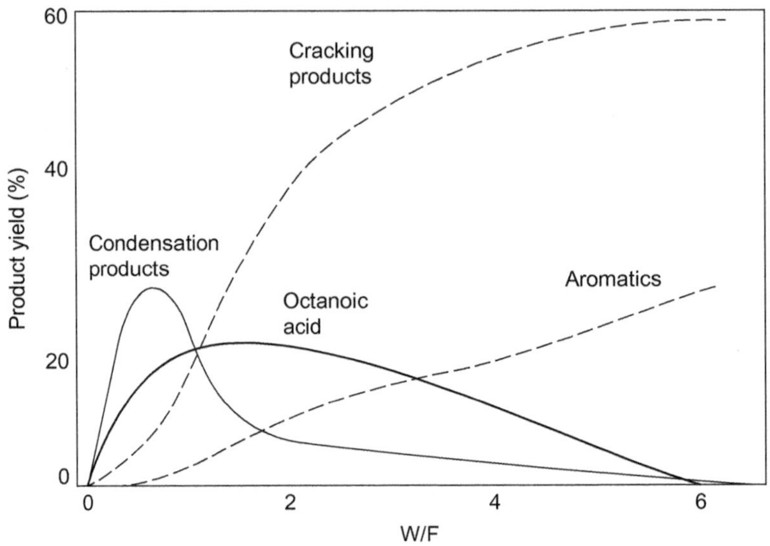

Fig. 18

Dependence of the product distribution on the contact time in the conversion of methyl octanoate on HZSM-5 zeolite. *Figure from reference Danuthai T, Jongpatiwut S, Rirksomboon T et al. Appl Catal A Gen 2009;361(1–2):99 with permission.*

The deoxygenation of methyl octanoate yielded hydrocarbons C_1–C_7 with significant amounts of aromatics. Octanoic acid and heavy products, such as 8-pentadecanone, were formed as primary products by acid-catalyzed hydrolysis and condensation, respectively (Fig. 18). The formation of a small amount of ethylbenzene and o-xylene at low conversions of methyl octanoate indicates that direct dehydrocyclization is possible.

We should also mention the reaction of benzene and phthalic anhydride resulting in anthraquinone (used in the synthesis of H_2O_2) on Ce-Y zeolite, though at a quite high temperature of 550°C [161].

Aromatics nitration on H-BEA or HZSM-5 zeolites using commercial HNO_3 in the vapor-phase process may be considered as a potentially ready-to-commercialize process [162], though the catalyst long-term stability may be a problem.

The above examples are mostly related to either acidic forms of zeolite-like materials or supported metals and complexes. The basic forms of zeolites are known to exhibit interesting properties in side chain alkylation of alkylaromatic hydrocarbons as well as in diverse condensation reactions (Knoevenagel, Michael, etc.). One example was given in the paper by Corma et al. [163] related to the condensation of benzaldehyde with diethyl malonate on alkali forms of X and Y zeolites. The activity increased with the size of the alkali cation, that is, with increasing basicity of the zeolite:

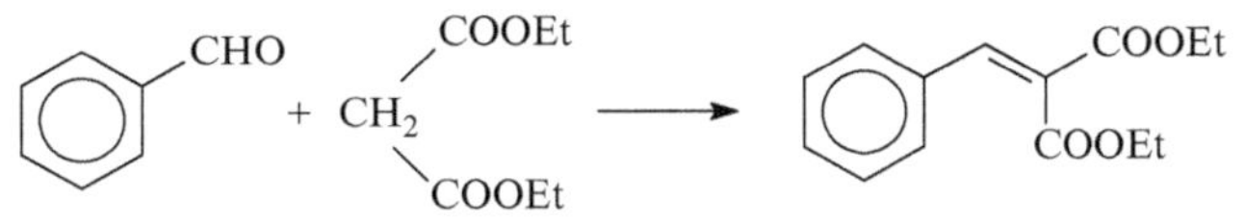

Zeolites can be used in catalytic processes not only as catalysts but also as dessicators or scavengers of compounds forming as the products (water, alcohols) or reactant concentrators. They can also act as carriers slowly releasing the reactant or as a nanoreactor for reaction/separation of a mixture of molecules with different sizes.

9 Reactions Related to the CO_2 Footprint (Conversion of Renewables, Biomass)

This part of the review gives a picture of the applications of zeolite-like materials in the conversion of renewable raw materials and other reactions aimed at decreasing the CO_2 environmental footprint. Biomass-derived oxygenates are promising future feedstocks for the chemical industry and fuel production. Fluid catalytic cracking of such feedstocks has the advantage of enabling deoxygenation without external hydrogen consumption [164]. Saturated alcohols, as a feedstock for FCC, yield the most valuable iso-alkanes. Fatty acids and triglycerides yield more aromatics, whereas the high-molecular-weight feedstock produces more di- and polyaromatics. Unsaturated feedstocks (vegetable oils and the like) result in the formation of aromatics (Fig. 19).

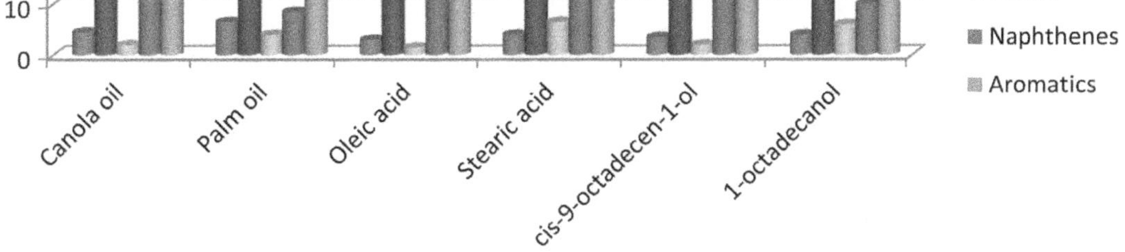

Fig. 19

Distribution of different classes of products obtained from vegetable oils and model compounds.

The catalyst acidity and porosity play an important role in the isomerization of linoleic acid in the case of mordenite and ZSM-5 (H^+, Ni^{2+} forms) [165]. The conversion of linoleic acid increased if weak or moderate Brønsted and Lewis acid sites are present together. The synergetic effect of Brønsted and Lewis sites causes the enhanced selectivity to the isomer cis-9,trans-11-CLA, whereas the higher selectivity to trans-10,cis-12-CLA was ensured by Brønsted acidity. The use of microwave activation for the Ni-exchange in the mordenite resulted in the formation of a new type of weak Lewis acid sites capable of increasing the conversion and the selectivity to cis-9,trans-11-CLA.

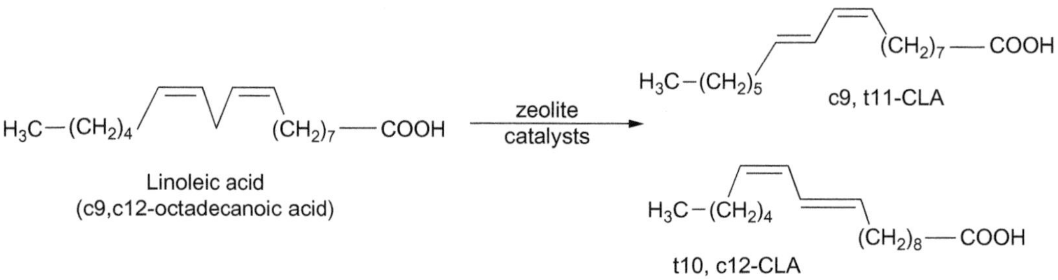

A commercial Beta-type zeolite was also used for upgrading unsaturated fatty acids [166]. The catalytic activity for the skeletal isomerization of oleic acid was attributed to the high Brønsted acidity of the zeolite. Addition of water (2–5 wt.%) to the fatty acid promoted the conversion. The methyl ester of oleic acid was more reactive than oleic acid on the zeolite catalysts. The carboxylic group of oleic acid shows a poisoning effect.

The conversion of rapeseed oil on Mo oxide and Ni-modified nanocrystalline ZSM-5 zeolites [167] resulted in the total deoxygenation of the vegetable oil. The major products were light olefins C_2–C_3 and aromatic hydrocarbons. Ni/HZSM-5 catalysts demonstrate a higher selectivity to light olefins, whereas Mo/HZSM-5 catalysts are more selective to aromatics:

The conversion of lignocellulosic biomass became one of the central problems of catalysis taking into account the large volumes of waste solids accumulated in forestry, agriculture, and pulp and paper industry, such as residual wood, saw dust or straw. Thus, the production of synthetic fuels and high added-value chemicals from biomass is a challenge of the present day. One of the approaches to this problem was developed within the Bioliq® concept [168]. The model studies and assessment of two different biomass-to-liquid concepts for the production of gasoline via DME synthesis and gasoline or diesel via Fischer-Tropsch synthesis were performed. The outcome of the techno-economic assessment is as follows: the total energy efficiencies from biomass to final products range between 38 and 39%, the production costs of gasoline via DME synthesis are about 0.82 US$/L, gasoline and diesel via FT synthesis can be produced at costs of about 0.88 US$/L. Thus, the production costs for biomass-derived fuels are 76 to 93% higher than the current market prices. However, the new technologies and catalysts with improved performances are coming up to be competitive to conventional refineries as the production costs are only 7 to 18% higher than the current market prices with coal as the main feedstock. The further advantages include the reduction of the CO_2 footprint; moreover, the implications of negative CO_2 emissions.

Furfural, one of the high-value chemicals, is produced by the dehydration of d-xylose derived from lignocellulosic biomass. Though the conversion of xylose to furfural on zeolite H-BEA is accompanied by side reactions and, as a result, the problems of separation, the proper design of the catalyst and the process scheme can improve the efficiency [169].

Hydrolysis of starch with parallel hydrogenation leading to sorbitol can effectively proceed on a dual-function zeolite catalyst (Ru/H-USY) at 180°C, 55 atm with an almost quantitative conversion and selectivity to sorbitol over 95% [170].

Catalytic transformation of levoglucosan (1-6-anhydro-β-d-glucopyranose), another carbohydrate derivative produced from biomass, on zeolite and mesoporous catalysts, including MCM-22 and MCM-41, at 573 K gives oxygenated liquid products, mainly aldehydes and furfurals (glycolaldehyde, formaldehyde, acetaldehyde, furfural, 5-methylfurfural, acetic acid) [171]. The yield of the liquid products was higher for MCM-41 than for MCM-22, except acetic acid predominantly formed on MCM-22. The deactivation via coking was more severe over the zeolite compared to the mesoporous material.

Glycerol, as a low-value product of biodiesel production from vegetable oils, is another target of the zeolite catalysis. Rare-earth forms of H-BEA zeolite (La^{3+}, Ce^{3+}, Nd^{3+}, and Eu^{3+}) demonstrated quite promising results in the etherification of glycerol with isobutylene to synthesize glycerol ethers, that is, products with a high-octane number to be used as blending additives to gasoline [172]. Nd/H-BEA zeolite exhibited the best catalytic performance. At the optimal isobutylene to glycerol ratio of 3:1 at 343 K, the conversion of 92% was found for glycerol that transformed into a mixture of mono-*tert*-butyl ethers of glycerol, di-*tert*-butyl ethers of glycerol and tri-*tert*-butyl ether of glycerol. The selectivities to the two latter products were 65 and 8%, respectively.

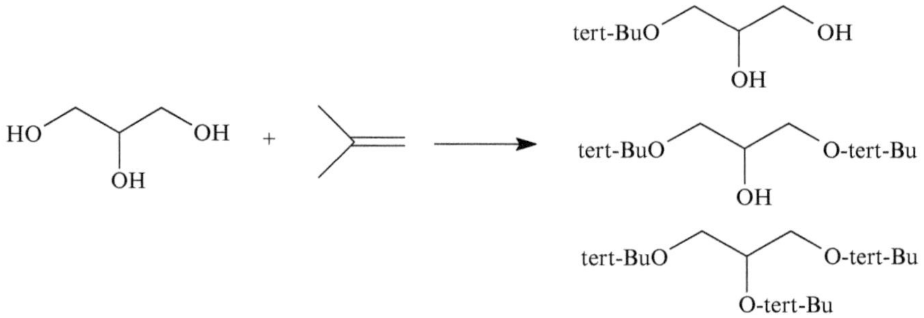

The conversion of glucose to aromatics [173] on a small pore ZK-5 and SAPO-34 materials, medium-pore Ferrierite, ZSM-23, MCM-22, SSZ-20, ZSM-11, ZSM-5, IM-5, and TNU-9, and large-pore SSZ-55, Beta zeolite, and Y zeolite yielded the aromatics, with the yield being a function of the pore size of the zeolite. Small-pore zeolites did not produce any aromatics, while oxygenated products were formed. Aromatic yields were highest in the medium-pore zeolites with pore sizes of 5.2–5.9 Å. High coke yield, low-aromatic yields, and low-oxygenate yields were observed with large-pore zeolites. Thus the large pores facilitate the formation of heavy deposits. The internal pore space and steric hindrance (the shape-selectivity effects) are crucial for aromatic production. Medium-pore zeolites with moderate internal pore space and steric hindrance (ZSM-5 and ZSM-11) exhibit the highest aromatic yield and the reduced amount of coke. In the case of the smaller-pore zeolites, the polycyclic aromatics may form by secondary reactions on the catalyst surface, either directly or from the smaller aromatics.

The advantages of the introduced porosity were demonstrated in the catalytic pyrolysis of low-density polyethylene, another abundant waste of the industry [174]. The porosity was created by desilication of the medium-pore ferrierite zeolite in sodium hydroxide solutions, resulting in enhanced accessibility and ultimately improved catalytic performance (Fig. 20). It was shown that compared to other frameworks (eg, MFI, MTW, MOR, and BEA), FER requires more severe conditions to extract silicon to create mesoporosity. The authors succeeded in increasing the mesopore surface area of the NaOH-treated ferrierite by a factor of 3–4 with respect to the parent zeolite, while preserving the crystallinity and acidity.

Another work in the area of high-density polyethylene pyrolysis was carried out using HZSM-5, HY and H-BEA zeolites at 500°C and atmospheric pressure, with light olefins and automotive fuel hydrocarbons being the products [175]. Products contain light olefins C_2–C_4, light alkanes $<C_4$, nonaromatic C_5–C_{11} compounds, single-ring aromatics and C_{11+} hydrocarbons, wax, and coke. HZSM-5 zeolite catalyst shows a high selectivity to light olefins, about 58%, whereas high yields of nonaromatic C_5–C_{11} products (45%) are obtained with H-BEA and HY zeolites. Wax was produced on HY and H-BEA zeolites.

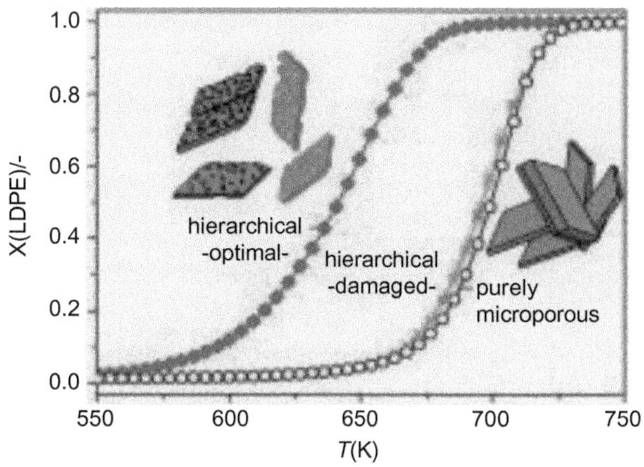

Fig. 20
Polyethylene conversion on desilicated zeolites. *Figure from reference Bonilla A, Baudouin D, Pérez-Ramírez J, J Catal 2009;265(2):170 with permission.*

10 New Technologies: Microwave, Plasma, Sonochemistry, Supercritical Fluids, Cascade Processes, Domino Reactions

10.1 Microwave-Assisted Processes

The existing technologies mostly reached the saturation level in their development and the further progress becomes possible, if at all, by using nontraditional approaches, such as the nonequilibrium conditions, gradient technologies, combination of endothermic and exothermic processes, ex situ or in situ application of electromagnetic activation modes, such as microwave activation, low-temperature plasma, and more.

The key advantage of energy transfer with microwave fields over convective methods for heating is the transfer of energy through radiation rather than heat transfer or convection. This ensures fast penetration of energy into the volume of materials transparent to microwave fields (like zeolites), that is, almost instantaneous heating (and cooling when the field is switched off) of materials. The spatial temperature distribution in the material is thus different from that observed under traditional convective or contact heating conditions. The most important differences related to the appearance of temperature gradients and nonequilibrium conditions are observed when a reaction medium or material (a catalyst) consists of several phases with different abilities to be heated by microwave radiation.

Numerous data show that, under microwave activation conditions, higher reaction rates are obtained than in thermally activated processes, the temperature of the catalytic reaction usually decreases by some 100–200 K, the yields and selectivities to final products increase, thereby suppressing the deactivation and side reactions. There are good grounds in certain instances for believing that catalytic processes occur under the action of microwave fields differently than under traditional thermal activation with convective or conductive heating. The energy of microwave radiation quanta is, however, insufficient for directly breaking chemical bonds in reacting molecules. A necessary condition for the action of the microwave field on a catalytic process is therefore the possibility of interaction of a solid (a catalyst or its separate components) with this field.

Among various mechanisms of the action of microwave field on solids, several classes of catalysts and processes that offer promise for use in practice can be identified. For instance, the Maxwell-Wagner interphase polarization mechanism can likely operate in mixed oxide catalysts (such as catalysts of partial oxidation based on vanadium and molybdenum oxides). For such catalysts, "nontrivial" and "nonthermal" effects can probably be expected.

Among heterogeneous catalysts, there are also many systems representing a deposited active metallic phase. Strictly, the Maxwell-Wagner mechanism can also be valid for metallic catalysts deposited on a carrier that is inert with respect to microwave fields. Heat exchange between a deposited conducting phase and a carrier transparent to microwave radiation is of great importance for heterostructures of the type metal + carrier transparent to electromagnetic field. If the characteristic time of thermal relaxation is larger than the reaction duration, such a system is of great interest for bifunctional catalysis with the separation of functions between a metal and a carrier active in the given case, for instance, a carrier possessing acid properties (the case of zeolites). This can be of obvious advantage for processes that require fast and selective catalyst heating. In these instances, a microwave field can be used to solve technological problems.

It is important to distinguish between thermal and nonthermal effects for revealing the real action of the microwave field on catalytic reactions. Currently, the procedure for selecting an appropriate catalyst for a heterogeneous gas-phase catalytic process that is ideally suitable for activation in microwave fields is as previously not quite clear. The selection is complicated because the catalyst should, first, have substantial catalytic activity and, secondly, possess certain electrophysical properties (dielectric/magnetic properties and conductivity). To obtain a maximum positive effect of microwave radiation on a catalytic system, comparatively inert components strongly interacting with electromagnetic fields are introduced into catalysts.

The advantages of microwave technologies can more effectively be used if catalytically and electromagnetically active catalyst components are placed on a substrate transparent to microwave fields. This is done to form superheated regions only in the volume or on the surface of the active catalyst phase, which decreases the temperature of the catalyst as a whole and the gas phase. This offers the possibility of suppressing side reactions and decreasing energy expenditures for heating inert materials that do not directly participate in the catalytic process.

Special attention should be paid to bifunctional catalysis, in which the different stages of the process take place on different components of the catalytic system. A typical example is transformations of hydrocarbons on metallic catalysts deposited on acid carriers. A metallic component (as a rule, a Pt family metal) activates the initial molecule with breaking bonds and/or charge transfer (dehydrogenation, hydrogenolysis), whereas carbon skeleton transformations (alkylation and isomerization) occur on acid centers of the carrier. Simultaneous hydrogenation/dehydrogenation, cyclization, aromatization, and other processes can occur on both complex catalyst components in various combinations. Microwave radiation-creating temperature gradients in a heterogeneous system whose components differently interact with electromagnetic fields can then cause substantial changes in the rate of the process and, which is especially important, distribution of reaction products [176].

Technologically, the advantages of the use of microwave activation of heterogeneous gas-phase catalytic reactions compared with thermal activation become obvious when fast heating or cooling and pulsed energy supply are required.

Effective use of microwave heating in the oxidation of alkenes and alcohols under the action of hydrogen peroxide was described in [177]. The catalyst was MCM-41-type mesoporous titanium silicate modified by organic compounds. It was recycled several times without activity loss.

A positive effect of microwave radiation was obtained in the oxidation of benzene to phenol with the use of nitrous oxide as a mild oxidizer as a very promising one-stage method for the production of phenols. Microwave activation of the zeolite catalyst of this reaction (Fe-ZSM-5) was used to decrease the contribution of coking [178]. Although some increase in selectivity with respect to phenol and a decrease in the rate of catalyst deactivation were observed, no substantial improvement of process characteristics was obtained.

Friedel-Crafts acylation of indole with acetic anhydride on tungstophosphoric acid modified H-BEA zeolite under microwave and conventional heating [179] showed better catalytic performance under microwave radiation. The reaction proceeded selectively to 3-acetylindole compared with conventional heating.

The acylation of anisole with decanoic acid on Ce-containing zeolite catalysts under conventional heating and microwave stimulation [180] exhibited increased reaction rates due to microwave enhancement, presumably related to selective desorption of water (by-product) from the catalyst.

Microwave heating was applied for the liquid-phase acylation of anisole by acetic anhydride or benzoyl chloride at 80°C, with the reaction rate increasing dramatically compared to a traditional heating [181]. In addition, a significantly shorter reaction time compared to the conventional heating was observed.

The use of microwaves at the stage of catalyst preparation is a well-known area of research. One of the recent examples is related to sulfonic acid supported on SBA-15 and BEA zeolite prepared under microwave irradiation [182] Microwave-assisted sulfonated catalysts showed higher selectivity toward di- and triethers of glycerol (83–91%) than those sulfonated by conventional heating. Noteworthy that the high selectivity to the bulky triether for BEA samples (32–36%) proves that microporosity is not a barrier for the formation of the triether.

Zeolites modified with transition metals were used as catalysts under microwave irradiation [183]. Two main dielectric mechanisms of microwave heating were determined in zeolites: rotational polarization phenomenon and interfacial polarization. The selective catalytic oxidation of styrene with H_2O_2 resulted in benzaldehyde formation.

Regioselective alkylation of naphthalene with alcohols occurs on H-mordenite ($SiO_2/Al_2O_3=240$) under microwave irradiation [184]. For the microwave-assisted reaction, high reaction rates and high selectivities for 2,6-dialkylnaphthalenes were found. In the reaction of 2-isopropylnaphthalene with isopropyl alcohol, the conversion and the selectivity were 43.5 and 66.4%, respectively. In di-*tert*-butylation of naphthalene with *tert*-butyl alcohol, the conversion and the selectivity reached 86.5 and 70.4%, respectively. The conversions and the selectivities were enhanced by microwave heating.

Solvent-free liquid-phase acylation of p-cresol with aliphatic carboxylic acids (acetic, propionic, butyric, hexanoic, octanoic, and decanoic acids) was accomplished using BEA zeolite under conventional and microwave heating to demonstrate the significant enhancement in activity in the case of the microwave heating [185]. Under conventional heating, the conversion was below 20%, while under microwave heating, the conversion was up to 50–80%. Ester formed through O-acylation and ortho-hydroxyketone formed through Fries rearrangement of the ester were the products. The conversion of carboxylic acid exhibits a maximum at the carbon number in the chain of 6. O-acylation occurred quite fast followed by slow conversion to ortho-hydroxyketone. Coke formation was suppressed on the catalyst used in the microwave mode unlike the conventional heating.

Solvent-free alkylation of imidazole with alkyl halides, such as 1-Br-butane, 2-Br-butane and 1-Br-hexane, was performed using basic zeolites (NaX zeolite with alkaline-earth cations—Ca, Sr and Ba) as catalysts [186]. Conversions up to 95% and selectivities of 100% were found at 700 W of the microwave power input in only 5 min.

10.2 Plasma-Stimulated Processes

Plasma-stimulated processes are usually characterized by low productivities, since low pressures (<100 Torr) are used, and enhanced energy consumption compared to the microwave processes.

Nonthermal plasma technology is a promising process for flue gas treatment. Microwave catalytic NO_x and SO_2 removal was performed using FeCu/zeolite as a catalyst [187]. The microwave plasma tests on FeCu/zeolite demonstrated the catalytic oxidation of 91.7% NO_x to nitrate and 79.6% SO_2 to sulfate. The efficiency of the microwave-assisted catalytic reduction of NO_x and SO_2 on FeCu/zeolite with NH_4HCO_3 as a reducing agent is as high as 95.8 and 93.4%, respectively.

10.3 Use of Supercritical Fluids

The use of supercritical fluids became a powerful approach in "green" chemistry. Both CO_2 and hydrocarbons can serve a supercritical medium, but the preferable choice is the use of the substrate of the catalytic reaction itself as a sc-fluid.

Hydrocarbon cracking (*n*-dodecane) was efficiently accomplished under supercritical conditions (425°C, 150 bar) on zeolites Y and ZSM-5 [188].

The catalytic cracking of *n*-dodecane on HZSM-5 zeolite at 400–450°C under supercritical and subcritical pressures (0.1–4.0 MPa) was also studied in [189]. The activity of the catalyst and its stabilization toward deactivation decrease with increasing pressure, and the catalyst preserves a substantially higher activity if the feed rate exceeds 4 ml/min under supercritical conditions. The contribution of supercritical extraction to the activity of the HZSM-5 catalyst increases with increasing hydrocarbon feed rates and decreasing catalytic activities, and reaches the maximum when the coke formation rate equals to the coke removal rate by a supercritical hydrocarbon.

The use of supercritical conditions is most perspective in the C_4 alkylation process where the lifetime of the catalyst is a burden. The comparison of the catalytic performances and stabilities of H-Beta and H-USY zeolites for isobutene alkylation with butenes in the supercritical isobutane phase [190] clearly shows that the zeolite structure strongly influences the performance and deactivation behavior of the zeolite in the alkylation process in supercritical (SC) conditions. H-Beta outperformed H-USY zeolite because of its lower deactivation and the stable quality of the alkylate, composed mainly of isoparaffins, whereas olefin di-/oligomerization was responsible for the H-USY deactivation. The stable activity of H-Beta was attributed to the effective extractive effect of the supercritical iC_4 media to "clean" the acid sites located on the external surface of the small crystallites of H-Beta or located at the pore mouth. On the contrary, the deactivation observed for H-USY was caused by olefin oligomerization inside the zeolite supercages. The supercritical isobutane media cannot prevent the oligomers formation inside the zeolite supercages and extract these oligomers.

n-Butane isomerization under the supercritical conditions on heteropolyacids ($H_3PW_{12}O_{40}$, $H_4SiW_{12}O_{40}$) supported on H-mordenite as catalysts at 13.8 MPa [191] resulted in the stable activity and suppression of the deactivation. Even the catalyst aged in the gas-phase isomerization can be reactivated by the supercritical media and the catalyst performance can attain the initial value of 20–25% conversion, when the *n*-butane density is in the vicinity of its critical point (Fig. 21).

The reaction of benzene (or other aromatic hydrocarbons) oxidation with N_2O as a mild oxidant also showed an outstanding performance when both benzene and N_2O were in the supercritical states ($T=400°C$, $P=100$ atm). The data obtained show that the process under supercritical conditions is characterized by two to three times higher productivity of the catalyst and prolonged lifetime. Furthermore, the catalyst deactivated in the gas-phase conditions (1 atm) can be regenerated by switching the supercritical regime on (Fig. 22).

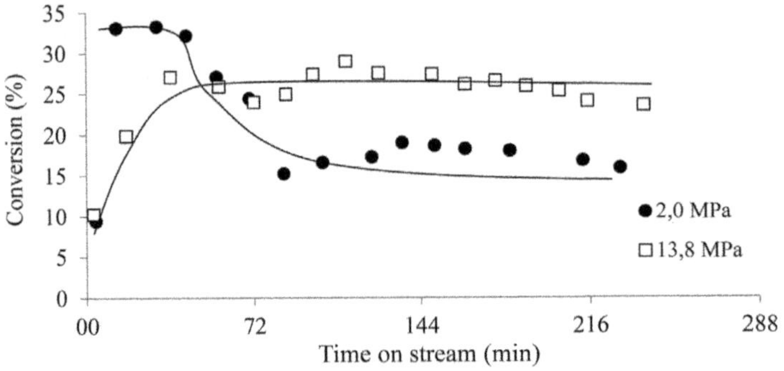

Fig. 21
Performance of H-mordenite in *n*-butane isomerization in gas-phase (2.0 MPa, 573 K) and supercritical conditions (13.8 MPa, 573 K).

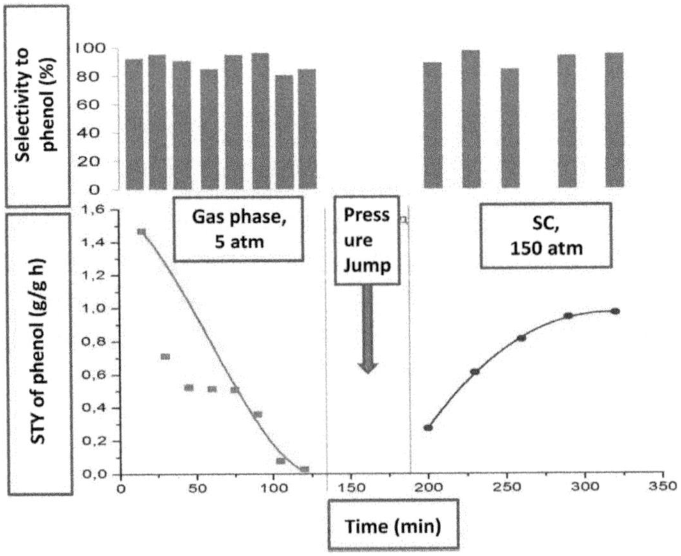

Fig. 22
Space-time yield of phenol and selectivity to phenol in gas phase and supercritical conditions ($VHSV=17\,h^{-1}$).

Fischer-Tropsch synthesis was carried out under supercritical conditions using *n*-butane as a medium using a hybrid catalyst comprising Co/SiO_2, HZSM-5, and Pd/SiO_2 [192]. The application of supercritical *n*-butane increased the yield of light isoparaffins significantly and suppressed the formation of methane. Under supercritical conditions, hydrogenolysis and isomerization reactions were accelerated. Since the optimum temperatures for FT and HZSM-5 catalysts are different, 513 K and over 573 K, respectively, two-step FT synthesis was considered. At the first step, Co/SiO_2 with a small amount of HZSM-5 catalyzed FT synthesis at 513 K, and the second step was catalyzed by a hybrid catalyst Pd/SiO_2 + zeolite for hydrogenolysis and isomerization at 573 K. The application of supercritical *n*-butane enhanced the isoparaffin selectivity and suppressed the methane yield. The production of heavy hydrocarbons C_9+ was inhibited in both gas and supercritical phase. The isoparaffin selectivity decreased in the gas phase with time-on-stream, but was very stable for the supercritical phase. HZSM-5 and H-BEA zeolites were the best choice for producing light isoparaffins. The two-step FT synthesis in supercritical *n*-butane was superior for producing light isoparaffins.

Alkylation of benzene with propene to cumene on H-BEA zeolites in supercritical conditions was a good example to prolong the catalyst lifetime (290–310°C, 5.7 MPa) [193].

HZSM-5 modified by tetraethyl orthosilicate was used for alkylation of toluene with ethylene in supercritical carbon dioxide [194]. The maximum yield and selectivity of *p*-ethyltoluene were 70 and 97%, respectively, at 623 K, 13.1 MPa, $WHSV = 5.0\,h^{-1}$ and a molar ratio of toluene to ethylene of 5.

Dialkylation of naphthalene with isopropanol to 2,6-diisopropylnaphthalene was catalyzed by HY zeolite modified by Fe^{3+}, Ni^{2+}, Co^{2+}, and Cu^{2+} under optimum supercritical conditions (50 bar, 220°C, isopropanol/naphthalene = 4, $WHSV = 18.8\,h^{-1}$, 6 h) [195]. Fe/HY revealed the highest selectivity, by increasing the 2,6/2,7-isomer ratio from 2.8 to 6.6. The selectivity to the desired 2,6-isomer changes in the order: Fe–HY > Ni–HY > Cu–HY > Co–HY > HY. The naphthalene conversion was over 73%, reaching 96%. Polyalkylated products were found on zeolites modified by Co and Cu, with a larger amount of coke being formed on these catalysts.

The positive effect of supercritical carbon dioxide on the transalkylation of diisopropylbenzene was found with benzene on zeolites BEA, Y, and mordenite under liquid, subcritical, and supercritical conditions [196]. The transalkylation rates were not increased with pressure. The use of supercritical CO_2 resulted in improvement in the product yield for BEA and MOR.

Supercritical CO_2 turned out to be a benign medium for the Friedel-Crafts hydroxyalkylation of anisole with paraformaldehyde [197]. Thus, the activity of mordenite in sc-CO_2 at 60°C is three times higher compared with the reaction in toluene at 110°C.

The transalkylation of diethylbenzene with benzene to produce ethylbenzene in supercritical conditions (sc-CO_2 and sc-benzene) on Beta zeolite [198] demonstrated that the supercritical conditions improved the catalyst performance in sc-substrate, but not in sc-CO_2. This effect was accounted for by the different densities and diethylbenzene diffusion coefficients in different media.

One more example concerns the use of Al- and Zr-containing mesoporous MMS-H materials for the esterification of glycerol with lauric acid in supercritical CO_2 [199]. The conversion up to 93% with the formation of equal amounts of lauric acid monoglyceride and diglyceride was observed in sc-CO_2 that was higher than that found in mesitylene and without a solvent. The coke formation was also suppressed in sc-media.

Isooctane synthesis was performed using ZSM-5 and ZSM-23 zeolites in sc-CO_2 via isobutene dimerization in comparison with sc-propane [200]. Except for dimerization, oligomerization and coke formation occur. The ZSM-23 catalyst exhibited the initial isobutene conversion of 80%. The conversions in sc-CO_2 and sc-propane were 56% after ~200 h and 32% after 120 h, respectively.

10.4 Tandem/Domino/Cascade Processes on Zeolites and Zeolite-Like Materials

The trend exists toward merging several steps in a one-pot process, and zeolites, especially, zeolite-like materials (MCM-41, SBA-15, MOFs), are among the key catalysts for such applications. A few examples were provided by Corma et al., for instance, the synthesis of complex amines from aldehydes, amines and acetylenic compounds [201]:

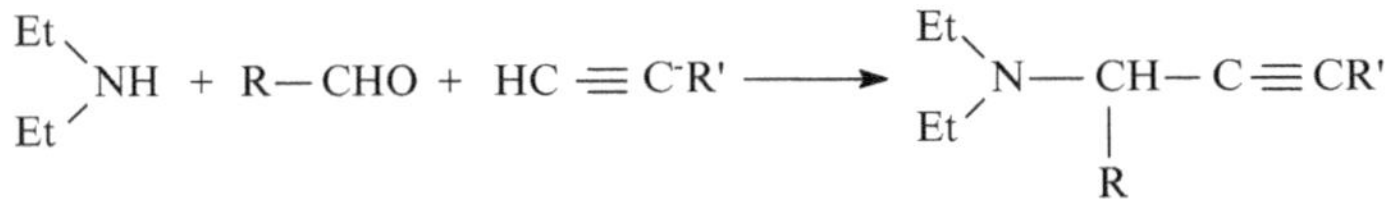

Typically, the reaction is catalyzed by Au(III), Au(I), or Au nanoparticles supported onto a mesoporous or microporous material.

A one-pot conversion of aromatic ketones into olefins was performed using a combination of two catalysts: the supported metal (Cu/SiO_2) catalyzes the hydrogenation steps, whereas the acidic form of zeolite (HZSM-5) is responsible for the dehydrogenation reaction [202]:

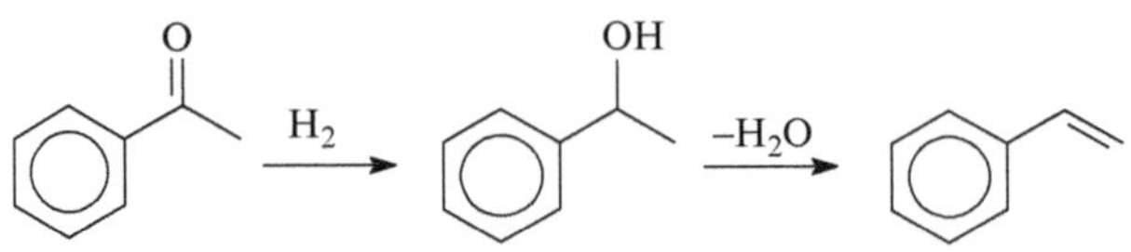

The space-time yield of styrene exceeds 70%, while the combination of two reactions in one reactor provides conditions for saving energy and solvents.

Another versatile approach to the synthesis of valuable products is outlined in [203]. The authors performed tandem reactions leading to substituted cyclopropanes using basic potassium form of Y zeolite:

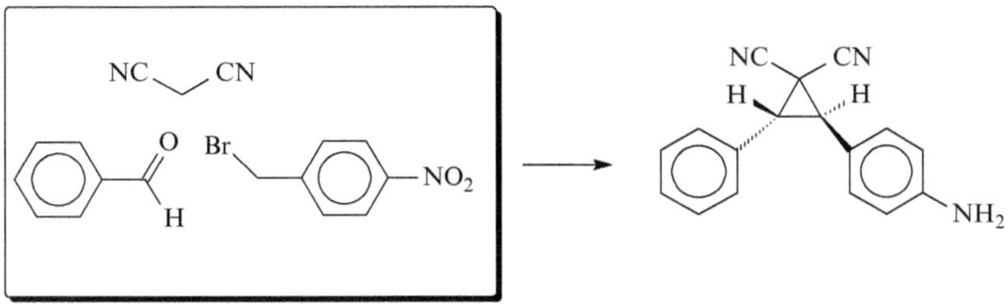

11 Conclusions

Thus, the progress in catalysis by zeolites cannot be stopped, and the last decade demonstrates we will take further steps toward greener and robust catalytic processes and technologies based on zeolites and related materials.

Acknowledgment

This work was supported by the Ministry of Science and Education of Russian Federation (Grant No. RFMEFI61615X0041).

References

[1] Flanigen EM. Pure Appl Chem 1980;52:2191.
[2] van Bekkum H, Flanigen EM, Jacobs PA, Jansen JC, editors. Introduction to zeolite science and practice. In: 2nd ed. Amsterdam-London-New-York-Oxford-Paris-Shannon-Tokyo: Elsevier; 2001.
[3] Martínez C, Corma A. Coord Chem Rev 2011;255(13–14):1558.
[4] Liu Z, Wang Y, Xie Z. Chin J Catal 2012;33(1):22.
[5] Masters AF, Maschmeyer T. Microporous Mesoporous Mater 2011;142(2–3):423.
[6] Martens JA, Thybaut JW, Denayer JFM, et al. Catal Today 2011;168(1):17.
[7] Holm MS, Taarning E, Egeblad K, Christensen CH. Catal Today 2011;168(1):3.
[8] Burd SD, Mazuik J. Hydrocarb Proc 1972;51:97.
[9] Bastiani R, Lam YL, Henriques CA, da Silva VT. Fuel 2013;107:680.
[10] Maheshwari S, Martínez C, Portilla MT, et al. J Catal 2010;272(2):298.
[11] O'Rear DJ. Ind Eng Chem Res 1987;26:2337.
[12] Burnens G, Bouchy C, Guillon E, Martens JA. J Catal 2011;282(1):145.
[13] Mota FM, Bouchy C, Guillon E, et al. J Catal 2013;301:20.
[14] de Jong KP, Mesters CMAM, Peferoen DGR, et al. Chem Eng Sci 1996;51(10):2053.
[15] Dalla Costa BO, Querini CA. Appl Catal A Gen 2010;385(1–2):144.
[16] McQueen D, Chiche BH, Fajula F, et al. J Catal 1996;161:587.
[17] Setiabudi HD, Jalil AA, Triwahyono S, et al. Chem Eng J 2013;217:300.
[18] Gagea BC, Lorgouilloux Y, Altintas Y, et al. J Catal 2009;265(1):99.
[19] Yang X, Wang JA, Chen L, et al. Catal Commun 2012;28:202.
[20] Wang JA, Zhou XL, Chen LF, et al. J Mol Catal A Chem 2009;299(1–2):68.

[21] Løften T, Gnep NS, Guisnet M, Blekkan EA. Catal Today 2005;100(3–4):397.
[22] Wan KT, Khouw CB, Davis ME. J Catal 1996;158:31.
[23] Cheung T-K, Jentoft FC, d'Itri JL, Gates BC. Chem Eng Sci 1997;52:4607.
[24] Ye F, Dong Z, Zhang H. Catal Commun 2009;10(15):2056.
[25] Mooiweer HH, de Jong KP, Kraushaar-Czarnetzki B, et al. In: Zeolites and related microporous materials: state of the Art: Elsevier; 1994. p. 2327.
[26] Xu W-Q, Yin YG, Suib SL, et al. J Catal 1996;163:232.
[27] Derouane EG, Vanderveken DJ. Appl Catal 1988;45:L15.
[28] Low DW, Tamm PW, Detz CM. Energy Progress 1987;7(4):215.
[29] Kubicka D, Kumar N, Venalainen T, et al. J Phys Chem B 2006;110(10):4937.
[30] Kubicka D, Kumar N, Maki-Arvela P, et al. Molecular sieves: from basic research to industrial applications, Parts A and B. Stud Surf Sci Catal, 158; 2005. p. 1669.
[31] Kubicka D, Kumar N, Maki-Arvela P, et al. J Catal 2004;227(2):313.
[32] Kubicka D, Kumar N, Maki-Arvela P, et al. J Catal 2004;222(1):65.
[33] Cusher NA. Handbook of petroleum refining processes. New York: McGraw-Hill; 1986.
[34] McVicker GB, Daage M, Touvelle MS, et al. J Catal 2002;210(1):137.
[35] Daage M, McVicker GB, Touvelle MS, et al. Stud Surf Sci Catal 2001;135:159.
[36] Galperin LB, Bricker JC, Holmgren JR. Appl Catal A Gen 2003;239(1-2):297.
[37] Sachtler WMH, et al. Adv Catal 1993;39:129.
[38] Dossi C, et al. J Catal 1994;145:377.
[39] Ostgard D, et al. J Catal 1992;133:342.
[40] Kubicka H, Okal J. Catal Lett 1994;25:157.
[41] Vaarkamp M, et al. J Catal 1995;151:330.
[42] Fu J, et al. Chem J Chin Univ-chin 1995;16:461.
[43] Martens JA, Mechanism of Isomerization and Hydrocracking of Long-Chain Paraffins on Large Pore and Shape Selective Zeolites, Ph. D. Thesis, Univ. Leuven, 1985.
[44] Brower D, et al. Recl Trav Chim Pays-Bas 1970;89:211.
[45] US Patent No. 5,382,730 (1995).
[46] Onyestyák G, Pál-Borbély G, Beyer HK. Appl Catal A Gen 2002;229(1–2):65.
[47] Castaño P, Gayubo AG, Pawelec B, et al. Chem Eng J 2008;140(1–3):287.
[48] Santikunaporn M, Alvarez WE, Resasco DE. Appl Catal A Gen 2007;325(1):175.
[49] Rabl S, Haas A, Santi D, et al. Appl Catal A Gen 2011;400(1–2):131.
[50] Rabl S, Santi D, Haas A, et al. Microporous Mesoporous Mater 2011;146(1–3):190.
[51] Santi D, Holl T, Calemma V, Weitkamp J. Appl Catal A Gen 2013;455:46.
[52] Bellussi G, Haas A, Rabl S, et al. Chin J Catal 2012;33(1):70.
[53] D'Ippolito SA, Gutierrez LB, Pieck CL. Appl Catal A Gen 2012;445–446:195.
[54] D'Ippolito SA, Gutierrez LB, Vera CR, Pieck CL. Appl Catal A Gen 2013;452:48.
[55] Mouli KC, Sundaramurthy V, Dalai AK. J Mol Catal A Chem 2009;304(1–2):77.
[56] Tang J-L, Xu Y-H, Wang X-Q. J Fuel Chem Technol 2012;40(6):721.
[57] Liu X, Smith KJ. Appl Catal A Gen 2008;335(2):230.
[58] Ardakani SJ, Liu X, Smith KJ. Appl Catal A Gen 2007;324:9.
[59] Ardakani SJ, Smith KJ. Appl Catal A Gen 2011;403(1–2):36.
[60] Wang L, Shen B, Fang F, et al. Catal Today 2010;158(3–4):343.
[61] Castaño P, Gutiérrez A, Pawelec B, et al. Appl Catal A Gen 2007;333(2):161.
[62] Ismagilov ZR, Yashnik SA, Startsev AN, et al. Catal Today 2009;144(3–4):235.
[63] Smiešková A, Hudec P, Kumar N, Salmi T, Murzin DYu, Jorík V. Appl Catal Gen 2010;377(1–2):83.
[64] Liu H, Wu S, Guo Y, et al. Fuel 2011;90(4):1515.
[65] Liu H, Yang S, Wu S, et al. Energy 2011;36(3):1582.
[66] Jin Z, Liu S, Qin L, et al. Appl Catal A Gen 2013;453:295.
[67] Liu H, Yang S, Hu J, et al. Fuel Process Technol 2012;96:195.
[68] Ngobeni MW, Carley AF, Scurrell MS, Nicolaides CP. J Mol Catal A Chem 2009;305(1–2):40.

[69] Zhang H, Kong A, Ding Y, et al. J Nat Gas Chem 2011;20(3):243.
[70] Cui Y, Xu Y, Lu J, et al. Appl Catal A Gen 2011;393(1–2):348.
[71] Ma S, Guo X, Zhao L, Scott S, Bao X. J Energy Chem 2013;22(1):1.
[72] Guo J, Lou H, Zheng X. J Nat Gas Chem 2009;18(3):260.
[73] Wong KS, Vazhnova T, Rigby SP, Lukyanov DB. Appl Catal A Gen 2013;454:137.
[74] Vazhnova T, Rigby SP, Lukyanov DB. J Catal 2013;301:125.
[75] Kumar MS, Holmen A, Chen D. Microporous and Mesoporous Mater 2009;126(1–2):152.
[76] Nawaz Z, Baksh F, Zhu J, Wei F. J Ind Eng Chem 2013;19(2):540.
[77] Vanhove D, Makambo P, Blanchard M. J Chem Soc Chem Commun 1979;14:605.
[78] Nijs HH, Jacobs PA. J Catal 1980;65:328.
[79] Mitsudo T, Boku H, Murachi S, et al. Chem Lett 1985;10:1463.
[80] Mukoma P, Hildebrandt D, Glasser DA. Ind Eng Chem Res 2006;45:5928.
[81] Lee YS, Koh DJ, Chung JS, Kim YG. Bull Chem Soc Jpn 1992;65:2884.
[82] Carli R, Blanchi CL, Ragaini V. Catal Lett 1995;44:49.
[83] Liu Z-W, Li X, Asami K, Fujimoto K. Catal Commun 2005;6:503.
[84] Martinez A, Lopez C, Peris E, Corma A. Stud Surf Sci Catal 2005;158:1327.
[85] Concepcion P, Lopez C, Martinez A, Puntes VF. J Catal 2004;228:321.
[86] US Patent No. 20100168258, 29.12.2009.
[87] Martinez A, Rollan J, Arribas MA, et al. J Catal 2007;249:162.
[88] US Patent No. 4304871, 19.08.1977.
[89] Zhao T-S, Chang J, Yoneyama Y, Tsubaki N. Ind Eng Chem Res 2005;44:769.
[90] He J, Yoneyama Y, Xu B, Nishiyama N, Tsubaki N. Langmuir 2005;21:1699.
[91] Bessel S. Appl. Catal. A. 1995;126:235.
[92] Chung JS, Yun HG, Koh DJ, Kim YG. J Mol Catal 1993;79:199.
[93] Tang Q, Wang Y, Zhang Q, Wan H. Catal Commun 2003;4:253.
[94] Glebov LS, Kliger GA. Usp Khim 1994;63:192.
[95] Lira E, Lopez CM, Oropeza F, et al. J Mol Catal 2008;281:146.
[96] Espinosa G, Domínguez JM, Morales-Pacheco P, et al. Catal Today 2011;166(1):47.
[97] Dalil M, Sohrabi M, Royaee SJ. J Ind Eng Chem 2012;18(2):690.
[98] Liu S, Gujar AC, Thomas P, et al. Appl Catal A Gen 2009;357(1):18.
[99] Kibby C, Jothimurugesan K, Das T et al., Catal Today, (2013) in press.
[100] Zonetti PC, Gaspar AB, Mendes FMT, et al. Fuel Process Technol 2010;91(5):469.
[101] Wang S, Yin Q, Guo J, et al. Fuel 2013;108:597.
[102] Kang S-H, Bae JW, Woo K-J, et al. Fuel Process Technol 2010;91(4):399.
[103] Prieto G, Martínez A, Concepción P, Moreno-Tost R. J Catal 2009;266(1):129.
[104] Xiong H, Zhang Y, Liew K, Li J. Fuel Process Technol 2009;90(2):237.
[105] Yang G, Xing C, Hirohama W et al., Catal Today, (2013) in press.
[106] Li X, He J, Meng M, Yoneyama Y, Tsubaki N. J Catal 2009;65(1):26.
[107] Jin Y, Yang R, Mori Y, et al. Appl Catal A Gen 2013;456:75.
[108] Liu K, Xie S, Liu S, et al. J Catal 2011;283(1):68.
[109] Weitkamp J. Acta Phys Chem 1985;31:271.
[110] Horsley JA. ChemTech October 1997;45.
[111] Lee GS, Maj JJ, Rocke SC, Garces JM. Catal Lett 1989;2:243.
[112] Chang CD, Silvestri AJ. Chemtech 1987;624.
[113] Wilson ST, Badger P. Microporous Mesoporous Mater 1999;29:117.
[114] Mier D, Aguayo AT, Gayubo AG, Olazar M, Bilbao J. Appl Catal A Gen 2010;383 (1–2):202.
[115] Simonetti DA, Carr RT, Iglesia E. J Catal 2012;285(1):19.
[116] Bellussi G, Mizia F, Calemma V, et al. Microporous Mesoporous Mater 2012;164:127.
[117] Fang Y, Tang J, Huang X, et al. Chin J Catal 2010;31(3):264.
[118] Li Y, Armor JN. Chem Commun 1997;2013.

[119] Espeel PH, de Peuter G, Tielen MC, Jacobs PA. J Phys Chem 1994;98:11588.
[120] Minachev Kh, Kharlamov V. Acta Phys Chem 1985;31:497.
[121] Mochida I, Hayata S, Kato A, Seiyama T. J Catal 1970;19:405.
[122] Centi G, Tvaruzkova Z, Trifiro F, et al. Appl Catal 1984;13:69.
[123] Suzuki E, Nakashiro K, Ono Y. Chem Lett 1988;953.
[124] Notari B. Stud Surf Sci Catal 1990;60:343.
[125] Roffa P, Leofanti G, Cesana A, et al. Chim Ind 1990;72:598.
[126] Bellussi G, Rigutto MS. In: van Bekkum H, Flanigen EM, Jacobs PA, Jansen JC, editors. Introduction to zeolite science and practice: Elsevier; 2001. p. 911.
[127] Bellussi G, Centi C, Perathoner S, et al. ACS Symp Ser 1993;523:281.
[128] Ramaswamy AW, Shivasanker S. Catal Lett 1993;22:239.
[129] Haruta M, Kobayashi T, Sano H, Yamada N. Chem Lett 1987;2:405.
[130] Huang J, Lima E, Akita T, et al. J Catal 2011;278:8.
[131] Chowdhury A, Bravo-Suarez JJ, Date M, et al. Angew Chem Int Ed 2006;45:412.
[132] Bond GC, Louis C, Thompson DT. Catalysis by gold. London: Imperial College Press; 2006. 366 p.
[133] Chretien S, Gordon MS, Metiu H. J Chem Phys 2004;121:3756.
[134] Sunil Sekhar AC, Sivaranjani K, Gopinath CS, Vinod CP. Catal Today 2012;198:92.
[135] Zahmakıran M, Özkar S. Mater Chem Phys 2010;121(1–2):359.
[136] Zhu KK, Hu JC, Richards R. Catal Lett 2005;100:195.
[137] Zheng J, Lin H, Wang Y-N, et al. J Catal 2013;297:110.
[138] Kalam A, Rahman L, Kumashiro M, Ishihara T. Catal Commun 2011;12(13):1198.
[139] Lin X, Hoel CA, Sachtler WMH, et al. J Catal 2009;265(1):54.
[140] Shen Z, Liu J, Xu H, et al. Appl Catal A Gen 2009;356(2):148.
[141] Malleswara Rao TV, Zahidi EM, Sayari A. J Mol Catal A Chem 2009;301(1–2):159.
[142] Rodrigues ACC, Monteiro JLF. Appl Catal A Gen 2009;362(1–2):185.
[143] Ratnasamy P, Srinivas D. Catal Today 2009;141(1–2):3.
[144] Adam F, Wong J-T, Ng E-P. Chem Eng J 2013;214:63.
[145] Salavati-Niasari M, Salimi Z, Bazarganipour M, Davar F. Inorg Chim Acta 2009;362(10):3715.
[146] Li X, Zhang D, Zhu X, et al. J Mol Catal A Chem 2013;372:121.
[147] Huang S, Liu S, Zhu Q, et al. Appl Catal A Gen 2007;323:94.
[148] Liu H, Zhang L, Li X, et al. J Nat Gas Chem 2009;18(3):331.
[149] Li X, Zhang W, Liu S, et al. J Mol Catal A Chem 2006;250(1–2):94.
[150] Li X, Zhang W, Liu S, et al. J Mol Catal A Chem 2009;313(1–2):38.
[151] Liu S, Li X, Xin W, et al. J Nat Gas Chem 2010;19(5):482.
[152] Borzatta V, Busca G, Poluzzi E, et al. Appl Catal A Gen 2004;257(1):85.
[153] Bhunia S, Sen R, Koner S. Inorg Chim Acta 2010;363(14):3993.
[154] Opanasenko M, Shamzhy M, Lamač M, Čejka J. Catal Today 2013;204:94.
[155] Eickelberg W, Hoelderich WF. J Catal 2009;263(1):42.
[156] Palkovits R, Schmidt W, Ilhan Y, et al. Microporous Mesoporous Mater 2009;117(1–2):228.
[157] Vaschetto EG, Monti GA, Herrero ER, et al. Appl Catal A Gen 2013;453:391.
[158] Ghiaci M, Aghaei H, Oroojeni M, et al. Catal Commun 2009;10(11):1486.
[159] Neri C, Buonomo F, Eur. Pat. 100,117, 1983.
[160] Danuthai T, Jongpatiwut S, Rirksomboon T, et al. Appl Catal A Gen 2009;361(1–2):99.
[161] Japan Patent No. 81,142,233 (1981).
[162] Kogelbauer A, Kouwenhoven HW. In: Sheldon RA, van Bekkum H, editors. Fine chemicals through heterogeneous catalysis. Weinheim: Wiley-VCH; 2001. p. 123.
[163] Corma A, Frones V, Martin-Aranda RM, et al. Appl Catal 1990;59:237.
[164] Černý R, Kubů M, Kubička D. Catal Today 2013;204:46.
[165] Cardó X, Bergadà O, Cesteros Y, Salagre P. Chem Eng J 2012;183:459.
[166] Ha L, Mao J, Zhou J, et al. Appl Catal A Gen 2009;356(1):52.

[167] Botas JA, Serrano DP, García A, et al. Catal Today 2012;195(1):59.
[168] Trippe F, Fröhling M, Schultmann F, et al. Fuel Process Technol 2013;106:577.
[169] Ferreira LR, Lima S, Neves P, et al. Chem Eng J 2013;215–216:772.
[170] Jacobs PA, Hinnekens H. Eur Pat 1989;329:923.
[171] Käldström M, Kumar N, Heikkilä T, et al. Biomass Bioenergy 2011;35(5):1967.
[172] Zhao W, Yi C, Yang B, et al. Fuel Process Technol 2013;112:70.
[173] Jae J, Tompsett GA, Foster AJ, et al. J Catal 2011;279(2):257.
[174] Bonilla A, Baudouin D, Pérez-Ramírez J. J Catal 2009;265(2):170.
[175] Elordi G, Olazar M, Lopez G, et al. J Anal Appl Pyrolysis 2009;85(1–2):345.
[176] Kustov LM, Sinev IV. J Phys Chem (Russ) 2010;84(10):1676.
[177] Balu AM, Hidalgo JM, Campelo JM, et al. J Mol Catal A Chem 2008;293:17.
[178] Gopalakrishnan S, Munch J, Herrmann R, et al. Chem Eng J 2006;120:99.
[179] Bai G, Li T, Yang Y, et al. Catal Commun 2012;29:114.
[180] Bond G, Gardner JA, McCabe RW, Shorrock DJ. J Mol Catal A Chem 2007;278(1–2):1.
[181] Winé G, Vanhaecke E, Ivanova S, et al. Catal Commun 2009;10(5):477.
[182] González MD, Cesteros Y, Llorca J, Salagre P. J Catal 2012;290:202.
[183] Jury FA, Polaert I, Estel L, Pierella LB. Appl Catal A Gen 2013;453:92.
[184] Yamashita H, Mitsukura Y, Kobashi H, et al. Appl Catal A Gen 2010;381(1–2):145.
[185] Chandra Shekar BM, Jai Prakash BS, Bhat YS. J Catal 2012;290:101.
[186] Costarrosa L, Ruiz-Martínez J, Rios RVRA, et al. Microporous Mesoporous Mater 2009;120(1–2):115.
[187] Wei ZS, Zeng GH, Xie ZR, et al. Fuel 2011;90(4):1599.
[188] Sicard M, Grill M, Raepsaet B, et al. Part B, Stud Surf Sci Catal 2008;174:1103.
[189] Xian X, Liu G, Zhang X, et al. Chem Eng Sci 2010;65(20):5588.
[190] Mota Salinas AL, Sapaly G, Ben Taarit Y, et al. Appl Catal A Gen 2008;336(1–2):61.
[191] Bogdan VI, Klimenko TA, Kustov LM, Kazansky VB. Appl Catal A Gen 2004;267(1–2):175.
[192] Yoneyama Y, Sun X, Zhao T, et al. Catal Today 2010;149(1–2):105.
[193] Tian Z, Qin Z, Wang G, et al. J Supercrit Fluids 2008;44(3):325.
[194] Yu CC, Tan CS. J Supercrit Fluids 2008;44(3):341.
[195] Hajimirzaee S, Leeke GA, Wood J. Chem Eng J 2012;207–208:329.
[196] Sotelo JL, Calvo L, Pérez-Velázquez A, et al. Appl Catal A Gen 2006;312:194.
[197] Álvaro M, Das D, Cano M, Garcia H. J Catal 2003;219(2):464.
[198] Rodríguez A, Uguina MA, Capilla D, Pérez-Velazquez A. J Supercrit Fluids 2008;46(1):57.
[199] Sakthivel A, Nakamura R, Komura K, Sugi Y. J Supercrit Fluids 2007;42(2):219.
[200] Koskinen R, Turunen H, Tiitta M, Keiski RL. Chem Eng J 2012;213:235.
[201] Zhang X, Llabrés i Xamena FX, Corma A. J Catal 2009;265(2):155.
[202] Bertero NM, Apesteguía CR, Marchi AJ. Catal Today 2011;172(1):171.
[203] Rama V, Kanagaraj K, Subramanian T, et al. Catal Commun 2012;26:39.

CHAPTER 9

Conversion of Biomass to Chemicals: The Catalytic Role of Zeolites

T. Ennaert, W. Schutyser, J. Dijkmans, M. Dusselier, B.F. Sels
Centre for Surface Chemistry and Catalysis, Leuven, Belgium

1 Introduction

Currently, fossil-based resources, such as petroleum and natural gas, are responsible for the production of the most important chemicals. To produce chemicals from fossil oil, the large hydrocarbon fraction is cracked and transformed into useful bulk chemicals like olefins (eg, ethylene and propylene) and aromatics. Central to these production processes—fluid catalytic cracking (FCC) being the most popular one—are zeolites where the unique strong Brønsted acidity of zeolites plays a crucial role. Functionalization of the cracked hydrocarbons with heteroatoms like oxygen, nitrogen, and sulfur is usually the next step in providing the desired chemical property to the molecule. Zeolites are catalytically active in these functionalization reactions as well, as has been demonstrated in many reviews [1–4], for instance, by the epoxidation of propylene and the hydroxylation of phenol to hydroquinone with Ti containing zeolites [5–8].

Natural gas requires an activation of the inert methane molecule, usually at high temperature in the presence of water, and promoted Ni-based catalysts, leading to syngas. Depending on the conditions and catalysts, a variety of chemicals could arise from the syngas, such as alcohols, olefins, and alkanes using Fischer-Tropsch, but also methanol can be produced from syngas. Zeolites such as Zeolite Socony Mobil-5 (ZSM-5) are often used in upgrading methanol to aromatics and olefins according to the methanol to gasoline (MTG) and methanol to olefin (MTO) technologies [9].

Unfortunately, easily accessible oil and gas fields are becoming depleted and the CO_2 emissions from fossil oil are influencing our climate. To reduce the emission effects and to secure the energy supply, valorization of biomass is stimulated globally. Governments take the opportunity to locally support the agricultural economy and try to become energy independent. Driven by energy and environmental needs, conversion of biomass to chemicals is also intensively practiced. First-generation chemicals are presently produced from high-value sugars and vegetable oils, but they are competing with the food chain. A higher demand

Zeolites and Zeolite-like Materials. http://dx.doi.org/10.1016/B978-0-444-63506-8.00010-0

of corn and vegetable oils already increases their prices in the United States and Europe, followed by additional rises due to speculation. Second-generation bio-derived chemicals and fuels, based on cheaper and abundant lignocellulosic feedstock, are therefore currently under development [10,11]. Due to the differences between both feedstocks (fossil oil vs. biomass), a new set of challenges in the development of active, selective, and stable catalysts and catalytic processes arise [12,13]. However, in an extension to the success of zeolites in petrochemistry [14], zeolites are gaining popularity in biomass processes as they have already shown excellent performances in a wide range of biomass-related reactions.

2 Biomass Feedstock and Their Current Chemical Uses

2.1 Carbohydrates

Carbohydrates is the most important class of chemicals in the biomass feedstock. Carbohydrates can be divided into monosaccharides, disaccharides, and polysaccharides. The most familiar monosaccharides are glucose, fructose, galactose and mannose (hexoses) and xylose, arabinose and ribose (pentoses) (Fig. 1). Sucrose (Fig. 3), a disaccharide of fructose and glucose, present in sugar beet and sugarcane and lactose (galactose-glucose), present in milk, are two well-known disaccharides.

The group of polysaccharides comprises starch, cellulose, hemicellulose, and inulin. Starch, an energy source of plants, is available from crops such as potatoes and cereals (wheat and corn). It is composed of two components: amylose and amylopectine (Fig. 2). Amylose is a linear chain of glucose units linked via α-1.4-glycosidic bonds, while the glucose units in amylopectine are connected via both α-1.4- and α-1.6-glycosidic bonds. The latter ones define the branched structure of amylopectine. Cellulose (Fig. 3) is a major structural component in the fibrous material of cell walls, where it is associated with hemicellulose and lignin. Cellulose polymers (1 MDa) contain β-1.4-glycosidic bonds between the glucose units. The different orientation of the ether bonds in cellulose in comparison with starch has a huge impact on the occurrence of the polysaccharide: the amorphous state of starch ensures that the chemical bonds are highly accessible, while cellulose contains denser, microcrystalline domains with hindered access to the ether bonds. Cellulose is largely used

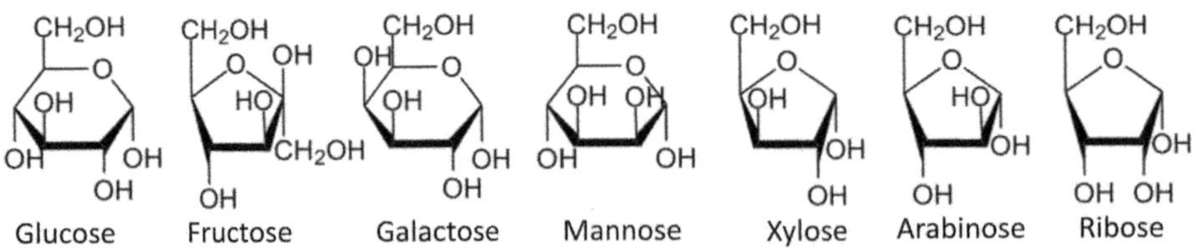

Fig. 1
Most abundant sugar monomers of the hexose and pentose family.

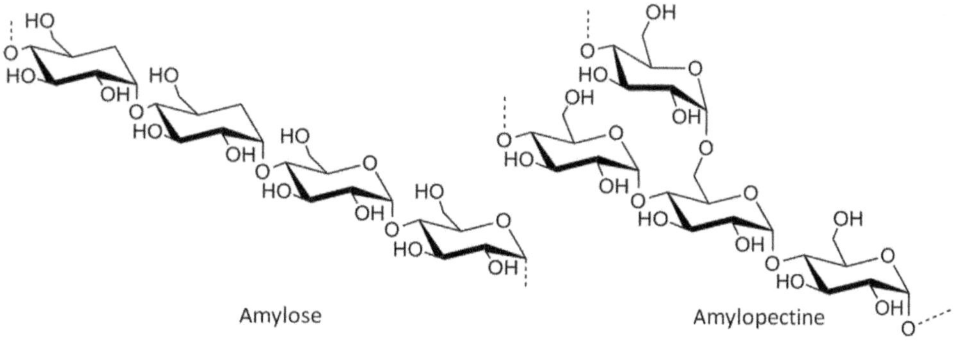

Fig. 2
Chemical structure of amylose and amylopectine.

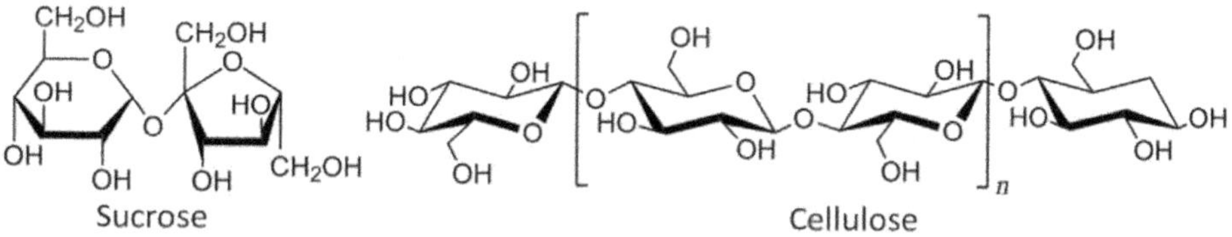

Fig. 3
Chemical structure of sucrose and cellulose.

in the paper and textile industry, while starch is consumed mainly for bioethanol production. Other applications of starch are binders in the paint and paper industry (mainly amylopectine) and use in film and adhesive applications (mainly amylose).

Hemicellulose (30 kDa) is the glue in the cellulose fibers; it has shorter, branched polymer chains built up from a variety of hexoses and pentoses. Inulin (10 kDa) is a heterogeneous branched polymer consisting of 3–100 monomers, mainly fructose, linked via β-2. 1-glycosidic bonds. Alike starch, hemicellulose and inulin are easy to digest via hydrolysis, in contrast to the more resistant nature of cellulose.

Glucose (starch and cellulose), fructose (inulin), xylose (xylan), and sucrose are the actual carbohydrate raw materials for current industrial applications because they are available in tonnage scale at reasonable price. Glucose lies at the origin of sorbitol and its dehydrated forms such as sorbitans and isosorbide. Glucose (and starch) can also be fermented to L-lactic acid, from which a biodegradable and biocompatible polylactic acid (PLA) polymer is formed [15]. Xylose, the major pentose, is hydrogenated to xylitol for food applications. Pentose streams from hemicellulose are used for the industrial production of furfural, a common solvent, by acid treatment of, for example, sugarcane bagasse. The first industrial furfural production process was established by Quaker Oats in 1921 [16]. In this process, oat hulls were converted into furfural using sulfuric acid and high pressure steam to strip out the furfural. The major part of the produced furfural was converted with hydrogen to furfuryl

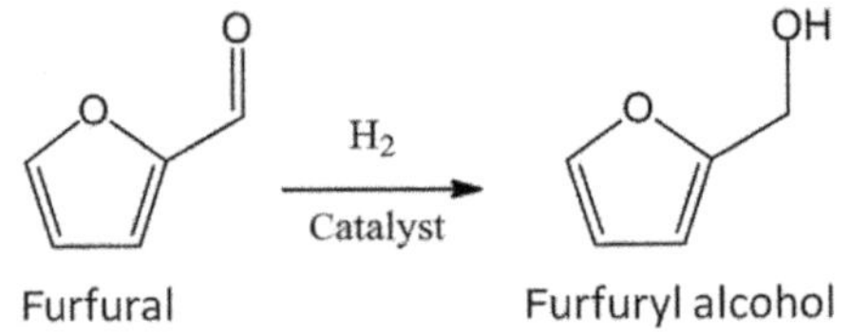

Scheme 1
Hydrogenation of furfural to furfuryl alcohol.

alcohol (Scheme 1). The latter finds application in the production of resins for bonding foundry sands for molds and in composite materials for the car and plane industry.

2.2 Lignin

The third component of lignocellulose, lignin, provides rigidity to the structure and antibacterial properties to the plant cell wall. Lignin is a complex, three-dimensional structure of propyl phenol units, namely coumaryl, confineryl, and sinapyl alcohol (Fig. 4), which are bound together through a variety of carbon-carbon and ether linkages. The chemical use of lignin is a great challenge due to high heterogeneity in the molecule.

Overall, lignocellulosic biomass contains 40–50 wt% cellulose, 25–40 wt% hemicellulose, and 10–25 wt% lignin. Large amounts (several $Gt\,a^{-1}$) of lignocellulose may be produced via dedicated crops and short rotation wood, while there is up to $5\,Gt\,a^{-1}$ waste from agricultural and forestry activities.

2.3 Vegetable Oils

Vegetable oils are esters of glycerol with three fatty acids (triacylglycerol or triglyceride (TAG)), ranging in length between 6 and 24 carbons (Fig. 5), 16–20 carbons being most abundant. They may contain a low to high value of unsaturation in the fatty

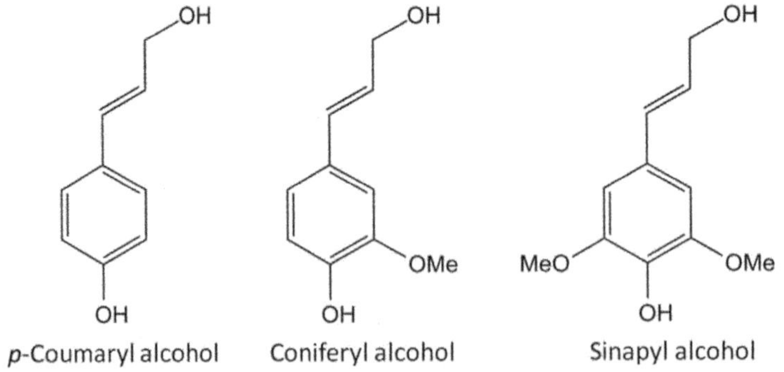

Fig. 4
Chemical structure of the three propyl phenol monomers of lignin.

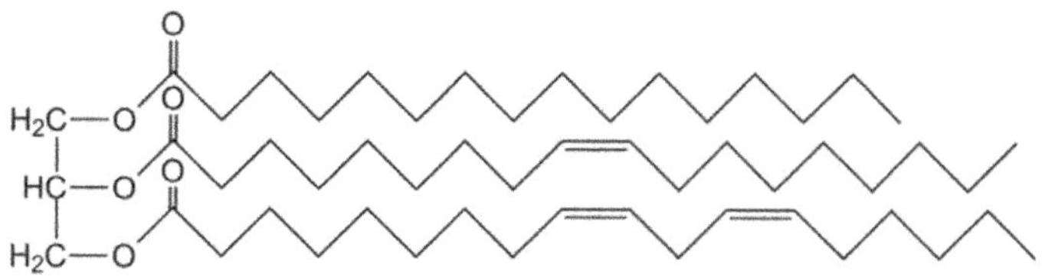

Fig. 5
Chemical structure of a triacylglycerol molecule; the left part contains glycerol with esters of palmitic acid (top fatty acid), oleic acid (middle fatty acid), and linoleic acid (bottom fatty acid).

acid hydrocarbon chain, located at specific positions and always in *cis* geometry. The variety in chain length and the amount of double bonds determine the occurrence of the fatty product (liquid to solid), while the latter also determines the oxygen (and light) stability. Both properties are important in the food industry, for instance for the production of dressings, margarines, and shortenings. Of all chemicals present in the biomass feedstock, triacylglycerols are in terms of structure the most comparable with that of fossil hydrocarbons, except for the ester functionality. It has the highest C to H ratio and it is therefore regarded as a valuable replacer of diesel. Trans-esterification of the oil with methanol in the presence of a catalyst leads to methyl fatty acids, which are used as biodiesel or blended in fossil diesel (B2–B20 with 2–20% biodiesel). Sources of vegetable oil are a variety of vegetables such as soybean oil, palm, rapeseed, castor and sunflower, and animal raw materials, while waste streams from the food industry are available as well.

Main chemical uses of the bio-based oleochemicals are plastic additives in polymers (in the form of epoxides and polyols) and as building blocks of polymers (in the form of dicarboxylic acids like azelaic acid, obtained via ozonolysis) such as polyamides (textile) and polyurethanes (isolation material and adhesives). The lubricant sector uses oleochemicals as a biodegradable alternative to the conventional mineral oil products with applications in the automobile and hydraulic systems. Hardening of the vegetable oil, that is, hydrogenation of the double bonds, is often applied for oxidative stability reasons. A wide range of products, such as surfactants, emollients, emulsifiers, and waxes, in the home and personal care applications contain oleochemicals. Polyunsaturated oils, including linseed oil and soybean oil, are used as drying oils in paints and varnishes.

Glycerol is a by-product of the biodiesel production: 1 ton biodiesel production produces 100 kg of glycerol. Glycerol is a 1.2.3-triol that is used in personal care and pharmaceutical products and as feed additive, but there are also a variety of derived chemicals such as propanediols (monomers in polyester and polyurethane materials), oligoglycerols (cosmetics and biodegradable lubricants), epichlorohydrin (polymers), and glycerol carbonate (solvent).

3 General Overview of the Major Biomass Conversion Strategies

Many strategies exist for the conversion of biomass to chemicals, fuels, and materials. Generally, these strategies can be divided into three main approaches: microbial, thermochemical, and chemocatalytic processes. The first and also the oldest method is the microbial conversion of biomass, which has been used by humans for millennia, such as for agriculture and brewing of beer. It involves the use of microorganisms, like bacteria and yeasts, for the conversion of sugars to simple acids and alcohols. Although this approach has often proved its elegance, we will not discuss the microbial strategy in this chapter. Comprehensive reviews can be found in other literature [17–21].

Vegetable oil and sugars are relatively simple in use as a chemical feedstock without severe treatment. In contrast, the usage of porous materials for direct second-generation biomass (lignocellulose) conversion is less straightforward for reasons of accessibility. Therefore, biomass needs to be broken down into smaller molecules. To accomplish this, thermochemical treatment under severe conditions or chemocatalytic hydrolysis under mild conditions are the two preferred options [22–25]. Thermal biomass conversion processes include gasification and pyrolysis. Gasification concerns burning the biomass (mainly cheap untreated lignocellulose) in air at temperatures from 800 to 1000°C to produce synthesis gas, a mixture of CO and H_2 [26]. The product is mainly relevant for power generation. However, the syngas may be converted to methanol or hydrocarbons via the Fischer-Tropsch synthesis. Note that a water gas shift reaction is required to adjust the H_2/CO ratio. Depending on the biomass composition, about one third to half the carbon is lost due to CO_2 formation, while oxygen is removed as CO_2 and water. Due to the high losses of C and O, this process suffers from relative low atom efficiency. Gasification of wood chips into syngas and subsequent Fischer-Tropsch has been demonstrated by CHOREN in a joint venture with Shell.

Pyrolysis of biomass is carried out under inert atmosphere and forms, depending on the residence time and temperature, char, oil, and gas. Pyrolysis with long residence time at low temperature (400°C) produces a black solid (charcoal), while fast pyrolysis at high temperature (500°C) favors the formation of a black liquor (bio-oil). The short contact times (<2 s at ca. 500°C) thus maximize the liquid yield. Fast pyrolysis is preferred by the chemical industry because of the relative ease of handling liquids. However, bio-oil produced by pyrolysis of bulk biomass contains more than 400 different components like carboxylic acids, ketones, aldehydes, sugars, furans, (substituted) phenols, aromatics, and tar (Table 1). Separation of useful chemicals from this complex pool is very difficult. As an alternative, pyrolysis can also be used as a first step for generating heat or electricity, followed by combusting the pyrolytic products. Excellent papers and reviews that describe fast pyrolysis in more detail are available [27–32].

Table 1: Composition of typical bio-oil [22].

Component	wt%
Acetic acid	1–32
Formic acid	1–20
Glyceraldehyde	1–13
Furfuryl alcohol	1–5
2-Hydroxyacetone	2–8
Syringol	1–5
Phenol	0–4
Methanol	1–3
Polyaromatics (tars)	2–7

Especially when the production of chemicals is pursued, the three main components of lignocellulosic (cellulose, hemicellulose, and lignin) are better separated prior to pyrolysis. Next to a separate pyrolysis of the (hemi)cellulose and lignin fractions, other strategies such as upgrading of the bio-oils (via cracking or hydrodeoxygenation), upgrading of pyrolysis vapors, or integrating catalysts in the fast pyrolysis (catalytic fast pyrolysis) are required to adjust the product composition toward lean oxygen content to become compatible with conventional transportation fuels. During all these processes, zeolites can play a prominent role.

The severe conditions of gasification and pyrolysis are not a necessity to solubilize lignocellulose. The ether linkages in the (hemi)cellulose and lignin part are sensitive to chemocatalytic hydrolysis below 250°C in the presence of enzymes or (soluble) acids and bases. In this approach, lignocellulose is best fractionated in its components (cellulose, hemicellulose, and lignin) prior to conversion. This separation entails a certain cost, which is a current concern in future biorefineries. Distillation, as it is done in crude oil refineries, is not possible because of the low volatility and stability of the biomass-related chemicals. Therefore, other techniques such as solvent treatments (with acid and base) and extraction, membrane separation, and chromatography will be required to separate the chemicals. However, most of these technologies are still under development. Once established, biomass fractionation and separation have the advantage of access to a purified fraction, which allows a more practical handling and a more selective formation of chemicals. Cellulose and hemicellulose can subsequently be converted to simple sugars, such as glucose and pentoses. Their chemical transformation to platform chemicals may be considered as analogous to the first-generation sugar feedstock. Nice examples come from cellulose (eg, glucose to bioethanol [33–37]) and hemicellulose (furfural [38–41]). However, controlled lignin depolymerization to aromatic alcohols is less obvious and this problem is still under investigation [42].

Use of heterogeneous catalysis is preferred above soluble acid and bases because of technologic advantages and fewer corrosion issues. However, solid acids and bases are generally too large to assist the depolymerization of the genuine biomass fractions via hydrolysis, especially in the case of crystalline cellulose and lignin. Many chemical and mechanical pretreatment techniques such as milling, irradiation, steam, acid, basic, ammonia, organosolv, and ultrasonic treatments are therefore under study to provide better accessibility of the ether linkages [43–45,23,46–51]. Some of the techniques are applied while performing the hydrolysis reaction. Once cellulose is solubilized, the sugars and oligomers may be treated in the first-generation technology platform, developed for starch and sugar conversion. Due to the complex and unknown structure of lignin, hydrolysis pathways to chemicals are currently less evolved.

The following parts provide more detail of the usefulness of zeolites in the valorization of both first- and second-generation biomass. Although not intending to be comprehensive, the examples attempt to emphasize the strength and weakness of zeolite properties in the different chemical reactions with biomass-related feedstocks.

4 Zeolites in Carbohydrate Chemistry

The transformation of woody biomass into drop-in transportation fuels or biofuels is a promising approach for a more sustainable energy production. However, integrated into a biorefinery, this will lead to a high-volume but low-margin process. To be competitive with petrochemical refineries, the production process should be cost-effective, that is, full use of the whole feedstock with an as-high-as-possible atom efficiency, combined with a maximum exploitation of the chemical functionalities present in the biomass. This implies that a biorefinery should also produce high-value chemicals next to the low-functionalized biofuels. In our opinion, the sugar fraction of biomass, that is, (hemi)cellulose, is extremely suitable for the production of such high-value chemicals due to the abundant chemical functionalities present in sugars.

Starting from polysaccharides such as (hemi)cellulose, a high variety of chemicals can be produced. Therefore, a careful selection of the targeted molecules is necessary. In this view, Dusselier, Mascal, and Sels recently published an assessment with the top chemical opportunities for sugars in a biorefinery [52]. They defined a functionality index (F:C), which is a measure of the amount and type of functional groups normalized to the number of carbon atoms. Fig. 6 shows the atom economy of some biomass primary products and derivatives, calculated from their respective polysaccharides or monosaccharides, versus the F:C. This figure clearly shows that biomass components and derivatives are characterized by a high F:C, while typical petroleum-derived (and O deficient) molecules such as alkanes, alkenes, and alkanols have a low F:C. We think that the most suitable targets are the molecules that retain a high amount of their F:C with respect to the substrate (indicated by the blue box in the graph). However, this is our suggestion and accounts only for the chemical's structure and functionality while it disregards the economic situation. The market for a given chemical

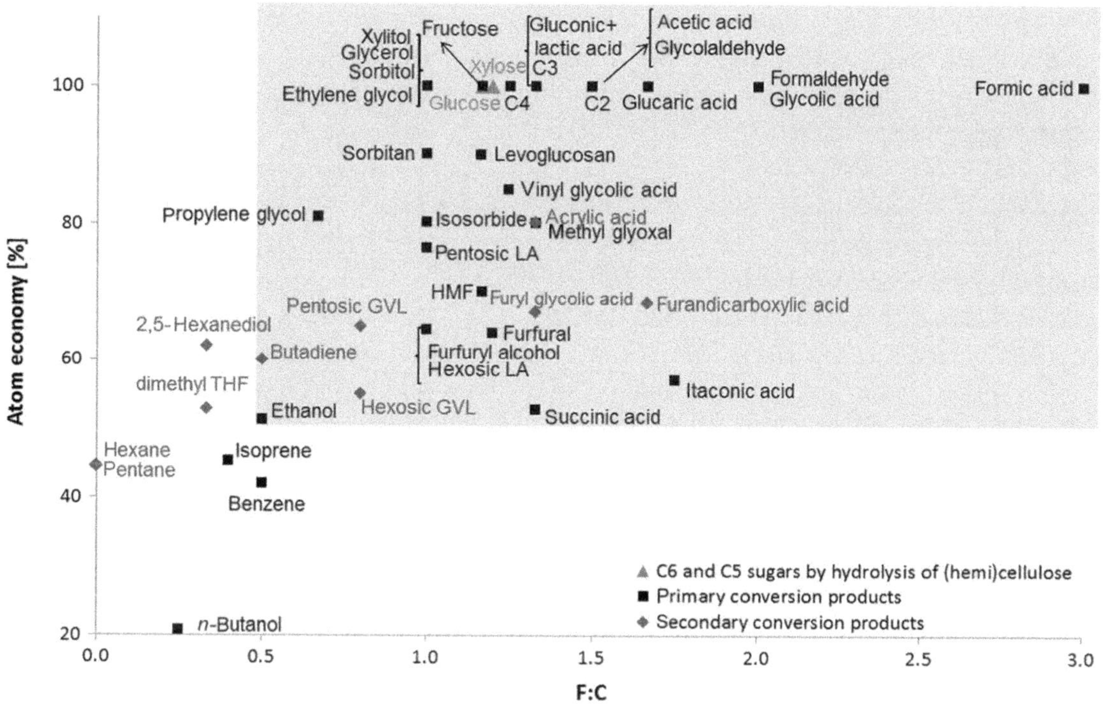

Fig. 6

Graphical representation of the atom economy vs. F:C for common carbohydrates and their derivatives. *Graph and data are reproduced with permission of Springer (Based on Dusselier M, Mascal M, Sels B. In: Nicholas KM, editor. Selective catalysis for renewable feedstocks and chemicals, vol. 353. Springer International Publishing; 2014. p. 1–40).*

ultimately will determine its value. This means that molecules with a low F:C value can still be very valuable when there is high demand. So in practice, every targeted molecule or reaction requires its own assessment.

The chemical transformation routes toward most of these "high-interest" molecules can be divided into only five different reaction types: isomerization, retro-aldol, dehydratation/rehydratation, hydrogenation, and oxidation. In the next section we will systematically discuss the potential of zeolites and ordered porous materials for each of these reaction types (except oxidations). Finally, we will discuss the potential of glycerol as a substrate in zeolite-based conversions.

4.1 Isomerization

The isomerization of glucose into fructose for the production of high-fructose corn syrups (HFCS), a mixture of glucose and fructose, which is 1.3 times sweeter than sucrose, is the largest biocatalytical process in the world. In the next decades, the isomerization of sugars probably will gain importance since the isomerization of glucose to fructose is considered to be an essential step for the selective and efficient production of fuels and chemicals from biomass [3]. Fructose, the keto-isomer of glucose, is more susceptible toward chemical reactions, probably due to a slightly higher linear occurrence [53]. Now, biocatalytic isomerization is industrially carried out with immobilized xylose isomerase enzyme. Although high yields are obtained by this method, the enzymatic process entails several drawbacks (eg, prereaction purification, narrow operating pH and temperatures, activity decay, etc.) causing difficulties with the integration in a multistep process. Therefore, stable inorganic isomerization catalysts have been explored. Isomerization of monosaccharides can be carried out with insoluble bases. Moreau et al. presented high-fructose selectivity (up to 86%) but only at low glucose conversion (maximum 25%) for zeolites with moderate basicity like NaX and KX. However, the authors observed a significant leaching of the metal cation leading to decreasing glucose conversions in successive runs [54]. More recently, large pore zeolites containing Lewis acid sites, comprising tin or titanium, proved to be very efficient isomerization catalysts (especially Sn-based catalysts) [55,56]. Moliner et al. were able to convert a 10 wt% glucose in water solution using Sn-beta zeolite with a glucose/Sn ratio of 50/1: 46% glucose, 31% fructose, and 9% mannose were obtained after 30 min at 110°C, which is close to the thermodynamic equilibrium.

Only seven sugars (viz. D-glucose, D-fructose, D-mannose, D-galactose, D-xylose, L-arabinose, and D-ribose) are found in nature in sufficiency to allow their exploitation on a commercial scale. All other C6 and C5 sugars are referred to as "rare sugars." Nevertheless, owing to their low natural existence, the latter sugars can be used in a wide range of useful applications in medicine (anticancer and antiviral drugs, antiinflammatory agents), food industry (low-calorie sweeteners), or chemical synthesis (chiral building blocks in natural products synthesis). To meet these requirements, natural sugars can undergo isomerization reactions to rare sugars. An example hereof is the isomerization of xylose with Sn-beta into xylulose and lyxose (27% and 11% yield respectively at 60% xylose conversion). When Brønsted acids are added to the Sn catalyst, furfural can be formed in a single pot (*vide infra*) [57].

By varying the reaction parameters (catalyst or solvent) the product distribution of an isomerization reaction can be influenced (Scheme 2). For instance, the main product of Sn-beta catalyzed glucose isomerization shifts from fructose to mannose if sodium tetraborate is added [58]. The same exercise can be made with xylose and arabinose where respectively lyxose and ribose are mainly formed in the presence of borate salts [58]. Changing the reaction solvent from water to methanol causes a similar shift in selectivity without the need for a cocatalyst, mannose becomes the main reaction product in the isomerization of glucose. Using of Ti-beta instead of Sn-beta for the isomerization of glucose causes the formation of another rare sugar, sorbose, albeit with low yields, in both water and methanol [59]. Finally Gounder and Davis demonstrate that low-defect zeolites, which are much more hydrophobic than highly defective zeolites, catalyze sugar isomerizations with considerable higher initial turnover rates. They reported glucose to fructose isomerization with initial turnover rates

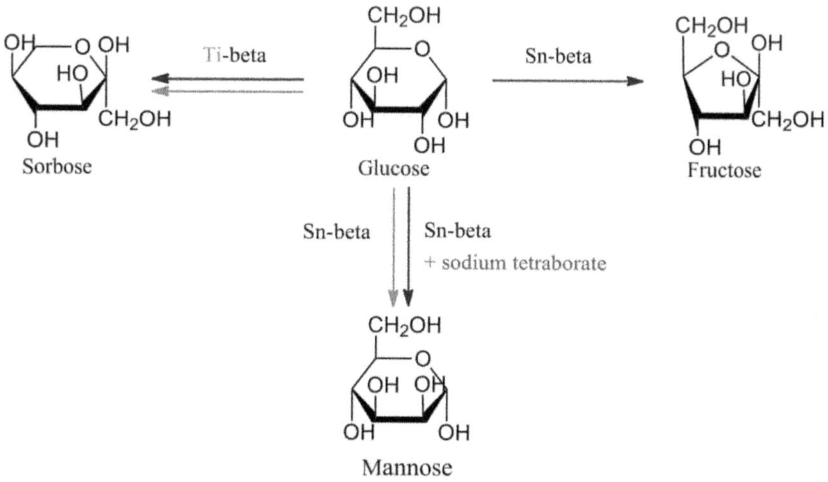

Scheme 2
Overview of the different products obtained by varying reaction conditions of the isomerization of glucose by metal incorporated beta, aqueous conditions *(blue arrow; dark gray in print versions)*, or alcoholic media *(green arrow; light gray in print versions)*.

10–30 times higher and even disaccharide isomerization (lactose into lactulose) 10^3–10^4 factors higher than with more hydrophilic zeolites. This higher isomerization rate is caused by weaker interactions of solvent molecules with the Lewis acid metal in hydrophobic zeolites, enabling faster sorption of sugars onto the active sites which eventually leads to higher turnover rates [60].

4.2 Retro-Aldol—Lactic Acid Formation

Retro-aldol chemistry is another option to convert carbohydrates to chemicals. Such a reaction with hexoses yields two trioses, viz. glyceraldehyde and 1.3-dihydroxyacetone, which are very interesting C_3 building blocks. For example, they can be converted to lactic acid or its esters. Lactic acid and alkyl lactates find applications in cosmetics, the food industry, and in green solvents. Besides they are also a vital building blocks for biodegradable PLA polymer synthesis [15,61]. In 2005 Hayashi et al. reported the remarkably fast and selective conversion of these trioses into alkyl lactates in alcoholic media with the use of soluble Sn halides [62]. These catalysts distinctively promoted the conversion of trioses to alkyl lactates while all other screened metal salts displayed little or no activity and selectivity toward alkyl lactates. A 10 mol% of $Sn^{II}Cl_2$ was, for instance, able to convert a solution of 2.5 mmol dihydroxyacetone (DHA) in methanol with a yield of 89% to methyl lactate in 3 h at 90°C, whereas Al^{III} and Cr^{III} based halides displayed only 62% and 50%, respectively [62]. Further catalytic superiority was demonstrated with $Sn^{IV}Cl_4{\cdot}5H_2O$: higher turnovers per Sn atom (8.8 mol $mol_{Sn}^{-1}\,h^{-1}$) were found with the Sn^{IV} catalyst, yielding 90% ethyl lactate or *n*-butyl lactate in 1 h at 90°C. The exceptional activity of Sn was tentatively attributed to the strong interaction of the Lewis acid Sn^{IV} with the carbonyl and alcohol group of the triose substrate, owing to the known tendency of Sn to coordinate OH bonds rather than to form Sn–O–C bonds [62]. Sels and coworkers were the first to report the zeolite-catalyzed conversion of trioses; moreover this was the first heterogeneous catalyzed conversion of trioses to alkyl lactates [63]. They reported acidic Ultra-Stable Y (USY) zeolites to be effective for isomerization and esterification of dihydroxyacetone in alcoholic media yielding moderate amounts of alkyl lactate.

Following this pioneering work, detailed analysis and isotope labeling experiments show the requirement of both Lewis and Brønsted acid sites (Scheme 3). The latter preferably catalyzes the rate-determining dehydration of trioses into pyruvaldehyde, while the Lewis acid assists the further rearrangement according to a hydride shift (Meerwein-Ponndorf-Verley or intramolecular Cannizzaro reaction type) into alkyl lactate [64]. This gives rise to the existence of a subtle balance between the amount and strength of the Brønsted and Lewis acid sites: strong Brønsted acidity is unwanted as it converts pyruvaldehyde into undesired dialkyl acetal. The formation of lactic acid is possible when the alcoholic solvent is replaced by water. The main disadvantages of working in water are lower rates, yields and selectivities.

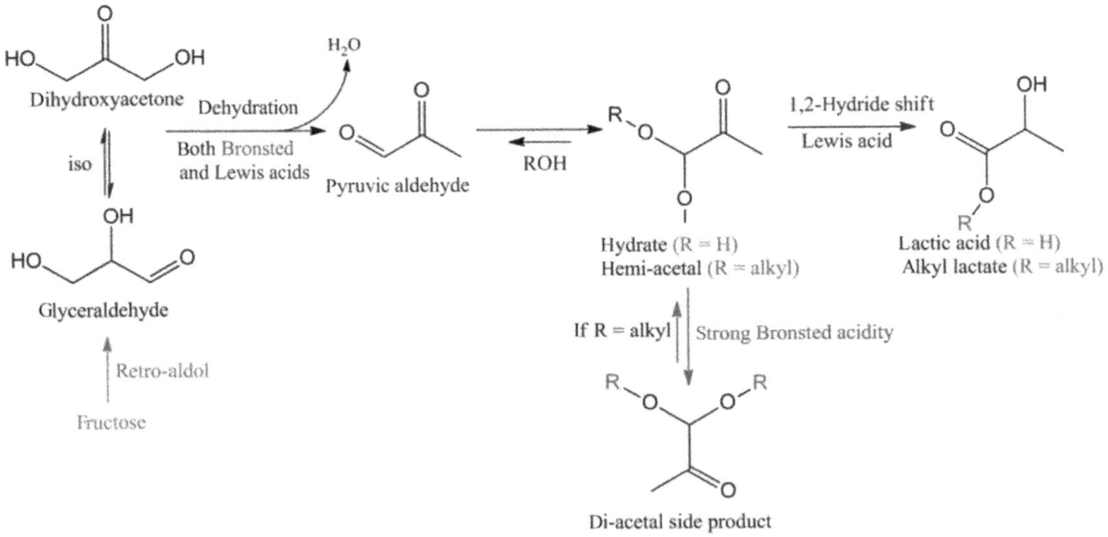

Scheme 3

Reaction network of the formation of lactic acid and alkyl lactate starting from triose sugars in aqueous conditions (R = H) or in alcoholic media (R = alkyl). *Reproduced with permission of the Royal Society of Chemistry (Modified from Dusselier M, Van Wouwe P, Dewaele A, Makshina E, Sels BF. Lactic acid as a platform chemical in the biobased economy: the role of chemocatalysis. Energy Environ Sci 2013;6:1415–42).*

Moreover, the stability of the catalyst in an organic acid, that is, lactic acid, and in liquid water at high temperatures becomes an important concern [65,66].

Besides the classical Si and Al-containing zeolites, there is an ongoing search toward other Lewis acidic zeolites and other microporous materials suitable for this reaction. Taarning et al. made a comparison between beta zeolites substituted with Al, Sn, Zr, and Ti. As with the homogeneous catalysts, this comparison pointed to Sn-based catalysts as extremely selective for this reaction. The authors obtained lactic acid and lactate yields of 90% and >99% at 125 and 80°C, respectively, with an Si:Sn ratio of 125 [67]. Since then, numerous reports were published including Sn-montmorillonite [68], mesoporous Sn-MCM-41 (Mobil Composition of Matter) [69,70], Sn-MFI (Mordenite Framework Inverted) [70,71], Sn-deAl-beta [72], Sn-SBA-15 (Santa Barbara Amorphous type material) [70], Sn-MWW (Zeolite Framework Type M-22) [73], desilicated MFI [74], and alkaline-assisted galliated USY, ferrierite, mordenite, MFI, and beta [75,76].

In search of other Lewis acidic heterogeneous catalysts, Li et al. reported Ga_2O_3 and Sn-MCM-41 mesoporous materials [69]. The latter proved to be as selective and faster (per Sn) than Sn-beta due the presence of both weak protonic silanols and isolated Sn sites. Direct proof of the elegant bifunctional catalytic mechanism for alkyl lactate synthesis from trioses has recently been reported by de Clippel and Dusselier et al. [77]. Based on their earlier experiences [78–80], they proposed a simple carbon-silica composite design to alter the number of Brønsted acid sites (provided by a carboxylic acid and phenolic functionalized carbon phase) independently of that of the Lewis acid sites (provided by Sn grafted onto the MCM-41 silica pore walls). A quantitative DHA conversion into ethyl lactate was obtained with a composite containing 15% of carbon, which was pyrolysed at 500°C and post-treated at 300°C under oxygen flow. With respect to the parent Sn-MCM-41, this composite achieved a sevenfold increased Sn-based initial turnover frequency of 289 h^{-1}. The bifunctional mechanism was recently further explored using zeolites containing both Brønsted acid Al^{3+} atoms, as well as Lewis acidic Sn^{4+} atoms inside a beta zeolite framework. It was demonstrated that each of the active sites catalyze a specific step in the reaction mechanism; the Brønsted acid Al^{3+} sites are used for the dehydration of DHA into the intermediate pyruvic aldehyde, whereas Sn is used in the subsequent hydrogen shift to the alkyl lactate [81].

Holm et al. were the first to report the one-step conversion of sugars into alkyl lactates [82]. After contacting sucrose with a Sn-beta zeolite in methanol 64% methyl lactate was obtained. The catalyst was reusable for at least six consecutive runs without loss of activity. The authors also reported the use of several other sugars, like fructose and glucose but yields never exceeded 45%. They even noticed that by using glycol aldehyde, a C2 sugar, as substrate a 16% yield of methyl lactate could be obtained, pointing to the appearance of an aldol condensation [83]. Nevertheless, the alkyl lactate was nonselectively produced as a by-product, more research is needed to find out the applicability of C2 sugars as a starting point for lactic acid production. By comparing the different mono- and disaccharides used, it is clear that sucrose is by far the best substrate, likely because of its higher hydrolysis resistance due to the unique 1.2-ether bond. Low fructose and glucose working concentrations are important to avoid competitive side reactions such as methylation to stable methyl sugars. The reaction pathway involves isomerization of glucose (if sucrose or glucose is used), retro-aldol, dehydration, and the final hydride shift, which were all catalyzed by one catalyst in the same pot. If one considers the reaction pathway starting from glucose, this catalyst is thus able to mimic a one-pot synthetic glycolysis [84]. de Clippel and Dusselier et al. also tested their bifunctional Sn-based carbon-silica composite materials for the conversion of hexoses [77]. The authors noticed the same trends and reached a 45% methyl lactate yield from sucrose in methanol at 155°C after 20 h.

Conversions of various pentoses in alcoholic media into methyl lactate have been demonstrated for the Sn-beta catalyst. Yields are somewhat lower when compared with reactions with hexoses, but more glycol aldehyde (dimethyl acetal) is formed [83]. Their formation is in agreement with the retro-aldol reaction pathway to form one triose and one glycol aldehyde, followed by the internal Cannizzarro reaction with the triose (Scheme 3). Surprisingly, interesting molecules were formed as by-products such as methyl vinylglycolate and methyl-2-hydroxy-4-methoxybutanoate [85]. Their formation was studied in more detail whereby their synthesis pathways based on retro-Michael reactions could been identified. The vinylglycolate in particular has potential as a building block for functional polyesters [86].

Unfortunately, the large-scale application of Sn-beta zeolite is hampered by a time-consuming (up to 40 days) and environmentally unfriendly (use of HF) synthesis procedure. Alternative Sn-based catalysts or alternative synthesis routes to Sn-beta are required. New synthesis methods have been explored extensively in the last years. Several protocols have been described to synthesize the Sn-containing beta zeolite in shorter times [87–89] or by using post-processing procedures on commercially available beta zeolites [72,90,91].

4.3 Dehydratation/Rehydratation

Dehydration of sugars such as hexoses and pentoses has been studied for a long time. The first study reporting on this topic dates from 1840, where the formation of formic acid and (the later known) levulinic acid from sucrose were described [92]. Later, Düll [93] and Kiermayer [94] published independently of each other the synthesis of 5-hydroxymethyl furfural (5-HMF), the dehydration product of glucose and fructose. In the next century, there was an ever-increasing interest in 5-HMF, which is called "the sleeping giant": it is not commercially available on a large scale because it is less straightforward to produce 5-HMF with high yields; nevertheless 5-HMF is an extremely important platform chemical for the furan chemistry as it is a starting material for the production of, among others, levulinic acid, 2.5-furandicarboxylic acid, and 2.5-furandicarbaldehyde. A summary of more chemicals and

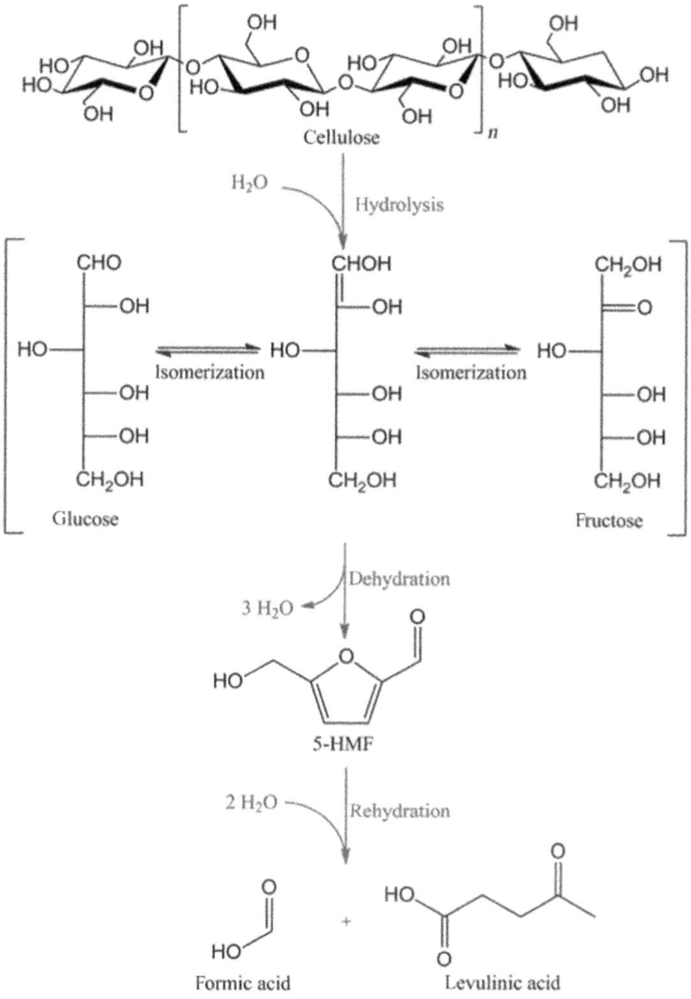

Scheme 4

Mechanism for converting cellulose into 5-HMF and levulinic acid. *Reproduced with permission of the Royal Society of Chemistry (Based on Van de Vyver S, Thomas J, Geboers J, Keyzer S, Smet M, Dehaen W, et al. Catalytic production of levulinic acid from cellulose and other biomass-derived carbohydrates with sulfonated hyperbranched poly(arylene oxindole)s. Energy Environ Sci 2011;4:3601–10).*

their production can be found elsewhere [95]. The formation pathway of 5-HMF and levulinic acid from cellulose, glucose, and fructose is presented in Scheme 4.

Herein, we limit ourselves to the description of the production of 5-HMF from carbohydrates with zeolites or other porous materials. More extensive reviews about the production of 5-HMF can be found in literature [96,97]. Glucose dehydration to 5-HMF has been reported over zeolite Y, but the yields from glucose are low, generally between 10% and 25% [98–100] due to both a low glucose conversion and low selectivities to 5-HMF. The low selectivities are caused by side reactions due to the presence of water. For example, the intermolecular condensation between partially dehydrated intermediates leads to the formation of undesired soluble polymers or insoluble humins [53,101]. Besides, water can hydrolyze 5-HMF into levulinic acid and formic acid [53,101,102]. From this point of view, the use of other solvents like DMSO [101,103] or ethanol [104] was proposed. If ethanol is used, 5-HMF will be etherified into 5-ethoxymethylfurfural (5-EMF) [104], which potentially can be used as a biofuel alternative. When performing the dehydration under biphasic conditions, 5-HMF is protected in the organic part from rehydration. Numerous reports have surfaced in the literature studying the use of biphasic solvent systems, including water-methyl isobutyl ketone (MIBK) [105–107], water-toluene [98], water-THF [108], and water-1-butanol [108].

The general use of fructose or (poly)fructans, like inulin, as feedstock can be a solution to improve the low conversion. In contrast to glucose, which forms very stable ring structures, fructose rings are less stable, making the enolization rate comparatively higher [53]. Because enolization is believed to be a key step in the dehydration of hexoses [102], the dehydration of fructose can be more easily performed than that of glucose [53,102]. Zeolites have been tested for the dehydration of fructose or inulin to 5-HMF [98,101,103,105–107,109,110]. The best yields were again obtained when a biphasic solvent system was applied. Mordenite with a Si/Al ratio of 11 was found to be the best catalyst in water-MIBK at 165°C, yielding 74% HMF, while yields were significantly lower with ZSM-5 and beta [106,107]. Due to the partition coefficient, large quantities of MIBK were required; therefore also other solutions were proposed to facilitate the 5-HMF separation, like the use of a continuous setup to extract 5-HMF [106], water removal by mild evacuation conditions [101], or the use of a new solid-liquid-liquid reactor setup [111]. In DMSO, H-beta produced 97% HMF yield when water is continuously removed during the reaction. Therefore, the zeolite performs as effectively as Amberlyst-15, showing nearly 100% 5-HMF yield when the dehydration was employed in DMSO at 120°C [101,103].

Unfortunately, polyfructans are also more expensive than cellulose and starch. To circumvent this problem, researchers combined the dehydration with the previously discussed isomerization of glucose to fructose with Lewis acid catalysts like Sn-beta. Nikolla et al. converted glucose to 5-HMF with Sn-beta in the presence of hydrochloric acid [108]. After the Sn-beta-catalyzed isomerization of glucose to fructose, the soluble Brønsted acid converts fructose to 5-HMF. Lew et al. applied the same approach but used the solid Brønsted acid catalyst Amberlyst-131 [104].

As presented in Scheme 4, 5-HMF is prone to rehydration and rearrangement to levulinic acid and formic acid. Formic acid may be regarded as a hydrogenation agent or an H_2 equivalent, while levulinic acid is a particularly useful chemical intermediate that can be catalytically converted to a variety of useful chemicals and fuels [112]. Although direct reaction of fructose to levulinic acid has not been very successful with soluble acids and commercial resins in early days [103], the use of zeolite was proposed, as the pores and cages could trap the intermediate 5-HMF [105,106,113], and subsequently converting 5-HMF further to levulinic acid with the aid of the strong acidity. First attempts with HY zeolite showed a levulinic acid yield of 43% from fructose at 140°C in water with glucose and 5-HMF as main by-products [109]. The use of glucose as substrate leads to somewhat lower yields: ±20% and 36% with aluminum-pillared montmorillonite [114] and MFI zeolite [115] as catalysts respectively.

Furfural, the dehydration product of xylose, as first described by Newth in the mid-20th century [116], is currently produced on a large scale (200,000 ton a^{-1}) [117]. Scheme 5 presents the proposed mechanism for furfural formation. Furfural can be converted, for

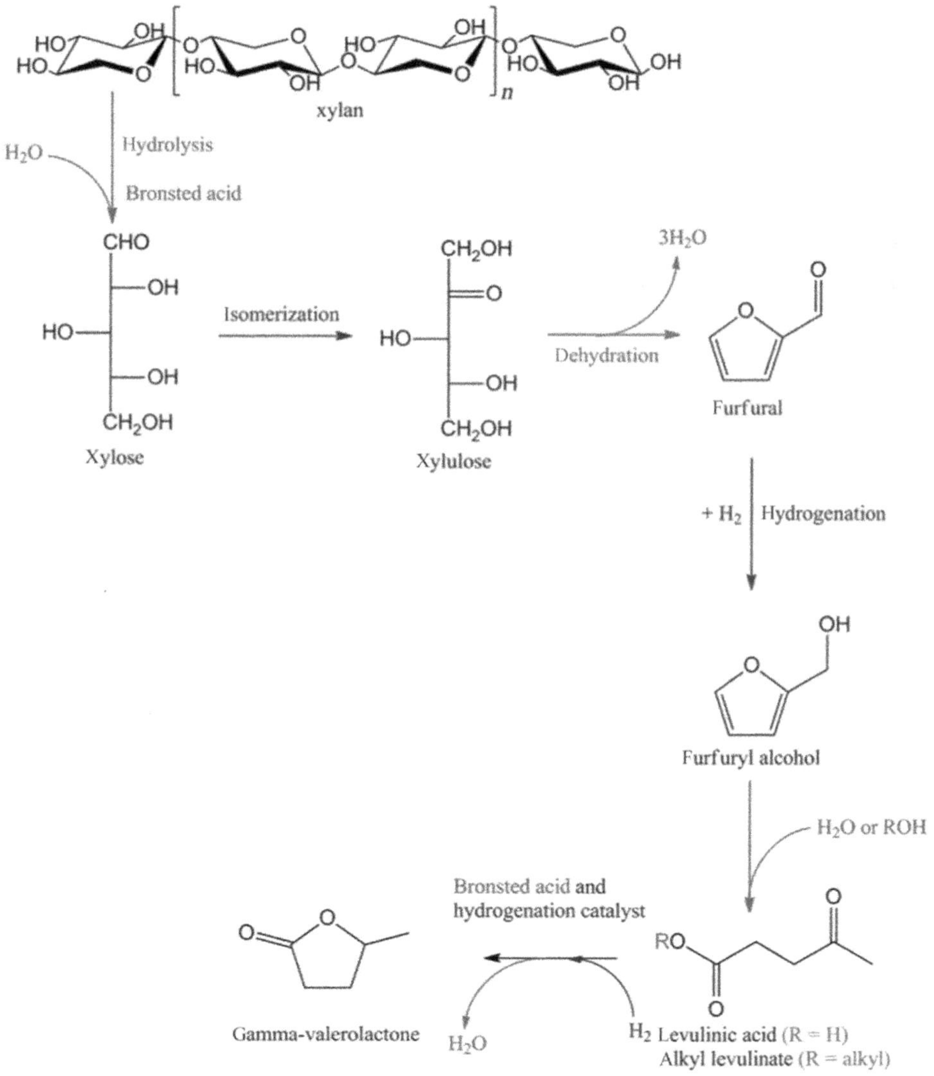

Scheme 5

Reaction scheme for converting xylan into furfural, furfuryl alcohol, levulinic acid, and gamma-valerolactone.

example, into furfuryl alcohol and 2-methylfuran [118]. The latter compound has recently been suggested to produce a high-quality diesel according to the so-called Sylvan process, whereby two molecules of 2-methyl furan are condensed with an aldehyde or ketone to yield C_{12}^{+} oxygenated intermediates, which are finally hydrodeoxygenated [119].

While the traditional dehydration of xylose employs corrosive soluble acids, the use of solid acids, such as zeolites, have been proposed to design a more sustainable process. For example, xylose is dehydrated in water by ZSM-5, yielding 46% furfural at 200°C [120]. Another approach was reported by Choudhary et al., in which they tried to integrate the earlier mentioned Sn-beta catalyst in the xylose dehydration [57]. The Lewis acidic Sn-beta catalyzes the isomerization of xylose to xylulose, while an additional added Brønsted catalyst, like HCl or Amberlyst, provides the dehydration to furfural. However, yields never exceeded 15%. A possible reason for these low yields could be the presence of a Lewis acid. Weingarten et al. showed that Lewis acid sites are detrimental for high furfural selectivity as they catalyze side reactions between xylose and furfural to humins [121]. Dhepe and Sahu reported the one-pot synthesis of furfural (12% yield) starting from hemicellulose with H-USY at 170°C [122]. Higher yields were obtained when water was replaced with aprotic polar solvents. MCM-41 and H-mordenite have been tested for the xylose dehydration in DMSO at 140°C, yielding 75% and 39% respectively [123–125]. Biphasic conditions were evaluated as well in presence of USY [125–128], SAPO-11 [129], beta [125,127,128,130], mordenite [125,126,128,131], MCM-41 [123,124,132], SBA-15 [133], TUD-1 [130], ferrierite [125,128], ZSM-5 [125,128], MCM-22 [134] in various mixtures like water/MIBK [124,126], water/toluene [123–126,129–131,133–142], water/1-butanol [132], water/*p*-xylene [127], water/DMSO/acetone [143] with furfural yields ranging from 20% to 98%. Mordenite and beta were the best zeolites, yielding 80% furfural in water. Green solvents, like gamma-valerolactone, have been used to replace the traditional solvents [144]. The main problem with solid acid catalysts is the stability of the zeolite catalyst in hot liquid water [66,145–148], and coke formation, especially when the sugar concentration is high (>20 wt%). More research about these issues is insurmountable before a sustainable solid acid-based catalyst system can be used on the long term.

Furfural, and so xylose, can be an alternative to glucose as feedstock for the production of levulinic acid (Scheme 5). This route has a higher atom economy (AE) than the route starting from glucose as there is no loss of carbon (76% AE for pentosic levulinic acid vs. 64% for hexosic levulinic acid). Moreover, when xylose is used as feedstock for the production of C5 molecules, glucose can exclusively be used for the production of C6 molecules, which promotes a more optimal use of lignocellulose. This approach was recently explored by Lange et al. The researchers investigated the conversion of furfuryl alcohol (hydrogenated furfural) toward ethyl levulinate with several ion-exchange resins based on sulfonated polystyrene frameworks and compared their activity with H_2SO_4. Amberlyst 46, a resin containing benzene sulfonic acid groups, was able to follow the performance (on H^+ basis) of H_2SO_4. However, all resins show the formation of by-products, which foul and deactivate the catalyst [149]. Unfortunately, these resins were not stable through regeneration of the catalyst by coke burn off. Therefore, a series of acidic zeolites (ZSM-5 with different Si/Al ratios, ZSM-12, ZSM-23, beta, mordenite, USY) were evaluated as well. This exercise pointed out that ZSM-5, with a Si/Al ratio of 30, was by far the most active of all investigated zeolites [149]. Although the higher formation of by-products like diethyl ether, this zeolite nearly approaches the intrinsic activity of the best ion-exchange resins. Levulinic acid can be converted into gamma-valerolacton (Scheme 5), a molecule with applications as solvent, perfume, and as food additive [149]. In 2013, the group of Roman-Leshkov tried to form gamma-valerolacton in one pot starting from furfural. In this reaction, furfural undergoes respectively a transfer hydrogenation and a hydrolytic ring-opening toward alkyl levulinate, followed by a second transfer hydrogenation, and finally a lactonization to form gamma-valerolacton. With the Lewis acid Zr-beta catalyzing the transfer-hydrogenation steps and an aluminosilicate with MFI topology and nanosheet morphology, they obtained up to 78% of gamma-valerolacton under optimized conditions [150].

4.4 Hydrogenation Coupled With Hydrolysis

Hydrogenation of D-glucose yields D-sorbitol, a valuable platform molecule for renewable chemicals and fuels [151]. This reaction is conventionally done by nickel-based catalysts like Raney nickel. Such catalysts suffer from metal leaching and deactivation due to the accumulation of by-products or organic impurities. Since ruthenium does not show leaching, some authors proposed ruthenium catalysts supported on carbon or oxygen-based supports [152–154]. Even though zeolites have proven to be excellent carriers for metal catalysts [155,156], there are only a few reports in the literature where zeolites are used as support for metals catalyzing hydrogenation of sugars. Zhang et al. presented a system with Ru nanoparticles impregnated on mesoporous MCM-41, which obtained D-sorbitol with 83% yield at nearly complete conversion. This catalyst outperforms other glucose hydrogenation systems such as Pd/C, Ni, Al-Ni, and Ru/C. Unfortunately, recycling tests demonstrate a decrease in catalyst performance, from 83% to 68% yield [157]. Recently Mishra et al. published an HY-zeolite-based catalyst (Ru/HY) with a very high selectivity (up to 99%) that remains active for at least four runs [158]. These authors were also able to hydrogenate C5 sugars (ie, xylose) with the same catalytic system. However, the very high performance during the first run (98% selectivity at complete conversion) decreases slightly after regeneration (92% selectivity at 92% conversion) [159].

When cellulose is used as feedstock, the main reaction to consider is cellulose hydrolysis to liberate glucose monomers. However, due to its structure, cellulose is water insoluble and extremely resistant to chemical and enzymatic attack, making hydrolysis very slow, all the more since the low thermostability of glucose limits reaction temperature. Hydrolysis results of mechanically treated cellulose with zeolites were found slow and unselective when compared with other solid acid catalysts such as sulfonated silica/carbon-like catalysts. Typically, a few percentages of glucose yield were observed with zeolites, while up to 60% of glucose yield was achieved with the composite catalyst in water at 423 K [160]. Assistance of ionic liquids (as solvent) and microwave heating improved the glucose hydrolysis yield with zeolites to 37% [161].

Hydrolysis of cellulose in water thus requires higher temperatures to break down the crystalline nature of the cellulose, but glucose is not stable in high concentration at such elevated temperatures. A promising strategy to avoid this problem, known as hydrolytic hydrogenation, is the combination of hydrolysis with a simultaneous hydrogenation that converts the formed sugars to more stable sugar alcohols, enabling higher process temperatures [162–167]. In many cases related compounds such as sorbitans and isosorbide can also be formed through acid dehydration of sugar alcohols [168].

This bifunctional pathway was pioneered halfway through the 20th century by Russian researchers who used combinations of mineral acids and mostly Ru-based hydrogenation catalysts [169,170]. Though mineral acids proved to be active and selective hydrolysis catalysts, reactor corrosion and the need for extensive neutralization proved to be prohibitive for large-scale applications. Therefore, research attention shifted to the use of solid acids. The first successes were reported for conversion of oligosaccharides over Ru catalysts supported on montmorillonite and active carbon [171]. However, heavy Ru sintering considerably shortened catalyst lifetime. This problem was solved by supporting Ru catalysts on H-USY zeolites. With this system, a stable performance for at least 10 consecutive runs was reported, after which the catalyst could be regenerated to its initial activity by simple re-reduction of the Ru [172]. Furthermore, the use of a highly acidic support allowed the conversion of water-soluble polysaccharides such as starch. Full conversion of 50 wt% starch feeds into >99% sugar alcohols in 1 h at 453 K were reported. Some years later, other researchers explored this system for the hydrolytic hydrogenation of sucrose [173].

The water-insoluble nature of cellulose impedes effective hydrolysis by intrazeolitic acid sites, as was shown by Fukuoka, who reported glucose yields under 4% from cellulose after 24 h at 463 K with a range of common zeolites [162]. Later, a solution to this accessibility problem was reported: by adding trace amounts of mineral acids (typically about 35 ppm) to a Ru/H-USY zeolite (Si/Al = 30), full cellulose conversion to 86% hexitols was achieved at 463 K. The minute concentration of mineral acid assists hydrolysis of cellulose to short-soluble oligomers, which in turn are easily adsorbed by the zeolite and hydrolyzed into glucose. Glucose is rapidly hydrogenated into sugar alcohols over the intraporous Ru particles. The low concentration of mineral acid should limit reaction corrosion and neutralization waste. Recently, Ennaert et al. optimized this system using a H-USY zeolite with a much lower Si/Al ratio (2.6) loaded with Ru resulting in unseen nearly quantitative hexitol yields (up to 95%) [66]. Similar reductive splitting of cellulose in hot water under

H_2 pressure has been applied successfully with Ni supported on ZSM-5, Pt promoted Ni-beta, and Ru or Ir on beta [174–177], but the total product yield did not surpass Ru/H-USY. Obtaining high yields with such bifunctional catalytic systems require an optimal balance between the two catalytic functions (ie, the acid sites and the redox sites). When the system exhibits too much redox activity, typical Ru catalyzed hydrogenolysis products will be formed (like ethylene glycol, propylene glycol, 1.2-propanediol, glycerol, C4 and C5 polyols with varying degrees of hydroxylation). On the other side, when the amount of active Ru sites is too low, the acid function becomes dominant over the metal function resulting in loss of glucose through thermal and acid degradation. Fig. 7 nicely illustrates this optimal balance for the Ru/H-USY system of ref [178].

Liu et al. used the same pathway to produce hexane from cellulose. With an Ir-ReO_x/SiO_2 catalyst combined with a H-ZSM-5 zeolite in a biphasic system (H_2O—dodecane), they were able to convert ball-milled cellulose and microcrystalline cellulose to hexane in one pot with yields up to 83% and 78%, respectively [179]. The catalytic reaction proceeds via the formation of sorbitol from cellulose; afterward it undergoes hydrogenolysis into hexane. Independently

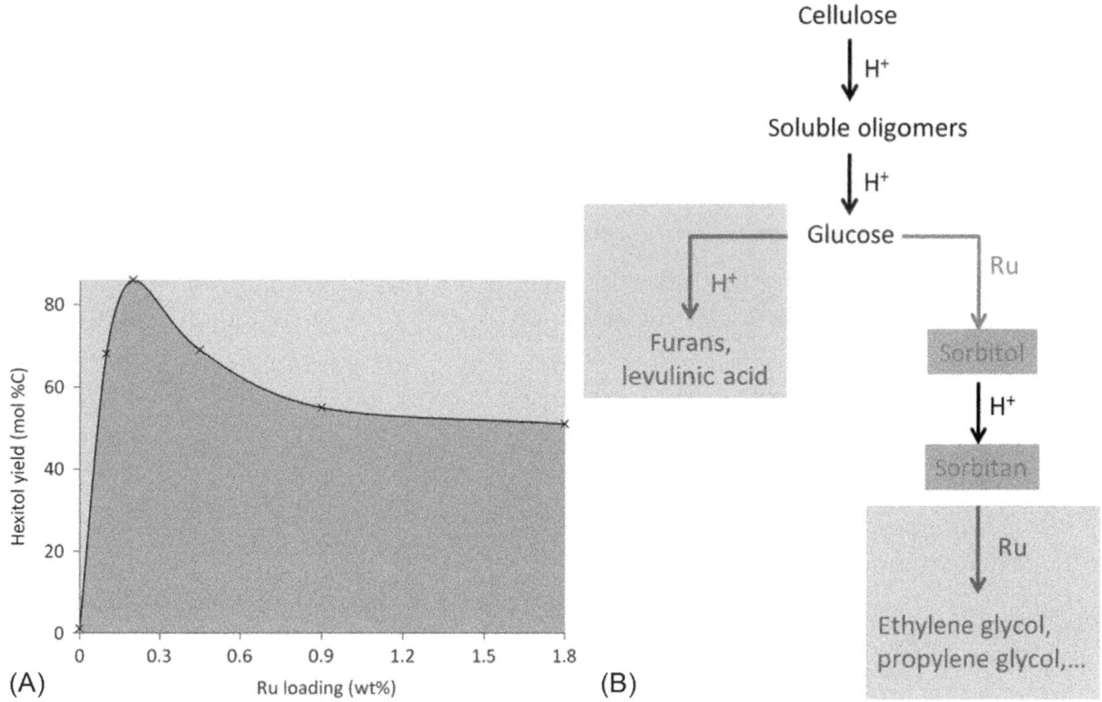

Fig. 7

(A) Optimal Ru-loading balance for the hydrolytic hydrogenation of cellulose, data used from Ref. [178] and (B) reaction scheme for cellulose conversion into hexitols.

from this group, the group of Sels also developed a biphasic (H_2O—decane) strategy for the direct production of hexane from cellulose with tungstosilicic acid and a modified Ru/C as catalysts [180]. Different from the approach of Liu et al., this catalytic system, due to the modification of Ru/C, promotes the formation of 5-HMF instead of sorbitol as an intermediate. It is essential in this system that the acidity (tungstosilicic acid) is only present in the aqueous phase and the redox activity (modified Ru/C) in the organic phase [180].

The hydrolytic hydrogenation of cellulose suffers from the same main drawback as most of the sugar reactions in aqueous phase, namely the stability of the zeolite catalyst in hot liquid water. Preliminary studies showed that the stability of zeolites in hot liquid water depends on the topology. Dense topologies like mordenite and MFI are relative stable in condensed water up to at least 250°C, in contrast, the more open faujasite and beta structures undergo extensive transformations [145,146]. Within one topology, the nature of the charge compensating cation has also a great influence on the stability. Lutz et al. showed that NaY stays nearly unchanged up to 250°C while the decomposing of HY already starts at 130°C [145]. Buhl et al. found that the stability of a series of metal-exchanged Linde Type A (LTA) zeolites followed the basicity of the metal hydroxides [181]. However, when zeolites are used in Brønsted-catalyzed reactions, one is limited to the protonated form of the corresponding zeolite. Ravenelle et al. have investigated the stability of some H-USY zeolites in liquid water between 150 and 200°C. The stability of these zeolites, in contrast to their steam stability, increased with decreasing Si/Al ratio [146]. Recently, Ennaert et al. found that the stability of H-USY zeolites depends on two main parameters: the framework Si/Al ratio and the extra-framework Al content [66]. Al in the framework protects the zeolite lattice by counteracting the hydrolysis of the framework. Besides, extra-framework Al at the outer surface of the zeolite prevents the solubilization of the framework due to its low water solubility. They concluded that mildly steamed zeolites (with a high framework and extra-framework Al content) stabilize over time after undergoing an initial period of transformation, whereby additional extra-framework Al is formed at the outer surface creating a protective layer around the zeolite crystals [66].

Zeolite-based catalysts are only moderately active in the hydrogenation of C5 sugars, with typical yields up to 30–40% [182,183]. Compared to other solid catalysts (sulfonated resins or silica), zeolites are initially less active, but they are more stable resulting in better performances in consecutives runs [182]. The performance of zeolites can be increased by undergoing posttreatments. Zhou et al. published a sugar yield of 55% after treating a H-USY zeolite with oxalic acid. This treatment induces macro/mesopores, which enhance the accessibility of the acidic sites and facilitate the formed sugars to diffuse out of the catalyst [184]. When the authors sulfonate these hierarchical zeolites, a remarkable sugar yield of 78% was obtained [185]. Despite the amorphous nature of hemicellulose, the hydrolytic hydrogenation of hemicellulose structures has been found to be less successful, only a few to a maximum of 20% sugar alcohols has been obtained [186–188]. Yet, biorefineries will only be competitive with petrorefineries when the entire feedstock can be selectively used. Therefore, it is necessary to find solutions for the, at the moment, low applications of hemicellulose for the production of chemicals. Due to the less recalcitrant structure of hemicellulose in comparison with cellulose, it is our strong believe that it should be possible to obtain higher yields of sugar alcohols starting from hemicellulose. However, this will require further investigation.

4.5 Catalytic Conversion of Glycerol

The ready availability of glycerol as a by-product of the soap and biodiesel industry recently encouraged researchers to study the use of glycerol as a chemical commodity. Several review articles have already been written on the subject [189–191], with particular focus on glycerol oxidation, dehydration, hydrogenation, selective esterification, chlorination, and polyol formation. By subjecting glycerol to these catalytic transformations, products ranging from acrolein and acetol to lactic acid, as well as various hydrogenolysis products like propanediols can be obtained.

Acrolein has a large number of uses, for example, as an intermediate for the production of acrylic acid and DL-methionine, as well as for the synthesis of various fine chemicals. Whereas the traditional industrial route involves oxidation of propene over complex multicomponent $BiMoO_x$-based catalysts, glycerol of renewable origin seems to be an attractive alternative for future acrolein production. The dehydration of glycerol can occur both in the gas and liquid phase. Although the highest yields in liquid phase dehydration are obtained with soluble acids or salts [192–197], some promising results were already achieved with zeolites. de Oliveira et al. reported 89% conversion with 99.5% selectivity after 10 h with large pore molecular sieves possessing mild or strong acidic sites [198]. In gas phase dehydration at temperatures of 330–360°C, it was shown that zeolites H-ZSM-11, H-ZSM-5, H-ZSM-22, and H-beta convert glycerol almost quantitatively into acrolein with yields up to 83% [199–209]. A major problem is the formation of cokes. Dalla Costa et al. reported a decreasing selectivity of acrolein with increasing time on stream due to an obstruction of the zeolite pores and deactivation of the catalytic sites. After a short time, only the catalytic sites on the external surface are available for the dehydration [199]. Alternatively to the commonly used fixed-bed reactors, Corma et al. also demonstrated the opportunity of reacting gas phase glycerol/water mixtures with zeolites in a continuous fluidized-bed reactor [210]. At 623 K and contact times of 0.5–2 s, they were able to obtain 61% acrolein yield with a ZSM-5 based catalyst. The current acrolein selectivity for zeolite-catalyzed dehydrations are slightly lower than those obtained when using less acidic metal oxides.

Another example of glycerol valorization with zeolites relies on the hydrogenolysis of glycerol to 1.2-propanediol via bifunctional heterogeneous catalysis [211]. The mechanism involves zeolite-catalyzed dehydration, followed by metal-mediated hydrogenation.

A sustainable way to circumvent the need of externally added hydrogen involves the integration of a glycerol-based hydrogen production via catalytic aqueous-phase reforming. D'Hondt et al. presented a chemical route for the hydrogenolysis of glycerol in the presence of Pt-impregnated NaY zeolites in the absence of external hydrogen supply [212]. Under nonoptimized process conditions, preliminary results with glycerol under inert atmosphere are reported to yield up to 55% 1.2-propanediol at 85% conversion. A bimodal size distribution of the Pt particles is crucial to balance aqueous reforming and hydrogenation toward maximum 1.2-propanediol yields and high H_2 efficiency [213].

Purushothaman et al. suggested an alternative route to produce alkyl lactates from biomass [214]. Instead of using cellulose, they proposed a one-pot transformation of glycerol to methyl lactate. Gold nanoparticles supported on a commercially available USY zeolite, with both Lewis and Brønsted acid sites, was active in this chemocatalytic conversion. The authors reported yields up to 75% methyl lactate at nearly quantitative conversion [214].

By a selective oxidation of glycerol, valuable products such as glyceric acid, tartronic acid, or ketomalonic acid can be formed. This reaction is commonly done by using Pd or Pt particles on a carbon support in the presence of dioxygen under basic conditions [215–220]. It was recently shown that gold particles are highly active and selective for this reaction [215–217,219–225]. Moreover, Au shows an improved resistance against overoxidation in the presence of O_2 as oxidant under liquid-phase oxidation conditions [216,218,220,226,227]. The disadvantage of using Au particles is the need for a base to improve the kinetics and reduce catalyst deactivation, which results in the formation of salts instead of free carboxylic acids. Working under acidic conditions, and the resulting direct formation of acids, is desirable from an industrial point of view. This problem can be conquered by the use of zeolites. Villa et al. reported the quantitative conversion of glycerol with a selectivity toward glyceric acid of 81% without the use of basic conditions. They used a bimetallic AuPt catalyst supported on an acidic H-mordenite zeolite. Due to the use of bimetallic AuPt nanoparticles, leaching of the metals is avoided and the catalyst lifetime is improved [228].

5 Fast Pyrolysis and Catalytic Upgrading

5.1 Introduction

As mentioned in part 2, fast pyrolysis is a very effective method to produce a liquid fuel from biomass in high yields (up to 70–80%) [32,31,229]. However, the produced bio-oil has some major drawbacks like the high oxygen content, corrosiveness due to the high acidity and the presence of phenols and the relative poor product stability. The highly oxygenated nature of the bio-oils results in a low calorific value of the oil and in a low miscibility with hydrocarbons, which hampers their integration in conventional refinery units. The poor stability is due to the presence of reactive oxygenates, like aldehydes, ketones, and phenolics, which are very prone to condensation and oligomerization. A consequence of the poor oil stability is the impossibility to separate by distillation [32,31,229]. For further application of the fast pyrolysis oils, the quality should be improved, that is, reducing the undesirable properties. This can be done through physical (filtration, solvent addition, emulsification, etc.) or chemical methods [31]. In this part, we will limit the discussion to the chemocatalytic methods which use zeolites or other porous materials.

The goal of catalytic upgrading is rejecting oxygen (deoxygenation) as a combination of CO, CO_2, and H_2O to produce a hydrocarbon product with an increased hydrogen and decreased oxygen content. It involves processes such as hydrotreatment/hydrodeoxygenation, catalytic cracking, gasification to syngas, and esterification, which are executed downstream from the fast pyrolysis reactor. Besides, it also includes catalytic pyrolysis, a process in which the biomass and the catalyst are physically mixed, so the catalytic upgrading takes place during pyrolysis [32,31,230]. Downstream upgrading processes such as catalytic cracking and hydrotreatment/hydrodeoxygenation (upgrading under hydrogen atmosphere) can be used for the upgrading of the resulting pyrolysis bio-oil, but can also be integrated in the pyrolysis process. During the latter way, the pyrolysis vapors are fed to the catalytic bed before they are condensed into an oil product [32,31,230]. Zeolites play a central role in catalytic pyrolysis and upgrading of pyrolysis bio-oils by removing part of the oxygen content, and increasing the amount of aromatic and olefinic compounds [24,25,32,31,230–232].

To begin, we discuss the upgrading of bio-oils by means of cracking and hydrodeoxygenation. This will be followed by upgrading pyrolysis vapors and catalytic pyrolysis. Finally, a short overview of catalytic fast pyrolysis of lignin will close this part of the chapter.

5.2 Zeolite Cracking

5.2.1 The effective hydrogen-to-carbon ratio of the feed

Chen et al. found that upgrading hydrogen-deficient compounds, such as acetic acid and methyl acetate, was difficult due to rapid catalyst aging [233]. They introduced an indicator for the degree of hydrogen deficiency of biomass-derived compounds, namely the "effective hydrogen-to-carbon ratio," which is defined as $\frac{H}{C_{eff}} = \frac{H-2O}{C}$ with H, O, and C the moles of hydrogen, oxygen, and carbon. This ratio is 0, 0.33, and 0.67 for glucose, sorbitol, and glycerol respectively, while the ratio for petroleum-derived feeds ranges from slightly over 2 (alkanes) to 1 (aromatics) [234]. Huber et al. performed the catalytic conversion of 10 biomass-derived feeds, that is, glucose, sorbitol, glycerol, tetrahydrofuran, methanol, and various hydrogenated bio-oil fractions, with different H/C_{eff} ratios over a ZSM-5 catalyst at 600°C and found the combined aromatic and olefin yield to increase and the coke yield to decrease with increasing H/C_{eff} ratio of the feed [235]. A feed with an H/C_{eff} ratio of at least 1.2 is recommended for aromatics and olefins production. According to Dao et al., the addition of methanol significantly increases the hydrocarbon yield from furfural during catalytic cracking with H-ZSM-5 at 400°C, but not that of carbohydrates like glycerol, glucose, and fructose [236]. Mentzel and Holm demonstrated by upgrading mixtures of a wide range of oxygenates and methanol with ZSM-5 that the H/C_{eff} ratio is not the only factor defining the upgrading process since structural isomers with identical H/C_{eff} ratios showed significant differences with respect to catalyst lifetime and product selectivity [237]. The functionalities in the oxygenates have a pronounced effect on the reactivity and molecules capable of dissociating CO or CO_2 generally experience higher conversion capacities.

5.2.2 Transformation of actual bio-oils

Direct cracking of bio-oils in a fixed-bed reactor has been studied over various catalysts, viz. H-ZSM-5, H-Y, H-mordenite, silicalite, and silica-alumina in the temperature range of 290–410°C [238–240]. H-ZSM-5 was the best catalyst yielding 28 wt% of hydrocarbons like toluene, xylenes, and trimethylbenzenes [239]. The hydrocarbon yield further decreased in the order H-Y (14 wt%) > silica-alumina (13 wt%) > silicalite (5 wt%) > H-mordenite (4 wt%), with significant differences in product distribution: ZSM-5 mainly produces aromatics, while the less acidic silica-alumina and the non-acidic silicalite mainly form aliphatics. This indicates that a high catalyst acidity is necessary for aromatics production. Apart from reaching the highest hydrocarbon yields, H-ZSM-5 provides the least coke formation [241].

Owing to catalyst deactivation, the content of oxygenates increases with time-on-stream. Vitolo et al. studied the regeneration of ZSM-5 zeolites after upgrading of wood pyrolysis oil, by removing the coke deposits in air at 500°C and found that the effectiveness of the catalysts toward aromatics is reduced by continuous regeneration [242]. The loss of activity was mainly ascribed to the disappearance of acid sites caused by the thermal regeneration cycles, which is suggested to proceed through dehydroxylation of the Bronsted acid sites, with formation of Lewis acid sites instead. These results are in line with observations of Gayubo et al. [243]. The researchers demonstrated an irreversible decrease in acidity due to dealumination in water. Besides, they also observe reversible catalyst deactivation by cokes deposition, both thermal coke and catalytic coke. The latter is formed during the reaction where the oxygenates in the mixture react with each other with synergistic effects. The degradation of pyrolytic lignin and the condensation of lignin-derived phenolics are the main contributors to the formation of thermal coke. The need for a preliminary separation of certain components (aldehydes, oxygenated phenols, furfural, and the respective derivatives) is recommended [244].

As described in the previous section, methanol addition increases the H/C_{eff} ratio, resulting in an increased hydrocarbon yield. Moreover, co-feeding of methanol (around 70 wt%) also decreases the catalytic deactivation. Methanol inhibits condensation reactions between different products and consequently minimizes the pyrolytic lignin deposition within and outside the catalyst particles [245]. Next to methanol addition, water addition is effective for bio-oil stabilization and reducing pyrolytic lignin deposition on the catalyst; however, in contrast to the former method, it rejects a part of the bio-oil that is not water soluble (most of the lignin-derived components) and only the aqueous phase is processed. So the disadvantages of water treatment are a lower total hydrocarbon yield due to a smaller converted fraction and a reduced catalyst activity/stability due to the high water content. In an effort to further enhance the catalytic transformation of a bio-oil/methanol (50:50 wt%) feed, a thermal step at 400°C was introduced prior to catalytic conversion with the objective to remove the pyrolytic lignin [246]. In this way, only the vapors from the thermal step are subjected to catalytic transformation, which is thereby similar to the upgrading of pyrolysis vapors (*vide infra*). Through this way, a lignin-like side-product is obtained that can be used for further valorization.

An alternative upgrading strategy was proposed by Huber et al., which comprises two hydroprocessing steps and one zeolite cracking step [247]. The hydroprocessing steps, viz. hydrogenation with Ru/C at 125°C followed by hydrogenation with Pt/C at 250°C, increase the H/C_{eff} ratio of the pyrolysis oil, producing polyols and alcohols. In the zeolite conversion step, which uses ZSM-5 at 600°C, these hydrogenated products are transformed to light olefins and aromatic hydrocarbons in a yield as much as three times higher than that produced with the pure pyrolysis oil. The yield of aromatics and olefins is proportional to the amount of hydrogen added to the pyrolysis oil in the hydroprocessing steps and can thereby be adjusted depending on the market values of the substrates and products. The cokes production during the zeolite upgrading step is significantly decreased due to the hydroprocessing steps. Moreover, since a fluidized-bed reactor is used for the zeolite cracking step, the coked catalyst can be regenerated by burning off the coke and recycling it back to the reactor. Vasalos et al. demonstrated the potential of a FCC type of riser technology for the conversion of a feed of the heavy fraction of a hydrotreated bio-oil mixed with vacuum gas oil over an USY zeolite, where the catalyst was regenerated with heat recuperation for the endothermic cracking process [248].

5.3 Zeolite-Assisted Hydrodeoxygenation of Bio-Oils

Next to zeolite cracking, hydrodeoxygenation has been suggested as an effective method for bio-oil upgrading [31,249–252]. In hydrodeoxygenation or hydrotreating, oxygen is removed by hydrogen addition at high hydrogen pressures (up to 200 bar), but mostly at lower temperatures than zeolite cracking (200–400°C). It involves a range of reactions like decarbonylation, decarboxylation, dehydration, cracking, alkylation and dealkylation, isomerization, hydrogenolysis, and hydrogenation whereby alkanes, aromatics, oxygen-containing organic compounds, water, and gases (CH_4, CO, and CO_2) are produced. A wide range of catalysts have been studied for hydrodeoxygenation, including conventional CoMo- and NiMo-based sulfide catalysts, noble metal (Pt, Pd, Ru, Rh, etc.) catalysts, base metal (Ni, Fe, Cu, etc.) catalysts, and metal nitride, carbide, and phosphide catalysts [249–252].

In comparison to zeolite upgrading, hydrodeoxygenation generally results in a lower coke and higher bio-oil yield. The bio-oil is also more deoxygenated and saturated, and so has a higher heating value [249]. However, the use of hydrogen increases the cost of hydrodeoxygenation. Limiting the hydrogen consumption should therefore be a primary goal.

As we have mentioned, pyrolysis bio-oils can contain large amounts of lignin-derived phenolic compounds. These phenols have by themselves many interesting applications (resins, adhesives, plasticizers), but the separation of phenols from bio-oils is very difficult [31,249]. An alternative is converting them into cycloalkanes and large cyclic hydrocarbons to produce molecules of appropriate mass for use as gasoline or diesel fractions. A large number of papers and reviews have focused on hydrodeoxygenation of lignin-derived phenolic compounds, where the most studies are performed with model compounds like phenol, guaiacol, cresols, and anisole [249–252]. A common strategy for hydrodeoxygenation is to use bifunctional catalysis, that is, combining an acid, in solid or aqueous form, with a metal catalyst. Due to their acidic nature and the possibility to perform bifunctional catalysis, zeolites could play a prominent role in these reactions.

Jones et al. studied the conversion of phenol over Pt/Y zeolite in a fixed-bed reactor at 200°C under 40 bar H_2 pressure [253]. They observed high yields of cyclohexane with the zeolite-based catalyst, while silica and alumina support materials formed particularly cyclohexanol. A bifunctional reaction mechanism was proposed in which the acid sites would be responsible for oxygen removal through dehydration, and the metal functions would perform the hydrogenation reactions (see Scheme 7). Next to cyclohexane, a range of bicyclic compounds were also observed as products, which were thought to originate from the alkylation of phenol with cyclohexanol or from aldol condensation of cyclohexanone (see Scheme 6). The zeolite morphology influenced the product selectivity: bicylohexyl, for example, was analyzed up to

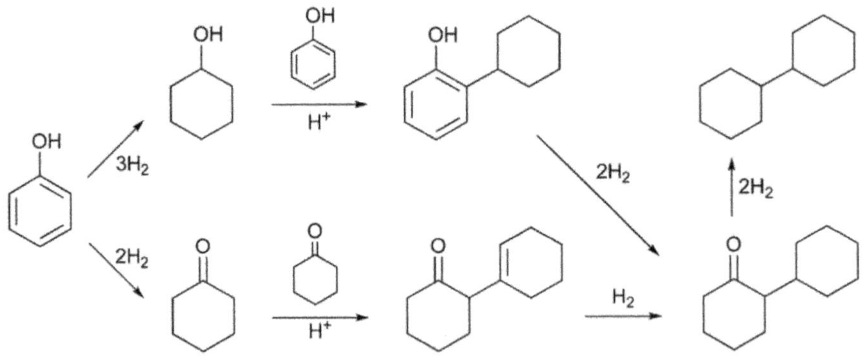

Scheme 6

Reaction pathway for the transformation of phenol into bicyclics. *Reprinted with permission from the Royal Society of Chemistry (From Hong D-Y, Miller SJ, Agrawal PK, Jones CW. Hydrodeoxygenation and coupling of aqueous phenolics over bifunctional zeolite-supported metal catalysts. Chem Commun 2010;46:1038–40).*

4% in reactions with Pt/beta and Pt/Y, while almost none was observed in reactions with Pt/ZSM-5 in accordance with the shape-selective effect of the pore sizes.

Resasco et al. explored the hydrodeoxygenation of anisol over 1 wt% Pt/Beta in a fixed-bed reactor at 400°C and 1 bar H_2, which yielded benzene, toluene, and xylenes [254]. The Brønsted acidity in the zeolite is responsible for the transalkylation of the methyl to the phenolic ring, giving phenol, cresols, and xylenols as products and so reduces the loss of carbon as methane. The metal function catalyzes demethylation, hydrodeoxygenation, and hydrogenation. Pt/SiO_2, a catalyst with only a metal function, produced only traces of methylated products, with 70% benzene, 15% methane, and 13% cyclohexane at complete anisol conversion, while the bifunctional Pt/beta yielded 50% benzene, 25% toluene, 10% xylene, 5% methane, and almost no saturated products. Use of a bifunctional catalyst is thereby very beneficial. Moreover, the presence of the metal also improves the catalyst stability with a moderate reduction in coke formation.

Lercher et al. have thoroughly studied the aqueous phase catalytic hydrodeoxygenation and hydrodealkylation of lignin-derived phenolics and dimers with a range of bifunctional catalyst systems consisting of metal and acid functions, like Pd, Pt, Rh, or Ru on carbon and with H_3PO_4 [255,256], Raney Ni and Nafion/SiO_2 [257], Pd/C combined with zeolites like ZSM-5 [258] and beta [259,260], and Ni supported on ZSM-5 [261–263], resulting in the selective production of cycloalkanes. They indicated that hydrodeoxygenation of phenolics proceeds through multiple steps consisting of hydrogenation, hydrolysis, and dehydration, as shown for phenol hydrodeoxygenation in Scheme 7 [255,256]. As bio-oils contain a lot of water, it was suggested that the conversion of the phenols might be preferably performed in aqueous conditions, which implies an increasing importance of the catalyst stability in aqueous conditions. A catalyst system comprised of Pd/C and ZSM-5 was able to convert 4-*n*-propylphenol to cycloalkanes at a lower temperature, viz. 150°C instead of 200°C, than Pd/C in combination with other solid Brønsted acid catalysts such as zirconia or Nafion/SiO_2 [258].

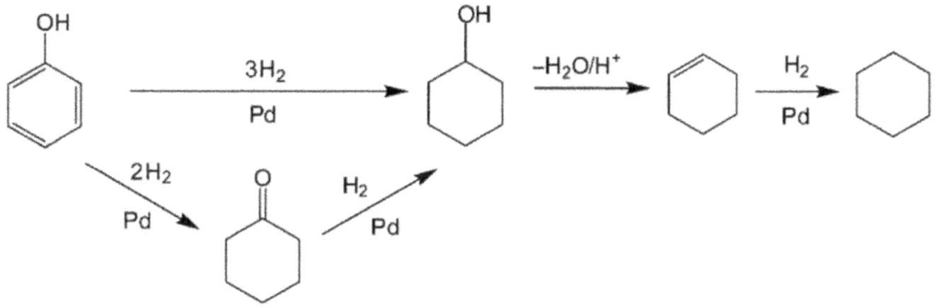

Scheme 7

Reaction pathway for the hydrodeoxygenation of phenol over Pd/C in the presence of an acid. *Reprinted with permission from Wiley-VCH Verlag (From Zhao C, Kou Y, Lemonidou AA, Li X, Lercher JA. Highly selective catalytic conversion of phenolic bio-oil to alkanes. Angew Chem Int Ed 2009;48:3987–90).*

This indicates a lower apparent activation energy and a higher reaction rate for cyclohexanol dehydration for Pd/C combined with ZSM-5. This catalytic system was also shown to be effective for the hydrodeoxygenation of more substituted methoxy-phenolics, with near-quantitative conversions to cycloalkanes and methanol at 200°C. By depositing 20 wt% Ni on ZSM-5, a catalyst containing both metal and acid sites was obtained, which exhibited a high activity and selectivity in the hydrodeoxygenation of a range of substituted phenolics toward cycloalkanes and aromatics at 250°C, without coke formation [261]. For all phenolic compounds, the combined selectivity for cycloalkanes, aromatics, and methanol was 100%. The catalyst was also effective for the conversion of phenolic dimers, furan-like compounds, and even a *n*-hexane-extracted crude bio-oil obtained from pyrolysis of pine trees. The bio-oil extract contained mainly phenols, furans, ketones, and aldehydes. The catalyst was able to cleave both aryl-ether and aryl-alkyl ether linkages (C–O–C), while the C–C bonds in dimeric compounds were mainly preserved. No notable change in activity was observed in subsequent runs with intermediate calcination and reduction of the catalyst. Ni leaching was very low (5 ppm), pointing to the high stability of the Ni particles on Ni/H-ZSM-5 in hot water [261].

In phenol hydrodeoxygenation with Ni/H-ZSM-5, the rate of the different steps increased in the following order: phenol hydrogenation $<$ cyclohexanone hydrogenation $<$ cyclohexanol dehydration $\ll$ cyclohexene hydrogenation [263]. The rate-determining step, which is the phenol hydrogenation, was accelerated up to five times by using an Ni-supported ZSM-5 catalyst with 19.3 wt% γ-Al_2O_3 binder. Besides, this catalyst also showed a higher overall phenol hydrodeoxygenation activity. The Ni nanoparticles interact strongly with alumina through NiO. The average nickel particle size was considerably smaller for the catalyst containing γ-Al_2O_3 (8.8 vs. 35 nm), resulting in a much higher concentration of accessible Ni atoms [262]. However, the γ-Al_2O_3 binder increased the amount of Lewis acid sites on the catalysts, which stabilizes cyclohexanone and inhibits its hydrogenation [263]. The catalyst deactivated by the transformation of Brønsted acid sites to Lewis acid sites and by Ni sintering and leaching during reaction, although the leaching was negligible for the catalysts with γ-Al_2O_3 binder. Furthermore, in situ IR analysis of adsorbed species indicated that the catalyst containing γ-Al_2O_3 exhibits a higher adsorption capacity for phenol, cyclohexanone, and cyclohexanol due to strong adsorption on the γ-Al_2O_3 [262].

The hydroalkylation of phenol for bicycloalkanes production, which was already observed by Jones et al. [253], was more thoroughly explored by Lercher et al. using a combination of Pd/C and beta in water at 200°C [259]. During reaction, phenol reacts selectively with the in situ generated cyclohexanol or cyclohexene on Brønsted acid sites (see Scheme 8). Other acid catalysts such as amorphous silica-alumina, sulfated zirconia, Amberlyst 15, ZSM-5, Y, and so on were used in the reaction, but the results were similar to those of Jones et al. [253], bicycloalkanes were only produced in the presence of zeolite beta and Y, with zeolite beta giving the highest yield. This indicates the necessity for a zeolite porous network with

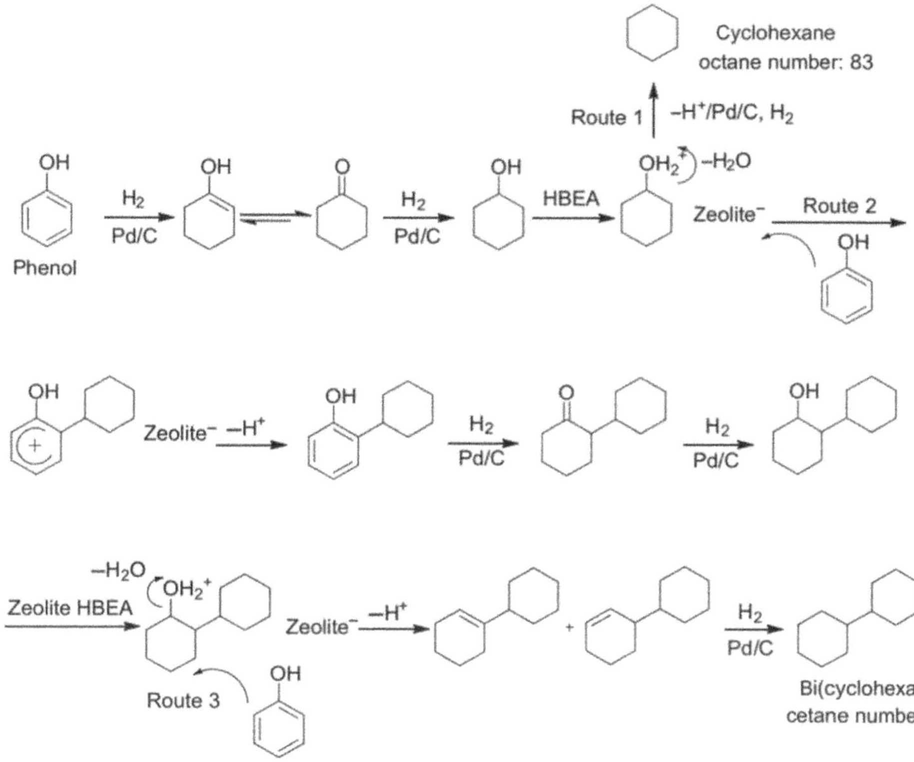

Scheme 8

Reaction pathway from phenol toward bicyclohexane [259]. *Reprinted with permission from Elsevier (From Zhao C, Camaioni DM, Lercher JA. Selective catalytic hydroalkylation and deoxygenation of substituted phenols to bicycloalkanes. J Catal 2012;288:92–103).*

the sufficient pore size to attain high activity and hydroalkylation selectivity, especially when reactions are performed with bulkier substrates like 4-*n*-propylphenol or 4-*n*-propylguaiacol. In comparison with other hydrogenation catalysts such as Pt/C, Rh/C, and Ru/C, Pd/C yielded the most bicycloalkanes, which is explained by its lower hydrogenation activity. Selective hydroalkylation, which involves hydrogenation, dehydration, and alkylation as competing reactions, requires the phenol hydrogenation and the cyclohexanol dehydration rates to be relatively slow compared to phenol alkylation. The optimal acid-to-metal site ratio for hydroalkylation was found to be around 21–22 for the combination of Pd/C and beta. La-incorporation in zeolite beta resulted in a more selective hydroalkylation, probably due to the higher rate of alkylation compared to dehydration [260]. It is known that dehydration is solely catalyzed by Brønsted acid sites, whereas alkylation can be achieved in the presence of La^{3+}. By deposition of Pd on zeolite beta, higher bicycloalkane yields could be obtained. For example, a 0.1 wt% Pd/beta catalyst achieved a bicycloalkane yield of 76 C% starting from 4-*n*-propylphenol [259].

In a follow-up study, the zeolite-catalyzed conversion of cyclohexanol in water was investigated with in situ magic-angle-spinning NMR spectroscopy [264]. Using 1-^{13}C-hexanol as a substrate, it was shown that significant migration of the hydroxyl group of cyclohexanol and the double bond in cyclohexene occurs during dehydration. It is suggested that the cyclohexyl cation undergoes a 1.2-hydride shift competitive with rehydration and deprotonation. Also, trace amounts of dicyclohexyl ether and the secondary product cyclohexyl-1-cyclohexene were detected during reaction. Further, beta is shown to be a much more active catalyst than phosphoric acid for the dehydration of cyclohexanol, with a dehydration rate 100 times higher per acid site.

5.4 Upgrading of Pyrolysis Vapors

During pyrolysis of biomass, often in a fluidized-bed reactor, the substrate thermally decomposes into vapors that can subsequently be driven over a zeolite catalyst bed, where the vapors are converted to hydrocarbons like aromatics, olefins, and a variety of partially deoxygenated compounds, next to CO, CO_2, water, and cokes. Afterward, the aromatics, olefins, deoxygenated products, and water are condensed into an upgraded bio-oil. Examination of the catalytic properties of catalysts in the upgrading of pyrolysis vapors is regularly performed in a fixed-bed configuration, in which the biomass substrate and catalyst bed are separated and the pyrolysis vapors are driven over the catalyst bed by an inert gas flow.

Use of zeolites like ZSM-5 for the upgrading of pyrolysis vapors was pioneered already in the 1980s to 1990s by Diebold and Scahill [265] and Horne and Williams [266–269]. Diebold and Scahill found that these vapors are very reactive with ZSM-5, producing methyl benzenes such as toluene and xylenes. Although the catalyst deactivated relatively quickly due to coke's deposition, it could be partly regenerated in hot air [265]. Williams and Horne saw that the ZSM-5 upgrading step resulted in a much lower oil yield compared to the noncatalyzed run, but a very large increase in the formation of aromatics, with mainly monoaromatics and beside also significant amounts of polyaromatics [266–269]. Although deoxygenation of phenolics is very challenging [270,271], aromatics were proposed to originate from both nonphenolic as well as phenolic compounds [266]. As the catalysis run time increased, the formation of monoaromatics was reduced and formation of oxygenated products increased due to catalyst deactivation [267]. Regeneration of the catalyst showed a decreasing effectiveness of the catalyst for aromatics production [266]. In a following study, methanol was co-fed in the upgrading step, as this increased the oxygen removal from the vapors as water and a corresponding lower production of CO and CO_2 [268]. Methanol addition significantly raised the formation of alkylated phenolic and aromatic compounds and reduced the formation of polyaromatic compounds.

In literature, the difference in catalytic behavior of a wide range of zeolites (H-ZSM-5, a partially exchanged Na-ZSM-5, H-Y, H-beta, H-mordenite, H-ferrierite) was extensively studied [269,272,273]. Both ZSM-5 catalysts emerge as the best catalysts with the highest yields of hydrocarbon products and lowest coke formation [269]. The aromatics yield tends to decrease in the order ZSM-5 > H-beta > H-mordenite > H-ferrierite/HY [273]. Stefanidis et al. demonstrated that, in comparison to a range of amorphous catalysts such as alumina, zirconia/titania, and magnesium oxide, ZSM-5 is more suitable for the reduction of undesirable compounds and production of aromatics in the upgrading of pyrolysis vapors from beech wood [274]. The excellent performance of ZSM-5 is attributed to the important role of its medium pore size [269]. Besides, Park et al. pointed out that ZSM-5 is more efficient than Y zeolites due to the proper distribution of strong acid sites [275].

Next to microporous zeolite catalysts, mesoporous materials with a uniform pore size distribution in the range of 2–10 nm were also used, with varying success. For example, pine pyrolysis vapors were upgraded with ZSM-5 and with two mesoporous zeolites, viz. mesoporous MFI zeolites synthesized using an amphiphilic organosilane (meso-MFI) and mesoporous ZSM-5 (MMZ_{ZSM-5}) synthesized via a combination of a top-down and bottom-up approach, in which the first step involved NaOH-treatment and the second step used the cationic surfactant cetyltrimethylammonium as the structure-directing agent [276,277]. The meso-MFI catalyst exhibited the best activity in deoxygenation and aromatization in the upgrading step, with high selectivity for aromatics such as benzene, toluene, and xylenes (BTX). Besides zeolites, also other mesoporous materials were tested. When pyrolysis vapors of a beech wood-derived feedstock and miscanthus were contacted over Al-MCM-41 at 500 °C, reduced bio-oil yields and increased production of water and cokes were found, similar to zeolite upgrading [278]. The catalytic transformation step decreased the undesirable fraction of acids, carbonyls, and heavy compounds, like high molecular weight phenolics, while increasing the amount of phenols and hydrocarbons. However, the amount of unwanted polyaromatic compounds also increased. A low silica-to-alumina ratio of MCM-41, indicating a high surface acid site density, was found to have a positive effect on phenol yield and composition [278]. Iliopoulou et al. tested Al-MCM-41 catalysts for the upgrading of pyrolysis vapors from a beech wood–derived feedstock and stated that the combination of the large surface area, tubular mesopores, and mild acidity of these materials provide the desired environment for controlled conversion of the bulky lignocellulosic molecules [279]. Comparing MCM-41 with ZSM-5 showed higher phenols and hydrocarbons formation for the mesoporous MCM-41 [231]. Triantafyllidis et al. compared two mesoporous aluminosilicate materials assembled from zeolite beta seeds ($MSU-S_{beta}$) with a different mesopore structure with an Al-MCM-41 catalyst in the same reaction and found that the $MSU-S_{beta}$ materials led to a significantly lower organic phase and higher aromatic (both mono- and polyaromatic), coke and char yield than the Al-MCM-41 catalyst, which was proposed to be due to the higher strength of the acid sites of $MSU-S_{beta}$ [280]. Park et al. performed the upgrading of pyrolysis vapors of miscanthus over Al-MCM-41, Al-MCM-48, H-ZSM-5, and a mesoporous MFI (meso-MFI), and found similar results, that is, higher phenolics production for the mesoporous catalysts

(MCM-48, H-ZSM-5, and meso-MFI) than the microporous ZMS-5, but observed a higher aromatics production for ZSM-5 and also meso-MFI, due to their higher acidity [281].

The incorporation of metals in the zeolites was found to affect the obtained product distribution. When gallium is incorporated in ZSM-5, a further increase of the bio-oil yield and aromatics selectivity could be observed [275]. Gallium incorporation was also beneficial for the above-mentioned meso-MFI zeolite, resulting in an increased organic oil yield and resistance to coke formation. The Ga-loading played an important role in the selectivity for aromatics, with an increase in selectivity for small Ga-loadings (1 wt%), but a significant decrease for high loadings (5 wt%) due to loss of acidity [276,277]. Besides gallium incorporation, the effect of incorporating transition metals was also extensively investigated. Aho et al. tested Fe-incorporated beta, Y and ferrierite zeolites in the upgrading of pine wood pyrolysis vapors at 450°C [282]. The incorporation of Fe into zeolites induced a decrease of the amount of methoxy-substituted phenols and an increase of the methyl-substituted phenol concentration. However, similar to other studies, a reduction of the catalyst acidity was observed in regeneration cycles. Iliopoulou et al. studied the influence of Ni- and Co-incorporation in ZSM-5 in the catalytic upgrading of pyrolysis vapors from a beech wood–derived feedstock [283]. The Ni-modified catalyst showed a lower organic liquid yield and a higher gas production compared to Co/ZSM-5. Both metal-modified catalysts enhanced the production of aromatics and phenols. An interesting observation was the in situ reduction of the supported metal oxides during the pyrolysis reaction, which might favor H-transfer reactions for aromatics production. A Co-modified commercial ZSM-5 catalyst was recently tested in a pilot scale circulating fluidized-bed pyrolyzer for the upgrading of pyrolysis vapors [284]. Co-incorporation significantly increased the CO_2/CO ratio in the gas phase, indicating a change in the decarboxylation/decarbonylation mechanism by metal impregnation, probably due to a change in the acidic properties of the catalyst. The enhancement in aromatics and phenols production due to Co-incorporation was even more pronounced in the pilot-scale experiments than in the bench-scale experiments. Stöcker et al. synthesized, in addition to Al-MCM-41 (*vide supra*), three other metal-containing mesoporous samples, viz. Cu-Al-MCM-41, Fe-Al-MCM-41, and Zn-Al-MCM-41, whereby Fe-Al-MCM-41 and Cu-Al-MCM-41 performed the best in terms of phenols production. Metal-incorporation induced increased concentrations of hydrogen in the gas phase, especially for Cu-Al-MCM-41 [278].

5.5 Catalytic Fast Pyrolysis

During the *in-bed* catalytic fast pyrolysis, the pyrolysis is carried out in the direct presence of the zeolite, that is the substrate and the catalyst are in physical contact during pyrolysis. Compared to catalytic upgrading of pyrolysis vapors, the zeolite impact in catalytic fast pyrolysis is much higher due to a better contact of the substrate and intermediates with the zeolite. A disadvantage of catalytic fast pyrolysis is the higher coke deposition on the catalyst, resulting in catalyst deactivation. Catalytic fast pyrolysis is a single-step process, which is preferably carried out in a fluidized-bed reactor similar to a FCC, but alike the upgrading of pyrolysis vapors, fundamental research such as the examination of catalyst properties is commonly performed in fixed-bed setup in which the substrate and catalyst are present in a single bed. During the process, the solid biomass is first thermally converted to volatile organics, gases and solid coke. The organics then enter the zeolite catalyst where they are converted into aromatics, partially deoxygenated compounds (glycolaldehyde, acetic acid, furanic compounds, etc.), CO, CO_2, water, and coke. A range of reactions proceed inside the zeolite, like dehydration, decarbonylation, decarboxylation, isomerization, oligomerization, and dehydrogenation. The aim of catalytic fast pyrolysis is to maximize the selectivity toward aromatics and to suppress the undesired coke formation, which can be formed through both thermal and catalytic reactions.

Huber et al. used a pyroprobe analytical pyrolizer for the catalytic pyrolysis of cellulose, cellobiose, glucose, and xylitol [234,285]. Catalytic fast pyrolysis with ZSM-5 at 600°C showed the aromatics yield to be higher for xylitol (47%) than for cellulose, cellobiose, and glucose (approximately 32%), what was explained by the higher H/C_{eff} ratio of xylitol. All substrates showed coke yields over 30%. The aromatic product distribution was largely the same for all substrates, with the selectivity decreasing in the order naphthalene ≫ toluene > xylenes > benzene > substituted benzene/indane, suggesting that conversion of the substrates proceeds through a similar intermediate. Brown et al. performed the catalytic pyrolysis of the individual lignocellulose components in an analytic pyrolyzer with ZSM-5 at 600°C and found the aromatic yield to decrease in the order cellulose (29%) > hemicellulose (19%) ≫ lignin (7%) [286]. The coke yield increased in the order cellulose (31%) > hemicellulose (40%) > lignin (59%), with cellulose and hemicellulose primarily giving rise to catalytic coke and lignin to thermal coke. It was proposed that there is no significant interaction between the lignocellulose components during catalytic fast pyrolysis, since the catalytic pyrolysis of a lignocellulose substrate, namely switchgrass, gave a product distribution that was approximately the additive contribution of the three individual components. They reported that dehydration is the dominant oxygen removal mechanism during pyrolysis and that decarbonylation was favored over decarboxylation.

A high aromatics selectivity, however, requires proper catalyst selection. Zhang et al. studied the fast pyrolysis of corncob in absence and presence of a catalyst (ie, ZSM-5) [287]. The presence of the catalyst increased the yields of noncondensable gas, water, and coke, while decreasing the liquid and char yields. The catalyst induced a decrease of the oxygen content of the liquid fraction by more than 25%. These studies indicate the importance of a catalyst during biomass pyrolysis. The most important catalytic parameters affecting the product distribution are pore structure and acid site type. This was demonstrated by testing silicalite, a material with the same pore structure as ZSM-5 but with intrinsic different acid sites, and silica-alumina, an amorphous material with Brønsted acid sites, in the catalytic pyrolysis of glucose [234,285]. Where ZSM-5 showed high selectivity toward aromatics, silicalite and silica-alumina produced primarily coke and only small amounts of aromatics, indicating the need for Brønsted acid sites and a crystalline pore structure for the production of aromatics. Furthermore, since beta and Y zeolites also produced large amounts of coke in the conversion of glucose, it was suggested that catalytic fast pyrolysis requires the correct pore size—a shape-selective process. In a following study, the influence of zeolite pore size and shape selectivity on the catalytic fast pyrolysis of glucose was investigated by testing a range of zeolites with varying pore size and shape, viz. small pore ZK-5, SAPO-34, medium pore ferrierite, ZSM-23, MCM-22, SSZ-20, ZSM-11, IM-5, TNU-9, beta zeolite, SSZ-55, and Y zeolite [288]. A clear influence of the zeolite pore size on the product distribution was observed, with small pore zeolites yielding mainly oxygenated products, CO, CO_2, coke, and no aromatics, while large pore zeolites producing a large amount of coke and only small amounts of aromatics and oxygenated products. Medium pore zeolites with pore sizes in the range of 5.2–5.9 Å, like ZSM-5 and ZSM-11, showed the highest aromatics production and the least amount of coke. The high aromatics yield for ZSM-5 was proposed to be due to an appropriate combination of pore size and cavity dimensions in this material. Next to the pore structure, the framework silica-to-alumina ratio of the zeolite also plays an important role in the product distribution, as was demonstrated by testing different ZSM-5 zeolites [289]. A maximum aromatic yield from glucose was obtained with a ZSM-5 zeolite with a silica-to-alumina ratio of 30.

To investigate the potential of hierarchical zeolites in the catalytic pyrolysis of glucose, a mesoporous ZSM-5 synthesized by a surfactant mediated method was tested, but showed a slight increase in coke formation and a decrease in monoaromatics yield compared to unmodified ZSM-5 [289]. On the other hand, Li et al. investigated the catalytic fast pyrolysis of beech wood in a pyroprobe analytical pyrolizer with mesoporous ZSM-5 catalysts prepared by desilication with NaOH solutions [290]. The creation of mesopores improved the diffusion and cracking of bulky oxygenates such as lignin-derived syringols. The desilicated zeolites gave higher aromatic yields and produced less coke. Carefully controlling the desilication of the zeolite is, however, important since desilication with a too high NaOH concentration resulted in a lower aromatic yield due to a considerable loss of microporosity in the zeolite. A zeolite treated with 0.3 M NaOH solution gave the best results, namely, an aromatic and coke yield of 30% and 40% respectively, compared to 24% and 44% for the nontreated ZSM-5.

Similar to the upgrading of pyrolysis vapors, the effect of metal incorporation was extensively studied. Hicks et al. tested ZSM-5, hierarchical ZSM-5 (meso-ZSM-5, synthesized using a dry-gel method followed by steam-assisted crystallization), and their Ce-incorporated analogues in the catalytic pyrolysis of glucose at 600°C [291]. ZSM-5 and Ce/ZSM-5 yielded mainly aromatics, while meso-ZSM-5 and especially Ce/meso-ZSM-5 exhibited product distributions greatly shifted from aromatics to oxygenated chemicals like aldehydes, furans, and ketones. Also, Ce/meso-ZSM-5 produced less coke and increased CO production through decarbonylation reactions than ZSM-5. Moreover it was shown to be capable of performing ketonization of carboxylic acids. It was suggested that the presence of Ce in the zeolite framework and the reduced acidity were responsible for the shift in selectivity and decrease in coke formation compared to ZSM-5. The performance of the spray-dried ZSM-5 in catalytic fast pyrolysis of pine wood in a fluidized-bed reactor could be significantly improved by Ga incorporation into the zeolite framework, with an increase in aromatic yield from 15% to 23% at 550°C [292]. The Ga/ZSM-5 catalyst appears to be a bifunctional catalyst, with Ga species facilitating decarbonylation and olefin aromatization, and the zeolite promoting other reactions such as oligomerization. Li et al. performed the catalytic fast pyrolysis of a range of substrates, viz. rice husk, sawdust, sugarcane bagasse, cellulose, hemicellulose and lignin, in a flowing fixed-bed system with ZSM-5 containing 6 wt% La at 600°C with the aim to produce light olefins [293]. The incorporation of La was shown to significantly increase the olefins yield during pyrolysis and was suggested to be due to a moderate increase in the amount of acid sites of medium strength. The olefins yield decreased in the order cellulose > hemicellulose > sugarcane bagasse > rice husk > sawdust > lignin. Substrates with a higher content of cellulose and hemicellulose yield more olefins than substrates with higher lignin contents. From sugarcane bagasse, an olefin yield of 21% was obtained. In another study, Li et al. demonstrated that Mg-incorporation into a ZSM-5 catalyst significantly increased the production of light olefins and reduced coke formation during catalytic conversion of bio-oil [294]. This was also attributed to a moderate increase in the amount of medium-strength acid sites.

The reaction mechanism of catalytic fast pyrolysis of carbohydrate feedstocks was investigated by Huber et al. with glucose as a model compound [295]. Glucose conversion involves first rapid thermal decomposition through fragmentation and dehydration reactions (thermal or pyrolysis reactions), followed by conversion of the intermediates within the zeolite pores (catalytic reactions). The thermal reactions can also occur on the catalyst active surface and are thus promoted by the presence of a catalyst. Fragmentation to smaller oxygenates (glyceraldehyde, hydroxyacetone, and hydroxyaldehyde) is favored at lower temperatures, while higher temperatures favor glucose dehydration to anhydrosugars, like levoglucosan and furanic compounds. Both the fragmentation and dehydration products are possible intermediates in the production of aromatics. However, independently from the temperature, the conversion to aromatics occurs through the formation of a hydrocarbon pool. A detailed reaction mechanism for catalytic fast pyrolysis of glucose is shown in Scheme 9. The major competing reaction to aromatics production is the formation of coke, which is most likely formed by polymerization of the furans on the external catalyst surface. For selective monoaromatics production, the concentration of oxygenates should be kept low to avoid the formation of coke and polyaromatics. However, the catalytic conversion step is significantly slower than the initial thermal step, therefore a high catalyst to feed ratio is essential for selective aromatics production.

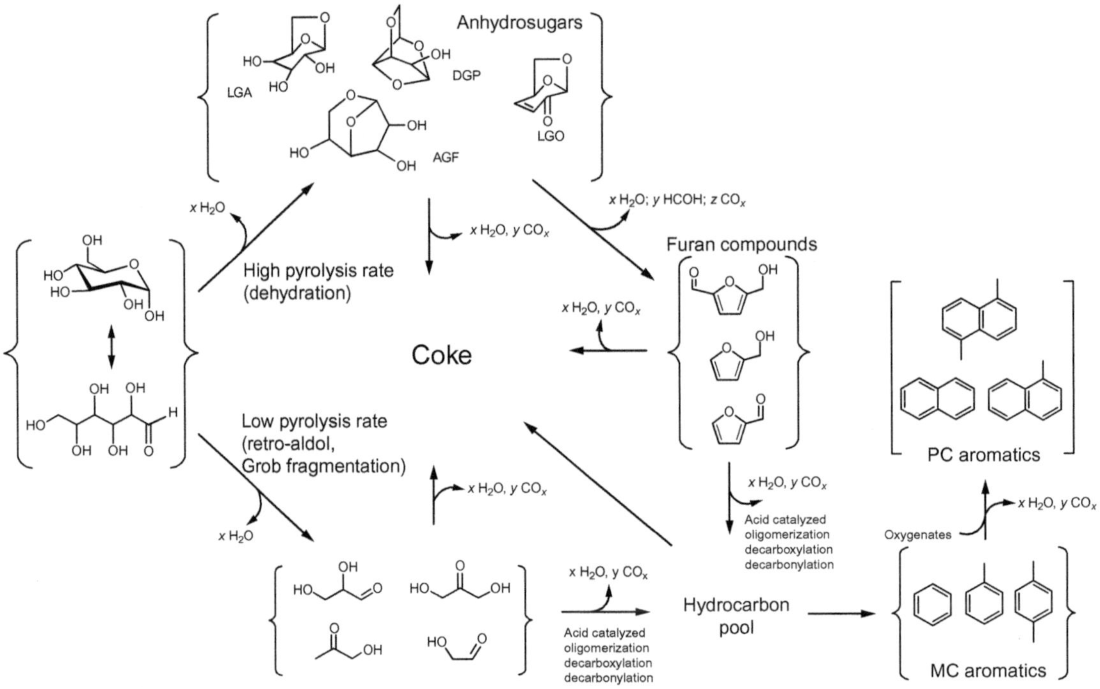

Scheme 9

Reaction scheme of the catalytic fast pyrolysis of glucose over ZSM-5. *Reprinted with permission from Elsevier (From Carlson TR, Jae J, Lin Y-C, Tompsett GA, Huber GW. Catalytic fast pyrolysis of glucose with HZSM-5: the combined homogeneous and heterogeneous reactions. J Catal 2010;270:110–24).*

Huber et al. performed the catalytic fast pyrolysis of pine wood with spray-dried ZSM-5 in a bubbling fluidized-bed reactor with co-feeding of different alcohols, viz. methanol, 1-propanol, 1-butanol, and 2-butanol, to adjust the H/C_{eff} ratio of the feed [296]. However, the optimal conditions for conversion of pine wood and the alcohols were different. Pine wood reached its maximum petrochemical yield (aromatics + C_2–C_4 olefins + C_5 compounds) of 24% at 600°C, while the maximum yield of 81% from methanol was reached at 450°C. Co-feeding of pine wood and the alcohols in the process was shown to enhance the petrochemical yield with a positive synergistic effect of the two feeds, but the ratio of the feeds should be tuned for optimal results. From the investigated alcohols, methanol showed the best results. Isotopic studies further indicated that the carbon of the pine wood and the alcohol is mixed within the zeolite and incorporated in the products.

In summary, catalytic fast pyrolysis is a promising technology once the main drawbacks, such as severe coking, deep cracking, and hydrothermal stability of the zeolite are solved. No commercial scale of a catalytic biomass pyrolysis has been demonstrated until now. The University of Massachusetts, one of the research experts in terms of catalytic pyrolysis, licensed their technology to startup Anellotech to produce renewable biogasoline and aromatics/olefins [297]. Better insight in the pore architecture, linked with the acid sites (location, number, and strength), should make new developments of better catalysts possible, that allow overall higher liquid yields and better control over the bio-oil composition in function of the applications (phenols, aromatics, polyaromatics, etc.).

5.6 Catalytic Fast Pyrolysis of Lignin

While entire biomass and cellulose fractions have been extensively investigated, catalytic fast pyrolysis of lignin, the third most abundant source of carbon, has been less extensive explored. Generally, lignin produces more cokes and char under these reaction conditions [286,298]. In fact, it has been suggested that the cokes formed during catalytic fast pyrolysis of biomass is mainly due to the lignin part. Removal of lignin prior to pyrolysis is therefore advantageous. Yet, lignin fractions (and insoluble parts in pyrolytic bio-oils) may be upgraded as well with catalytic fast pyrolysis in the presence of zeolites.

As described in the previous section, Brown et al. obtained an aromatic yield of 7% in the catalytic fast pyrolysis of milled wood lignin with ZSM-5 at 600°C, which was considerable lower than the yields obtained from cellulose and hemicellulose [286]. Also, the coke yield from lignin (59%) was much higher. Huber et al. used the lignin residue obtained after H_2SO_4 hydrolysis of maple wood as a substrate for pyrolysis in a fluidized-bed reactor with spray-dried ZSM-5 at 600°C [298]. Next to lignin, this substrate also contains humic residue, a compound also composed of phenolic entities. An aromatic and olefin yield of only 2% and 1% was obtained, compared to respective yields of 15% and 7% from pure maple wood. The coke yield from the solid residue (69%) was also much higher than the coke yield from maple wood (32%).

The catalytic pyrolysis of lignin is studied over a wide range of catalysts, like ZSM-5 [299–304], beta [300,304], MCM-41 [300–302], SBA-15 [300], solid phosphoric acid [301,302], Co-Mo/Al_2O_3 [301,302], Y [304], ferrierite [304] and mordenite [304]. Regardless the used catalyst, an almost complete removal of the aliphatic hydroxyl groups, a 44–85% decrease in carboxyl groups and up to 80% decrease in methoxy groups in the bio-oil can be observed compared to a noncatalytic run [304]. Besides, adding a catalyst significantly lowered the average molecular weight of the pyrolysis products. From the aforementioned catalysts, the highest aromatics yield is obtained with ZSM-5. The mixture of formed aromatics is mainly composed of toluene, xylenes, naphthalene, and benzene [300–302]. Using a different zeolite does not only affect the yield of aromatics, but even the composition of the formed products can strongly be influenced. Ferrierite and mordenite, for example, produces mainly heavy oils, while ZSM-5, Y, and beta produce more light oils [304]. Zhao et al. and Boateng et al. investigated the influence of the nature of the counter cation of ZSM-5 on the catalytic activity and found the aromatics yield to decrease in the order Na-ZSM-5 > H-ZSM-5 > K-ZSM-5 [300–302].

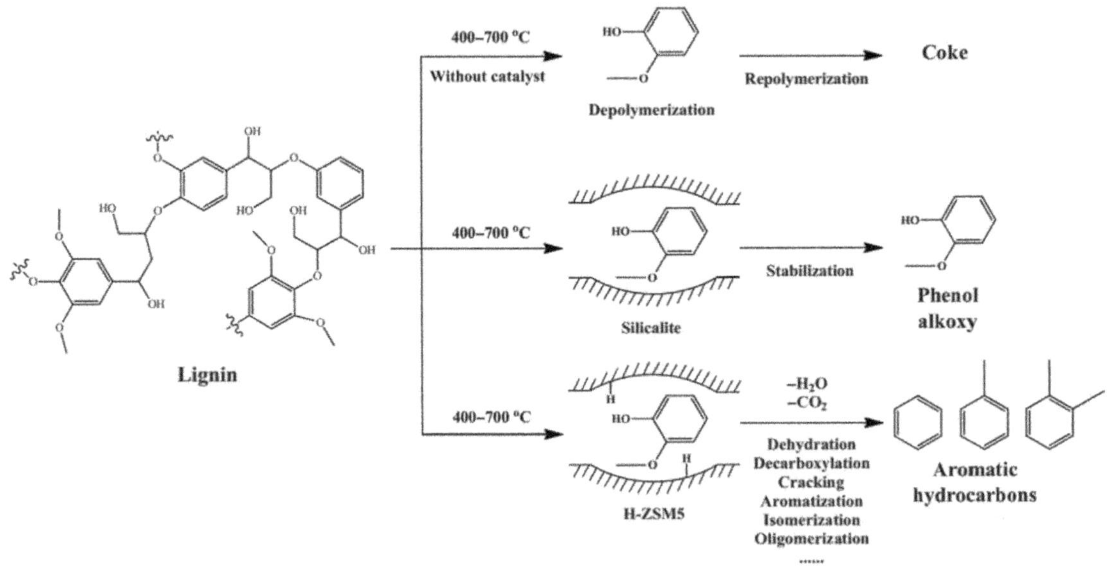

Scheme 10

Catalytic and noncatalytic fast pyrolysis of lignin. *Reprinted with permission from Elsevier (From Ma Z, Troussard E, van Bokhoven JA. Controlling the selectivity to chemicals from lignin via catalytic fast pyrolysis. Appl Catal A 2012;423:130–6).*

Ma et al. ascribed a dual role to the catalyst, with the acid sites converting the depolymerized intermediates into desirable and more stable products while the catalyst surface prevents repolymerization and coke formation (Scheme 10) [305]. A high catalyst-to-substrate ratio was shown to increase the liquid product yield. They also studied the effect of the Si/Al ratio and the acidity of the catalyst. Decreasing the silica-to-alumina ratio and thus increasing the amount of acid sites of ZSM-5 increased the production of aromatics. ZSM-5 with an intermediate silica-to-alumina ratio showed the highest phenols production (14 wt% of the liquid product). This was in line with the results of Li et al. [303].

Regarding catalyst pore size, it was stated that large pore sizes enable large molecules to fit in and be stabilized or converted [305]. USY, the catalyst with the largest pores and highest acid site density, yielded the highest amount of liquid product, viz. 75 wt%, and the highest amount of aromatics (40 wt% of the liquid phase) [305]. Yu et al. investigated the role of zeolite shape selectivity in the catalytic pyrolysis of lignin at 650°C and proposed that thermal distortions of the zeolite pore structure under the high-temperature conditions of pyrolysis effectively enlarge the crystallographically determined pore sizes of zeolites by 0.25–0.34 nm, thus enabling bulky lignin-derived oxygenates to enter the pores [306]. Too bulky compounds such as syringols could, however, not be effectively converted by ZSM-5 and mordenite.

As a conclusion, some promising results regarding aromatics production from lignin via catalytic pyrolysis are reported, although most studies obtain only low aromatics yields

(<8%). The presence of ZSM-5 decreased the yield of oxygenates, increased the yield of aromatics, and significantly lowered the average molecular weight of the pyrolysis products. The aromatic yield can further be optimized by increasing the acid density of ZSM-5, so decreasing the silica-to-alumina ratio, and increasing the catalyst-to-substrate ratio.

6 Zeolite Catalysis in Oleochemistry

6.1 Hydrogenation of Fatty Acid Methyl Esters and Vegetable Oils

The selective hydrogenation of vegetable oils is used in industry to stabilize oils against autoxidation and/or to obtain fat products with a certain melting profile [307]. However, during this hydrogenation process isomerization also occurs at the catalyst surface, leading to the formation of *trans* isomers. Although these *trans* fatty acids are associated with positive melting properties, the *cis/trans* isomerization should be limited as some of these *trans* isomers, especially elaidate (C18:1 *trans*-9), are considered as a risk factor in coronary heart diseases [308]. Therefore, governments have been forcing the industry to limit the content of *trans* fatty acids in food products. It is important to note that the debate concerning *trans* fatty acids in foods is still active, as proven by the recent change in the EU regulation [309]. Unfortunately, the literature reveals that low *cis/trans* isomerization is almost always associated with a lower hydrogenation selectivity, implying that considerable levels of unstable polyunsaturated fatty acids are not converted to more stable ones, leading to an unstable fat product and/or that high levels of trisaturates are formed, giving rise to undesirable physical properties [310]. Therefore, hydrogenated oils and fats have increasingly been replaced by interesterified and fractionized samples, especially in Europe. These alternative processes gained in popularity as they do not produce *trans* fatty acids. Also, when using oils low in polyunsaturated fatty acids and high in C14 and C16 saturated fatty acid chains, as palm or coconut oil, these alternative processes yield fat products with desirable physical properties [311–314].

Recently there has been a renewed interest in the use of hydrogenated fats, providing these fats contain only low amounts of *trans* fatty acids, show a good stability, and have desirable physical properties. The high quantity of cholesterol-raising saturated fatty acids as palmitic acid (C16:0) in interesterified bakery shortenings [315], sustainability concerns of customers, and the increasing vegetable oil prices, explain this renewed interest in hydrogenated fats. Up to now, hydrogenation is the only process that selectively converts thermally unstable polyunsaturated fatty acids to more stable ones. Hence, using this process which is common to polyunsaturated oils, for soybean and rapeseed oils, can obtain stable end products.

Catalyst design is crucial to obtain a hydrogenation process showing both high hydrogenation selectivities and low *cis/trans* isomerization. For example, the catalyst porosity, the metal dispersion, and the accessibility of the triglyceride molecules to the active metal centers significantly influence the activity, selectivity, and the degree of *cis/trans* isomerization [310].

It is known that zeolites act as molecular sieves, excluding the access of molecules with a diameter too large to enter the pores. Different zeolite topologies were tested to find a pore system that is able to distinguish between the linear *trans* chain and the bended *cis* chain. ZSM-5 zeolite appears to be the best choice: the quasi-linear *trans* isomer is sorbed more preferably. This preference was further investigated in relation to various compositional parameters. Loading the ZSM-5 zeolite with Pt clusters within the crystal matrix allows the selective hydrogenation of the *trans* isomer, while leaving the *cis* isomer almost unreacted [316]. Obviously, the location of Pt is very critical. Only the Pt particles encapsulated in the intracrystalline space of the zeolite crystal will selectively hydrogenate the *trans* isomer (Fig. 8). Various synthesis procedures with varying compositions and heat pretreatments were investigated to obtain well-dispersed Pt within the zeolite matrix, as was demonstrated with scanning and transmission electron microscopic images. By using various activation methods, it was shown that the calcination step is crucial as it allows the complete transformation of the Pt-precursor into Pt^{2+}, which can be seen by the development of a greenish sample color. The critical conditions require application of a low heating rate (0.3°C/min), a high oxygen flow (120 mL/min.g), and an elevated calcination temperature (350°C).

Besides the position of the Pt clusters, the zeolite particle size is another important parameter for selective hydrogenation of the *trans* fatty acid methyl ester isomer. Larger crystals have a higher ratio of internal to external surface and consequently a lower ratio of external to internal Pt clusters, which induce an improving hydrogenation selectivity. Also the Si/Al ratio influences the catalytic system. A Pt/ZSM-5 catalyst with an Si/Al ratio up to 80–150 results in further improvement of the reaction selectivity based on shape selectivity of diffusional origin.

The shape-selective Pt/ZSM-5 catalyst was also tested in the hydrogenation of model triacylglycerols. In agreement with the results obtained with the fatty acid methyl esters, a

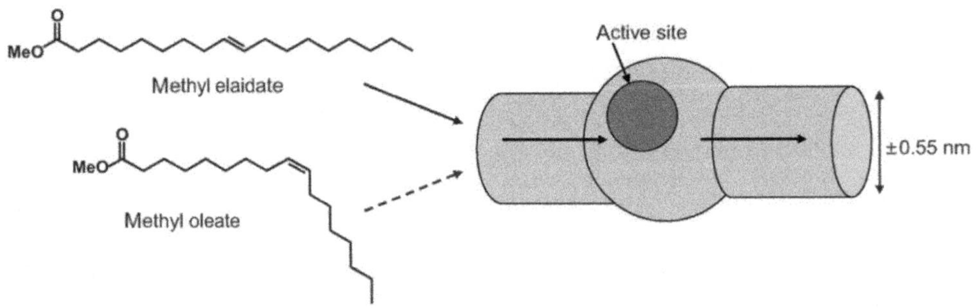

Fig. 8
Shape-selective hydrogenation of methyl oleate (*cis* isomer) and methyl elaidate (*trans* isomer) over Pt/ZSM-5. *Reprinted with permission from Wiley-VCH Verlag (From Philippaerts A, Jacobs PA, Sels BF. Is there still a future for hydrogenated vegetable oils? Angew Chem Int Ed 2013;52:5220–6).*

preference for the hydrogenation of *trans* over *cis* fatty acid chains is also seen. Moreover, a second, new type of shape selectivity was observed: the chain on the central position of the glycerol backbone is preferably hydrogenated [317]. This observation points to the existence of a very specific adsorption mode of the triacylglycerol molecules on the catalyst surface, viz. in a tuning fork configuration. Surprisingly, Pt located in micropores of ZSM-5 zeolite also shows a preferred stepwise reduction to the mono-unsaturated level as illustrated in the hydrogenation of trilinoleate. This behavior is unusual for Pt-catalysts and thus should be associated with the tuning fork sorption mode of triacylglycerols on the zeolite external surface in combination with an improved affinity of more polar chains (trienes > dienes > monoenes) for the polar pore mouths of MFI [317].

Finally, this concept was tested in the hydrogenation of vegetable oils, viz. soybean oil. It was shown that a single hydrogenation process of soybean oil at low temperature and high hydrogen pressure, using a shape-selective Pt/ZSM-5 catalyst, yields products that are thermally stable, have a high nutritive value, and have the desired plasticity required for shortening applications [318]. The hydrogenated fat samples contain high levels of oleic acid (C18:1*c*) and linoleic acid (C18:2), very low levels of *trans* fatty acids and low levels of cholesterol-raising fatty acids such as myristic (C14:0) and palmitic acid (C16:0). The fairly selective conversion of polyunsaturates, viz. linolenic acid (C18:3), results in an oxidatively stable fat product. In addition, the regioselectivity induced by the Pt/ZSM-5 catalyst results in a mixture of intermediate melting TAGs without formation of large amounts of distasteful tristearates. This unique fat composition leads to low contents of undesirable beta crystals in the solid product, more preferable melting properties and a quite flat solid fat content curve in the temperature region from 10 to 40°C, indicative of high plasticity.

6.2 Conjugation of Linoleic Acid and Vegetable Oils Rich in Linoleic Acid

Conjugated vegetable oils are of high interest in industry as the conjugated double bonds are highly active in polymerization reactions. Traditionally conjugated oils are added to paints, varnishes, and inks to improve their drying properties [319,320]. Furthermore, conjugated linoleic acids (CLAs) are also used in the production of bioplastics [321]. In recent years various positive health effects have been attributed to CLAs. However, from the different possible CLA isomers, only for the *c*9,*t*11, *t*10,*c*12, and *t*9,*t*11 CLA isomers beneficial health effects are known [322–324].

Currently, CLAs are mainly produced by conjugation of oils high in linoleic acid, viz. soybean and safflower oil, or alternatively by dehydration of castor oil [325,326]. Unfortunately, these processes require the use of homogeneous alkali bases, solvents, and neutralizing acids and hence do not show the characteristics of a sustainable process. Heterogeneous catalysts are more attractive as they can be easily recovered by filtration. However, literature reveals that it is not easy to design a catalyst with high CLA selectivity as the competing hydrogenation reaction limits the CLA yield [327]. Another drawback is the low productivity of the heterogeneous processes described in literature.

Highly dispersed Ru/USY catalysts with a high Si/Al ratio show high productivities of CLA from methyl linoleate [328]. Additionally, almost no hydrogenated products are formed as the reactions are performed "under an" inert atmosphere. Conjugation seems to be performed by means of hydrogen transfer between the fatty acids and the active Ru metal clusters on the catalyst, that is, according to a classical Horiuti-Polanyi mechanism, since *c*9,*t*11 and *t*10,*c*12 CLA isomers were shown to be primary products (Scheme 11). The latter are converted consecutively into *trans*, *trans* isomers with double bonds at the same and finally other positions (*t*9,*t*11). Also, the selectivity for the primary products (desirable isomers) increases, when the size of the charge compensating cations is raised ($H^+ < Na^+ < Cs^+$).

The obtained isomerization productivities are much higher compared to those of other heterogeneous processes found in literature and are comparable to the homogeneous process, today used industrially [327]. Moreover, the turnover frequency for Ru/Cs-USY is one order of magnitude larger than that of the best heterogeneous catalyst found in literature, viz. Ru/C. This extraordinary isomerization activity of Ru in Ru/Cs-USY can be explained by the small size of the Ru clusters. In addition to the high activity, also the CLA selectivity benefits from the small Ru cluster size. The biologically most active CLA isomers, viz. *c*9,*t*11, *t*10,*c*12 and *t*9,*t*11, were the main products, especially at low catalyst concentration.

As very small Ru clusters are necessary to obtain such high activity and selectivities, also here the synthesis method of the catalyst is crucial. Catalyst characterization with CO-chemisorption,

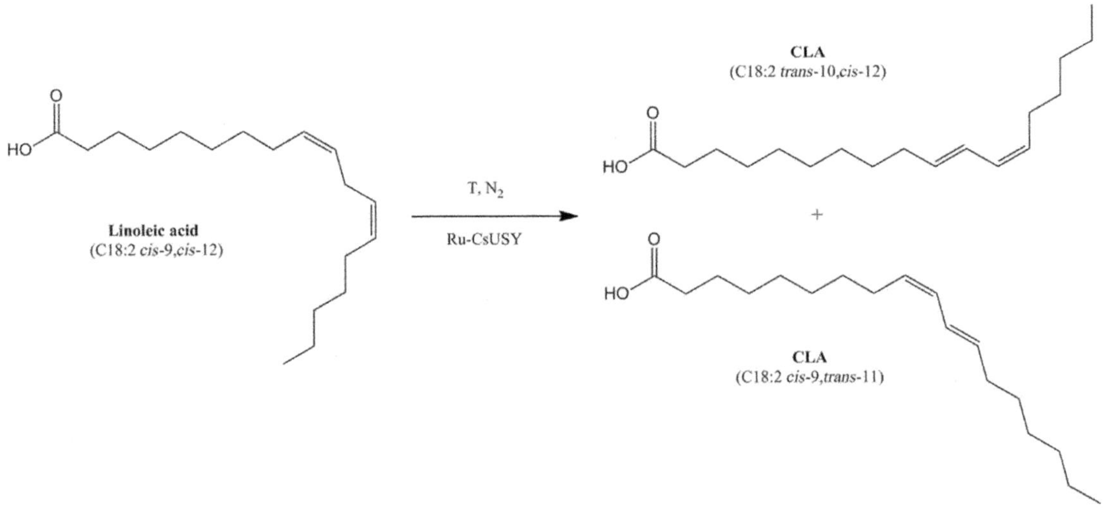

Scheme 11
Conjugation of linoleic acid over Ru/USY zeolite to CLAs [328].

extended X-ray absorption fine structure (EXAFS) measurements and transmission electron microscopy (TEM) analysis reveal highly dispersed RuO_2 species in Ru/Cs-USY. It is assumed that such clusters in reaction conditions are transformed into highly dispersed Ru metal clusters via a reaction of autoxidation of methyl linoleate by Ru–O [328].

The Ru/USY catalyst was also tested in the isomerization of pure vegetable oils. In order to obtain high activities in the conjugation of bulky triglyceride molecules, a simple one-step method to prepare hierarchically structured USY zeolites was developed [329]. By contacting a commercial steamed USY zeolite with diluted aqueous NH_3 solution, a considerable amount of fairly uniform large micropores (1.5–2 nm) in addition to the original large mesopores (20–30 nm; due to steaming) and micropores (0.7 nm, inherent to Y zeolite) is created, without distorting the crystal morphology. These additional large micropores facilitate transport of the bulky triglyceride molecules to and from the active centers of the catalyst. Surprisingly, also the Ru dispersion benefits from this hierarchical structure. With EXAFS measurements almost no Ru-Ru interactions could be observed, implying the presence of atomically dispersed Ru. By using these hierarchical Ru/USY zeolites in the hydrogen-free conjugation of safflower oil, it was observed that not only the activity, but also the selectivity for the beneficial CLA isomers improves due to the more open pore architecture. As an example, the isomerization of safflower oil at 180°C under N_2, yielded after 2 h 75 mg CLA/g oil with a very high specific yield, viz. 377 g_{CLA}/g_{metal} h, pointing to a very efficient process. The level of the desirable *c*9,*t*11 and *t*10,*c*12 isomers at this point was almost 55 mg CLA/g oil. Moreover, as no H_2 or another H-donor is used in the catalytic process, almost no hydrogenation was observed, while only very low amounts of undesirable C18:1 *trans* are present in the conjugated oil [328].

6.3 Production of Branched Fatty Acids

Saturated branched fatty acids, such as isostearic acid, are important intermediates in various industrial products, including lubricants, surfactants, detergents, cosmetics, biodiesel, and hydraulic fluids. Saturated branched fatty acids and their alkyl esters have important advantages over natural linear fatty acids: they have an excellent oxidative stability compared to unsaturated fatty acids and have a more desirable liquidity compared to linear saturated fatty acids. Furthermore, branched saturated fatty acids show a desirable low viscosity and improved formulation flexibility [330]. As naturally occurring branched fatty acids are rare, catalytic skeletal isomerization of common linear fatty acids is required.

Currently, branched fatty acids are commercially produced as a by-product from dimer fatty acid production using clay catalysts. However, this process suffers from a rather poor selectivity, yielding complex mixtures of both alkyl-branched and dimeric fatty acids [331]. Although skeletal isomerization of linear hydrocarbons has been a common practice in the petrochemical industry, the direct catalytic isomerization of fatty acids or esters has turned out much more challenging. Most commercial solid acids fail to produce branched products from fatty acids due to the strong inhibiting effect of the carboxylate group [332]. More recently, acid zeolites have been found to be particularly effective catalysts for the production of branched fatty acids [330]. Yields of monomeric branched fatty acids around 80% can be obtained, starting from a wide range of feeds, applying a nitrogen atmosphere and elevated temperatures, typically 250°C. The obtained unsaturated branched fatty acids can be easily saturated in a subsequent hydrogenation step.

This exceptional performance of acidic zeolites compared to other solid acids can be explained by their hydrophobic surface, favoring the aliphatic part of a fatty acid molecule over the carboxylic acid group and thus reducing the poisoning effect of the carboxylate group [332]. Furthermore, it is believed that the micropores of zeolites are large enough for a fatty acid to diffuse through the pore system toward the active sites and allow skeletal isomerization, but small enough to hinder the formation of oligomeric products [333].

The skeletal isomerization of unsaturated fatty acids is a Brønsted acid catalyzed reaction and follows the conventional carbenium ion mechanism. The zeolitic Brønsted acid sites provide protons to form carbenium ions from adsorbed fatty acids that contain unsaturated double bonds. Therefore, the activity of the acid zeolite catalyst is highly dependent on the acid density. For instance, excellent performance in the production of isostearic acids from oleic acid is obtained with an acid ferrierite zeolite catalyst with low Si/Al ratio in the presence of small amounts of deionized water. This catalytic system gave quantitative conversion and >85% selectivity [333]. This high activity of ferrierite zeolite compared to other acid zeolite catalysts with comparable Si/Al ratios can be attributed to its high density of Brønsted acid sites. However, it has been demonstrated that the accessibility of the acid sites is also crucial. For instance, among various commercial beta zeolites with comparable Si/Al ratios, those containing significant amounts of mesopores, were found to be much more active in the isomerization of oleic acid and more stable against coke formation [332].

7 Conclusions

This chapter explored the current state of the use of zeolites in biomass conversion strategies. From this, it can be concluded that zeolites, in an extension to their success in petrochemistry, can also play a pivotal role in the conversion of biomass. The success of zeolites in petrochemistry is closely linked to their strong Bronsted acidity in a crystalline matrix with porosity of molecular dimensions which could be one by one easily modified. The examples in this chapter showed that these properties are also very decisive in biomass transformations.

However, the different chemical composition of fossil feedstocks versus biomass-based feedstocks poses a new set of challenges in the development of active, selective, and stable zeolite catalysts. In contrast to petroleum resources, biomass feedstocks generally consist of highly oxygenated compounds with bulky and high molecular weight components such as (hemi)cellulose, lignin, and fatty acids resulting in the need of sufficiently accessible active sites by an interconnected pore system which can be obtained by hierarchization. The highly oxygenated nature of the components leads to an enhanced cokes production requiring regeneration processes to retain the high selectivities. Besides, hierarchization of the zeolites can counteract or delay the coke formation due to a better mass transport: reagents reach faster the active sites and the products can be more easily diffused out of the zeolite particle that obviate the possibility to undergo secondary reactions leading to coke formation or other unwanted products. In addition, biomass conversion usually takes place in condensed phase, often in hot aqueous conditions. In order to convert the biomass compounds efficiently, active sites should be accessible and stable in aqueous conditions at high temperatures.

It has been illustrated that specific strategies exist to adapt the zeolite properties to the specificity of biomass in comparison with the needs for valorizing fossil feeds. Nonetheless, there is still some work left in exploring the complete chemistry of some zeolitic biomass transformation reactions. Provided there is continued research interest, we believe that the refinery success of zeolites can be extended to biomass feeds in integrated biorefineries.

References

[1] Espeel P, Parton R, Toufar H, Martens JA, Hölderich WF, Jacobs PA. In: Weitkamp J, Puppe L, editors. Catalysis and Zeolites—Fundamentals and Applications. Berlin: Springer Verlag; 1999. p. 377–436.
[2] Hölderich WF, van Bekkum H. In: van Bekkum H, Flanigen EM, Jansen JC, editors. Studies in surface science and catalysis, vol. 58. Amsterdam: Elsevier; 1991. p. 631–726.
[3] Venuto PB. Organic catalysis over zeolites: a perspective on reaction paths within micropores. Microporous Mater 1994;2:297–411.
[4] De Vos DE, Jacobs PA. Zeolite effects in liquid phase organic transformations. Microporous Mesoporous Mater 2005;82:293–304.
[5] Taramasso M, Perego G, Notari B. Google Patents; 1983.
[6] Notari B. In: Grobet PJ, Mortier WJ, Vansant EF, Schulz-Ekloff G, editors. Studies in surface science and catalysis, vol. 37. Amsterdam: Elsevier; 1988. p. 413–25.
[7] BASF. HPPO—a groundbreaking technology for more environmental compatibility and efficiency, 2014. http://www.basf.com [accessed 08.09.14].
[8] IHS. Hydroquinone from phenol using Ti-Si catalyst, 2014. http://www.ihs.com [accessed 08.09.14].
[9] Stöcker M. Methanol-to-hydrocarbons: catalytic materials and their behavior. Microporous Mesoporous Mater 1999;29:3–48.
[10] Klass DL. Biomass for renewable energy, fuels, and chemicals. San Diego: Academic Press; 1998.
[11] Lange J-P. In: Centi G, Van Santen RA, editors. Catalysis for renewables. Weinheim: John Wiley & Sons, Ltd.; 2007. p. 21–51.
[12] Jacobs PA, Dusselier M, Sels BF. Will zeolite-based catalysis be as relevant in future biorefineries as in crude oil refineries? Angew Chem Int Ed 2014;53:8621–6.
[13] Kubička D, Kubičková I, Čejka J. Application of molecular sieves in transformations of biomass and biomass-derived feedstocks. Catal Rev 2013;55:1–78.
[14] Vermeiren W, Gilson JP. Impact of zeolites on the petroleum and petrochemical industry. Top Catal 2009;52:1131–61.
[15] Dusselier M, Van Wouwe P, Dewaele A, Jacobs PA, Sels BF. Shape-selective zeolite catalysis for bioplastics production. Science 2015;349:78–80.

[16] Zeitsch KJ. The chemistry and technology of furfural and its many by-products. Amsterdam: Elsevier; 2000.
[17] Bugg TDH, Ahmad M, Hardiman EM, Rahmanpour R. Pathways for degradation of lignin in bacteria and fungi. Nat Prod Rep 2011;28:1883–96.
[18] Ghosh P, Singh A. Physicochemical and biological treatments for enzymatic/microbial conversion of lignocellulosic biomass. Adv Appl Microbiol 1993;39:295–333.
[19] Sharma S, Reena, Kumar A, Mittal P. In: Babu V, Thapliyal A, Patel GK, editors. Biofuels production. John Wiley & Sons, Inc., Weinheim; 2013. p. 119–45.
[20] Villas-Bôas SG, Esposito E, Mitchell DA. Microbial conversion of lignocellulosic residues for production of animal feeds. Anim Feed Sci Technol 2002;98:1–12.
[21] Williams KC. Microbial conversion systems for increasing the efficiency and cost effectiveness of ethanol production from lignocellulose. Davis: University of California; 2013.
[22] Kersten SRA, van Swaaij WPM, Lefferts L, Seshan K. In: Centi G, Van Santen RA, editors. Catalysis for renewables. John Wiley & Sons, Ltd., Weinheim; 2007. p. 119–45.
[23] Lin Y-C, Huber GW. The critical role of heterogeneous catalysis in lignocellulosic biomass conversion. Energy Environ Sci 2009;2:68–80.
[24] Perego C, Bosetti A. Biomass to fuels: the role of zeolite and mesoporous materials. Microporous Mesoporous Mater 2011;144:28–39.
[25] Taarning E, Osmundsen CM, Yang X, Voss B, Andersen SI, Christensen CH. Zeolite-catalyzed biomass conversion to fuels and chemicals. Energy Environ Sci 2011;4:793–804.
[26] Brown RC. In: Kamm B, Gruber PR, Kamm M, editors. Biorefineries—industrial processes and products. John Wiley & Sons, Ltd., Weinheim; 2008. p. 227–52.
[27] Bridgwater AV. Catalysis in thermal biomass conversion. Appl Catal A 1994;116:5–47.
[28] Bridgwater AV, Peacocke GVC. Fast pyrolysis processes for biomass. Renew Sust Energ Rev 2000;4:1–73.
[29] Kersten SRA, Wang X, Prins W, van Swaaij WPM. Biomass pyrolysis in a fluidized bed reactor. Part 1: Literature review and model simulations. Ind Eng Chem Res 2005;44:8773–85.
[30] Mohan D, Pittman CU, Steele PH. Pyrolysis of wood/biomass for bio-oil: a critical review. Energy Fuel 2006;20:848–89.
[31] Bridgwater AV. Review of fast pyrolysis of biomass and product upgrading. Biomass Bioenergy 2012;38:68–94.
[32] Butler E, Devlin G, Meier D, McDonnell K. A review of recent laboratory research and commercial developments in fast pyrolysis and upgrading. Renew Sust Energ Rev 2011;15:4171–86.
[33] Tolan JS. In: Kamm B, Gruber PR, Kamm M, editors. Biorefineries—industrial processes and products. Wiley-VCH Verlag GmbH, Weinheim; 2008. p. 193–208.
[34] de Souza ROMA, Miranda LSM, Luque R. Bio(chemo)technological strategies for biomass conversion into bioethanol and key carboxylic acids. Green Chem 2014;16:2386–405.
[35] Kumar V, Dhall P, Kumar R, Kumar A. In: Babu V, Thapliyal A, Patel GK, editors. Biofuels production. John Wiley & Sons, Ltd., Weinheim; 2013. p. 85–118.
[36] Sousa-Aguiar EF, Appel LG, Zonetti PC, Fraga AdC, Bicudo AA, Fonseca I. Some important catalytic challenges in the bioethanol integrated biorefinery. Catal Today 2014;234:13–23.
[37] Vohra M, Manwar J, Manmode R, Padgilwar S, Patil S. Bioethanol production: feedstock and current technologies. J Environ Chem Eng 2014;2:573–84.
[38] Gomez Bernal H, Bernazzani L, Raspolli Galletti AM. Furfural from corn stover hemicelluloses. A mineral acid-free approach. Green Chem 2014;16:3734–40.
[39] Gürbüz EI, Wettstein SG, Dumesic JA. Conversion of hemicellulose to furfural and levulinic acid using biphasic reactors with alkylphenol solvents. ChemSusChem 2012;5:383–7.
[40] Kim ES, Liu S, Abu-Omar MM, Mosier NS. Selective conversion of biomass hemicellulose to furfural using maleic acid with microwave heating. Energy Fuels 2012;26:1298–304.
[41] Sánchez C, Serrano L, Andres MA, Labidi J. Furfural production from corn cobs autohydrolysis liquors by microwave technology. Ind Crops Prod 2013;42:513–9.

[42] Zakzeski J, Bruijnincx PCA, Jongerius AL, Weckhuysen BM. The catalytic valorization of lignin for the production of renewable chemicals. Chem Rev 2010;110:3552–99.
[43] Literature review of physical and chemical pretreatment processes for lignocellulosic biomass. Wageningen: Energy Research Centre of the Netherlands; 2010.
[44] Jørgensen H, Kristensen JB, Felby C. Enzymatic conversion of lignocellulose into fermentable sugars: challenges and opportunities. Biofuels Bioprod Biorefin 2007;1:119–34.
[45] Kumar P, Barrett DM, Delwiche MJ, Stroeve P. Methods for pretreatment of lignocellulosic biomass for efficient hydrolysis and biofuel production. Ind Eng Chem Res 2009;48:3713–29.
[46] Menon V, Rao M. Trends in bioconversion of lignocellulose: biofuels, platform chemicals & biorefinery concept. Prog Energy Combust Sci 2012;38:522–50.
[47] Mosier N, Wyman C, Dale B, Elander R, Lee YY, Holtzapple M, et al. Features of promising technologies for pretreatment of lignocellulosic biomass. Bioresour Technol 2005;96:673–86.
[48] Sun Y, Cheng J. Hydrolysis of lignocellulosic materials for ethanol production: a review. Bioresour Technol 2002;83:1–11.
[49] Taherzadeh MJ, Karimi K. Pretreatment of lignocellulosic wastes to improve ethanol and biogas production: a review. Int J Mol Sci 2008;9:1621–51.
[50] Sánchez ÓJ, Cardona CA. Trends in biotechnological production of fuel ethanol from different feedstocks. Bioresour Technol 2008;99:5270–95.
[51] Kamm B, Gruber PR, Kamm M. Biorefineries—industrial processes and products. John Wiley & Sons, Ltd., Weinheim; 2010.
[52] Dusselier M, Mascal M, Sels B. In: Nicholas KM, editor. Selective catalysis for renewable feedstocks and chemicals, vol. 353. Springer International Publishing, Switzerland; 2014. p. 1–40.
[53] Kuster BFM. 5-Hydroxymethylfurfural (HMF). A review focussing on its manufacture. Starch-Stärke 1990;42:314–21.
[54] Moreau C, Durand R, Roux A, Tichit D. Isomerization of glucose into fructose in the presence of cation-exchanged zeolites and hydrotalcites. Appl Catal A 2000;193:257–64.
[55] Moliner M, Román-Leshkov Y, Davis ME. Tin-containing zeolites are highly active catalysts for the isomerization of glucose in water. Proc Natl Acad Sci U S A 2010;107:6164–8.
[56] Román-Leshkov Y, Moliner M, Labinger JA, Davis ME. Mechanism of glucose isomerization using a solid Lewis acid catalyst in water. Angew Chem Int Ed 2010;49:8954–7.
[57] Choudhary V, Pinar AB, Sandler SI, Vlachos DG, Lobo RF. Xylose isomerization to xylulose and its dehydration to furfural in aqueous media. ACS Catal 2011;1:1724–8.
[58] Gunther WR, Wang Y, Ji Y, Michaelis VK, Hunt ST, Griffin RG, et al. Sn-beta zeolites with borate salts catalyse the epimerization of carbohydrates via an intramolecular carbon shift. Nat Commun 2012;3:1109.
[59] Gounder R, Davis ME. Titanium-beta zeolites catalyze the stereospecific isomerization of D-glucose to L-sorbose via intramolecular C5–C1 hydride shift. ACS Catal 2013;3:1469–76.
[60] Gounder R, Davis ME. Monosaccharide and disaccharide isomerization over Lewis acid sites in hydrophobic and hydrophilic molecular sieves. J Catal 2013;308:176–88.
[61] Dusselier M, Van Wouwe P, Dewaele A, Makshina E, Sels BF. Lactic acid as a platform chemical in the biobased economy: the role of chemocatalysis. Energy Environ Sci 2013;6:1415–42.
[62] Hayashi Y, Sasaki Y. Tin-catalyzed conversion of trioses to alkyl lactates in alcohol solution. Chem Commun 2005;2716–8.
[63] Janssen KPF, Paul JS, Sels BF, Jacobs PA. Glyoxylase biomimics: zeolite catalyzed conversion of trioses. Stud Surf Sci Catal 2007;170B:1222–7.
[64] Pescarmona PP, Janssen KPF, Delaet C, Stroobants C, Houthoofd K, Philippaerts A, et al. Zeolite-catalysed conversion of C3 sugars to alkyl lactates. Green Chem 2010;12:1083–9.
[65] West RM, Holm MS, Saravanamurugan S, Xiong J, Beversdorf Z, Taarning E, et al. Zeolite H-USY for the production of lactic acid and methyl lactate from C3-sugars. J Catal 2010;269:122–30.
[66] Ennaert T, Geboers J, Gobechiya E, Courtin CM, Kurttepeli M, Houthoofd K, et al. Conceptual frame rationalizing the self-stabilization of H-USY zeolites in hot liquid water. ACS Catal 2015;5:754–68.

[67] Taarning E, Saravanamurugan S, Holm SM, Xiong J, West RM, Christensen CH. Zeolite-catalyzed isomerization of triose sugars. ChemSusChem 2009;2:625–7.
[68] Wang J, Masui Y, Onaka M. Conversion of triose sugars with alcohols to alkyl lactates catalyzed by Brønsted acid tin ion-exchanged montmorillonite. Appl Catal B 2011;107:135–9.
[69] Li L, Stroobants C, Lin K, Jacobs PA, Sels BF, Pescarmona PP. Selective conversion of trioses to lactates over Lewis acid heterogeneous catalysts. Green Chem 2011;13:1175–81.
[70] Osmundsen CM, Holm MS, Dahl S, Taarning E. Tin-containing silicates: structure-activity relations. Proc R Soc A 2012;468:2000–16.
[71] Lew CM, Rajabbeigi N, Tsapatsis M. Tin-containing zeolite for the isomerization of cellulosic sugars. Microporous Mesoporous Mater 2012;153:55–8.
[72] Hammond C, Conrad S, Hermans I. Simple and scalable preparation of highly active Lewis acidic Sn-β. Angew Chem Int Ed 2012;51:11736–9.
[73] Guo Q, Fan F, Pidko EA, van der Graaff WNP, Feng Z, Li C, et al. Highly active and recyclable Sn-MWW zeolite catalyst for sugar conversion to methyl lactate and lactic acid. ChemSusChem 2013;6:1352–6.
[74] Dapsens PY, Mondelli C, Pérez-Ramírez J. Highly selective lewis acid sites in desilicated MFI zeolites for dihydroxyacetone isomerization to lactic acid. ChemSusChem 2013;6:831–9.
[75] Dapsens PY, Kusema BT, Mondelli C, Pérez-Ramírez J. Gallium-modified zeolites for the selective conversion of bio-based dihydroxyacetone into C1–C4 alkyl lactates. J Mol Catal A Chem 2014;388–389:141–7.
[76] Dapsens PY, Menart MJ, Mondelli C, Perez-Ramirez J. Production of bio-derived ethyl lactate on GaUSY zeolites prepared by post-synthetic galliation. Green Chem 2014;16:589–93.
[77] de Clippel F, Dusselier M, Van Rompaey R, Vanelderen P, Dijkmans J, Makshina E, et al. Fast and selective sugar conversion to alkyl lactate and lactic acid with bifunctional carbon-silica catalysts. J Am Chem Soc 2012;134:10089–101.
[78] de Clippel F, Harkiolakis A, Ke X, Vosch T, Van Tendeloo G, Baron GV, et al. Molecular sieve properties of mesoporous silica with intraporous nanocarbon. Chem Commun 2010;46:928–30.
[79] de Clippel F, Harkiolakis A, Vosch T, Ke X, Giebeler L, Oswald S, et al. Graphitic nanocrystals inside the pores of mesoporous silica: synthesis, characterization and an adsorption study. Microporous Mesoporous Mater 2011;144:120–33.
[80] de Clippel F, Khan AL, Cano-Odena A, Dusselier M, Vanherck K, Peng L, et al. CO2 reverse selective mixed matrix membranes for H2 purification by incorporation of carbon-silica fillers. J Mater Chem A 2013;1:945–53.
[81] Dijkmans J, Dusselier M, Gabriëls D, Houthoofd K, Magusin PCMM, Huang S, et al. Cooperative catalysis for multistep biomass conversion with Sn/Al beta zeolite. ACS Catal 2015;5:928–40.
[82] Holm MS, Saravanamurugan S, Taarning E. Conversion of sugars to lactic acid derivatives using heterogeneous zeotype catalysts. Science 2010;328:602–5.
[83] Holm MS, Pagan-Torres YJ, Saravanamurugan S, Riisager A, Dumesic JA, Taarning E. Sn-beta catalysed conversion of hemicellulosic sugars. Green Chem 2012;14:702–6.
[84] Lobo RF. Synthetic glycolysis. ChemSusChem 2010;3:1237–40.
[85] Dusselier M, Van Wouwe P, de Clippel F, Dijkmans J, Gammon DW, Sels BF. Mechanistic insight into the conversion of tetrose sugars to novel α-hydroxy acid platform molecules. ChemCatChem 2013;5:569–75.
[86] Dusselier M, Van Wouwe P, De Smet S, De Clercq R, Verbelen L, Van Puyvelde P, et al. Toward functional polyester building blocks from renewable glycolaldehyde with Sn cascade catalysis. ACS Catal 2013;3:1786–800.
[87] Chang C-C, Wang Z, Dornath P, Je Cho H, Fan W. Rapid synthesis of Sn-beta for the isomerization of cellulosic sugars. RSC Adv 2012;2:10475–7.
[88] Kang Z, Zhang X, Liu H, Qiu J, Han W, Yeung KL. Factors affecting the formation of Sn-beta zeolites by steam-assisted conversion method. Mater Chem Phys 2013;141:519–29.
[89] Kang Z, Zhang X, Liu H, Qiu J, Yeung KL. A rapid synthesis route for Sn-beta zeolites by steam-assisted conversion and their catalytic performance in Baeyer-Villiger oxidation. Chem Eng J 2013;218:425–32.

[90] Dijkmans J, Gabriels D, Dusselier M, de Clippel F, Vanelderen P, Houthoofd K, et al. Productive sugar isomerization with highly active Sn in dealuminated [small beta] zeolites. Green Chem 2013;15:2777–85.
[91] Li P, Liu G, Wu H, Liu Y, Jiang J-g, Wu P. Postsynthesis and selective oxidation properties of nanosized Sn-beta zeolite. J Phys Chem C 2011;115:3663–70.
[92] Mulder GJ. Untersuchungen über die Humussubstanzen. J Prakt Chem 1840;21:203–40.
[93] Düll G. Action of oxalic acid on inulin. Chem-Ztg 1895;19:216–7.
[94] Kiermayer J. A derivative of furfuraldehyde from laevulose. Chem-Ztg 1895;19:1003–5.
[95] Hu L, Zhao G, Hao W, Tang X, Sun Y, Lin L, et al. Catalytic conversion of biomass-derived carbohydrates into fuels and chemicals via furanic aldehydes. RSC Adv 2012;2:11184–206.
[96] Karinen R, Vilonen K, Niemelä M. Biorefining: heterogeneously catalyzed reactions of carbohydrates for the production of furfural and hydroxymethylfurfural. ChemSusChem 2011;4:1002–16.
[97] van Putten R-J, van der Waal JC, de Jong E, Rasrendra CB, Heeres HJ, de Vries JG. Hydroxymethylfurfural, a versatile platform chemical made from renewable resources. Chem Rev 2013;113:1499–597.
[98] Lima S, Antunes MM, Fernandes A, Pillinger M, Ribeiro MF, Valente AA. Acid-catalysed conversion of saccharides into furanic aldehydes in the presence of three-dimensional mesoporous Al-TUD-1. Molecules 2010;15:3863–77.
[99] Lourvanij K, Rorrer GL. Reactions of aqueous glucose solutions over solid-acid Y-zeolite catalyst at 110–160°C. Ind Eng Chem Res 1993;32:11–9.
[100] Lourvanij K, Rorrer GL. Reaction rates for the partial dehydration of glucose to organic acids in solid-acid, molecular-sieving catalyst powders. J Chem Technol Biotechnol 1997;69:35–44.
[101] Shimizu K-i, Uozumi R, Satsuma A. Enhanced production of hydroxymethylfurfural from fructose with solid acid catalysts by simple water removal methods. Catal Commun 2009;10:1849–53.
[102] Van de Vyver S, Thomas J, Geboers J, Keyzer S, Smet M, Dehaen W, et al. Catalytic production of levulinic acid from cellulose and other biomass-derived carbohydrates with sulfonated hyperbranched poly(arylene oxindole)s. Energy Environ Sci 2011;4:3601–10.
[103] Nakamura Y, Morikawa S. The dehydration of D-fructose to 5-hydroxymethyl-2-furaldehyde. Bull Chem Soc Jpn 1980;53:3705–6.
[104] Lew CM, Rajabbeigi N, Tsapatsis M. One-pot synthesis of 5-(ethoxymethyl)furfural from glucose using Sn-BEA and amberlyst catalysts. Ind Eng Chem Res 2012;51:5364–6.
[105] Moreau C, Durand R, Pourcheron C, Razigade S. Preparation of 5-hydroxymethylfurfural from fructose and precursors over H-form zeolites. Ind Crops Prod 1994;3:85–90.
[106] Moreau C, Durand R, Razigade S, Duhamet J, Faugeras P, Rivalier P, et al. Dehydration of fructose to 5-hydroxymethylfurfural over H-mordenites. Appl Catal A 1996;145:211–24.
[107] Ordomsky VV, van der Schaaf J, Schouten JC, Nijhuis TA. The effect of solvent addition on fructose dehydration to 5-hydroxymethylfurfural in biphasic system over zeolites. J Catal 2012;287:68–75.
[108] Nikolla E, Román-Leshkov Y, Moliner M, Davis ME. "One-pot" synthesis of 5-(hydroxymethyl)furfural from carbohydrates using tin-beta zeolite. ACS Catal 2011;1:408–10.
[109] Jow J, Rorrer GL, Hawley MC, Lamport DTA. Dehydration of d-fructose to levulinic acid over LZY zeolite catalyst. Biomass 1987;14:185–94.
[110] Moreau C, Durand R, Duhamet J, Rivalier P. Hydrolysis of fructose and glucose precursors in the presence of H-form zeolites 1. J Carbohydr Chem 1997;16:709–14.
[111] Rivalier P, Duhamet J, Moreau C, Durand R. Development of a continuous catalytic heterogeneous column reactor with simultaneous extraction of an intermediate product by an organic solvent circulating in countercurrent manner with the aqueous phase. Catal Today 1995;24:165–71.
[112] Leonard RH. Levulinic acid as a basic chemical raw material. Ind Eng Chem 1956;48:1330–41.
[113] Kruger JS, Choudhary V, Nikolakis V, Vlachos DG. Elucidating the roles of zeolite H-BEA in aqueous-phase fructose dehydration and HMF rehydration. ACS Catal 2013;3:1279–91.
[114] Lourvanij K, Rorrer GL. Dehydration of glucose to organic acids in microporous pillared clay catalysts. Appl Catal A 1994;109:147–65.

[115] Zeng W, Cheng D-G, Zhang H, Chen F, Zhan X. Dehydration of glucose to levulinic acid over MFI-type zeolite in subcritical water at moderate conditions. React Kinet Mech Catal 2010;100:377–84.

[116] Newth FH. In: Claude SH, Sidney MC, editors. Advances in carbohydrate chemistry, vol. 6. New York: Academic Press; 1951. p. 83–106.

[117] Lichtenthaler FW, editor. Ullmann's encyclopedia of industrial chemistry. John Wiley & Sons, Ltd., Weinheim; 2000.

[118] Sitthisa S, Resasco D. Hydrodeoxygenation of furfural over supported metal catalysts: a comparative study of Cu, Pd and Ni. Catal Lett 2011;141:784–91.

[119] Corma A, de la Torre O, Renz M. Production of high quality diesel from cellulose and hemicellulose by the Sylvan process: catalysts and process variables. Energy Environ Sci 2012;5:6328–44.

[120] O'Neill R, Ahmad MN, Vanoye L, Aiouache F. Kinetics of aqueous phase dehydration of xylose into furfural catalyzed by ZSM-5 zeolite. Ind Eng Chem Res 2009;48:4300–6.

[121] Weingarten R, Tompsett GA, Conner Jr WC, Huber GW. Design of solid acid catalysts for aqueous-phase dehydration of carbohydrates: the role of Lewis and Brønsted acid sites. J Catal 2011;279:174–82.

[122] Dhepe PL, Sahu R. A solid-acid-based process for the conversion of hemicellulose. Green Chem 2010;12:2153–6.

[123] Dias AS, Lima S, Pillinger M, Valente AA. Acidic cesium salts of 12-tungstophosphoric acid as catalysts for the dehydration of xylose into furfural. Carbohydr Res 2006;341:2946–53.

[124] Dias AS, Pillinger M, Valente AA. Dehydration of xylose into furfural over micro-mesoporous sulfonic acid catalysts. J Catal 2005;229:414–23.

[125] Kim S, You S, Kim Y, Lee S, Lee H, Park K, et al. Dehydration of D-xylose into furfural over H-zeolites. Korean J Chem Eng 2011;28:710–6.

[126] Moreau C, Durand R, Peyron D, Duhamet J, Rivalier P. Selective preparation of furfural from xylose over microporous solid acid catalysts. Ind Crops Prod 1998;7:95–9.

[127] Sahu R, Dhepe PL. A one-pot method for the selective conversion of hemicellulose from crop waste into C5 sugars and furfural by using solid acid catalysts. ChemSusChem 2012;5:751–61.

[128] You SJ, Park ED. Effects of dealumination and desilication of H-ZSM-5 on xylose dehydration. Microporous Mesoporous Mater 2014;186:121–9.

[129] Lima S, Fernandes A, Antunes M, Pillinger M, Ribeiro F, Valente A. Dehydration of xylose into furfural in the presence of crystalline microporous silicoaluminophosphates. Catal Lett 2010;135:41–7.

[130] Lima S, Antunes MM, Fernandes A, Pillinger M, Ribeiro MF, Valente AA. Catalytic cyclodehydration of xylose to furfural in the presence of zeolite H-beta and a micro/mesoporous Beta/TUD-1 composite material. Appl Catal A 2010;388:141–8.

[131] Lessard J, Morin J-F, Wehrung J-F, Magnin D, Chornet E. High yield conversion of residual pentoses into furfural via zeolite catalysis and catalytic hydrogenation of furfural to 2-methylfuran. Top Catal 2010;53:1231–4.

[132] Zhang J, Zhuang J, Lin L, Liu S, Zhang Z. Conversion of D-xylose into furfural with mesoporous molecular sieve MCM-41 as catalyst and butanol as the extraction phase. Biomass Bioenergy 2012;39:73–7.

[133] Shi X, Wu Y, Yi H, Rui G, Li P, Yang M, et al. Selective preparation of furfural from xylose over sulfonic acid functionalized mesoporous SBA-15 materials. Energies 2011;4:669–84.

[134] Antunes MM, Lima S, Fernandes A, Pillinger M, Ribeiro MF, Valente AA. Aqueous-phase dehydration of xylose to furfural in the presence of MCM-22 and ITQ-2 solid acid catalysts. Appl Catal A 2012;417–418:243–52.

[135] Agirrezabal-Telleria I, Larreategui A, Requies J, Güemez MB, Arias PL. Furfural production from xylose using sulfonic ion-exchange resins (Amberlyst) and simultaneous stripping with nitrogen. Bioresour Technol 2011;102:7478–85.

[136] Dias AS, Lima S, Carriazo D, Rives V, Pillinger M, Valente AA. Exfoliated titanate, niobate and titanoniobate nanosheets as solid acid catalysts for the liquid-phase dehydration of d-xylose into furfural. J Catal 2006;244:230–7.

[137] Dias AS, Pillinger M, Valente AA. Mesoporous silica-supported 12-tungstophosphoric acid catalysts for the liquid phase dehydration of D-xylose. Microporous Mesoporous Mater 2006;94:214–25.

[138] García-Sancho C, Agirrezabal-Telleria I, Güemez MB, Maireles-Torres P. Dehydration of d-xylose to furfural using different supported niobia catalysts. Appl Catal B 2014;152–153:1–10.
[139] Lima S, Pillinger M, Valente AA. Dehydration of D-xylose into furfural catalysed by solid acids derived from the layered zeolite Nu-6(1). Catal Commun 2008;9:2144–8.
[140] Pholjaroen B, Li N, Wang Z, Wang A, Zhang T. Dehydration of xylose to furfural over niobium phosphate catalyst in biphasic solvent system. J Energy Chem 2013;22:826–32.
[141] Sádaba I, Lima S, Valente AA, López Granados M. Catalytic dehydration of xylose to furfural: vanadyl pyrophosphate as source of active soluble species. Carbohydr Res 2011;346:2785–91.
[142] Suzuki T, Yokoi T, Otomo R, Kondo JN, Tatsumi T. Dehydration of xylose over sulfated tin oxide catalyst: influences of the preparation conditions on the structural properties and catalytic performance. Appl Catal A 2011;408:117–24.
[143] Daengprasert W, Boonnoun P, Laosiripojana N, Goto M, Shotipruk A. Application of sulfonated carbon-based catalyst for solvothermal conversion of cassava waste to hydroxymethylfurfural and furfural. Ind Eng Chem Res 2011;50:7903–10.
[144] Gürbüz EI, Gallo JMR, Alonso DM, Wettstein SG, Lim WY, Dumesic JA. Conversion of hemicellulose into furfural using solid acid catalysts in γ-valerolactone. Angew Chem Int Ed 2013;52:1270–4.
[145] Lutz W, Toufar H, Kurzhals R, Suckow M. Investigation and modeling of the hydrothermal stability of technically relevant zeolites. Adsorption 2005;11:405–13.
[146] Ravenelle RM, Schüβler F, D'Amico A, Danilina N, van Bokhoven JA, Lercher JA, et al. Stability of zeolites in hot liquid water. J Phys Chem C 2010;114:19582–95.
[147] Zapata PA, Faria J, Ruiz MP, Jentoft RE, Resasco DE. Hydrophobic zeolites for biofuel upgrading reactions at the liquid-liquid interface in water/oil emulsions. J Am Chem Soc 2012;134:8570–8.
[148] Zapata PA, Huang Y, Gonzalez-Borja MA, Resasco DE. Silylated hydrophobic zeolites with enhanced tolerance to hot liquid water. J Catal 2013;308:82–7.
[149] Lange J-P, van de Graaf WD, Haan RJ. Conversion of furfuryl alcohol into ethyl levulinate using solid acid catalysts. ChemSusChem 2009;2:437–41.
[150] Bui L, Luo H, Gunther WR, Román-Leshkov Y. Domino reaction catalyzed by zeolites with Brønsted and Lewis acid sites for the production of γ-valerolactone from furfural. Angew Chem Int Ed 2013;52:8022–5.
[151] Huber GW, Chheda JN, Barrett CJ, Dumesic JA. Production of liquid alkanes by aqueous-phase processing of biomass-derived carbohydrates. Science 2005;308:1446–50.
[152] Crezee E, Hoffer BW, Berger RJ, Makkee M, Kapteijn F, Moulijn JA. Three-phase hydrogenation of D-glucose over a carbon supported ruthenium catalyst—mass transfer and kinetics. Appl Catal A 2003;251:1–17.
[153] Hoffer BW, Crezee E, Mooijman PRM, van Langeveld AD, Kapteijn F, Moulijn JA. Carbon supported Ru catalysts as promising alternative for Raney-type Ni in the selective hydrogenation of D-glucose. Catal Today 2003;79–80:35–41.
[154] Eisenbeis C, Guettel R, Kunz U, Turek T. Monolith loop reactor for hydrogenation of glucose. Catal Today 2009;147:S342–6.
[155] Martens JA, De Vos DE, Kirschhock CEA, Pescarmona PP, Sels BF, Vankelecom IFJ. Forty years of designing catalytic and adsorptive sites in FAU type zeolites at K.U. Leuven. Top Catal 2009;52:1119–30.
[156] Sartipi S, Makkee M, Kapteijn F, Gascon J. Catalysis engineering of bifunctional solids for the one-step synthesis of liquid fuels from syngas: a review. Catal Sci Technol 2014;4:893–907.
[157] Zhang J, Lin L, Zhang J, Shi J. Efficient conversion of D-glucose into D-sorbitol over MCM-41 supported Ru catalyst prepared by a formaldehyde reduction process. Carbohydr Res 2011;346:1327–32.
[158] Mishra DK, Dabbawala AA, Park JJ, Jhung SH, Hwang J-S. Selective hydrogenation of D-glucose to D-sorbitol over HY zeolite supported ruthenium nanoparticles catalysts. Catal Today 2014;232:99–107.
[159] Mishra DK, Dabbawala AA, Hwang J-S. Ruthenium nanoparticles supported on zeolite Y as an efficient catalyst for selective hydrogenation of xylose to xylitol. J Mol Catal A Chem 2013;376:63–70.
[160] Van de Vyver S, Peng L, Geboers J, Schepers H, de Clippel F, Gommes CJ, et al. Sulfonated silica/carbon nanocomposites as novel catalysts for hydrolysis of cellulose to glucose. Green Chem 2010;12:1560–3.

[161] Zhang Z, Zhao ZK. Solid acid and microwave-assisted hydrolysis of cellulose in ionic liquid. Carbohydr Res 2009;344:2069–72.
[162] Fukuoka A, Dhepe PL. Catalytic conversion of cellulose into sugar alcohols. Angew Chem Int Ed 2006;45:5161–3.
[163] Geboers J, Van de Vyver S, Carpentier K, de Blochouse K, Jacobs P, Sels B. Efficient catalytic conversion of concentrated cellulose feeds to hexitols with heteropoly acids and Ru on carbon. Chem Commun 2010;46:3577–9.
[164] Palkovits R, Tajvidi K, Procelewska J, Rinaldi R, Ruppert A. Hydrogenolysis of cellulose combining mineral acids and hydrogenation catalysts. Green Chem 2010;12:972–8.
[165] Palkovits R, Tajvidi K, Ruppert AM, Procelewska J. Heteropoly acids as efficient acid catalysts in the one-step conversion of cellulose to sugar alcohols. Chem Commun 2011;47:576–8.
[166] Van de Vyver S, Geboers J, Dusselier M, Schepers H, Vosch T, Zhang L, et al. Selective bifunctional catalytic conversion of cellulose over reshaped Ni particles at the tip of carbon nanofibers. ChemSusChem 2010;3:698–701.
[167] Zheng M-Y, Wang A-Q, Ji N, Pang J-F, Wang X-D, Zhang T. Transition metal-tungsten bimetallic catalysts for the conversion of cellulose into ethylene glycol. ChemSusChem 2010;3:63–6.
[168] Op de Beeck B, Geboers J, Van de Vyver S, Van Lishout J, Snelders J, Huijgen WJJ, et al. Conversion of (ligno)cellulose feeds to isosorbide with heteropoly acids and Ru on carbon. ChemSusChem 2013;6:199–208.
[169] Sharkov VI. Production of polyhydric alcohols from wood polysaccharides. Angew Chem Int Ed 1963;2:405–9.
[170] Balandin A, Vasyunina N, Chepigo S, Barysheva G. Hydrolytic hydrogenation of cellulose. Dokl Akad Nauk SSSR 1959;128:941–4.
[171] Sabadie J, Barthomeuf D, Charcosset H, Descotes G. Catalytic hydrogenolysis of oligosaccharides. Bull Soc Chim Fr 1981;288–92.
[172] Jacobs PA, Hinnekens H. Single-step catalytic process for the direct conversion of polysaccharides to polyhydric alcohols. EP0329923; 1989.
[173] Barbosa CMBM, Falabella E, Mendes MJ, Lima NM, Abreu CAM. Kinetic evaluation of hydrogenation of sucrose over ruthenium containing Y zeolites. React Kinet Catal Lett 1999;68:291–8.
[174] Liang G, Cheng H, Li W, He L, Yu Y, Zhao F. Selective conversion of microcrystalline cellulose into hexitols on nickel particles encapsulated within ZSM-5 zeolite. Green Chem 2012;14:2146–9.
[175] Liang G, He L, Cheng H, Li W, Li X, Zhang C, et al. The hydrogenation/dehydrogenation activity of supported Ni catalysts and their effect on hexitols selectivity in hydrolytic hydrogenation of cellulose. J Catal 2014;309:468–76.
[176] Negoi A, Triantafyllidis K, Parvulescu VI, Coman SM. The hydrolytic hydrogenation of cellulose to sorbitol over M (Ru, Ir, Pd, Rh)-BEA-zeolite catalysts. Catal Today 2014;223:122–8.
[177] Shrotri A, Tanksale A, Beltramini JN, Gurav H, Chilukuri SV. Conversion of cellulose to polyols over promoted nickel catalysts. Catal Sci Technol 2012;2:1852–8.
[178] Geboers J, Van de Vyver S, Carpentier K, Jacobs PA, Sels BF. Efficient hydrolytic hydrogenation of cellulose in the presence of Ru-loaded zeolites and trace amounts of mineral acid. Chem Commun 2011;47:5590–2.
[179] Liu S, Tamura M, Nakagawa Y, Tomishige K. One-pot conversion of cellulose into *n*-hexane over the Ir-ReO_x/SiO_2 catalyst combined with HZSM-5. ACS Sustainable Chem Eng 2014;2:1819–27.
[180] Op de Beeck B, Dusselier M, Geboers J, Holsbeek J, Morre E, Oswald S, et al. Direct catalytic conversion of cellulose to liquid straight-chain alkanes. Energy Environ Sci 2015;8:230–40.
[181] Buhl JC, Gerstmann M, Lutz W, Ritzmann A. Hydrothermal stability of the novel zeolite type LSX in comparison to the traditional 13X modification. Z Anorg Allg Chem 2004;630:604–8.
[182] Cara PD, Pagliaro M, Elmekawy A, Brown DR, Verschuren P, Shiju NR, et al. Hemicellulose hydrolysis catalysed by solid acids. Catal Sci Technol 2013;3:2057–61.
[183] Vilcocq L, Castilho PC, Carvalheiro F, Duarte LC. Hydrolysis of oligosaccharides over solid acid catalysts: a review. ChemSusChem 2014;7:1010–9.

[184] Zhou L, Shi M, Cai Q, Wu L, Hu X, Yang X, et al. Hydrolysis of hemicellulose catalyzed by hierarchical H-USY zeolites—the role of acidity and pore structure. Microporous Mesoporous Mater 2013;169:54–9.
[185] Zhou L, Liu Z, Shi M, Du S, Su Y, Yang X, et al. Sulfonated hierarchical H-USY zeolite for efficient hydrolysis of hemicellulose/cellulose. Carbohydr Polym 2013;98:146–51.
[186] Faba L, Kusema BT, Murzina EV, Tokarev A, Kumar N, Smeds A, et al. Hemicellulose hydrolysis and hydrolytic hydrogenation over proton- and metal modified beta zeolites. Microporous Mesoporous Mater 2014;189:189–99.
[187] Käldström M, Kumar N, Tenho M, Mokeev MV, Moskalenko YE, Murzin DY. Catalytic transformations of birch kraft pulp. ACS Catal 2012;2:1381–93.
[188] Kusema BT, Faba L, Kumar N, Mäki-Arvela P, Díaz E, Ordóñez S, et al. Hydrolytic hydrogenation of hemicellulose over metal modified mesoporous catalyst. Catal Today 2012;196:26–33.
[189] Katryniok B, Paul S, Capron M, Dumeignil F. Towards the sustainable production of acrolein by glycerol dehydration. ChemSusChem 2009;2:719–30.
[190] Pagliaro M, Ciriminna R, Kimura H, Rossi M, Della Pina C. From glycerol to value-added products. Angew Chem Int Ed 2007;46:4434–40.
[191] Sels B, D'Hondt E, Jacobs P. In: Centi G, Van Santen RA, editors. Catalysis for renewables. Wiley-VCH Verlag GmbH & Co. KGaA, Weinheim; 2007. p. 223–55.
[192] Groll H, Hearne G. Process of converting a polyhydric alcohol to a carbonyl compound. US2042224; 1936.
[193] Hoyt HE, Manninen TH. Production of acrolein from glycerol. US2558520; 1951.
[194] Lehr V, Sarlea M, Ott L, Vogel H. Catalytic dehydration of biomass-derived polyols in sub- and supercritical water. Catal Today 2007;121:121–9.
[195] Ott L, Bicker M, Vogel H. Catalytic dehydration of glycerol in sub- and supercritical water: a new chemical process for acrolein production. Green Chem 2006;8:214–20.
[196] Ramayya S, Brittain A, De Almeida C, Mok W, Antal Jr MJ. Acid-catalysed dehydration of alcohols in supercritical water. Fuel 1987;66:1364–71.
[197] Watanabe M, Iida T, Aizawa Y, Aida TM, Inomata H. Acrolein synthesis from glycerol in hot-compressed water. Bioresour Technol 2007;98:1285–90.
[198] de Oliveira AS, Vasconcelos SJS, de Sousa JR, de Sousa FF, Filho JM, Oliveira AC. Catalytic conversion of glycerol to acrolein over modified molecular sieves: activity and deactivation studies. Chem Eng J 2011;168:765–74.
[199] Dalla Costa BO, Peralta MA, Querini CA. Gas phase dehydration of glycerol over, lanthanum-modified beta-zeolite. Appl Catal A 2014;472:53–63.
[200] Dubois JL, Duquenne C, Hoelderlich W. Process for dehydrating glycerol to acrolein. WO 2006087083; 2006.
[201] Jia C-J, Liu Y, Schmidt W, Lu A-H, Schüth F. Small-sized HZSM-5 zeolite as highly active catalyst for gas phase dehydration of glycerol to acrolein. J Catal 2010;269:71–9.
[202] Kasuga H, Kirishiki M, Matsunami E, Okada M, Okuno M, Takahashi T. Production method of acrolein. WO2007132926; 2007.
[203] Kasuga H, Okada M. Method for producing acrolein from glycerol. JP2008137950; 2008.
[204] Kim YT, Jung K-D, Park ED. Gas-phase dehydration of glycerol over ZSM-5 catalysts. Microporous Mesoporous Mater 2010;131:28–36.
[205] Li X-Z. Process for preparing acrolein by glycerin dewatering. CN101070276; 2007.
[206] Zhao H, Zhou CH, Wu LM, Lou JY, Li N, Yang HM, et al. Catalytic dehydration of glycerol to acrolein over sulfuric acid-activated montmorillonite catalysts. Appl Clay Sci 2013;74:154–62.
[207] Zhou C-J, Huang C-J, Zhang W-G, Zhai H-S, Wu H-L, Chao Z-S. In: Dongyuan Zhao SQYT, Chengzhong Y, editors. Studies in surface science and catalysis, vol. 165. Amsterdam: Elsevier; 2007. p. 527–30.
[208] Zhuang A, Zhang C, Wen S, Zhao X, Wu T. Method for preparing acroleic acid by using glycerol as raw material. CN101225039; 2008.
[209] Hoang TQ, Zhu XL, Danuthai T, Lobban LL, Resasco DE, Mallinson RG. Conversion of glycerol to alkyl-aromatics over zeolites. Energy Fuel 2010;24:3804–9.

[210] Corma A, Huber GW, Sauvanaud L, O'Connor P. Biomass to chemicals: catalytic conversion of glycerol/water mixtures into acrolein, reaction network. J Catal 2008;257:163–71.
[211] Nakagawa Y, Tomishige K. Heterogeneous catalysis of the glycerol hydrogenolysis. Catal Sci Technol 2011;1:179–90.
[212] D'Hondt E, Van de Vyver S, Sels BF, Jacobs PA. Catalytic glycerol conversion into 1,2-propanediol in absence of added hydrogen. Chem Commun 2008;6011–2.
[213] Van de Vyver S, D'Hondt E, Sels BF, Jacobs PA. Preparation of Pt on NaY zeolite catalysts for conversion of glycerol into 1,2-propanediol. Stud Surf Sci Catal 2010;175:771–4.
[214] Pazhavelikkakath Purushothaman RK, van Haveren J, Melián-Cabrera I, van Eck ERH, Heeres HJ. Base-free, one-pot chemocatalytic conversion of glycerol to methyl lactate using supported gold catalysts. ChemSusChem 2014;7:1140–7.
[215] Brandner A, Lehnert K, Bienholz A, Lucas M, Claus P. Production of biomass-derived chemicals and energy: chemocatalytic conversions of glycerol. Top Catal 2009;52:278–87.
[216] Carrettin S, McMorn P, Johnston P, Griffin K, Kiely CJ, Hutchings GJ. Oxidation of glycerol using supported Pt, Pd and Au catalysts. Phys Chem Chem Phys 2003;5:1329–36.
[217] Garcia R, Besson M, Gallezot P. Chemoselective catalytic oxidation of glycerol with air on platinum metals. Appl Catal A 1995;127:165–76.
[218] Ketchie WC, Fang Y-L, Wong MS, Murayama M, Davis RJ. Influence of gold particle size on the aqueous-phase oxidation of carbon monoxide and glycerol. J Catal 2007;250:94–101.
[219] Kimura H, Tsuto K, Wakisaka T, Kazumi Y, Inaya Y. Selective oxidation of glycerol on a platinum-bismuth catalyst. Appl Catal A 1993;96:217–28.
[220] Porta F, Prati L. Selective oxidation of glycerol to sodium glycerate with gold-on-carbon catalyst: an insight into reaction selectivity. J Catal 2004;224:397–403.
[221] Jérôme F, Pouilloux Y, Barrault J. Rational design of solid catalysts for the selective use of glycerol as a natural organic building block. ChemSusChem 2008;1:586–613.
[222] Liang D, Gao J, Wang J, Chen P, Hou Z, Zheng X. Selective oxidation of glycerol in a base-free aqueous solution over different sized Pt catalysts. Catal Commun 2009;10:1586–90.
[223] Luque R, Budarin V, Clark JH, Macquarrie DJ. Glycerol transformations on polysaccharide derived mesoporous materials. Appl Catal B 2008;82:157–62.
[224] Zhou C-H, Beltramini JN, Fan Y-X, Lu GQ. Chemoselective catalytic conversion of glycerol as a biorenewable source to valuable commodity chemicals. Chem Soc Rev 2008;37:527–49.
[225] Corma A, Iborra S, Velty A. Chemical routes for the transformation of biomass into chemicals. Chem Rev 2007;107:2411–502.
[226] Carrettin S, McMorn P, Johnston P, Griffin K, Hutchings GJ. Selective oxidation of glycerol to glyceric acid using a gold catalyst in aqueous sodium hydroxide. Chem Commun 2002;696–7.
[227] Demirel-Gülen S, Lucas M, Claus P. Liquid phase oxidation of glycerol over carbon supported gold catalysts. Catal Today 2005;102–103:166–72.
[228] Villa A, Veith GM, Prati L. Selective oxidation of glycerol under acidic conditions using gold catalysts. Angew Chem Int Ed 2010;49:4499–502.
[229] Mohan D, Pittman Jr CU, Steele PH. Pyrolysis of wood/biomass for bio-oil: a critical review. Energy Fuels 2006;20:848–89.
[230] Tan S, Zhang Z, Sun J, Wang Q. Recent progress of catalytic pyrolysis of biomass by HZSM-5. Chin J Catal 2013;34:641–50.
[231] Stocker M. Biofuels and biomass-to-liquid fuels in the biorefinery: catalytic conversion of lignocellulosic biomass using porous materials. Angew Chem Int Ed 2008;47:9200–11.
[232] Kubicka D, Kubickova I, Cejka J. Application of molecular sieves in transformations of biomass and biomass-derived feedstocks. Catal Rev Sci Eng 2013;55:1–78.
[233] Chen NY, Walsh DE, Koenig LR. Fluidized-bed upgrading of wood pyrolysis liquids and related-compounds. ACS Symp Ser 1988;376:277–89.
[234] Carlson TR, Vispute TR, Huber GW. Green gasoline by catalytic fast pyrolysis of solid biomass derived compounds. ChemSusChem 2008;1:397–400.

[235] Zhang H, Cheng Y-T, Vispute TP, Xiao R, Huber GW. Catalytic conversion of biomass-derived feedstocks into olefins and aromatics with ZSM-5: the hydrogen to carbon effective ratio. Energy Environ Sci 2011;4:2297–307.
[236] Dao LH, Haniff M, Houle A, Lamothe D. Reactions of model compounds of biomass-pyrolysis oils over ZSM-5 zeolite catalysts. ACS Symp Ser 1988;376:328–41.
[237] Mentzel UV, Holm MS. Utilization of biomass: conversion of model compounds to hydrocarbons over zeolite H-ZSM-5. Appl Catal A 2011;396:59–67.
[238] Adjaye JD, Bakhshi NN. Upgrading of a wood-derived oil over various catalysts. Biomass Bioenergy 1994;7:201–11.
[239] Adjaye JD, Bakhshi NN. Production of hydrocarbons by catalytic upgrading a fast pyrolysis bio-oil. 1. Conversion over various catalysts. Fuel Process Technol 1995;45:161–83.
[240] Sharma RK, Bakhshi NN. Catalytic upgrading of pyrolysis oil. Energy Fuels 1993;7:306–14.
[241] Adjaye JD, Bakhshi NN. Production of hydrocarbons by catalytic upgrading of a fast pyrolysis bio-oil. 2. Comparative catalyst performance and reaction pathways. Fuel Process Technol 1995;45:185–202.
[242] Vitolo S, Bresci B, Seggiani M, Gallo MG. Catalytic upgrading of pyrolytic oils over HZSM-5 zeolite: behaviour of the catalyst when used in repeated upgrading-regenerating cycles. Fuel 2001;80:17–26.
[243] Gayubo AG, Aguayo AT, Atutxa A, Prieto R, Bilbao J. Deactivation of a HZSM-5 zeolite catalyst in the transformation of the aqueous fraction of biomass pyrolysis oil into hydrocarbons. Energy Fuels 2004;18:1640–7.
[244] Gayubo AG, Aguayo AT, Atutxa A, Valle B, Bilbao J. Undesired components in the transformation of biomass pyrolysis oil into hydrocarbons on an HZSM-5 zeolite catalyst. J Chem Technol Biotechnol 2005;80:1244–51.
[245] Gayubo AG, Valle B, Aguayo AT, Olazar M, Bilbao J. Attenuation of catalyst deactivation by cofeeding methanol for enhancing the valorisation of crude bio-oil. Energy Fuels 2009;23:4129–36.
[246] Gayubo AG, Valle B, Aguayo AT, Olazar M, Bilbao J. Pyrolytic lignin removal for the valorization of biomass pyrolysis crude bio-oil by catalytic transformation. J Chem Technol Biotechnol 2010;85:132–44.
[247] Vispute TP, Zhang H, Sanna A, Xiao R, Huber GW. Renewable chemical commodity feedstocks from integrated catalytic processing of pyrolysis oils. Science 2010;330:1222–7.
[248] Lappas AA, Bezergianni S, Vasalos IA. Production of biofuels via co-processing in conventional refining processes. Catal Today 2009;145:55–62.
[249] Mortensen PM, Grunwaldt JD, Jensen PA, Knudsen KG, Jensen AD. A review of catalytic upgrading of bio-oil to engine fuels. Appl Catal A 2011;407:1–19.
[250] Wang H, Male J, Wang Y. Recent advances in hydrotreating of pyrolysis bio-oil and its oxygen-containing model compounds. ACS Catal 2013;3:1047–70.
[251] Saidi M, Samimi F, Karimipourfard D, Nimmanwudipong T, Gates BC, Rahimpour MR. Upgrading of lignin-derived bio-oils by catalytic hydrodeoxygenation. Energy Environ Sci 2014;7:103–29.
[252] Ruddy DA, Schaidle JA, Ferrell JRI, Wang J, Moens L, Hensley JE. Recent advances in heterogeneous catalysts for bio-oil upgrading via "ex situ catalytic fast pyrolysis": catalyst development through the study of model compounds. Green Chem 2014;16:454–90.
[253] Hong D-Y, Miller SJ, Agrawal PK, Jones CW. Hydrodeoxygenation and coupling of aqueous phenolics over bifunctional zeolite-supported metal catalysts. Chem Commun 2010;46:1038–40.
[254] Zhu X, Lobban LL, Mallinson RG, Resasco DE. Bifunctional transalkylation and hydrodeoxygenation of anisole over a Pt/HBeta catalyst. J Catal 2011;281:21–9.
[255] Zhao C, Kou Y, Lemonidou AA, Li X, Lercher JA. Highly selective catalytic conversion of phenolic bio-oil to alkanes. Angew Chem Int Ed 2009;48:3987–90.
[256] Zhao C, He J, Lemonidou AA, Li X, Lercher JA. Aqueous-phase hydrodeoxygenation of bio-derived phenols to cycloalkanes. J Catal 2011;280:8–16.
[257] Zhao C, Kou Y, Lemonidou AA, Li X, Lercher JA. Hydrodeoxygenation of bio-derived phenols to hydrocarbons using RANEY (R) Ni and Nafion/SiO_2 catalysts. Chem Commun 2010;46:412–4.
[258] Zhao C, Lercher JA. Selective hydrodeoxygenation of lignin-derived phenolic monomers and dimers to cycloalkanes on Pd/C and HZSM-5 catalysts. ChemCatChem 2012;4:64–8.

[259] Zhao C, Camaioni DM, Lercher JA. Selective catalytic hydroalkylation and deoxygenation of substituted phenols to bicycloalkanes. J Catal 2012;288:92–103.
[260] Zhao C, Song W, Lercher JA. Aqueous phase hydroalkylation and hydrodeoxygenation of phenol by dual functional catalysts comprised of Pd/C and H/La-BEA. ACS Catal 2012;2:2714–23.
[261] Zhao C, Lercher JA. Upgrading pyrolysis oil over Ni/HZSM-5 by cascade reactions. Angew Chem Int Ed 2012;51:5935–40.
[262] Zhao C, Yu Y, Jentys A, Lercher JA. Understanding the impact of aluminum oxide binder on Ni/HZSM-5 for phenol hydrodeoxygenation. Appl Catal B 2013;132:282–92.
[263] Zhao C, Kasakov S, He J, Lercher JA. Comparison of kinetics, activity and stability of Ni/HZSM-5 and Ni/Al2O3-HZSM-5 for phenol hydrodeoxygenation. J Catal 2012;296:12–23.
[264] Vjunov A, Hu MY, Feng J, Camaioni DM, Mei D, Hu JZ, et al. Following solid-acid-catalyzed reactions by MAS NMR spectroscopy in liquid phase-zeolite-catalyzed conversion of cyclohexanol in water. Angew Chem Int Ed 2014;53:479–82.
[265] Diebold J, Scahill J. Biomass to gasoline—upgrading pyrolysis vapors to aromatic gasoline with zeolite catalysis at atmospheric-pressure. ACS Symp Ser 1988;376:264–76.
[266] Williams PT, Horne PA. The influence of catalyst regeneration on the composition of zeolite-upgraded biomass pyrolysis oils. Fuel 1995;74:1839–51.
[267] Horne PA, Williams PT. The effect of zeolite ZSM-5 catalyst deactivation during the upgrading of biomass-derived pyrolysis vapors. J Anal Appl Pyrolysis 1995;34:65–85.
[268] Horne PA, Nugranad N, Williams PT. Catalytic coprocessing of biomass-derived pyrolysis vapors and methanol. J Anal Appl Pyrolysis 1995;34:87–108.
[269] Williams PT, Horne PA. The influence of catalyst type on the composition of upgraded biomass pyrolysis oils. J Anal Appl Pyrolysis 1995;31:39–61.
[270] Gayubo AG, Aguayo AT, Atutxa A, Aguado R, Bilbao J. Transformation of oxygenate components of biomass pyrolysis oil on a HZSM-5 zeolite. I. Alcohols and phenols. Ind Eng Chem Res 2004;43:2610–8.
[271] Horne PA, Williams PT. Reaction of oxygenated biomass pyrolysis model compounds over a ZSM-5 catalyst. Renew Energy 1996;7:131–44.
[272] Pattiya A, Titiloye JO, Bridgwater AV. Fast pyrolysis of cassava rhizome in the presence of catalysts. J Anal Appl Pyrolysis 2008;81:72–9.
[273] Mihalcik DJ, Mullen CA, Boateng AA. Screening acidic zeolites for catalytic fast pyrolysis of biomass and its components. J Anal Appl Pyrolysis 2011;92:224–32.
[274] Stefanidis SD, Kalogiannis KG, Iliopoulou EF, Lappas AA, Pilavachi PA. In-situ upgrading of biomass pyrolysis vapors: catalyst screening on a fixed bed reactor. Bioresour Technol 2011;102:8261–7.
[275] Park HJ, Dong J-I, Jeon J-K, Yoo K-S, Yim J-H, Sohn JM, et al. Conversion of the pyrolytic vapor of radiata pine over zeolites. Ind Eng Chem 2007;13:182–9.
[276] Lee HI, Park HJ, Park Y-K, Hur JY, Jeon J-K, Kim JM. Synthesis of highly stable mesoporous aluminosilicates from commercially available zeolites and their application to the pyrolysis of woody biomass. Catal Today 2008;132:68–74.
[277] Park HJ, Heo HS, Jeon J-K, Kim J, Ryoo R, Jeong K-E, et al. Highly valuable chemicals production from catalytic upgrading of radiata pine sawdust-derived pyrolytic vapors over mesoporous MFI zeolites. Appl Catal B 2010;95:365–73.
[278] Antonakou E, Lappas A, Nilsen MH, Bouzga A, Stocker M. Evaluation of various types of Al-MCM-41 materials as catalysts in biomass pyrolysis for the production of bio-fuels and chemicals. Fuel 2006;85:2202–12.
[279] Iliopoulou EF, Antonakou EV, Karakoulia SA, Vasalos IA, Lappas AA, Triantafyllidis KS. Catalytic conversion of biomass pyrolysis products by mesoporous materials: effect of steam stability and acidity of Al-MCM-41 catalysts. Chem Eng J 2007;134:51–7.
[280] Triantafyllidis KS, Iliopoulou EF, Antonakou EV, Lappas AA, Wang H, Pinnavaia TJ. Hydrothermally stable mesoporous aluminosilicates (MSU-S) assembled from zeolite seeds as catalysts for biomass pyrolysis. Microporous Mesoporous Mater 2007;99:132–9.

[281] Park HJ, Park K-H, Jeon J-K, Kim J, Ryoo R, Jeong K-E, et al. Production of phenolics and aromatics by pyrolysis of miscanthus. Fuel 2012;97:379–84.
[282] Aho A, Kumar N, Lashkul AV, Eranen K, Ziolek M, Decyk P, et al. Catalytic upgrading of woody biomass derived pyrolysis vapours over iron modified zeolites in a dual-fluidized bed reactor. Fuel 2010;89:1992–2000.
[283] Iliopoulou EF, Stefanidis SD, Kalogiannis KG, Delimitis A, Lappas AA, Triantafyllidis KS. Catalytic upgrading of biomass pyrolysis vapors using transition metal-modified ZSM-5 zeolite. Appl Catal B 2012;127:281–90.
[284] Iliopoulou EF, Stefanidis S, Kalogiannis K, Psarras AC, Delimitis A, Triantafyllidis KS, et al. Pilot-scale validation of Co-ZSM-5 catalyst performance in the catalytic upgrading of biomass pyrolysis vapours. Green Chem 2014;16:662–74.
[285] Carlson TR, Tompsett GA, Conner WC, Huber GW. Aromatic production from catalytic fast pyrolysis of biomass-derived feedstocks. Top Catal 2009;52:241–52.
[286] Wang K, Kim KH, Brown RC. Catalytic pyrolysis of individual components of lignocellulosic biomass. Green Chem 2014;16:727–35.
[287] Zhang H, Xiao R, Huang H, Xiao G. Comparison of non-catalytic and catalytic fast pyrolysis of corncob in a fluidized bed reactor. Bioresour Technol 2009;100:1428–34.
[288] Jae J, Tompsett GA, Foster AJ, Hammond KD, Auerbach SM, Lobo RF, et al. Investigation into the shape selectivity of zeolite catalysts for biomass conversion. J Catal 2011;279:257–68.
[289] Foster AJ, Jae J, Cheng Y-T, Huber GW, Lobo RF. Optimizing the aromatic yield and distribution from catalytic fast pyrolysis of biomass over ZSM-5. Appl Catal A 2012;423:154–61.
[290] Li J, Li X, Zhou G, Wang W, Wang C, Komarneni S, et al. Catalytic fast pyrolysis of biomass with mesoporous ZSM-5 zeolites prepared by desilication with NaOH solutions. Appl Catal A 2014;470:115–22.
[291] Neumann GT, Hicks JC. Novel hierarchical cerium-incorporated MFI zeolite catalysts for the catalytic fast pyrolysis of lignocellulosic biomass. ACS Catal 2012;2:642–6.
[292] Cheng Y-T, Jae J, Shi J, Fan W, Huber GW. Production of renewable aromatic compounds by catalytic fast pyrolysis of lignocellulosic biomass with bifunctional Ga/ZSM-5 catalysts. Angew Chem Int Ed 2012;51:1387–90.
[293] Huang W, Gong F, Fan M, Zhai Q, Hong C, Li Q. Production of light olefins by catalytic conversion of lignocellulosic biomass with HZSM-5 zeolite impregnated with 6 wt.% lanthanum. Bioresour Technol 2012;121:248–55.
[294] Hong C, Gong F, Fan M, Zhai Q, Huang W, Wang T, et al. Selective production of green light olefins by catalytic conversion of bio-oil with Mg/HZSM-5 catalyst. J Chem Technol Biotechnol 2013;88:109–18.
[295] Carlson TR, Jae J, Lin Y-C, Tompsett GA, Huber GW. Catalytic fast pyrolysis of glucose with HZSM-5: the combined homogeneous and heterogeneous reactions. J Catal 2010;270:110–24.
[296] Zhang H, Carlson TR, Xiao R, Huber GW. Catalytic fast pyrolysis of wood and alcohol mixtures in a fluidized bed reactor. Green Chem 2012;14:98–110.
[297] http://www.anellotech.com/.
[298] Bond JQ, Upadhye AA, Olcay H, Tompsett GA, Jae J, Xing R, et al. Production of renewable jet fuel range alkanes and commodity chemicals from integrated catalytic processing of biomass. Energy Environ Sci 2014;7:1500–23.
[299] Thring RW, Katikaneni SPR, Bakhshi NN. The production of gasoline range hydrocarbons from Alcell (R) lignin using HZSM-5 catalyst. Fuel Process Technol 2000;62:17–30.
[300] Zhao Y, Deng L, Liao B, Fu Y, Guo Q-X. Aromatics production via catalytic pyrolysis of pyrolytic lignins from bio-oil. Energy Fuels 2010;24:5735–40.
[301] Jackson MA, Compton DL, Boateng AA. Screening heterogeneous catalysts for the pyrolysis of lignin. J Anal Appl Pyrolysis 2009;85:226–30.
[302] Mullen CA, Boateng AA. Catalytic pyrolysis-GC/MS of lignin from several sources. Fuel Process Technol 2010;91:1446–58.
[303] Li X, Su L, Wang Y, Yu Y, Wang C, Li X, et al. Catalytic fast pyrolysis of Kraft lignin with HZSM-5 zeolite for producing aromatic hydrocarbons. Front Environ Sci Eng 2012;6:295–303.

[304] Ben H, Ragauskas AJ. One step thermal conversion of lignin to the gasoline range liquid products by using zeolites as additives. RSC Adv 2012;2:12892–8.
[305] Ma Z, Troussard E, van Bokhoven JA. Controlling the selectivity to chemicals from lignin via catalytic fast pyrolysis. Appl Catal A 2012;423:130–6.
[306] Yu Y, Li X, Su L, Zhang Y, Wang Y, Zhang H. The role of shape selectivity in catalytic fast pyrolysis of lignin with zeolite catalysts. Appl Catal A 2012;447:115–23.
[307] Patterson HBW. Hydrogenation of fats and oils: theory and practice. Urbana, IL: AOCS Press; 1994.
[308] Mozaffarian D, Katan MB, Ascherio A, Stampfer MJ, Willett WC. Trans fatty acids and cardiovascular disease. N Engl J Med 2006;354:1601–13.
[309] Labelling of foodstuffs, European Rule n° 1169/2011 Patent; 2011. p. 18–63.
[310] Philippaerts A, Jacobs PA, Sels BF. Is there still a future for hydrogenated vegetable oils? Angew Chem Int Ed 2013;52:5220–6.
[311] Chu BS, Ghazali HM, Lai OM, Che Man YB, Yusof S, Tee SB, et al. Comparison of lipase-transesterified blend with some commercial solid frying shortenings in Malaysia. J Am Oil Chem Soc 2001;78:1213–9.
[312] Khatoon S, Reddy SRY. Plastic fats with zero trans fatty acids by interesterification of mango, mahua and palm oils. Eur J Lipid Sci Technol 2005;107:786–91.
[313] Lee JH, Akoh CC, Lee K-T. Physical properties of trans-free bakery shortening produced by lipase-catalyzed interesterification. J Am Oil Chem Soc 2008;85:1–11.
[314] Siew WL, Cheah KY, Tang WL. Physical properties of lipase-catalyzed interesterification of palm stearin with canola oil blends. Eur J Lipid Sci Technol 2007;109:97–106.
[315] Katan MB, Zock PL, Mensink RP. Effects of fats and fatty acids on blood lipids in humans: an overview. Am J Clin Nutr 1994;60:1017S–22S.
[316] Philippaerts A, Paulussen S, Turner S, Lebedev OI, Van Tendeloo G, Poelman H, et al. Selectivity in sorption and hydrogenation of methyl oleate and elaidate on MFI zeolites. J Catal 2010;270:172–84.
[317] Philippaerts A, Paulussen S, Breesch A, Turner S, Lebedev OI, Van Tendeloo G, et al. Unprecedented shape selectivity in hydrogenation of triacylglycerol molecules with Pt/ZSM-5 zeolite. Angew Chem Int Ed 2011;50:3947–9.
[318] Philippaerts A, Breesch A, De Cremer G, Kayaert P, Hofkens J, Van den Mooter G, et al. Physical properties of nutritive shortenings produced from regioselective hardening of soybean oil with Pt containing zeolite. J Am Oil Chem Soc 2011;88:2023–34.
[319] Bradley TF. Drying Oils. US2350583; 1944.
[320] Sleeter RT. Method of conjugating double bonds in drying oils. US5719301; 1998.
[321] Lu Y, Larock RC. Novel polymeric materials from vegetable oils and vinyl monomers: preparation, properties, and applications. ChemSusChem 2009;2:136–47.
[322] Bhattacharya A, Banu J, Rahman M, Causey J, Fernandes G. Biological effects of conjugated linoleic acids in health and disease. J Nutr Biochem 2006;17:789–810.
[323] Coakley M, Johnson MC, McGrath E, Rahman S, Ross RP, Fitzgerald GF, et al. Intestinal bifidobacteria that produce trans-9, trans-11 conjugated linoleic acid: a fatty acid with antiproliferative activity against human colon SW480 and HT-29 cancer cells. Nutr Cancer 2006;56:95–102.
[324] Pariza MW, Park Y, Cook ME. The biologically active isomers of conjugated linoleic acid. Prog Lipid Res 2001;40:283–98.
[325] Saebo A, Skarie C, Jerome D, Haroldsson G. Conjugated linoleic acid compositions and methods of making same. US6410761 B1; 2002.
[326] Westfechtel AR, Albiez W, Zander L, Horlacher P. Method for the production of conjugated linoleic acids. US20060106238 A1; 2006.
[327] Philippaerts A, Goossens S, Jacobs PA, Sels BF. Catalytic production of conjugated fatty acids and oils. ChemSusChem 2011;4:684–702.
[328] Philippaerts A, Goossens S, Vermandel W, Tromp M, Turner S, Geboers J, et al. Design of Ru-zeolites for hydrogen-free production of conjugated linoleic acids. ChemSusChem 2011;4:757–67.
[329] Philippaerts A, Goossens S, Geboers J, Sels BF. Method for the production of conjugated polyunsatured fatty acids with heterogenous catalysts. WO2012068645 A3; 2012.

[330] Zhang Z, Dery M, Zhang S, Steichen D. New process for the production of branched-chain fatty acids. J Surfactant Deterg 2004;7:211–5.
[331] Biermann U, Metzger JO. Synthesis of alkyl-branched fatty acids. Eur J Lipid Sci Technol 2008;110:805–11.
[332] Zhang S, Zhang ZC. Skeletal isomerization of unsaturated fatty acids: the role of mesopores in HBeta zeolites. Catal Lett 2007;115:114–21.
[333] Ngo HL, Nuñez A, Lin W, Foglia TA. Zeolite-catalyzed isomerization of oleic acid to branched-chain isomers. Eur J Lipid Sci Technol 2007;109:214–24.

CHAPTER 10

Cu-Zeolite Selective Catalytic Reduction Catalysts for NO_x Conversion

R. Oord, B.M. Weckhuysen
Utrecht University, Utrecht, The Netherlands

1 Introduction

Selective catalytic reduction (SCR) of nitrogen oxides (NO_x) is an important technology to reduce environmentally harmful NO_x from exhaust gases, originating from mobile sources, such as cars and ships [1]. NO_x causes acid rain and is a major component in smog. The smog is usually found in big cities, where it is mainly produced by vehicles and power plants. This smog is highly toxic to humans. Forty percent of the total NO_x released into the atmosphere comes from road transport [2]. Newer catalyst formulations were needed because the old two-way exhaust catalysts, consisting of metals such as platinum, palladium, and rhodium supported on a metal oxide, did not meet the new emission standards, including EURO 5, or the 2010 US EPA standards for both light- and heavy-duty diesel trucks, especially on their capability to remove NO_x [3]. The SCR component is one of the catalysts in the typical exhaust emission catalytic system. The total system normally consists of a diesel oxidative catalyst (DOC), diesel particulate filter (DPF), and the NH_3-SCR system, an example of such system is schematically illustrated in Fig. 1.

In the NH_3-SCR reaction NO_x gases are reduced to N_2 and H_2O, by using NH_3 as a reducing agent. Depending on the operating conditions of the engine as well as of the exhaust system, different gas compositions are possible, which has an influence on how the NH_3-SCR reactions proceeds. The three most important reaction pathways of this process are [4]:

$$\text{The standard SCR reaction}: 4NH_3 + 4NO + O_2 \rightarrow 4N_2 + 6H_2O \quad (1)$$

$$\text{The fast SCR reaction}: 2NH_3 + NO + NO_2 \rightarrow 2N_2 + 3H_2O \quad (2)$$

$$\text{The } NO_2 \text{ SCR reaction}: 4NH_3 + 3NO_2 \rightarrow 3.5N_2 + 6H_2O \quad (3)$$

In the exhaust gases containing NO_x, NO is generally present, while only a small fraction of NO_2 is present. Therefore, reaction (1) is called the standard SCR reaction. The reaction of a 1 to 1 mixture of NO and NO_2 is a lot faster, and is therefore called the fast SCR

Zeolites and Zeolite-like Materials. http://dx.doi.org/10.1016/B978-0-444-63506-8.00011-2

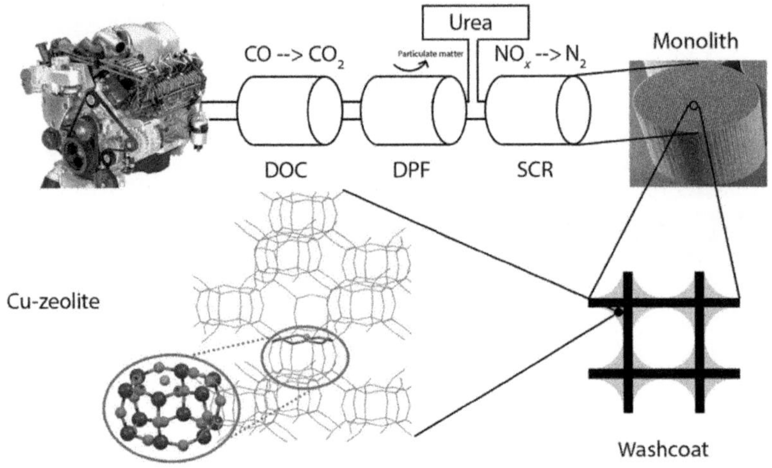

Fig. 1
Schematic of the exhaust catalyst catalytic system. The NH_3-SCR takes place in the selective catalytic reactor (SCR) catalyst component. The other components depicted are diesel oxidative catalyst (DOC), diesel particulate filter (DPF), urea, the source of NH_3. The active component, most commonly a Cu-exchanged zeolite, is placed as a wash coat on a ceramic monolith.

reaction (reaction (2)). If there is more NO_2 than NO present in the exhaust gases, the SCR reaction will be as in reaction (3) and only NO_2 will react [4]. Oxidation of NH_3 to NO or even NO_2 is possible, but Cu-ion exchanged zeolite catalysts are generally very selective for N_2 formation, in the right temperature window [4]. In this chapter we discuss briefly the different types of catalysts that are used for the NH_3-SCR reaction, with special emphasis on Cu-exchanged zeolites. We will discuss the differences in synthesis methods and their implications, as well as the location of Cu in the zeolite framework. Next we will put the focus on the differences between Cu-SAPO-34 and Cu-SSZ-13. After this we will move to the deactivation mechanisms for Cu-exchanged zeolites. For more details and background on both stationary and mobile catalysts we refer the interested reader to the book entitled *Urea-SCR Technology for deNOx After Treatment of Diesel Exhaust* [5]. For a more mechanistic view on Fe-exchanged zeolites and their role in NH_3-SCR we can recommend the review article by Brandenberger et al. [4].

2 Different Catalysts Used for NH_3-SCR

Different catalysts capable of performing the NH_3-SCR reaction are now being used [5]. For stationary applications, supported base metal catalysts, such as vanadium on titania, are used. However, for mobile applications, Cu- and Fe-exchanged zeolites are normally used, due to their good redox capacity. Zeolites are crystalline aluminosilicates consisting of repeating SiO_4 and AlO_4 tetrahedral subunits to form porous three-dimensional networks. The negative charge, which is a result of Al^{3+} replacing Si^{4+} in the zeolite framework, results in giving zeolites both an acid functionality and an ion exchange capacity. Cu-zeolite materials are known to be highly active in the SCR of NO_x with NH_3 (NH_3-SCR) [5]. In this section, we will discuss why different catalysts are being used and which drawbacks they have.

An SCR catalyst needs to be hydrothermally stable, even at elevated temperatures up to 650°C, since the DPF needs to be generated at temperatures above 600°C periodically. Vanadium-based catalysts, which are commonly employed in stationary applications, such as power plants, however, are not stable at these elevated temperatures, while Cu- and Fe-zeolite-based SCR catalysts are much more stable [6]. This is not a problem for stationary operations, since the temperature doesn't reach these temperatures during stationary operations.

Fe-zeolite catalysts are very resistant to the high temperatures and show good SCR activity. The drawback of these catalysts is that they perform less at relatively low temperatures (200–300°C) as the Cu-zeolite catalysts, which are common exhaust operation conditions. This makes Cu-zeolite catalysts the preferred catalyst for mobile diesel NO_x removal [7].

To show the difference in level of activity after aging, Ford Motor Company performed research in which they compared Fe- and Cu-exchanged zeolites as well as a vanadium-supported titania catalyst, normally used in stationary applications [8,9]. After aging for 64 h at 670°C to represent ~120 k miles on a diesel vehicle with a regenerating DPF, they concluded that the vanadium-supported titania catalyst is not the right catalyst formulation for mobile applications, since the activity is very low (in comparison with the Cu- and Fe-exchanged zeolites). A graphical representation of this set of experiments can be found in Fig. 2. This figure also illustrates the better performance of the Cu-zeolite based SCR catalyst material at lower temperatures.

3 Cu-Exchanged Zeolites

If we have a closer look at Cu-exchanged zeolites, one of the features with which we can differentiate types of zeolites is the largest pore size opening of the framework structure. These pores can be small (8-membered ring), medium (10-membered ring), large (12-membered ring), and extra-large (>12-membered ring). Examples of zeolite structures, which are mostly used for catalysis research, are shown in Fig. 3. Medium- and large-pore Cu-zeolites are less stable under severe hydrothermal conditions as their small pore counterparts [10,12–14]. The small pore Cu zeolites have an improved selectivity toward N_2 and a reduced selectivity toward N_2O formation. Small-pore Cu-zeolite SCR catalysts are currently used successfully in diesel engine-powered heavy vehicles. Most research done in the early days was performed on the medium- and large-pore zeolites like ZSM-5 (MFI), mordenite (MOR), zeolite Y (FAU), and beta (BEA). The last couple of years the research

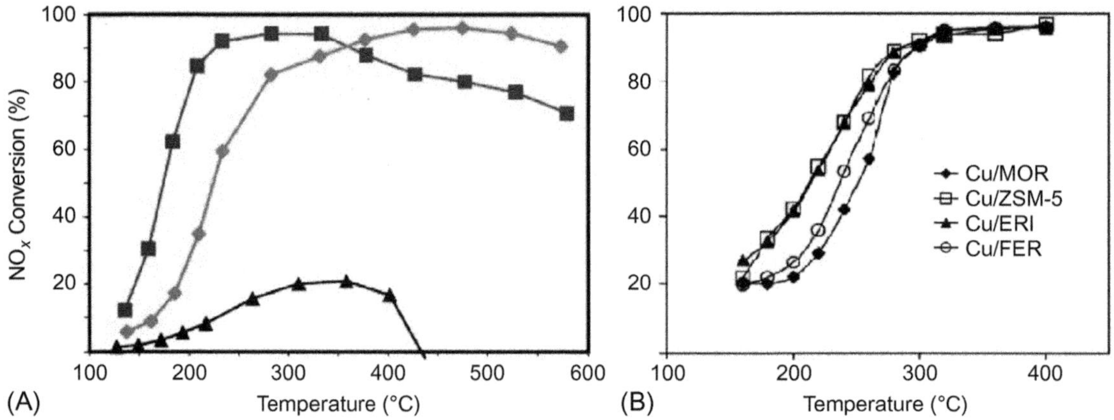

Fig. 2
(A) NO_x conversion over Cu-exchanged zeolites (■), Fe-exchanged zeolites (♦), and vanadium-based (▲) formulation, as a function of temperature. A round sample core of 1 in. diameter was taken from a wash-coated monolith obtained from a supplier and a gas space velocity of 30,000 h^{-1} [8,9]. *Reproduced from Li J, Chang H, Ma L, Hao J, Yang RT. Low-temperature selective catalytic reduction of NO_x with NH_3 over metal oxide and zeolite catalysts—a review. Catal. Today 2011;175:147–156. Copyright 2011, with permission of Elsevier.* (B) NO_x conversion over Cu/zeolites and a gas hourly space velocity of 75,000 h^{-1} [10]. *Reproduced from Sultana A, Nanba T, Sasaki M, Haneda M, Suzuki K, Hamada H. Selective catalytic reduction of NO_x with NH_3 over different copper exchanged zeolites in the presence of decane. Catal. Today 2011;164:495-499. Copyright 2011, with permission of Elsevier.*

has shifted toward small-pore zeolites; like the microporous aluminosilicate SSZ-13 and the microporous silico-aluminophosphate SAPO-34, which have both the CHA framework topology, and Erionite (ERI). The CHA structure belongs to the ABC-6 family of zeolite structures [15]. The framework is basically built by stacking layers of double 6-membered rings in the order AABBCC…, forming double 6-rings at each apex of the rhombic unit cell. This leads to the formation of large cavities, accessible through a three-dimensional 8-membered ring (8r) structure, with an aperture of 3.8 × 3.8 Å. The framework topology of the CHA structure type is rhombohedral, R3m, which has also been attributed to chabazite [16]. Fig. 3 shows the cage of the CHA topology viewed normal to the [001] plane. Instead, Erionite (ERI) is a small-pore zeolite, the biggest accessible pore in this system is also an 8-membered ring (3.6 × 5.1 Å), similar to chabazite. Other small-pore zeolites that are briefly researched in literature are: SSZ-52 (SFW), SSZ-39 (AEI, not a member of the family, but closely related to CHA structure) and SSZ-16. (AFX, stacking sequence AABBCCBB) all show good activity for the NH_3-SCR reaction, indicating that other units with d6r units might be promising as future DeNO_x catalysts [15,17,18]. Ferrierite (FER) has both 10-membered ring pores (in the c direction) having an aperture 5.4 × 4.2 Å, which are interconnected by 8-membered ring pores (in the b direction) having an aperture of 4.8 × 3.5 Å [19]. Fig. 2

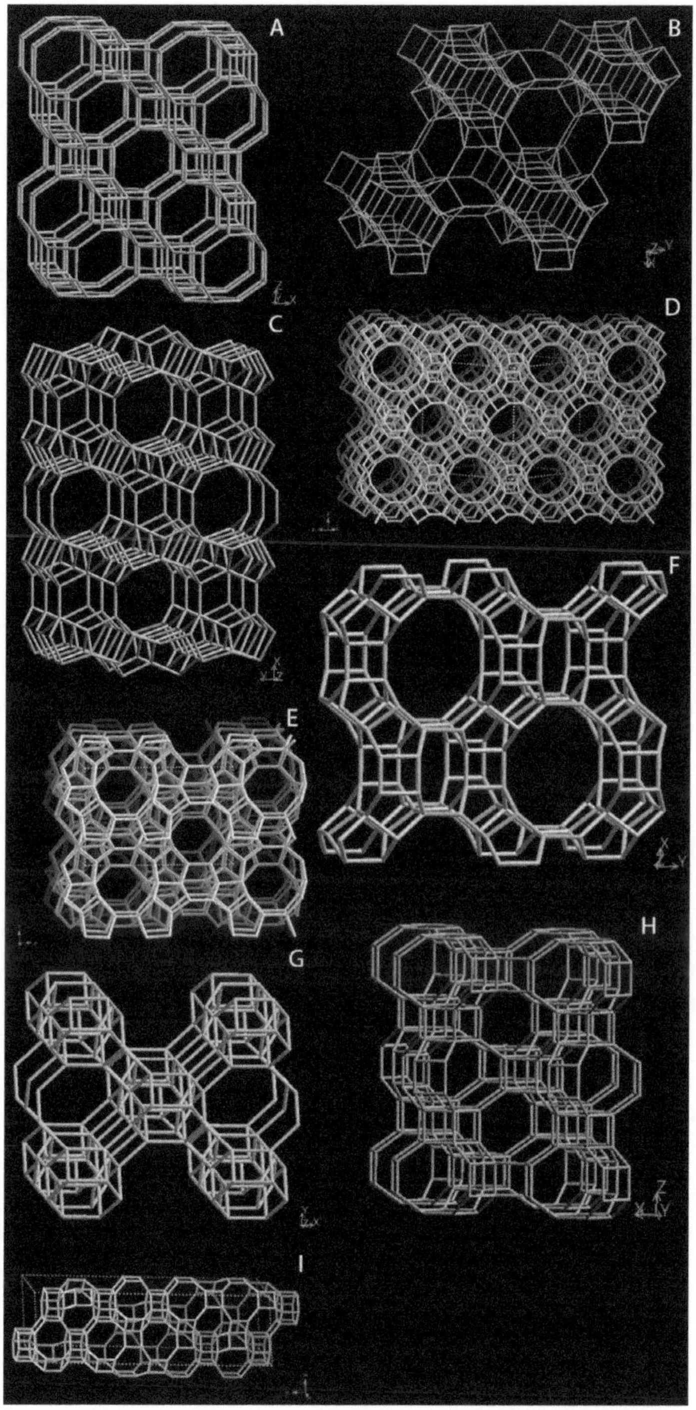

Fig. 3

Different types of framework topologies for the zeolites mentioned in this chapter, topologies are described by their three-letter code. A CHA B ERI C FER D FAU E MFI F MOR G AEI H AFX I SFW [11]. *Modified from IZA. IZA structural database, http://www.iza-structure.org/databases/. Copyright 2014, with permission of IZA structural database.*

shows the difference in NO_x conversion over small (ie, Cu-ERI), medium (ie, Cu-MFI), large (ie, Cu-MOR) pore zeolite materials as well as an in-between small and medium (ie, Cu-FER) zeolite material. It can be clearly seen that the small-pore zeolites show a higher activity at lower reaction temperatures.

Both the Brønsted acid sites as the copper sites inside the zeolite are important for the SCR reaction, the acid sides adsorb NH_3 to form NH_4^+ and the Cu is an essential redox center for the NH_3-SCR reaction. To get an optimal NH_3-SCR activity, the Cu exchange level should be well balanced, as both these functions are essential in the catalyst [20]. How the Cu exactly behaves during the SCR reaction is still under debate, but it is generally accepted that Cu should be in its ionic state. CuO clusters, which may appear in the zeolite, because of imperfect synthesis or hydrothermal deactivation, are more active in catalyzing NH_3 oxidation at high temperatures, yielding a negative effect on the selectivity for NO_x in NH_3-SCR [18,21].

The total mechanism of the NH_3-SCR reaction, for small-pore zeolites, is still not fully understood. One thing that is known is that the reaction pathways are different for both the low and high temperature ranges. According to Ma et al. [22] at low temperatures, ammonium nitrates from the reaction between surface NH_3 and nitrates are most important intermediates that are later reduced by gas-phase NO to yield nitrogen and water. At high temperatures, gas phase NO_2 is an important intermediate in the NH_3-SCR reaction. Pre-adsorbed NH_3 reacts gas phase NO_2 to form an intermediate after which it decomposes into nitrogen and water. Kinetic studies are difficult to perform as they are complicated due to the pore diffusion limitation, which is associated especially with the small pore structures of the CHA framework. Furthermore, studying the reaction mechanism under in situ conditions is even more complicated for this reaction system, as –OH, H_2O, and NH_3 all have strong ligand effects, which cause mobility and transient interactions among Cu^{2+} ions [23]. A few considerations for the total reaction mechanism, based on the work of Gao et al. [23] are: before full dehydration, isolated Cu^{2+} ions are very mobile; isolated Cu^{2+} ions do not catalyze NO oxidation, if the temperature remains under 400°C; Cu-SSZ-13 follows a Cu^{2+}/Cu^{+} redox mechanism. They conclude as well, that for different loadings, different kinds of Cu-dimers play a role in their catalytic system.

4 Differences in Synthesis Methods

Copper-exchanged zeolites can be prepared using different methods and yielding different results; an example of this can be found in Fig. 4 [24]. A few methods are being used in the literature today: conventional wet ion exchange, chemical vapor deposition (CVD), solid-state (SS) ion exchange, and a direct synthesis using a Cu containing precursor, a summary of which can be found in Table 1.

Conventional wet ion exchange is used in all types of fields to exchange metal ions into different kinds of zeolites. Normally, a zeolite material in its proton or ammonia form is

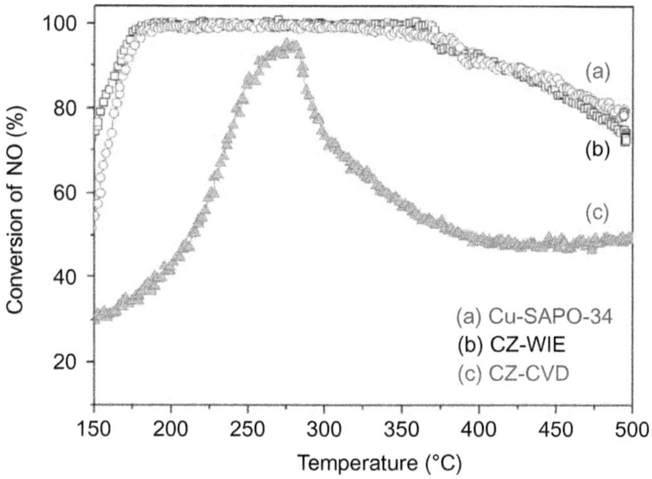

Fig. 4

Conversion of NO over (A) Cu-SAPO-34 (red); (B) CZ-WIE (black), and (C) CZ-CVD (gray). The gas hourly space velocity equals 100,000 h^{-1} [24]. *Reprinted from Deka U, Lezcano-Gonzalez I, Warrender SJ, Lorena Picone A, Wright PA, Weckhuysen BM, Beale AM. Changing active sites in Cu–CHA catalysts: deNO$_x$ selectivity as a function of the preparation method. Microporous Mesoporous Mater. 2013;166:144–152. Copyright 2013 with permission of Elsevier.*

Table 1: Advantages and disadvantages of different synthesis methodologies to prepare Cu-based SCR catalyst materials

Method	Advantages	Disadvantages
Wet ion exchange	- Easiest method experimentally - Copper in the desired cationic locations	- Yields lots of waste
Chemical vapor deposition	- Limits waste - Techniques already available industrially	- Develops CuAlO2 phase which causes bad selectivity
Solid-state ion exchange	- Limits waste - Easy experimental method	- High temperature needed - Sample needs to be dry to avoid damage to zeolite
Direct synthesis	- Limits waste - Good distribution over zeolite	- More expensive structure directing agents

added to a solution containing the desired metal ions, while continuously stirring. The temperature can be varied to get a better distribution over the zeolite. In general for small-pore zeolites, the temperature should be increased to increase the rate of diffusion [4]. After the ion exchange process is done, the sample should be filtered and washed properly. The risk of this procedure is over-exchange of the metal precursor, yielding metal or metal oxide clusters, but by correct timing and correct concentration of exchange liquid one can get very active catalysts [25]. CVD is a technique mostly used to cover flat surfaces with a few layers of metals. Generally, when correct salts, conditions, and flows are being used, it is also possible to use this method to deposit ions inside the zeolite pore structure. The salts most used for this procedure are chlorine salts. The resulting exchanged zeolite should be washed carefully, to remove all of the chlorine left behind with this procedure, as chlorine is expected to cause a higher mobility of metal species [26]. SS ion exchange is possible for Cu-zeolite systems [27] and is recently applied in the preparation for SCR catalysts [28]. In this method a Cu-containing precursor and the zeolite are mixed and consequently heated to high temperatures in a flow in order to get the metals mobile and inside the zeolite framework. Finally, the last method discussed is a direct synthesis using a Cu-containing precursor. Different groups have been working on this system [29–33]. Instead of using a normal NH_4-zeolite or H-zeolite, a Cu-containing precursor is used to synthesize the zeolite around it, immediately resulting in a Cu-zeolite. The advantage of this one-step synthesis is that after the process, further treatment is no longer necessary, thereby limiting the waste that is generated during ion exchange. By using Cu-containing structure-directing agents, the Cu distribution over crystal should be more homogeneous as a postsynthesis treatment, as the Cu is used as a charge-balancing molecule during synthesis [31].

Of all these methods, the most used method is the conventional wet ion exchange, it is by far the easiest method experimentally and yields copper in the desired cationic locations inside the zeolite. A very good alternative, once the synthesis method using Cu-structure directing agents is optimized to perfection, would be the direct synthesis route, since it limits waste and saves energy in the production of the Cu-materials. SS ion exchange is also an option since it limits waste and is experimentally easy. These synthesized materials have shown that they have excellent activity in the NO_x NH_3-SCR reaction, and also show a good hydrothermal stability [31]. Each of the above-mentioned methods seems to lead a different local environment for Cu inside the zeolite pore system, and not all of them are equally effective for the NH_3-SCR reaction [24]. The CVD method also seems to develop a $CuAlO_2$ phase next to the Cu^{2+} ions, which causes a poor deNO_x activity and high selectivity toward the formation of undesired N_2O [24].

In what will follow we will focus on the most studied small-pore zeolites in NO_x NH_3-SCR reaction, in particularly Cu-SSZ-13 and Cu-SAPO-34. We discuss here the precise Cu location and different deactivation pathways proposed.

5 Cu Location

Different research groups have been investigating the location of the Cu inside small-pore zeolite systems. Both Fickel et al. and Deka et al. have done extensive investigations on the location and redox state of the copper with respect to the zeolite framework. Fickel et al. [17]

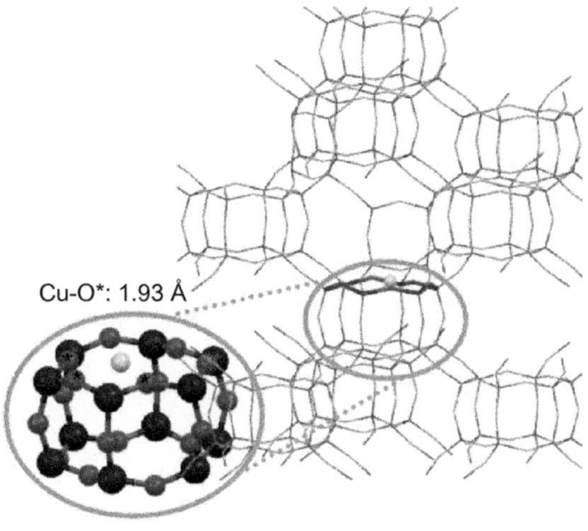

Fig. 5
Local environment of Cu^{2+} in zeolite SSZ-13, the proposed acid site for the NH_3-SCR reaction. The distance between the Cu ion and the oxygen's marked with an asterisk is 1.93 Å [36]. *Reprinted from Deka U, Lezcano-Gonzalez I, Weckhuysen BM, Beale AM. Local environment and nature of Cu active sites in zeolite-based catalysts for the selective catalytic reduction of NO_x. ACS Catal. 2013;3:413–427. Copyright 2013, with permission of American Chemical Society.*

used Rietveld refinement of synchrotron XRD at different temperatures on a Cu-SSZ-13 catalyst and concluded that copper within the SSZ-13 framework is located within the cage, just outside the 6-membered rings of the zeolite. Korhonen et al. [34] demonstrated with their XAFS data that the Cu^{2+} is coordinated by three oxygen atoms of the 6-membered ring. Deka et al. [35] confirmed this location also during NH_3-SCR, the ions located on the plane and slightly distorted from the center of the d6r subunits of SSZ-13, by using in situ XAFS/XRD. The exact location is visible in Fig. 5, the distance between the Cu ion and the oxygen's marked with an asterisk is 1.93 Å. When ion exchange levels reach close to a 100%, the Cu species still exist as isolated ion monomers [34].

Recent papers gave many new insights in other locations for Cu ions inside zeolite framework structures. Andersen et al. [37] concluded that the majority of the Cu^{2+} ions are in a specific site in the 8R of SSZ-13, as well as in the d6R, based on Rietveld and MEM analysis applied to synchrotron pXRD data, on a dehydrated Cu-SSZ-13. They used a system with Si/Al ratio of 15.5 and a Cu/Al ratio of 0.45. By DFT calculations they assigned this specific site in 8R to $[Cu^{2+}OH^-]^+$ close to a Al atom, and Cu^{2+} in the d6r. This was backed up by the work of Godiksen et al. [38] where they used EPR to study their Cu-SSZ-13 system (Si/Al = 14). They explain the loss of EPR signals upon dehydration due to EPR silent species, $[Cu^{2+}OH^-]^+$. Borfecchia et al. [39] concluded that the majority of the dehydrated Cu species are found preferentially in 8R units of the SSZ-13 framework, as tri-coordinated $[CuOH]^+$ species at O_2 activation, and bi-coordinated Cu^+ ions upon

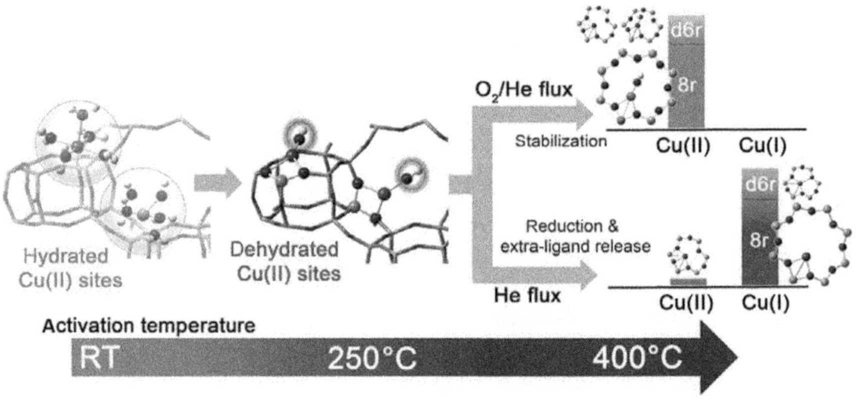

Fig. 6
Schematic representation of Cu-speciation in the Cu-SSZ-13 catalyst as a function of the dehydration temperature and conditions. In the O_2 -activated Cu-SSZ-13 no traces of Cu(I). A minor fraction (around 10%) of Cu(II) in the He-activated sample was detectable in HERFD XANES spectra. The right side of the scheme concerns the dominant structural components identified by XAS, XES, and FTIR in the Cu-SSZ-13 sample investigated in the present work [39]. *Reprinted from Borfecchia E, Lomachenko KA, Giordanino F, Falsig H, Beato P, Soldatov AV, Bordiga S, Lamberti C. Revisiting the nature of Cu sites in the activated Cu-SSZ-13 catalyst for SCR reaction. Chem. Sci. 2015;6:548–563. Copyright 2014, with permission of The Royal Society of Chemistry.*

He activation. A schematic representation can be found in Fig. 6. They used a Cu-SSZ-13 zeolite with an Si/Al ratio of 12 and a Cu/Al ratio of 0.44.

Verma et al. [40] using a Cu-SSZ-13 zeolite (Si/Al = 4.5, Cu/Al varies) came to slightly other conclusions. They used UV–vis, XAS, catalytic measurements, and a computational kinetic study of dry NO oxidation, to conclude that there are at least two Cu ion configurations within SSZ-13. For Cu/Al ratios below 0.2 the dominant Cu^{2+} ion is a hydrated isolated Cu^{2+} ion. Under dry NO oxidation, these species were converted into dehydrated isolated Cu^{2+} ions close by two framework Al sites, in the d6r unit. For Cu/Al = 0.2, all framework Al pairs in the d6r units were exchanged with dehydrated isolated Cu^{2+} ions. For Cu/Al > 0.2, Cu_xO_y ($x \geq 2$, $y \geq 1$) species inside the 8r unit, balanced by the Al in the 8r unit, play a major role in NO oxidation. Gao et al. [23], using their Cu-SSZ-13 system (Si/Al = 6, Cu/Al varies), studied with TPR, EPR, and catalytic testing with NO oxidation, NH_3-oxidation, and NH_3-SCR, the state of Cu inside the zeolite. They state that for Cu/Al ≤ 0.22, transient Cu-dimers are relevant active sites at reaction temperatures under 300°C, which are in equilibrium with monomeric Cu ion complexes. At temperatures above 350°C, at low Cu loadings, the Cu-dimers completely dissociate to isolated Cu^{2+} monomers. At loadings 0.032 ≤ Cu/Al ≤ 0.22 Cu^{2+} ion monomers next to the 8r become relevant sites for SCR. At Cu loadings above one Cu^{2+} ion per unit cell, stable Cu-dimers form, and they do not dissociate at temperatures above 350°C. Borfecchia et al. [39] however, state that they could not find a significant fraction of dimeric oxo-Cu species (Cu_xO_y ($x \geq 2$, $y \geq 1$)) and according to them these species are very unlikely.

The major difference between these studies is the Si/Al ratio in the parent zeolite, which apparently has a large influence on the Cu location and coordination inside the zeolite framework, since a single conclusion, which backs up everything is not yet found.

Both Cu-SSZ-13 and Cu-SAPO-34 are more stable during the NH_3-SCR reaction, as the larger pore Cu-zeolites, resulting in an increased activity and selectivity after aging, as can be seen in Fig. 7 [41,42]. Several studies have been carried out to determine the exact location and state of the Cu ions in both Cu-SSZ-13 and Cu-SAPO-34 catalysts. According to Ma et al. [22,43] after hydrothermal treatment at high temperatures, Cu-SSZ-13 was proven to be less stable as Cu-SAPO-34, based on catalyst performance testing. Wang et al. [28] also came to this conclusion, for SS ion exchanged Cu-SSZ-13 and Cu-SAPO-34. The difference can be clearly seen in Fig. 8. Hydrothermal treatment at 800°C caused a significant reduction in the NH_3-SCR performance for Cu-SSZ-13, accompanied by a loss in crystallinity and a loss of active Cu-sites. Cu-SAPO-34 did, however, not lose this activity at these conditions [28]. This difference in hydrothermal stability is not yet properly explained in current literature. Deka et al. did not detect major differences in Cu locations for Cu-SSZ-13 or Cu-SAPO-34 in their extensive study [24]. The chemistry of the framework is different for silico-aluminophosphates (SAPO), and it is reported that SSZ-13 shows a higher acid strength than its SAPO analog SAPO-34[44].

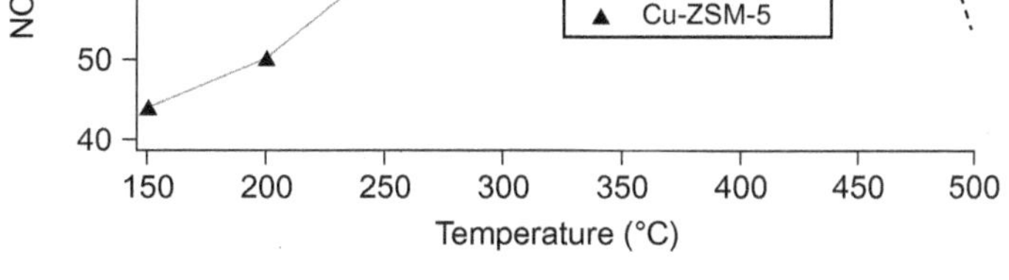

Fig. 7

NH_3-SCR activity after the 14-h hydrothermal treatment of Cu-SSZ-13, Cu- SAPO-34, Cu-SSZ-16, and Cu-ZSM-5 [41]. *Reprinted from Fickel DW, D'Addio E, Lauterbach JA, Lobo RF. The ammonia selective catalytic reduction activity of copper-exchanged small-pore zeolites. Appl. Catal. B 2011;102:441–448. Copyright 2011, with permission of Elsevier.*

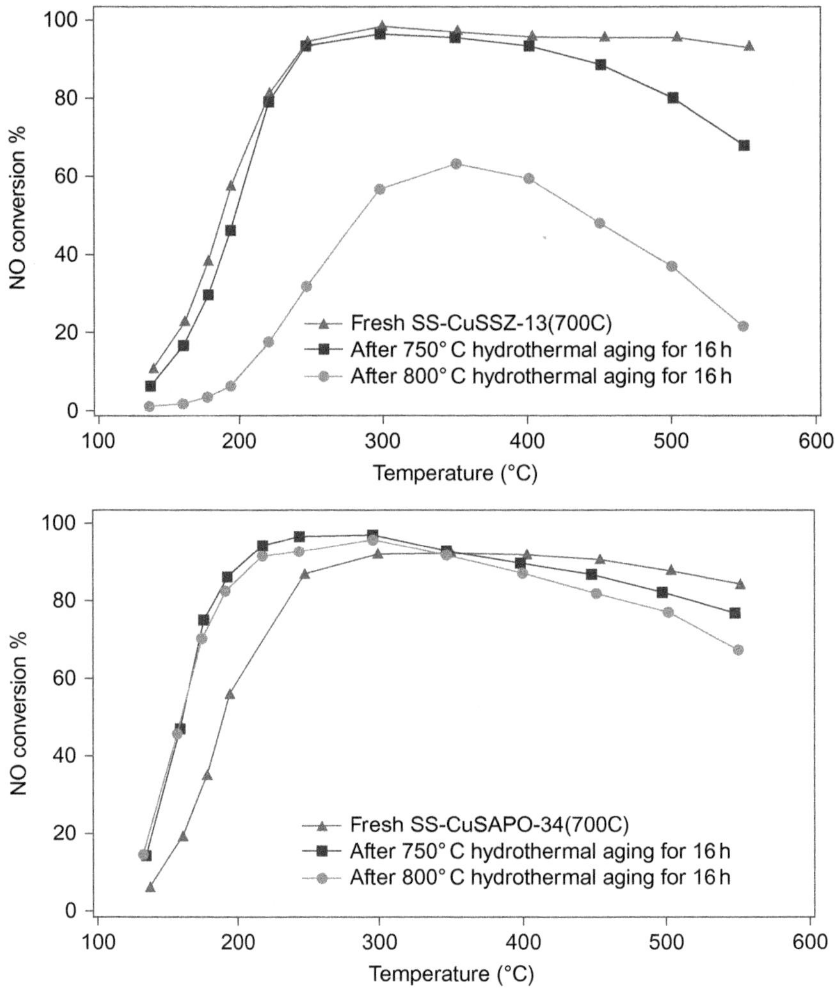

Fig. 8

CR performance of SS-Cu-SSZ-13 (left), and SS-Cu-SAPO-34 (right) before and after hydrothermal aging for 16 h [28]. *Reprinted from Wang D, Jangjou Y, Liu Y, Sharma MK, Luo J, Li J, Kamasamudram K, Epling WS. A comparison of hydrothermal aging effects on NH_3-SCR of NO_x over Cu-SSZ-13 and Cu-SAPO-34 catalysts. Appl. Catal. B 2015;165:438–445. Copyright 2014, with permission of Elsevier.*

6 Deactivation

Deactivation of zeolite-based SCR systems can be divided into two main categories: hydrothermal deactivation by steam in the exhaust system or chemical deactivation by different chemicals found in an exhaust stream of typical fossil fuel engine. Steam is present in the exhaust, since it is a by-product of burning fossil fuels, while different chemicals might be present due to lubrication oil, contaminants in the fuel, or they may be originating from a different catalyst in the exhaust system.

6.1 Hydrothermal Deactivation

The main drawback of using zeolites is decreased activity at high temperatures due to hydrothermal deactivation. Especially for the medium- and large-pore zeolites, at temperatures exceeding 700°C and in the presence of steam, the catalysts deactivate quickly and show lower SCR activity [12–14]. The main cause of this deactivation is dealumination of the zeolite framework, the aluminum inside the framework becomes unstable and is removed from the tetrahedral framework locations [13,45–47]. Since aluminum atoms positioned in tetrahedral framework locations are the Brønsted acid sites in the zeolites, the Brønsted acidity is also lost. Furthermore, less charge compensation is needed due to the removal of aluminum, so the Cu also starts to migrate, and forms CuO or copper aluminate clusters [48,49]. For most larger pore zeolites, Cu exchanged zeolites are less stable as the normal zeolites, as can be seen by in situ XRD [5]. For smaller pore zeolites, like SSZ-13, this seems to be not the case. In situ XRD also shows the difference between NH_4-SSZ-13 and Cu-SSZ-13. Clearly the zeolite retains its crystal structure longer, at more severe hydrothermal temperatures [17].

Small-pore zeolites are much less susceptible to this hydrothermal deactivation, although H_2O can still enter the pores easily, the kinetic diameter of H_2O being 2.65 Å, and the pore diameter of the largest pore being the 8-membered ring with dimensions of 3.8 × 3.8 Å [16]. Fickel at al. [41] hypothesized, in order to get defects in the zeolite because of dealumination, $Al(OH)_3$ units have to leave the zeolite framework, and out of the pores. They estimate the kinetic diameter of $Al(OH)_3$ at 5.03 Å, making it impossible for the $Al(OH)_3$ to escape the pores and locking it inside. They also propose based on other literature about the postsynthesis alumination treatments [50,51], that at low temperatures, the $Al(OH)_3$ is reincorporated into the zeolite framework [41]. Because of these reasons Cu-SSZ-13 displays an improved hydrothermal stability in comparison with bigger pore Cu-zeolite catalysts for NH_3-SCR. The mechanism that deactivates SSZ-13 is still similar to other zeolites, namely dealumination and structural collapse.

6.2 Other Types of Deactivation

Another type of deactivation found under automotive conditions is the storage of hydrocarbons, and sequentially coking of these hydrocarbon molecules inside the zeolite pore system. Hydrocarbons are always present in the exhaust stream, but even more so during cold starts of the engine. The buildup of hydrocarbons inside the zeolite can burn during the regeneration cycle of the DPF filter, creating more heat locally inside the zeolite pore system, which can in turn thermally destroy the zeolite. One of the reasons Cu-SSZ-13 is more stable for this deactivation has to do with the small-pore openings of this zeolite. For hydrocarbon molecules with a kinetic diameter larger than 3.8 Å (methane) it is very difficult to enter the pores, making diffusion very slow [13].

Other types of deactivation are based on the chemical deactivation of the Cu zeolite system. Chemical deactivation of NH_3-SCR catalysts is due to contaminants from oil and fuel and chemicals from different catalysts in the exhaust system, like Pt, originating from the DOC [52–54]. The volatized Pt from the DOC produce severe deactivation of the SCR catalyst, where it is known to oxidize NH_3 toward NO_x and N_2O [52–54]. Other impurities such as phosphorous, calcium, magnesium, and zinc originate lubrication oil in the engine, where they are added as additives [54]. Lezcano-Gonzalez et al. [54] have performed extensive research on the deactivation mechanisms of Ca, Zn, Pt, and P. The main influence on the catalytic performance of Ca and Zn is a decrease in NO conversion, and an increased NH_3 slip. For Pt there is also a decrease in NO conversion, but also a strong influence on the N_2 selectivity and an increase in NO oxidation activity. P is responsible for a complete suppression of the catalytic activity. The main deactivation mechanisms are different for all investigated poisons. For Ca, it is due to pore blocking and filling by CaO and CuO aggregates. For Zn it is also pore blocking and filling, by ZnO and CuO aggregates, but also the decrease of Cu^{2+} sites. Pt species itself promote NO and NH_3 oxidation, and as a secondary effect it also fills the pores with Pt species. P disturbs the zeolite framework, by reacting with the Al species in the framework, it is also responsible for the decrease in the amount of Cu^{2+} ions and the formation of CuO. Furthermore, it forms polyphosphates and CuO, which block the pores of the zeolite system (Table 2) [54].

7 Conclusions and Look into the Future

Compared to medium and large-pore Cu-exchanged zeolite catalysts, small-pore Cu-exchanged zeolite catalysts show an excellent and improved activity, hydrothermal stability, and selectivity toward N_2. Because of the small pores, the small-pore Cu-exchanged zeolite catalysts also show a better resistance against hydrocarbon poisoning, as well as dealumination effects. Different synthesis procedures, however, might result in differences in catalyst performance, so it is key to pick the right synthesis procedure, while taking care of waste treatment. At the moment, these catalysts are already employed in heavy-duty diesel trucks on a commercial level, however, a lot more work is needed to fully understand and employ these new Cu exchanged small-pore zeolites. A full mechanistic catalytic cycle, for example, is still not yet established. New vehicle emission regulations could be expected in the future, and the low temperature activity for Cu-SSZ-13 and Cu-SAPO-34 has to be improved. As mentioned, pore diffusion causes difficulties for both Cu-SSZ-13 and Cu-SAPO-34. To overcome these problems more studies into the synthesis of very small zeolite crystals or mesopores and modified zeolites is necessary. The full mechanism of the NH_3-SCR reaction inside the Cu-SSZ-13 framework is still not known as more research is needed in the field of kinetic studies.

Table 2: Overview of the deactivation mechanisms observed for Ca, Zn, Pt, P, hydrocarbons (HC) and steam for Cu-based zeolite NO_x SCR catalyst materials [54]

Deactivation factor	Influence on catalytic performance	Main deactivation mechanism	Secondary deactivation mechanism
Ca	- Decrease in NO conversion - Increased NH_3 slip	- Pore blocking or filling by CaO and CuO aggregates	- Decrease of isolated Cu^{2+} ions and CuO formation
Zn	- Decrease in NO conversion - Increased NH_3 slip	- Pore blocking or filling by ZnO and CuO aggregates - Decrease of isolated Cu^{2+} ions and CuO formation	
Pt	- Decrease in NO conversion - Strong influence on N_2 selectivity - Increase in NO oxidation activity	- Pt promotes NO and NH_3 oxidation	- Pore blocking or filling by Pt species
P	- Complete suppression of catalytic activity	- Decrease of isolated Cu^{2+} ions and CuO formation - Pore blocking or filling by polyphosphates and CuO aggregates - Disruption of zeolite framework	
HC	- Decrease in NO conversion - Generation of local heat while burning them off	- Thermal destruction of zeolite framework - Pore blocking or filling by HC species (coking)	
Steam	- Decrease in NO conversion	- Dealumination of zeolite framework - Loss of Brønsted acidity - Decrease of isolated Cu^{2+} ions and CuO formation	

Data adapted from Lezcano-Gonzalez I, Deka U, van der Bij HE, Paalanen P, Arstad B, Weckhuysen BM, Beale AM. Chemical deactivation of Cu-SSZ-13 ammonia selective catalytic reduction (NH_3-SCR) systems. Appl. Catal. B 2014;154–155:339–349. Copyright 2014, with permission of Elsevier.

Acknowledgments

The authors acknowledge financial support of a NWO-CW Top grant.

References

[1] Air Quality in Europe, European Environment Agency. 2012.
[2] Klingstedt F, Arve K, Eraenen K, Murzin DY. Toward improved catalytic low-temperature NO_x removal in diesel-powered vehicles. ChemInform 2006;37:273–282.

[3] Granger P, Parvulescu VI. Catalytic NO_x abatement systems for mobile sources: from three-way to lean burn after-treatment technologies. Chem Rev 2011;111:3155–207.
[4] Brandenberger S, Kröcher O, Tissler A, Althoff R. The state of the art in selective catalytic reduction of NO_x by ammonia using metal-exchanged zeolite catalysts. Catal Rev 2008;50:492–531.
[5] Chen H-Y, Nova I, Tronconi E. Urea-SCR technology for deNO_x after treatment of diesel exhausts. New York: Springer; 2014. p. 123–47.
[6] Madia G, Elsener M, Koebel M, Raimondi F, Wokaun A. Thermal stability of vanadia-tungsta-titania catalysts in the SCR process. Appl Catal B 2002;39:181–90.
[7] Colombo M, Nova I, Tronconi E. A comparative study of the NH_3-SCR reactions over a Cu-zeolite and a Fe-zeolite catalyst. Catal Today 2010;151:223–30.
[8] Cavataio G, Girard J, Patterson J, Montreuil C, et al. Laboratory testing of urea-SCR formulations to meet tier 2 bin 5 emissions. SAE Tech Pap 2007-01-1575 2007, http://dx.doi.org/10.4271/2007-01-1575.
[9] Li J, Chang H, Ma L, Hao J, Yang RT. Low-temperature selective catalytic reduction of NO_x with NH_3 over metal oxide and zeolite catalysts—a review. Catal Today 2011;175:147–56.
[10] Sultana A, Nanba T, Sasaki M, Haneda M, Suzuki K, Hamada H. Selective catalytic reduction of NO_x with NH_3 over different copper exchanged zeolites in the presence of decane. Catal Today 2011;164:495–9.
[11] IZA. IZA structural database, http://www.iza-structure.org/databases/.
[12] Palella BI, Cadoni M, Frache A, Pastore HO, Pirone R, Russo G, et al. On the hydrothermal stability of CuAPSO-34 microporous catalysts for N_2O decomposition: a comparison with CuZSM-5. J Catal 2003;217:100–6.
[13] Ye Q, Wang L, Yang RT. Activity, propene poisoning resistance and hydrothermal stability of copper exchanged chabazite-like zeolite catalysts for SCR of NO with ammonia in comparison to Cu/ZSM-5. Appl Catal A 2012;427–428:24–34.
[14] Kwak JH, Tran D, Burton SD, Szanyi J, Lee JH, Peden CHF. Effects of hydrothermal aging on NH_3-SCR reaction over Cu/zeolites. J Catal 2012;287:203–9.
[15] Xie D, McCusker LB, Baerlocher C, Zones SI, Wan W, Zou X. SSZ-52, a zeolite with an 18-layer aluminosilicate framework structure related to that of the DeNO_x catalyst Cu-SSZ-13. J Am Chem Soc 2013;135:10519–24.
[16] IZA. Chabazite series, http://www.iza-online.org/natural/Datasheets/Chabazite/chabazite.htm.
[17] Fickel DW, Lobo RF. Copper coordination in Cu-SSZ-13 and Cu-SSZ-16 investigated by variable-temperature XRD. J Phys Chem C 2010;114:1633–40.
[18] Moliner M, Franch C, Palomares E, Grill M, Corma A. Cu-SSZ-39, an active and hydrothermally stable catalyst for the selective catalytic reduction of NO_x. Chem Commun 2012;48:8264–6.
[19] IZA, Ferrierite series, http://www.iza-online.org/natural/Datasheets/Ferrierite/ferrierite.htm.
[20] Mihai O, Widyastuti C, Kumar A, Li J, Joshi S, Kamasamudram K, et al. The effect of NO_2/NO_x feed ratio on the NH_3-SCR system over Cu–zeolites with varying copper loading. Catal Lett 2014;144:70–80.
[21] Delahay G, Coq B, Kieger S, Neveu B. The origin of N_2O formation in the selective catalytic reduction of NO_x by NH_3 in O_2 rich atmosphere on Cu-faujasite catalysts. Catal Today 1999; 54:431–8.
[22] Ma L, Cheng Y, Cavataio G, McCabe RW, Fu L, Li J. In situ DRIFTS and temperature-programmed technology study on NH_3-SCR of NO_x over Cu-SSZ-13 and Cu-SAPO-34 catalysts. Appl Catal B 2014;156–157:428–37.
[23] Gao F, Walter ED, Kollar M, Wang Y, Szanyi J, Peden CHF. Understanding ammonia selective catalytic reduction kinetics over Cu/SSZ-13 from motion of the Cu ions. J Catal 2014;319:1–14.
[24] Deka U, Lezcano-Gonzalez I, Warrender SJ, Lorena Picone A, Wright PA, Weckhuysen BM, et al. Changing active sites in Cu–CHA catalysts: DeNO_x selectivity as a function of the preparation method. Microporous Mesoporous Mater 2013;166:144–52.
[25] Giordanino F, Vennestrom PNR, Lundegaard LF, Stappen FN, Mossin S, Beato P, Bordiga S, Lamberti C. Characterization of Cu-exchanged SSZ-13: a comparative FTIR, UV–vis, and EPR study with Cu-ZSM-5 and Cu-[small beta] with similar Si/Al and Cu/Al ratios. Dalton Trans 2013;42:12741–61.

[26] Grant A, Ranney J, Campbell C, Evans T, Thornton G. The influence of chlorine on the dispersion of Cu particles on Cu/ZnO(0001) model catalysts. Catal Lett 2000;65:159–68.
[27] King ST. Oxidative carbonylation of methanol to dimethyl carbonate by solid-state ion-exchanged CuY catalysts. Catal Today 1997;33:173–82.
[28] Wang D, Jangjou Y, Liu Y, Sharma MK, Luo J, Li J, et al. A comparison of hydrothermal aging effects on NH_3-SCR of NO_x over Cu-SSZ-13 and Cu-SAPO-34 catalysts. Appl Catal B 2015;165:438–45.
[29] Gao F, Walter ED, Washton NM, Szanyi J, Peden CHF. Synthesis and evaluation of Cu/SAPO-34 catalysts for NH_3-SCR 2: Solid-state ion exchange and one-pot synthesis. Appl Catal B 2015;162:501–14.
[30] Martínez-Franco R, Moliner M, Concepcion P, Thogersen JR, Corma A. Synthesis, characterization and reactivity of high hydrothermally stable Cu-SAPO-34 materials prepared by "one-pot" processes. J Catal 2014;314:73–82.
[31] Martínez-Franco R, Moliner M, Thogersen JR, Corma A. Efficient one-pot preparation of Cu-SSZ-13 materials using cooperative OSDAs for their catalytic application in the SCR of NO_x. ChemCatChem 2013;5:3316–23.
[32] Ren L, Zhu L, Yang C, Chen Y, Sun Q, Zhang H, Li C, Nawaz F, Meng X, Xiao F-S. Designed copper-amine complex as an efficient template for one-pot synthesis of Cu-SSZ-13 zeolite with excellent activity for selective catalytic reduction of NO_x by NH_3. Chem Commun 2011;47:9789–91.
[33] Xie L, Liu F, Ren L, Shi X, Xiao F-S, He H. Excellent performance of one-pot synthesized Cu-SSZ-13 catalyst for the selective catalytic reduction of NO_x with NH_3. Environ Sci Technol 2014;48:566–72.
[34] Korhonen ST, Fickel DW, Lobo RF, Weckhuysen BM, Beale AM. Isolated Cu^{2+} ions: active sites for selective catalytic reduction of NO. Chem Commun 2011;47:800–2.
[35] Deka U, Juhin A, Eilertsen EA, Emerich H, Green MA, Korhonen ST, et al. Confirmation of isolated Cu^{2+} Ions in SSZ-13 zeolite as active sites in NH_3-selective catalytic reduction. J Phys Chem C 2012;116: 4809–18.
[36] Deka U, Lezcano-Gonzalez I, Weckhuysen BM, Beale AM. Local environment and nature of Cu active sites in zeolite-based catalysts for the selective catalytic reduction of NO_x. ACS Catal 2013;3:413–27.
[37] Andersen CW, Bremholm M, Vennestrom PNR, Blichfeld AB, Lundegaard LF, Iversen BB. Location of Cu^{2+} in CHA zeolite investigated by X-ray diffraction using the Rietveld/maximum entropy method. IUCrJ 2014;1:382–6.
[38] Godiksen A, Stappen FN, Vennestrøm PNR, Giordanino F, Rasmussen SB, et al. Coordination environment of copper sites in Cu-CHA zeolite investigated by electron paramagnetic resonance. J Phys Chem C 2014;118:23126–38.
[39] Borfecchia E, Lomachenko KA, Giordanino F, Falsig H, Beato P, Soldatov AV, et al. Revisiting the nature of Cu sites in the activated Cu-SSZ-13 catalyst for SCR reaction. Chem Sci 2015;6:548–63.
[40] Verma AA, Bates SA, Anggara T, Paolucci C, Parekh AA, Kamasamudram K, et al. NO oxidation: a probe reaction on Cu-SSZ-13. J Catal 2014;312:179–90.
[41] Fickel DW, D'Addio E, Lauterbach JA, Lobo RF. The ammonia selective catalytic reduction activity of copper-exchanged small-pore zeolites. Appl Catal B 2011;102:441–8.
[42] Kwak JH, Tonkyn RG, Kim DH, Szanyi J, Peden CHF. Excellent activity and selectivity of Cu-SSZ-13 in the selective catalytic reduction of NO_x with NH_3. J Catal 2010;275:187–90.
[43] Ma L, Cheng Y, Cavataio G, McCabe RW, Fu L, Li J. Characterization of commercial Cu-SSZ-13 and Cu-SAPO-34 catalysts with hydrothermal treatment for NH_3-SCR of NO_x in diesel exhaust. Chem Eng J 2013;225:323–30.
[44] Bordiga S, Regli L, Lamberti C, Zecchina A, Bjørgen M, Lillerud KP. FTIR adsorption studies of H_2O and CH_3OH in the isostructural H-SSZ-13 and H-SAPO-34: Formation of H-bonded adducts and protonated clusters. J Phys Chem B 2005;109:7724–32.
[45] Aramburo LR, Karwacki L, Cubillas P, Asahina S, de Winter DAM, Drury MR, et al. The porosity, acidity, and reactivity of dealuminated zeolite ZSM-5 at the single particle level: the influence of the zeolite architecture. Chemistry 2011;17:13773–81.
[46] Triantafillidis CS, Vlessidis AG, Nalbandian L, Evmiridis NP. Effect of the degree and type of the dealumination method on the structural, compositional and acidic characteristics of H-ZSM-5 zeolites. Microporous Mesoporous Mater 2001;47:369–88.

[47] In: van Bokhoven JA, Danilina N, editors. Zeolites and catalysis. Weinheim: Wiley-VCH Verlag GmbH & Co. KGaA; 2010. p. 283–300.
[48] Vennestrøm PNR, Janssens TVW, Kustov A, Grill M, Puig-Molina A, Lundegaard LF, et al. Influence of lattice stability on hydrothermal deactivation of Cu-ZSM-5 and Cu-IM-5 zeolites for selective catalytic reduction of NO_x by NH_3. J Catal 2014;309:477–90.
[49] Zhang Y, Flytzani-Stephanopoulos M. Hydrothermal stability of cerium modified Cu-ZSM-5 catalyst for nitric oxide decomposition. J Catal 1996;164:131–45.
[50] Chen LY, Ping Z, Chuah GK, Jaenicke S, Simon G. A comparison of post-synthesis alumination and sol–gel synthesis of MCM-41 with high framework aluminum content. Microporous Mesoporous Mater 1999;27:231–42.
[51] Song Y, Sun C, Shen W, Lin L. Hydrothermal post-synthesis of HZSM-5 zeolite to enhance the coke-resistance of Mo/HZSM-5 catalyst for methane dehydroaromatization reaction: reconstruction of pore structure and modification of acidity. Appl Catal A 2007;317:266–74.
[52] Chen X, Currier N, Yezerets A, Kamasamudram K. Mitigation of platinum poisoning of Cu-zeolite SCR catalysts. SAE Int J Engines 2013;6:856–61.
[53] Jen H-W, Girard JW, Cavataio G, Jagner MJ. Detection, origin and effect of ultra-low platinum contamination on diesel-SCR catalysts. SAE Int J Fuels Lubr 2009;1:1553–9.
[54] Lezcano-Gonzalez I, Deka U, van der Bij HE, Paalanen P, Arstad B, Weckhuysen BM, et al. Chemical deactivation of Cu-SSZ-13 ammonia selective catalytic reduction (NH_3-SCR) systems. Appl Catal B 2014;154–155:339–49.

Index

Note: Page numbers followed by *f* indicate figures, *t* indicate tables and *s* indicate schemes.

D

E

F

G

H